DYNAMIC
PLASTICITY

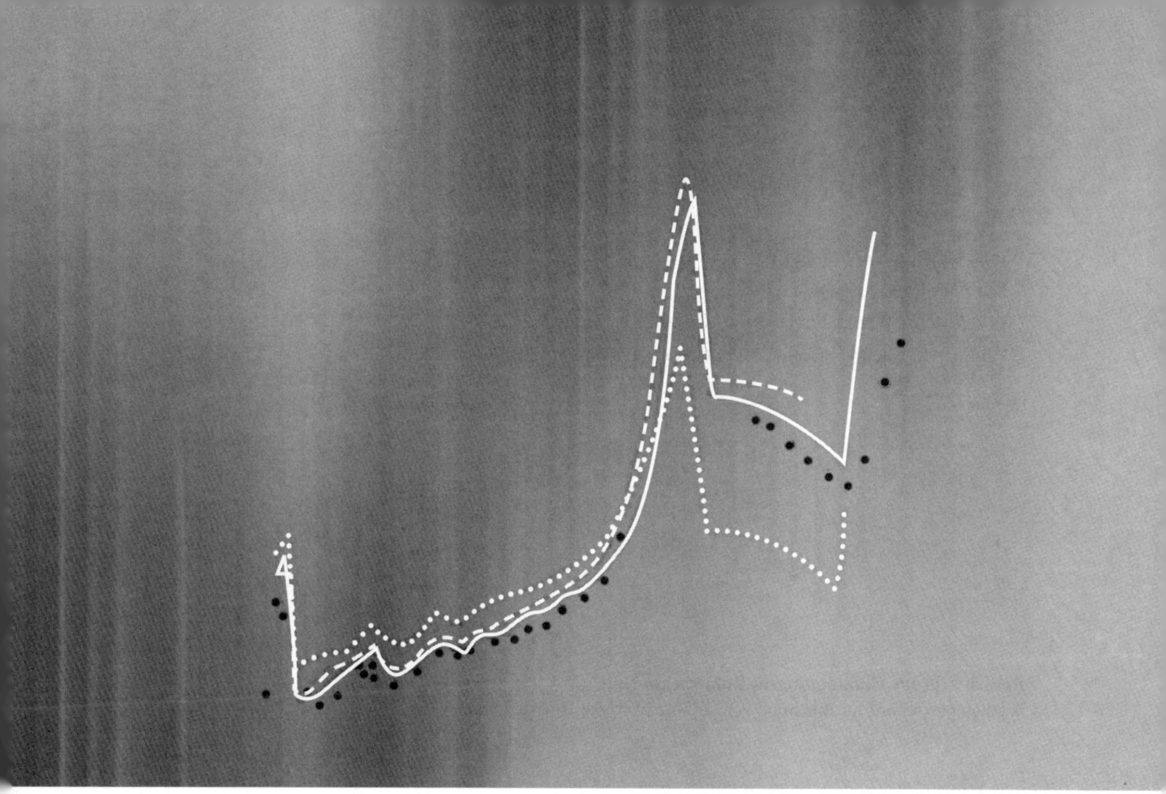

DYNAMIC PLASTICITY

N D Cristescu

University of Florida, USA

World Scientific

NEW JERSEY · LONDON · SINGAPORE · BEIJING · SHANGHAI · HONG KONG · TAIPEI · CHENNAI

Published by

World Scientific Publishing Co. Pte. Ltd.

5 Toh Tuck Link, Singapore 596224

USA office: 27 Warren Street, Suite 401-402, Hackensack, NJ 07601

UK office: 57 Shelton Street, Covent Garden, London WC2H 9HE

British Library Cataloguing-in-Publication Data
A catalogue record for this book is available from the British Library.

DYNAMIC PLASTICITY

ISBN-13 978-981-256-747-5
ISBN-10 981-256-747-X

Printed in Singapore by World Scientific Printers (S) Pte Ltd

IN MEMORIAM

I would like to mention, with all the sadness possible,
the memory of my wife Cornelia, which disappeared meantime,
and which helped me always so much.

Preface

The present book is not simply a new addition of the book Dynamic Plasticity, initially published in 1967, a long time ago. Certainly this edition is not only a new version, containing the essential of the old book and what has been done meantime. Why again Dynamic Plasticity? Well because very many books published meantime on the subject are not mentioning the waves which are to be considered in Dynamic Plasticity. Also, generally, the plastic waves are slower than the elastic one. Thus, when considering a simple problem of propagation of waves in thin bars, for any loading at the end, the plastic waves are reached by the elastic ones, and will not propagate any more. Only a part of the bar is deforming plastically. Examples of this kind are very few.

I thought that this new version is too restrictive for the today students which know little of static plasticity, differential equations, dynamic elastic–plastic properties, etc. Therefore, I thought to write a simpler book, containing the main concepts of dynamic plasticity, but also something else. Thus I thought that this new version would contain the elementary concepts of static plasticity, etc., which would be useful to give. Also it would be good to give other problems, not directly related to dynamic plasticity. Thus I started with some classical problems on static plasticity, but only the simplest things, so that the readers would afterwards understand also the dynamic problems. Also, since in dynamic problems the soils and rocks played a fundamental role, I thought to write a chapter on rocks and soils. Then were expressed several chapters about dynamic plasticity, as propagation of elastic–plastic waves in thin bars, the rate influence and the propagation of waves in flexible strings. It is good to remember here that all problems related to dynamic problems, are to be considered using the mechanics of the wave propagation; without the wave propagation mechanics all results concerning constitutive equations, rate effect, etc. are only informative. Such problems are mentioned however in the book. We have presented mainly the different aspects on constitutive equations of materials, as resulting from dynamic problems. Rate effects are considered in this way. They have been used by a variety of authors. The same with the mechanics of flexible strings, presented afterwards. Not very many authors have considered till now the mechanics of deformable cables.

Therefore I thought to write a very simple book, which can be read by the students themselves, without any additional help. They can understand what "plasticity" is after all. Then several other problems have been presented. Not trying to remove the fundamentals, I have thought also to add some additional problems, which are in fact dynamic, though the inertia effect is disregarded. They are the stationary problems, quite often met in many applications. It is question obviously, about problems involving Bingham bodies, as wire drawing, floating with working plug, extrusion, stability of natural inclined plane, etc.

Further I have considered various problems of plastic waves, using various theories. Also the perforation problems, was presented, using various symmetry assumptions, or any other assumption made.

The last chapter is on hypervelocity impact. To keep it simple, I have given only very few information about. Thus I wished to show what hypervelocity is and how is it considered now.

Though the book is a very simple one, I wished to ask any author to disregard possible missing of some papers. All literature is certainly incomplete. One has done today much more than given here. It was impossible for me to mention "all" authors in this field.

N. D. Cristescu
cristesc@ufl.edu

Contents

Introduction

Theory of Plasticity studies the distribution of stresses and particle velocities (or displacements) in a plastically (irreversible) deformed body, when are known the external factors which have acted upon him and the history of variation of these factors. The theory was applied to metals to describe working processes both at cold (drawing, rolling, etc.) and warm (extrusion, forging, etc.), to describe term behavior (high and law) involving also temperatures, to short term behavior, to describe impact, shocks, perforation, etc. It was applied to geomaterials, as soils, rocks, sands, clays, etc., with the description of civil engineering applications as tunnels, wells, excavations of all sorts, etc. It was applied to other materials as concrete, asphalt, ceramics, ice, powder-like materials, various pastes, slurries, etc.

In the classical sense the Plasticity Theory is time independent. However a time dependent theory was also developed and called Viscoplasticity. Besides Rheology deals with any flow or deformation in which time is the main parameter.

From the point of view of formulation of problems, in plasticity one considers in some of the problems, as in elasticity, that the strains are small; whoever in some other problems the consideration of the problems are as in nonlinear fluid mechanics when the strain are finite.

1 Diagnostic Tests

These are the slow tests in compression or in tension ($\dot{\varepsilon} \leq 10^{-2}$ s^{-1}, say) so as the strain is uniform along the specimen. We denote by

$$\sigma_{\mathrm{PK}} = \frac{F}{A_0} \quad \text{and by} \quad \sigma_C = \frac{F}{A}$$

the Piola–Kirchhoff and the Cauchy's stresses. Here F is the total force applied axially to the specimen, and A the current area, and by A_0 the initial area of the cross section of the specimen.

We also denote by

$$\varepsilon_H = \ln \frac{l}{l_0} \quad \text{and by} \quad \varepsilon_c = \frac{l - l_0}{l_0}$$

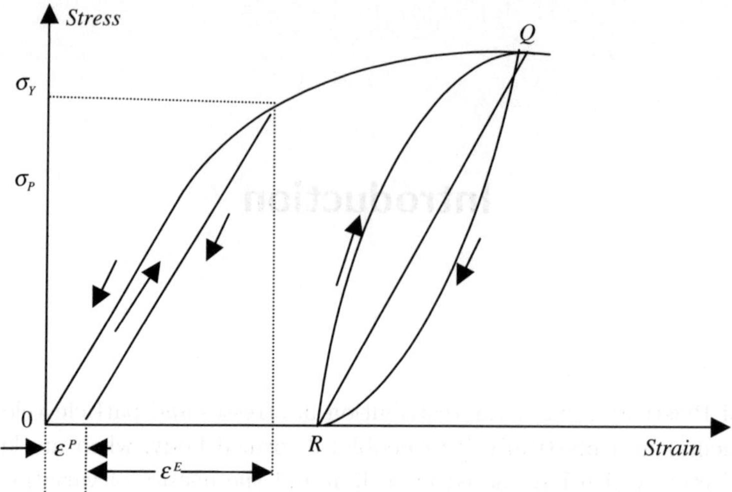

Fig. 1   Typical diagram of a diagnostic test.

the Henchy's strain or the Cauchy's strain. Here again l is the length of the working area of the specimen, and l_0 is the initial length of the same area. As a sign convention, $\sigma > 0$ in tension for metals, but it is a reverse convention for rocks and soils (see Fig. 1). σ_P is the proportionally limit of the specimen where we apply the Hooke's law $\sigma = E\varepsilon$ with E the Young's modulus which is constant, and independent on the loading rate and on the loading history. Up to σ_P we apply the Hooke's law in both loading und unloading. σ_Y is a conventional or offset yield limit defined by the permanent ε_Y, generally $0.1\% \doteq 0.5\%$ of the total strain. Essentially is that ε_Y is defined by a convention. Thus for $\varepsilon < \varepsilon_Y$ the unloading is perfectly elastic without hysteresis loop, as

$$\sigma = \sigma_Q + E(\varepsilon - \varepsilon_Q).$$

Thus we assume small strains and

$$\varepsilon = \varepsilon^E + \varepsilon^P.$$

In elasticity we apply the Hooke's law written as

$$\boldsymbol{\sigma} = \mathbf{C}[\boldsymbol{\varepsilon}] \quad \text{or} \quad \sigma_{ij} = C_{ijkl}\varepsilon_{kl}$$

where $\mathbf{C}$ is a fourth order tensor. We apply this law during loading and unloading and the natural reference configuration is the stress-free strain-free state. If we introduce the two deviators by:

$$\boldsymbol{\sigma}' = \boldsymbol{\sigma} - \frac{\operatorname{tr}\boldsymbol{\sigma}}{3}\mathbf{1} \quad \text{and} \quad \boldsymbol{\varepsilon}' = \boldsymbol{\varepsilon} - \frac{\operatorname{tr}\boldsymbol{\varepsilon}}{3}\mathbf{1},$$

the Hooke's law can be written

$$\boldsymbol{\sigma}' = 2G\boldsymbol{\varepsilon}' \quad \text{and} \quad \operatorname{tr}\boldsymbol{\sigma} = 3K\operatorname{tr}\boldsymbol{\varepsilon},$$

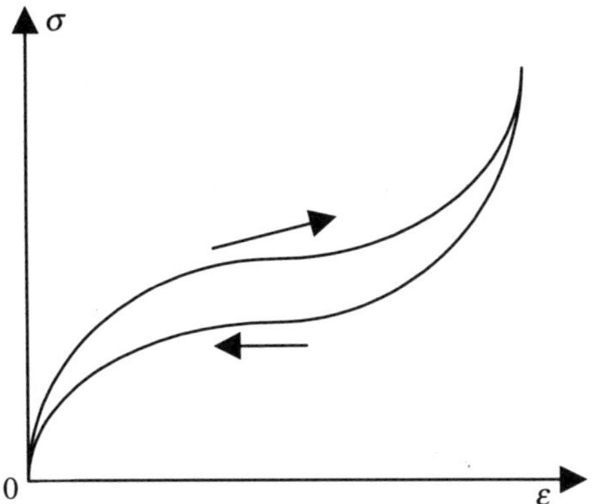

Fig. 2   Nonlinear elastic curve.

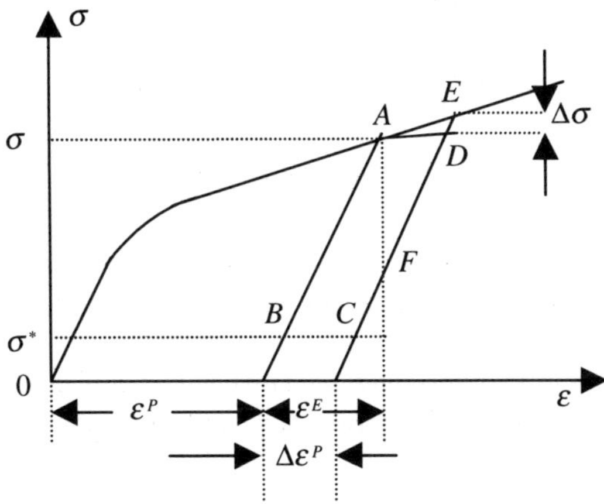

Fig. 3   Stress work per unit volume.

where G and K are the two elastic constants. The above relations are applied for any elastic isotropic body.

There are nonlinear elastic bodies as for instance rubber (see Fig. 2). The stress–strain curve is nonlinear but reversible. The unloading is according to some other law, exhibiting a significant hysteresis loop. We cannot describe this behavior by the Hooke's law but with a nonlinear law, giving a one-to-one correspondence. Since the material remembers his initial configuration, the reference configuration is the initial one.

For dissipative materials as the plastic ones, we can define an irreversible stress work per unit volume by:

$$W(T) = \int_0^T \sigma(t)\dot{\varepsilon}^P(t)\,dt\,.$$

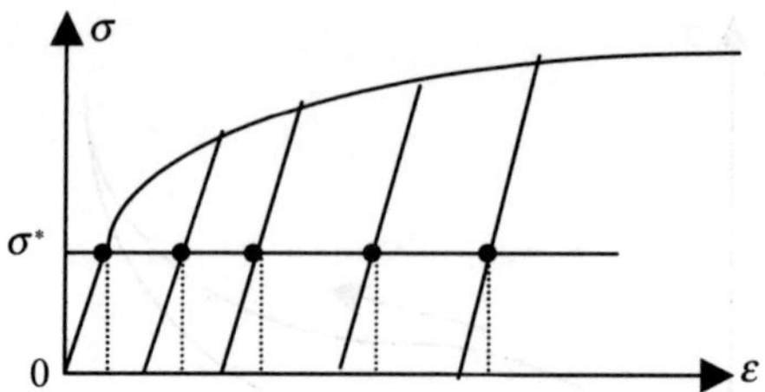

Fig. 4   Lack of one-to-one correspondence.

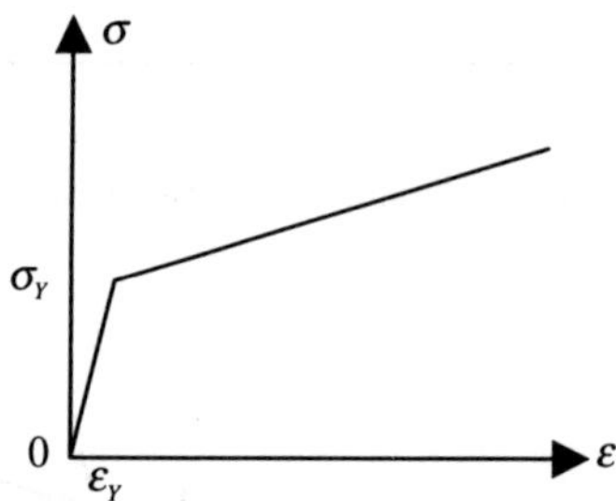

Fig. 5   Linear work-hardening.

That is shown in Fig. 3; it is the total irreversible area under the curve. The remaining area is the potential energy of deformation reversible for the reversible elastic materials (conservative).

In order to define work-hardening, we start with the Fig. 3. In a loading loop BAEDC producing the irreversible strain $\Delta\varepsilon^P$ and returning to the same stress σ^*, we define by:

$$(\sigma - \sigma^*)\Delta\varepsilon^P > 0 \quad \text{irreversibility},$$

$$\Delta\sigma\Delta\varepsilon^P > 0 \quad \text{stability}.$$

These two conditions are known as the Drucker's postulate and are used to define the plastic work-hardening.

Another postulate is due to Iliushin's; it says that the loading–unloading FAEDF must be positive.

In plasticity there is no one-to-one stress–strain correspondence. That is very clear in Fig. 4. The loading history must be known; to a single stress correspond several strains. Plasticity starts with unloading, as compared with nonlinear elastic behavior; and with the plastic strains which can develop only if $\sigma > \sigma_Y$.

The linear work-hardening is defined by two straight lines (Fig. 5):

$$\sigma = E\varepsilon \qquad \text{if } \sigma \leq \sigma_Y,$$

$$\sigma = \sigma_Y + E_1(\varepsilon - \varepsilon_Y) \quad \text{if } \sigma \geq \sigma_Y.$$

Here E_1 is the constant work-hardening parameter, and $E_1 \ll E$.

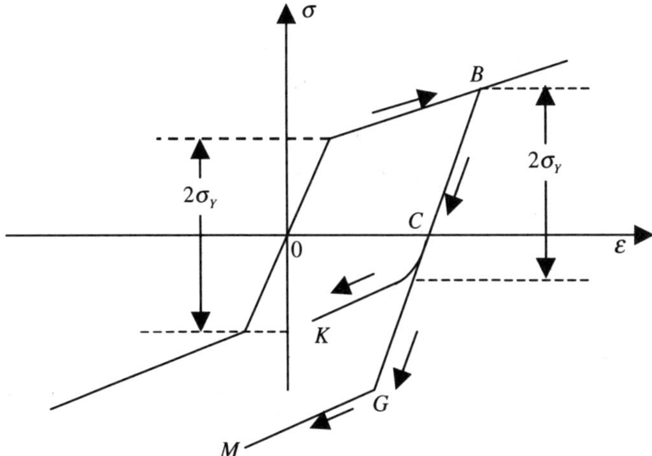

Fig. 6   Various hardening laws.

If the stress is increasing very slowly, in the so-called "soft" machines, one is observing some steps on the stress–strain curve. It is question of the so-called Savart–Masson effect, later rediscovered by Portevin–Le Chatelier effect. It was shown that this effect can by described by a rate-type constitutive equation (Suliciu [1981]). The viscosity coefficient has strong variation in some regions of the ε, σ plane that lie above the equilibrium curve.

For most materials, if σ is on a plastic state, then $-\sigma$ is also on the plastic state. As it is well known, there are a lot of materials which do not satisfy this condition. For rocks for instance, if σ_{Yt} is the yield stress in tension, then σ_{Yc} is in compression, and $|\sigma_{Yc}| \gg |\sigma_{Yt}|$. That is also for concrete, cast iron, soils, glass, powders, etc. That is called Bauschinger effect, discovered in 1886. In Fig. 6 it is along BCK, that is the segment $2\sigma_Y$ stays more or less constant during loading.

For metals the elastic domain has a constant size $2\sigma_Y$ during loading. If during unloading we follow $BCGM$ then the hardening is said to be <u>isotropic</u>. If during unloading we follow the path BCK we say that the hardening is <u>kinematic</u>. Generally, if the yield stress in one direction is diminished by a previous plastic deformation in the opposite direction we have a Bauschingr effect. It introduces anisotropy, sough it can be removed by annealing at high temperature. If we do a loading in a single direction, we cannot distinguish between the two hardening.

2   Tests Performed at Long and Short Term Intervals

If dx is a material element in current configuration, and dX in the initial configuration, we call $\lambda = dx/dX$ elongation. The rate of elongation is $D = \dot{\lambda}/\lambda \ (= \dot{\varepsilon}$ sometimes).

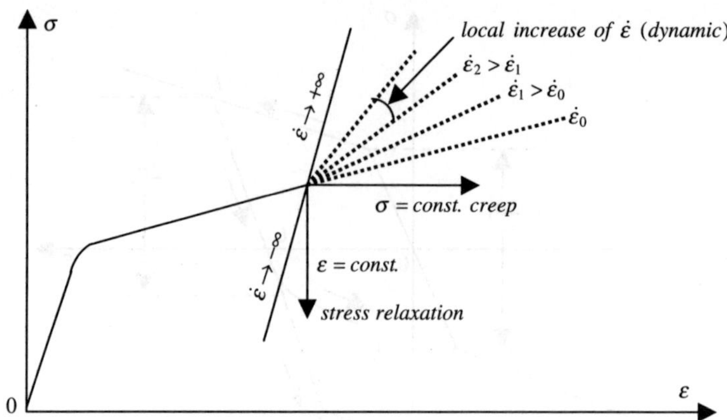

Fig. 7   Effect of change of rate of elongation.

Table 1   Variation of $\dot{\varepsilon}$.

$\dot{\varepsilon} < 10^{-20}$ s^{-1}	slow tectonic motion,
10^{-8} s$^{-1} \geq \dot{\varepsilon} \geq 10^{-12}$ s^{-1}	creep tests,
10^{-4} s$^{-1} \geq \dot{\varepsilon} \geq 10^{-1}$ s^{-1}	testing machines,
$\dot{\varepsilon} \simeq 1$ s^{-1}	hammer drop,
$\dot{\varepsilon} \simeq 10$ s^{-1}	the strain is not uniform, wave propagation is needed,
$\dot{\varepsilon} \simeq 10^2$ s^{-1}	metal drawing, air gun bullet,
$\dot{\varepsilon} \simeq 10^4$ s^{-1}	high speed impact, ballistics,
$\dot{\varepsilon} \simeq 10^6$ s$^{-1} - 10^7$ s^{-1}	high speed drawing of very fine wires, or very fast tests.

The change of rate of deformation is shown in Fig. 7. An increase of $\dot{\varepsilon}$ is raising the curves. But this raise in technically limited by the machine we have. For an additional increase, we need dynamic curves, with an local increase of $\dot{\varepsilon}$. This increase is done by elastic waves propagating with the velocity $c_0 = \sqrt{E/\rho}$. A table of approximate increase of $\dot{\varepsilon}$ is given in Table 1. This is a very approximate table of variation of $\dot{\varepsilon}$. For constant stress we have *creep*, but for constant strain we have *stress relaxation*. Any other intermediate variation of the strain rate is possible. For $|\dot{\varepsilon}| \to \infty$ we have very fast variation of the strain rate, impossible to realize practically.

In order to have a representation we take into account that mainly the plastic properties are influenced by the change of the rate of strain (see Fig. 8). A relationship was proposed by Ludwik from 1909, and is of the kind shown on Fig. 8. Thus we have for a fixed strain:

$$\sigma = \sigma_Y + \sigma_0 \ln \frac{\dot{\varepsilon}^P}{\dot{\varepsilon}_0^P} \quad \text{and} \quad \sigma > \sigma_Y > 0$$

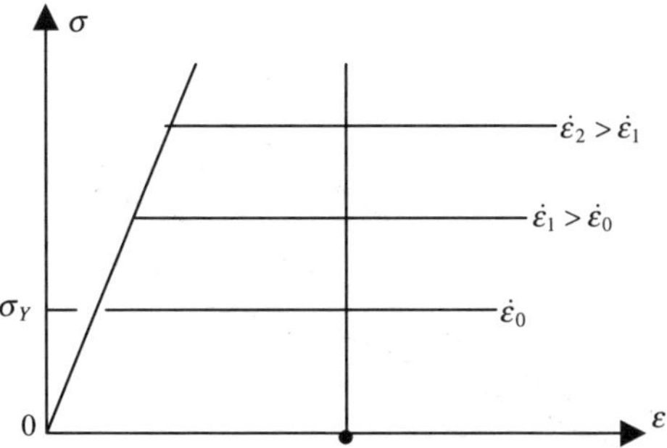

Fig. 8   Influence of the strain rate on the stress–strain curve.

with $\sigma_0 = $ constant. If the elastic strains are disregarded, the stress–strain curves are

$$\dot{\varepsilon} = \begin{cases} \dfrac{\sigma - \sigma_Y}{3\eta} & \text{if } \sigma > \sigma_Y\,, \\[2mm] 0 & \text{if } 0 \le \sigma \le \sigma_Y\,, \end{cases}$$

where $\sigma - \sigma_Y$ is the <u>overstress</u>. σ_Y is the yield stress for a conventional small $\dot{\varepsilon}_0$ obtained in very slow performed tests, when flow starts being possible. η is a viscosity coefficient; if two tests are performed with the strain rates $\dot{\varepsilon}_1$ and $\dot{\varepsilon}_2$ we have:

$$3\eta = \frac{\sigma_2 - \sigma_1}{\dot{\varepsilon}_2 - \dot{\varepsilon}_1}\,,$$

to determine η. If η is constant for any strain rates, the relation is linear, otherwise nonlinear. Since in this relation there is no strain, the reference configuration is the actual one.

For work-hardening materials (Malvern):

$$\dot{\varepsilon} = \begin{cases} \dfrac{\sigma - f(\varepsilon)}{3\eta} & \text{if } \sigma > f(\varepsilon)\,, \\[2mm] 0 & \text{if } 0 \le \sigma \le f(\varepsilon)\,. \end{cases}$$

The reference configuration is the initial one or a relative one, corresponding to the state when the test started (for geomaterials, for instance).

Let as give several examples. In Fig. 9 is given the stress–strain curves for schist. One can see that the influence of the strain rate is felt from the beginning. The hole curve is influenced, not only the plastic part. Also, the last points correspond to failure. Thus with an increase of loading rate the stress at failure is increased, while the strain at failure is decreased. Thus a theory of failure expressing in stresses only, would not work.

In order to see that the influence of the strain rate is not always to be seen always on the stress–strain curves. In Fig. 10 are given the curves for granite, obtained in

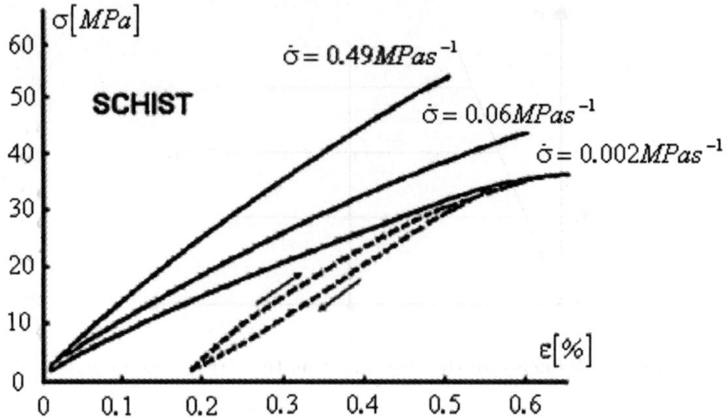

Fig. 9   Stress–strain curves for schist.

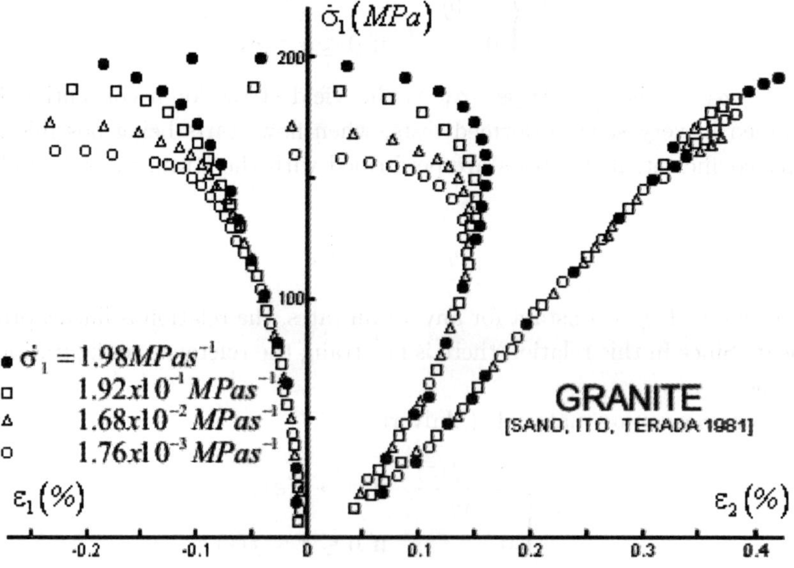

Fig. 10   Influence of the strain rate on granite.

triaxial tests (after Sano, Ito, Terada [1981]). One can see that the influence of the strain rate is practically zero on the axial stress strain curves. But the influence on the radial strain is remarkable on the other curves. Failure is the last point, but for the lowest radial curve is extended until -0.064, thus extending very much outside the figure. Concerning failure one can see that stress at failure is increased with the strain rate, while the strain is decreased. Thus failure would be impossible to describe with a condition written in stresses only.

3   Long-Term Tests

In order to find out the principal properties of creep and relaxation, one is doing long-terms tests. If the stress applied is relative small, the creep is transient, i.e., stabilizing after a certain time. In Fig. 11 is shown the creep curves for schist. One can see that the first curves correspond to transient creep only. The stresses are too small. The fact that the stresses are small, on the right side is given the $\dot{\varepsilon}-t$ curves. The curves corresponding to the transient case come quite fast at the origin ($\dot{\varepsilon} \to 0$). For higher stresses, we obtain a steady state creep, with $\dot{\varepsilon} = $ const. That is obtained generally for $\sigma \geq 0.6\sigma_c$, where σ_c is the short-term failure strengths. Temperature has a strong influence on creep. With increasing temperature the creep curves start at a lower stresses and extend very much.

When the applied stress is still increased, we arrive at the tertiary creep, where $\ddot{\varepsilon} > 0$. That means that soon failure will take place. From the right side of the diagram, the tertiary creep is not bringing the curves at zero.

From the very many laws existing, the most known is the Norton law where $\dot{\varepsilon} \sim \sigma^n$. Generally we describe creep by $\dot{\varepsilon} = \mathbf{f}(\boldsymbol{\sigma}, \boldsymbol{\varepsilon})$, the temperature being also influenced. For relaxation one is describing it with $\dot{\boldsymbol{\sigma}} = \mathbf{g}(\boldsymbol{\sigma}, \boldsymbol{\varepsilon})$, the temperature being again present.

Jugging only from the creep curves, it is difficult to see if they correspond to a viscoelastic model or to a viscoplastic one. The only things we can say is the difference in the history dependent principle. Figure 12 is showing the difference. If one is loading with the stress σ_1 and then increase it to σ_2, or applied from the beginning the stress σ_2, we obtain the same thing (at left) if there are no internal changes, or obtain something different (at right), with possibly internal changes. In the first case the loading is history independent, while in the second we have a history dependency. The history dependency is probably viscoplastic.

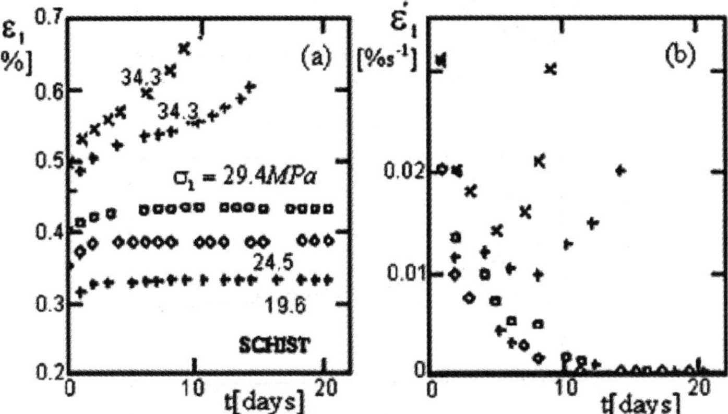

Fig. 11   Creep curves for schist.

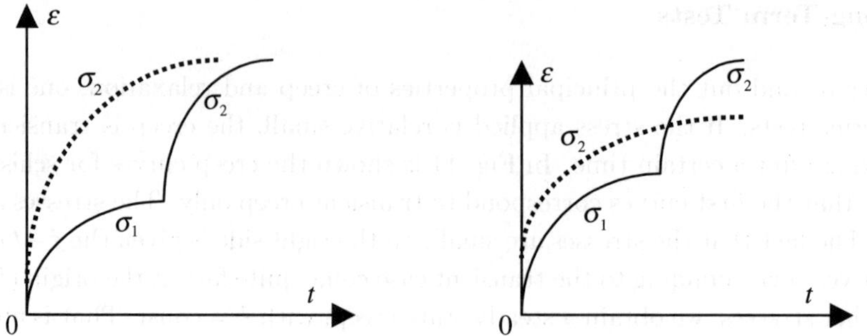

Fig. 12   Principle of history dependency.

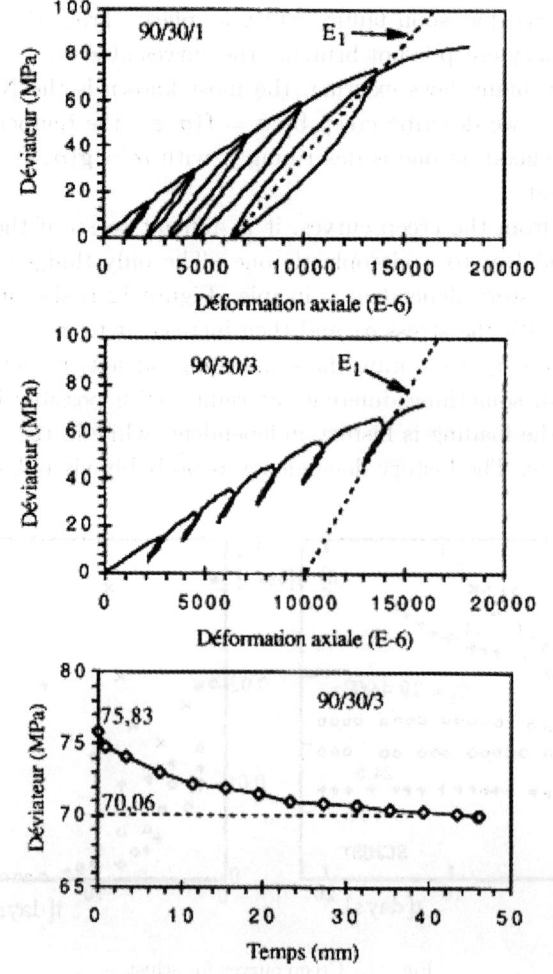

Fig. 13   The method used for the determination of the elastic parameters.

Thus we have to consider <u>plasticity</u> only if:

- the stresses $\sigma > \sigma_Y$,
- we obtain an irreversible strain,
- we have a history dependence,
- generally, we have a initial reference configuration,
- generally, the deviator relationship $(\boldsymbol{\sigma}' \sim \boldsymbol{\varepsilon}')$ is distinct from the volumetric $(\operatorname{tr} \boldsymbol{\sigma} \sim \operatorname{tr} \boldsymbol{\varepsilon})$.

Concerning the procedure used to measure the elastic parameters for various materials, for metals there are no problems. The problem is for the powder-like materials, or for rocks, for instance. The idea is that for such materials the hysteresis loop is quite important, so that if one is doing the unloading immediately, it is difficult to determine the elastic parameters directly. Figure 13 (after a PhD of Niandou [1994]) is showing such a problem. If one is trying to determine the elastic parameters directly immediately after loading, one obtains the hysteresis loops shown. One has to wait after each re-loading a number of minutes during which the rock is creeping. On the last figure shown one can see that the rate of deformation is decreasing wary much. Thus after a period of time if one is doing the unloading, no hysteresis loops are observed any more. This method is to separate the rheological properties from the unloading. That is the method proposed for rocks since 1988 (Cristescu). It was applied to many other materials since that time.

4 Temperature Influence

A temperature increases will decrease:

- the elastic parameters,
- the yield stress,
- the work-hardening modulus,
- the ultimate strength.

It is important to mention here the Zener–Hollomon parameter for metals:

$$Z = \dot{\varepsilon} \exp\left(\frac{\Delta H}{RT}\right),$$

where ΔH is the activation energy which may depend on temperature, R is a universal gas constant and T the absolute temperature. For $\varepsilon = \text{const.}$ the stress can be obtained from

$$\sigma = f(Z) = f\left(\dot{\varepsilon} \exp \frac{\Delta H}{RT}\right).$$

That is we can change the rate of strain or temperature according to this relation.

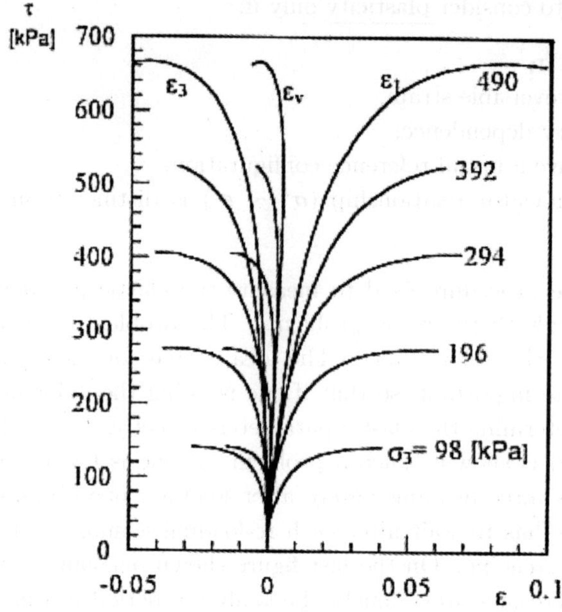

Fig. 14   Triaxial stress–strain curves for alumina powder.

5   The Influence of the Hydrostatic Pressure

It was shown that with an increase of pressure: the yield stress slight increases, the ultimate strength increases, the elastic modules G and K increase (increase of wave velocities), the work-hardening modulus increase and the ductility increases significantly. The last increase has lead to new methods of metal working, where an additional pressure was used. Generally, for metals the constitutive equations is written in deviators.

The above is not true for: rocks, soils, wood, cast iron, powder like materials, granular like materials, etc. For all these materials the pressure plays an important role. In Fig. 14 is shown the stress–strain curves for alumina powder, obtained in triaxial tests. One can see that all curves, the axial, diameter, and volumetric curves, are strongly dependent on the pressure (written along the curves).

6   Variation of Elastic Parameters with Plastic Strain (Metals)

During plastic deformation the elastic parameters vary. That has a significance for the expression of wave velocities (say). For cylindrical specimens subjected to uniaxial tests

$$\varepsilon_2 = \varepsilon_3 = -\nu\varepsilon_1$$

with ν the Poisson's ratio. For isotropic elastic materials $0 \leq \nu \leq 0.5$. If $\nu \approx 0$ no lateral strain is to be expected. For $\nu \approx 0.5$ we have an incompressible material $\varepsilon_{11} + \varepsilon_{22} + \varepsilon_{33} = 0$. For rubber $\nu = 0.47$, steel $0.25 \leq \nu \leq 0.30$, aluminum $0.26 \leq \nu \leq 0.36$, cork $\nu = 0.00$, polyethylene and paraffin $\nu \approx 0.5$.

In the plastic domain for aluminum ν is very close to 0.5, thus shoving plastic incompressibility. Thus the variation of the volume is elastic only: $\sigma = 3K\varepsilon^E$ and $\varepsilon^P = 0$.

This result is true for metals, only. For rocks, granular materials, etc. it is not.

Bibliography

Niandou H., 1994, Thése de Doctorat, Lab. de Mécanique de Lille URA 1441 CNRS.

Sano O., Itô I. and Terada M., 1981, Influence of strain rate on dilatancy and strength of oshima granite under uniaxial compression, *J. Geophys. Res.* **86**, B10, 9299–9311.

Suliciu I., 1981, On the Savart–Masson effect, *J. Applied Mech.* **48**, 426–428.

Chapter 1

Yield Conditions

1.1 Stresses

In order to understand well the problem we have to show several concepts which are helping. We choose an octhaedrical plane and its normal (see Fig. 1.1.1). For a point of stress $(\boldsymbol{\sigma}_1, \boldsymbol{\sigma}_2, \boldsymbol{\sigma}_3)$ we have the stress vector $\mathbf{t}_n$ which must be projected on the hydrostatic line and on a normal to the hydrostatic line. Since

$$\mathbf{t}_n = [n_1, n_2, n_3] \begin{bmatrix} \sigma_1 & 0 & 0 \\ 0 & \sigma_2 & 0 \\ 0 & 0 & \sigma_3 \end{bmatrix}, \tag{1.1.1}$$

its components are

$$\mathbf{t}_n = \mathbf{n} \cdot \mathbf{T} = \sigma_1 \frac{\mathbf{i}_1}{\sqrt{3}} + \sigma_2 \frac{\mathbf{i}_2}{\sqrt{3}} + \sigma_3 \frac{\mathbf{i}_3}{\sqrt{3}} \tag{1.1.2}$$

so that the projection of this vector on the hydrostatic line is

$$\mathbf{t}_n \cdot \mathbf{n} = \frac{\sigma_1}{3} + \frac{\sigma_2}{3} + \frac{\sigma_3}{3} = \sigma, \tag{1.1.3}$$

where σ is the mean stress. The projection on the hydrostatic line is now

$$(\mathbf{t}_n \cdot \mathbf{n})\mathbf{n} = \sigma \frac{\mathbf{i}_1}{\sqrt{3}} + \sigma \frac{\mathbf{i}_2}{\sqrt{3}} + \sigma \frac{\mathbf{i}_3}{\sqrt{3}}$$

with absolute value

$$|(\mathbf{t}_n \cdot \mathbf{n})\mathbf{n}| = |\sigma|. \tag{1.1.4}$$

The projection normal to the octhaedrical line is

$$\mathbf{t}_\tau = \mathbf{t}_n - (\mathbf{t}_n \cdot \mathbf{n})\mathbf{n}$$

$$= \left(\frac{\sigma_1}{\sqrt{3}} - \frac{\sigma}{\sqrt{3}} \right) \mathbf{i}_1 + \left(\frac{\sigma_2}{\sqrt{3}} - \frac{\sigma}{\sqrt{3}} \right) \mathbf{i}_2 + \left(\frac{\sigma_3}{\sqrt{3}} - \frac{\sigma}{\sqrt{3}} \right) \mathbf{i}_3$$

$$= \frac{1}{\sqrt{3}} (\sigma_1' \mathbf{i}_1 + \sigma_2' \mathbf{i}_2 + \sigma_3' \mathbf{i}_3)$$

15

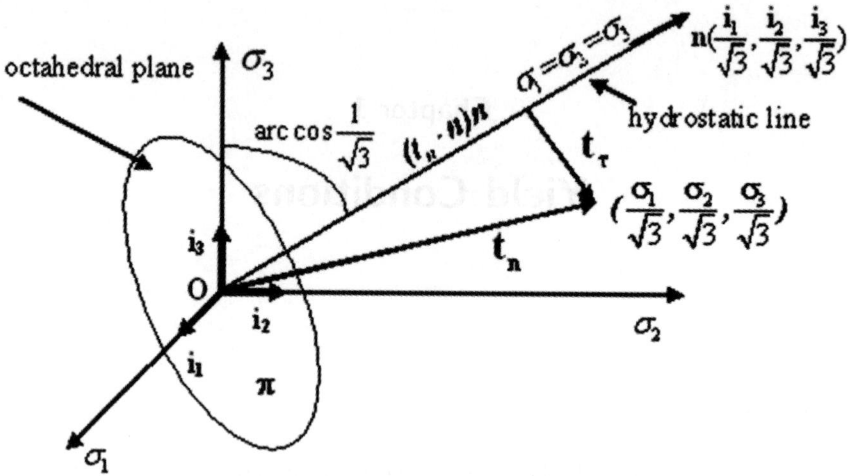

Fig. 1.1.1   Octahedral plane and all other associated concepts.

with the absolute value

$$|\mathbf{t}_\tau| = \tau_{\text{oct}} = \sqrt{\frac{2}{3}II_{\sigma'}}\,. \tag{1.1.5}$$

We got thus the interpretation of Roš and Eichinger [1926]: the absolute value of the vector tensor normal to the linear hydrostatic axes is equal to the square root of the second invariant of the deviatoric stress tensor. For notation we give

$$II_{\sigma'} = \frac{1}{2}[(\sigma_1')^2 + (\sigma_2')^2 + (\sigma_3')] = \frac{3}{2}\tau_{\text{oct}}^2 = \frac{1}{3}\bar{\sigma}^2 \tag{1.1.6}$$

and $\sigma_{ij}' = \sigma_{ij} - \sigma\delta_{ij}$ is the stress deviator.

Since the vector $\mathbf{t}_\tau$ in entirely in the octahedral plane, one can try to project it on various directions in this plane. Thus projecting on the π plane is giving (see Fig. 1.1.2):

$$\mathbf{t}_n|_\pi = \frac{\sigma_1}{\sqrt{3}}\sqrt{\frac{2}{3}}\mathbf{i}_1' + \frac{\sigma_2}{\sqrt{3}}\sqrt{\frac{2}{3}}\mathbf{i}_2' + \frac{\sigma_3}{\sqrt{3}}\sqrt{\frac{2}{3}}\mathbf{i}_3'\,.$$

Now projecting on the $\boldsymbol{\sigma}_3'$ direction we have,

$$\mathbf{t}_n \cdot (-\mathbf{i}_3') = \frac{\sigma_1}{\sqrt{3}}\sqrt{\frac{2}{3}}\frac{1}{2} + \frac{\sigma_2}{\sqrt{3}}\sqrt{\frac{2}{3}}\frac{1}{2} - \frac{\sigma_3}{\sqrt{3}}\sqrt{\frac{2}{3}}$$

$$= \frac{1}{3\sqrt{2}}(\sigma_1 + \sigma_2 - 2\sigma_3)$$

$$= -\frac{\sigma_3'}{\sqrt{2}}\,.$$

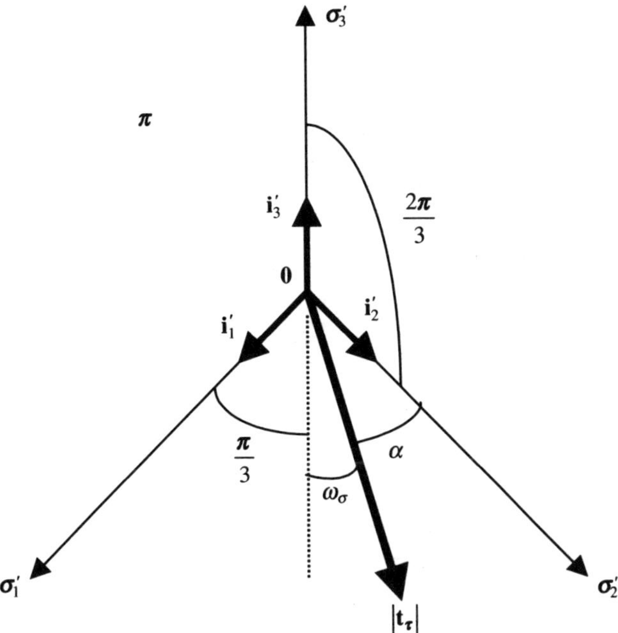

Fig. 1.1.2   Octahedral plane with the assumption $\sigma_2' > \sigma_1' > \sigma_3'$.

Thus we can write:

$$\mathbf{t}_n \cdot \mathbf{i}_2' = \tau_{\text{oct}} \cos\alpha = \frac{\sigma_2'}{\sqrt{2}} \,,$$

$$\mathbf{t}_n \cdot \mathbf{i}_3' = \tau_{\text{oct}} \cos\left(\alpha + \frac{2\pi}{3}\right) = \frac{\sigma_3'}{\sqrt{2}} \,,$$

$$\mathbf{t}_n \cdot \mathbf{i}_1' = \tau_{\text{oct}} \cos\left(\alpha - \frac{2\pi}{3}\right) = \frac{\sigma_1'}{\sqrt{2}} \,, \qquad (1.1.7)$$

$$\sigma_1'\sigma_2'\sigma_3' = III_{\sigma'} \,.$$

Taking into account that

$$\cos\alpha \cos\left(\alpha + \frac{2\pi}{3}\right)\cos\left(\alpha - \frac{2\pi}{3}\right) = \frac{1}{4}\cos 3\alpha \,,$$

we arrive at the relation

$$\cos 3\alpha = \left(\sqrt{\frac{3}{II_{\sigma'}}}\right)^3 \frac{III_{\sigma'}}{2} \,, \qquad (1.1.8)$$

$\alpha \in [0, \pi/3]$ defines the orientation in the octahedral plane of the stress vector $\mathbf{t}(\mathbf{x}, \mathbf{n})$ and is the angle with $\mathbf{i}_i'$ corresponding to σ_i' maximum. $\mathbf{t}_\tau$ is located between $\mathbf{i}_i'$ corresponding to σ_i' maxim and $\mathbf{i}_j'$ corresponding to σ_j' intermediate.

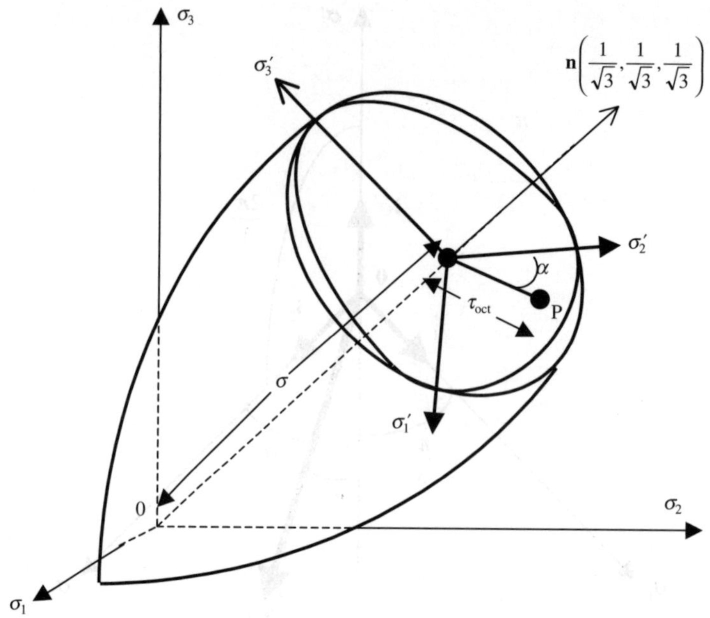

Fig. 1.1.3   Possible use of the invariants.

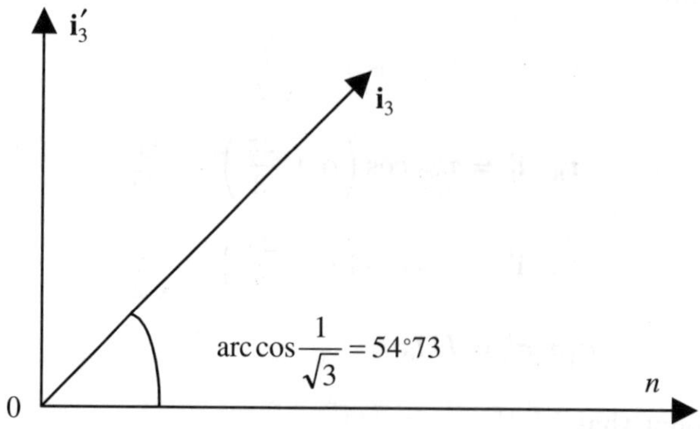

Fig. 1.1.4   Projection on the octahedral plane.

If we write

$$\frac{\sigma_2'/\sqrt{2}}{\sigma_3 - \sigma_1} - \sqrt{3}\frac{\cos\alpha}{\sin\alpha} = m,$$

we obtain the Lode parameter, which is defining the orientation in the octahedral plane.

From (1.1.7) we get:

$$\frac{\sigma_1 - \sigma_2}{2} = \sqrt{\frac{3}{2}}\tau_{\text{oct}} \sin\left(\alpha - \frac{\pi}{3}\right),$$

$$\frac{\sigma_2 - \sigma_3}{2} = \sqrt{\frac{3}{2}}\tau_{\text{oct}} \sin\left(\alpha + \frac{\pi}{3}\right), \tag{1.1.9}$$

$$\frac{\sigma_3 - \sigma_1}{2} = -\sqrt{\frac{3}{2}}\tau_{\text{oct}} \sin\alpha.$$

Since we have chosen $\sigma_2' > \sigma_1' > \sigma_3'$ we have:

$$\sigma_1 - \sigma_2 < 0 \quad \text{or} \quad \pi + \frac{\pi}{3} \le \alpha \le 2\pi + \frac{\pi}{3},$$

$$\sigma_2 - \sigma_3 > 0 \quad \text{or} \quad -\frac{\pi}{3} \le \alpha \le \pi - \frac{\pi}{3},$$

$$\sigma_3 - \sigma_1 < 0 \quad \text{or} \quad 0 \le \alpha \le \pi.$$

Therefore we have $0 \le \alpha \le \pi/3$ which is satisfying all the inequalities.

Since $\tau_{\max} = \tau_1 = (\sigma_2 - \sigma_3)/2$, we have:

$$\frac{\tau_{\max}}{\sqrt{(3/2)}\tau_{\text{oct}}} = \sin\left(\alpha + \frac{\pi}{3}\right)$$

and since we have to choose only half of the interval $0 \le \alpha \le \pi/6$ to get different values for the ratio, we get:

$$(0.866 =)\frac{\sqrt{3}}{2} \le \frac{\tau_{\max}}{\sqrt{II_{\sigma'}}} \le 1 \tag{1.1.10}$$

for $\alpha = \pi/6$ and $\alpha = 0$. The conclusion is that $\tau_{\max}$ and $\sqrt{II_{\sigma'}}$ is quite close to each other.

Thus a point of coordinates $\sigma_1, \sigma_2, \sigma_3$ can be replaced by a point σ, τ, α, introducing the three invariants: the first invariant of the stress, and the second and third invariants of the stress deviator (Fig. 1.1.3).

We have still show how a quantity is projected on the octahedral plane.

We have $\cos(\mathbf{i}_3, \mathbf{i}_3') = \sqrt{2/3}$, and just any quantity which is projected on the octahedral plane is to be multiplied by $\sqrt{2/3}$ (Fig. 1.1.4).

1.2 Yield Conditions

It is assumed that when deforming plastically all materials are satisfying a condition called yield condition. At least most metals satisfy this condition. The yield conditions satisfy a few general conditions. Let us describe them.

<u>Assumption 1.</u> We assume that for each material a <u>yield function</u> $f(\boldsymbol{\sigma})$ exists so that:

$$f(\boldsymbol{\sigma}) < 0 \quad \text{or if } f(\boldsymbol{\sigma}) = 0 \text{ and } \frac{\partial f}{\partial \sigma_{ij}}\dot{\sigma}_{ij} < 0 \text{ the material is elastic};$$

$$f(\boldsymbol{\sigma}) = 0 \quad \text{and} \begin{cases} \dfrac{\partial f}{\partial \sigma_{ij}}\dot{\sigma}_{ij} = 0 & \text{for loading in perfect plasticity}, \\[2mm] \dfrac{\partial f}{\partial \sigma_{ij}}\dot{\sigma}_{ij} > 0 & \text{for loading for work-hardening plasticity}. \end{cases}$$

In order to give an example let us consider the yield criteria of von Mises which can be written $f(\boldsymbol{\sigma}) := II_{\sigma'} - k^2$. Here k if it is constant the material is perfectly plastic; if however k is changing when the plastic deformation takes place, that is $k(\varepsilon^P)$, the material is work-hardening.

<u>Assumption 2.</u> For metals the yield is independent of the spherical part of the stress. Thus f depends on the stress deviator $f(\boldsymbol{\sigma})$. That means that the yield conditions are cylindrical in the stress space.

<u>Assumption 3.</u> The material is initially isotropic. Therefore $f(I_\sigma, II_\sigma, III_\sigma) = k^2$ the yield function depends on invariants only. There are no preferred directions; f does not depend on the orientation of the principal axes. Taking into account *the* previous assumption, f depends on the deviator invariants:

$$f(II_{\sigma'}, III_{\sigma'}) = k^2 \,. \tag{1.2.1}$$

<u>Assumption 4.</u> Since for most metals the curves σ–ε are symmetric with respect to the origin of axes, we must have $f(\boldsymbol{\sigma}) = f(-\boldsymbol{\sigma})$. Because $II_{\sigma'} = II_{(-\sigma')}$ the third deviator invariant must be involved in <u>even</u> powers.

<u>The Tresca [1868] Yield Condition.</u> It is, in the octhaedrical plane, a regular hexagon as shown in Fig. 1.2.1.

Thus the Tresca hexagon is define as $\tau_{\max} = k = const.$ It can be expressed in terms of invariants as described previously. Thus we can write

$$(\sigma_1' - \sigma_2') = 4k^2 \,,$$

$$\sigma_1'\sigma_2' + \sigma_2'\sigma_3' + \sigma_3'\sigma_1' = -II_{\sigma'} \,,$$

$$\sigma_1'\sigma_2'\sigma_3' = III_{\sigma'} \,,$$

$$\sigma_1' + \sigma_2' + \sigma_3' = 0 \,.$$

We can eliminate all the deviator stress components to get

$$\frac{27 II_{\sigma'}^3}{64k^6}\left[\frac{4}{27} - \frac{III_{\sigma'}^2}{II_{\sigma'}^3}\right] - \left[1 - \frac{3II_{\sigma'}}{4k^2}\right]^2 = 0,$$

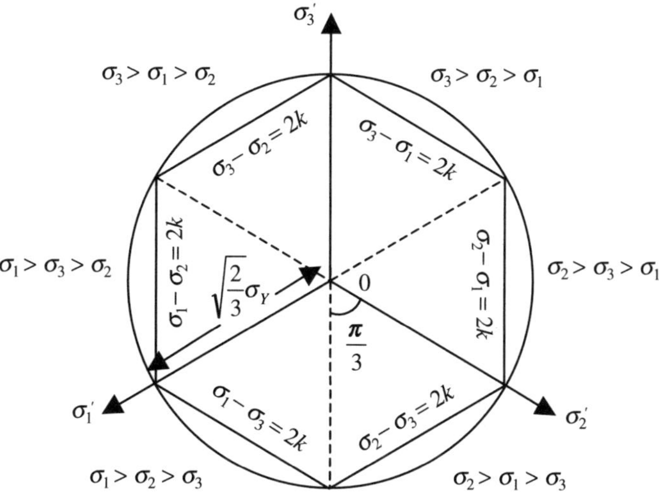

Fig. 1.2.1   The Tresca hexagon.

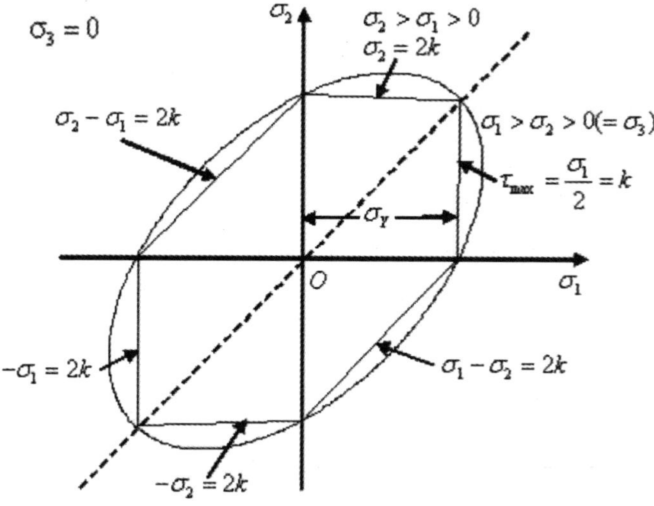

Fig. 1.2.2   The plane Tresca hexagon.

which is due to Reuss. That is a complicated way of expressing the Tresca yield condition, so that today it is expressed also as $\tau_{\max} = k$.

For the plane problems it is expressed in a simpler way as in Fig. 1.2.2.

For the Tresca condition the relationship between the yield stresses in shear and tension is $2\tau_Y = \sigma_Y$. The number $1/2$ is a little too small, experimentally it is ranging somewhere between 0.55 and 0.6.

The Mises [1913] Yield Condition. The von Mises yield condition is expressed as $II_{\sigma'} = k^2$, where

$$II_{\sigma'} = \frac{1}{6}[(\sigma_{11} - \sigma_{22})^2 + (\sigma_{22} - \sigma_{33})^2 + (\sigma_{33} - \sigma_{11})^2] + \sigma_{12}^2 + \sigma_{23}^2 + \sigma_{31}^2$$

$$= \frac{1}{2}[(\sigma_1')^2 + (\sigma_2')^2 + (\sigma_3')^2].$$

The von Mises yield condition is represented in the three axial stress space as a circular cylinder shown in the last two figures. For the plane case it is an ellipse. For the von Mises yield condition the relationship between the yield stress in shear and tension is $\tau_Y = \sigma_Y/\sqrt{3} \cong 0.557\sigma_Y$. That is a much better value. All the correspondences of the two yield condition and the tests are shoving a better fitting of the von Mises than that of Tresca. Also, any generalization of this condition to for anisotropic materials is a combination of the two conditions.

1.3 The Classical Constitutive Equation for Perfectly Plastic Materials

Saint-Venant–Mises Constitutive Equation

The first law of plasticity is due to Barré de Saint-Venant [1870]. He has done several assumptions which have been inspired from metal working theories but the meaning of the plastic flow was changed. They are:

$$\varepsilon^E = 0 \quad \text{the elastic part of the strain rate is negligible}$$

since $\varepsilon_V^P \approx 0$ plastic incompressibility is assumed, i.e., $\boldsymbol{D} = \boldsymbol{D}'$, $\boldsymbol{\sigma} = \lambda \boldsymbol{D}$ as in Newtonian viscosity, but λ is variable, λ is determined from a plasticity condition; squaring the above law we have

$$\boldsymbol{\sigma}' : \boldsymbol{\sigma}' = \lambda^2 \boldsymbol{D} : \boldsymbol{D} \quad \text{or} \quad \lambda = \frac{\sqrt{\boldsymbol{\sigma}' : \boldsymbol{\sigma}'}}{\sqrt{\boldsymbol{D} : \boldsymbol{D}}}$$

and for a von Mises plasticity condition $II_{\sigma'} = k^2 = conts.$ for perfect plasticity:

$$\lambda = \frac{k}{\sqrt{II_D}}.$$

Thus the constitutive equation is:

$$\boldsymbol{\sigma}' = \frac{k}{\sqrt{II_D}}\boldsymbol{D} \quad \text{and} \quad \text{tr}\,\boldsymbol{D} = 0 \quad \text{if } II_{\sigma'} = k^2 \quad \text{for } \underline{\text{perfectly plastic}},$$

$$\sqrt{II_{\sigma'}} < k \quad \text{or} \quad II_D = 0 \qquad\qquad\qquad \text{for } \underline{\text{rigid behavior}}.$$

The meaning of k follows from the yield conditions. For von Mises yield conditions

$$\frac{1}{2}\boldsymbol{\sigma}' : \boldsymbol{\sigma}' = \frac{1}{2}(\sigma_{11}'^2 + \sigma_{22}'^2 + \sigma_{33}'^2) + \sigma_{12}^2 + \sigma_{23}^2 + \sigma_{31}^2 = k^2$$

if all $\sigma_{ij}' = 0$ besides $\sigma_{12} \neq 0$ we obtain $\sigma_{12}^2 = k^2$ or, if the yield stress in pure shear is τ, $\tau = k$. In uniaxial tensile tests $\sigma_{11} \neq 0$ all other $\sigma_{ij} = 0$, we obtain $Y^2/3 = k^2$, or $k = Y/\sqrt{3}$, where Y is the yield stress in uniaxial tensile (or compressive) tests. Thus we arrive at the relation

$$\tau_Y = \frac{\sigma_Y}{\sqrt{3}},$$

for the yield stresses in pure shear and tensile, for the von Mises yield condition.

The classical constitutive equation seems to be a non-Newtonian fluid, but it is not, since it is time-independent (nonviscous). To show that we write the constitutive equation as

$$\frac{\boldsymbol{D}}{\sqrt{II_D}} = \frac{\boldsymbol{\sigma}'}{k},$$

and we change v with μv, with $\mu > 0$, that is we change the time. We have $\boldsymbol{D} \to \mu \boldsymbol{D}$, and $II_D \to \mu^2 II_D$ thus μ disappears from the constitutive equation.

Thus the main properties of the constitutive equation are time independent, the reference configuration is the present one (in the case we have no internal changes) $\boldsymbol{D}$ is involved as for fluids; it is used when the plastic deformation are significant, the elastic part of the strain rate are negligible, as in some metal working problems.

The Prandtl–Reuss Constitutive Equation

This is a constitutive equation developed by Prandtl [1924] for two-dimensional cases and extended by Reuss [1930] for the three-dimensional case. Assuming small deformations the constitutive equations are:

$$\dot{\boldsymbol{\varepsilon}} = \dot{\boldsymbol{\varepsilon}}^E + \dot{\boldsymbol{\varepsilon}}^P,$$

$$\dot{\boldsymbol{\varepsilon}}^{E'} = \frac{1}{2G}\dot{\boldsymbol{\sigma}}',$$

$$\mathrm{tr}\,\dot{\boldsymbol{\sigma}} = 3K\,\mathrm{tr}\,\dot{\boldsymbol{\varepsilon}},$$

$$\dot{\boldsymbol{\varepsilon}}^P = \frac{\lambda}{2G}\boldsymbol{\sigma}', \quad \mathrm{tr}\,\dot{\boldsymbol{\varepsilon}}^P = 0.$$

$$(1.3.1)$$

Thus the strain rates are additive, the reference configuration is usually the initial one, G is the shearing modulus, and K is the bulk modulus with $G = E/2(1 + v)$ and $K = E/3(1 - 2v)$. From there one receive

$$2G\dot{\varepsilon}_{ij}' = \dot{\sigma}_{ij}' + \lambda\sigma_{ij}'. \qquad (1.3.2)$$

In (1.3.2) we have 10 variables, 3 velocities, 6 stresses and λ. If we assume the von Mises yield condition $II_{\sigma'} = k^2$, we obtain either

$$\dot{\varepsilon}_{ij}' = \frac{\dot{\sigma}_{ij}'}{2G} + \frac{\sqrt{II_{\dot{\varepsilon}'^P}}}{k}\sigma_{ij}'$$

or

$$\dot{\varepsilon}'_{ij} = \frac{\dot{\sigma}'_{ij}}{2G} + \frac{\dot{W}^P}{2k^2}\sigma'_{ij}$$

with $\dot{W}^P = \sigma_{ij}\dot{\varepsilon}^P_{ij}$ the stress work per unit volume.

With all these the Prandtl–Reuss constitutive equation for perfectly plastic materials is

$$\left.\begin{aligned}\dot{\varepsilon}'_{ij} &= \frac{\dot{\sigma}'_{ij}}{2G} + \frac{\sqrt{II_{\varepsilon'}}}{k}\sigma'_{ij}\\[2mm]\operatorname{tr}\dot{\boldsymbol{\sigma}} &= 3K\,\operatorname{tr}\dot{\boldsymbol{\varepsilon}}\end{aligned}\right\} \quad \text{if } II_{\sigma'} = k^2 \quad \text{and} \quad \sigma'_{ij}\dot{\sigma}'_{ij} = 0\,,$$

$$\left.\begin{aligned}\dot{\varepsilon}'_{ij} &= \frac{\dot{\sigma}'_{ij}}{2G}\\[2mm]\operatorname{tr}\dot{\boldsymbol{\sigma}} &= 3K\,\operatorname{tr}\dot{\boldsymbol{\varepsilon}}\end{aligned}\right\} \quad \begin{aligned}&\text{if } II_{\sigma'} < k^2 \quad \text{or}\\[2mm]&\text{if } II_{\sigma'} = k^2 \quad \text{and} \quad \sigma'_{ij}\dot{\sigma}'_{ij} < 0\,.\end{aligned}$$

$$(1.3.3)$$

The relation $\sigma'_{ij}\dot{\sigma}'_{ij} = 0$ is called <u>consistency</u> condition. The reference configuration is either the actual one or the configuration after unloading from current configuration. Also, $\boldsymbol{\varepsilon}^E$ does not satisfy the Saint-Venant compatibility conditions, since there may be no single valued continuous displacement field that would take the whole body from deformed configuration to an unstressed configuration. Thus $\dot{\boldsymbol{\varepsilon}}^P$ is not explicitly integrable. It is easy to show that the Prandtl–Reuss constitutive equation is time-independent, i.e., they are nonviscous.

<u>The Hencky Constitutive Equation</u>

This is a constitutive equation developed by H. Hencky [1924], tough A. Nadai [1923] has used it for torsion problems, and afterwards it was used by A. A. Iliushin [1961]. The main assumption is that one can write it as for Hooke law, but the "plastic" parameters are variable, not constant. Thus:

$$\boldsymbol{\sigma}' = 2G_P\boldsymbol{\varepsilon}'\,,$$

$$\sigma = 3K_P\varepsilon\,.$$

But K_P coincides with K because the volume behavior is elastic. So it is only G which can be written:

$$2G_P = \frac{\sqrt{II_{\sigma'}}}{\sqrt{II_{\varepsilon'}}} = \frac{S}{E}\,,$$

if one uses the Sokolovsckii notation. Thus for work-hardening condition:

$$S = F(E)\,,$$

$$\boldsymbol{\sigma}' = \frac{F(E)}{E}\boldsymbol{\varepsilon}'\,,$$

$$\sigma = 3K\varepsilon\,,$$

$$(1.3.4)$$

or,

$$E = F^{-1}(S),$$

$$\varepsilon' = \frac{F^{-1}(S)}{S}\sigma',$$

$$\dot{\varepsilon} = \frac{\dot{\sigma}}{3K} \text{ for all cases.}$$

(1.3.5)

One has to observe that this constitutive equation was extensively used because it is easily reversible; it is written in finite form and called constitutive equation of "plastic deformation"; it was used when the strain are increasing continuously; for proportional loading paths $\sigma(X,t) = \lambda(t)\sigma^0(X)$, the Prandtl–Reuss constitutive equation coincides with the Hencky (Iliushin [1961]); the reference configuration is the initial one.

We can think to another variant of the Hencky constitutive equation by assuming:

$$\varepsilon = \varepsilon^E + \varepsilon^P,$$

ε^E satisfy the Hooke constitutive equation, the plastic strain satisfies

$$\varepsilon^{P'} = \left(\frac{F^{-1}(S)}{S} - \frac{1}{2G}\right)\sigma'.$$

(1.3.6)

If we differentiate the constitutive equation we can write:

$$\dot{\varepsilon}' = \begin{cases} \dfrac{F^{-1}(S)}{S}\dot{\sigma}' + \dfrac{d}{dt}\left(\dfrac{F^{-1}(S)}{S}\right)\sigma' & \text{for } E = F^{-1}(S) \text{ and } \dot{S} > 0 \quad \text{for loading}, \\[4mm] \dfrac{\dot{\sigma}'}{2G} & \text{for } \overbrace{E = F^{-1}(S) \text{ and } \dot{S} \leq 0}^{\text{for unloading}} \text{ or } \overbrace{E < F^{-1}(S)}^{\text{elastic}} \end{cases}$$

$$\dot{\varepsilon} = \frac{\sigma}{3K} \quad \text{for all cases.}$$

The Hencky's constitutive equation is obviously time independent.

1.4 Work-Hardening Materials

We assume for work-hardening materials small strains so that

$$\varepsilon = \varepsilon^E + \varepsilon^P.$$

The elastic rate of deformation components satisfy

$$\dot{\varepsilon}_{ij}^E = A_{ijkl}\dot{\sigma}_{kl}.$$

The yield function F exists which satisfy

$$F(\sigma_{ij}, \varepsilon_{kl}^P, \chi) = 0$$

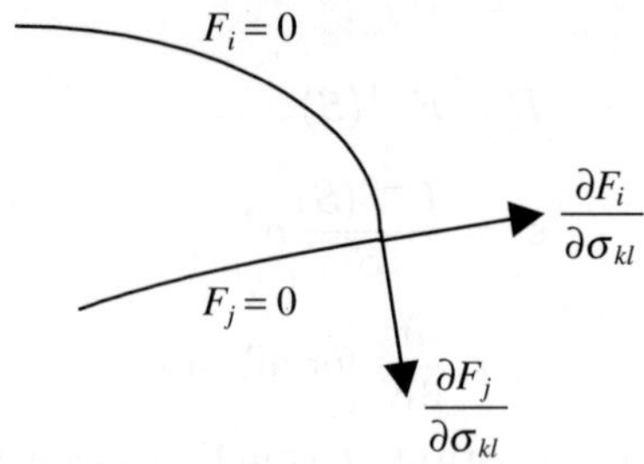

Fig. 1.4.1   The Koiter generalization.

with $F(0, \varepsilon_{ij}^P, \chi) < 0$ at reference configuration with ε_{ij}^P and χ fixed at a particle. χ is a scalar called work-hardening parameter.

If $F(\sigma_{ij}, \chi)$, i.e., ε_{ij}^P is not explicitly involved the work-hardening is called isotropic. If however, $F(\sigma_{ij}, \varepsilon_{kl}^P)$ the work-hardening is inducing anisotropy.

The plastic rate of deformation is given by:

$$
\dot{\varepsilon}_{ij}^P =
\begin{cases}
0 & \text{for}
\begin{cases}
F(\sigma_{ij}, \varepsilon_{kl}^P, \chi) < 0 \quad \text{or} \\[2mm]
F(\sigma_{ij}, \varepsilon_{kl}^P, \chi) = 0 \quad \text{and} \quad \dfrac{\partial F}{\partial \sigma_{ij}}\dot{\sigma}_{ij} \le 0\,. \\
\end{cases} \\[6mm]
B_{ij}(\sigma_{kl}, \varepsilon_{mn}^P, \chi, \dot{\sigma}_{ij}) & \text{for } F(\sigma_{ij}, \varepsilon_{kl}^P, \chi) = 0 \quad \text{and} \quad \dfrac{\partial F}{\partial \sigma_{ij}}\dot{\sigma}_{ij} > 0\,.
\end{cases}
$$

The initial data are defined by $\varepsilon^P(0) = \varepsilon_0^P$, and $\chi(0) = \chi_0$, thus a reference configuration in assumed, not necessarily the stress-free, strain-free configuration.

Since it is assumed that both at time t and $t + \Delta t$ the yield condition is satisfied during a loading, we have

$$
\frac{\partial F}{\partial \sigma_{ij}}\dot{\sigma}_{ij} + \frac{\partial F}{\partial \varepsilon_{ij}^P}\dot{\varepsilon}_{ij}^P + \frac{\partial F}{\partial \chi}\dot{\chi} = 0\,,
$$

which shows that all the increments are not independent. This is the consistency condition.

Because when approaching the yield condition by continuity, we must have:

$$
\lim B_{ij}(\sigma_{kl}, \varepsilon_{mn}^P, \chi, \dot{\sigma}_{ij}) = 0\,,
$$

$$
\frac{\partial F}{\partial \sigma_{ij}}\dot{\sigma}_{ij} \to 0\,,
$$

which is the continuity condition. That means that only the normal component of the increment of the stress tensor is giving a plastic increment of the strain field. Thus we have

$$
\dot{\varepsilon}_{ij}^P = \lambda(\sigma_{kl}, \dot{\sigma}_{mn})\frac{\partial F}{\partial \sigma_{ij}}
$$

and we call the constitutive law <u>associated</u> to the yield condition. That is true for most metals, but not true for rocks or soils, for instance. For such materials we have to introduce a <u>plastic potential</u> following the idea of von Mises from 1928:

$$\dot{\varepsilon}_{ij} = \mu(\sigma_{kl}, \dot{\sigma}_{mn})\frac{\partial H}{\partial \sigma_{ij}},$$

where H is the plastic potential. For such materials the constitutive equation is called <u>nonassociated</u>.

There are yield conditions which have several constitutive equation meeting at a point. For instance the Tresca yield condition is of this kind (see Fig. 1.4.1). In this case we use the idea of Koiter [1953]:

$$\dot{\varepsilon}_{ij}^{P} = \lambda_k \frac{\partial F_k}{\partial \sigma_{ij}}$$

with the constants λ_k defined by:

$$\lambda_k > 0 \quad \text{if} \quad F_k(\sigma_{mn}) = 0 \quad \text{and} \quad
\begin{cases}
\dfrac{\partial F_k}{\partial \sigma_{ij}}\dot{\sigma}_{ij} = 0 & \text{for perfect plasticity} \\[3mm]
\dfrac{\partial F_k}{\partial \sigma_{ij}}\dot{\sigma}_{ij} > 0 & \text{for working-hardening}
\end{cases}$$

$$\lambda_k = 0 \quad \text{if}
\begin{cases}
F_k(\sigma_{mn}) < 0 \quad \text{or} \\[3mm]
F_k(\sigma_{mn}) = 0 \quad \text{and}
\begin{cases}
\dfrac{\partial F_k}{\partial \sigma_{mn}}\dot{\sigma}_{mn} < 0 & \text{for perfect plasticity} \\[3mm]
\dfrac{\partial F_k}{\partial \sigma_{mn}}\dot{\sigma}_{mn} \leq 0 & \text{for working-hardening}.
\end{cases}
\end{cases}$$

Thus, only at the corner are involved several values of λ_k.

1.5 Isotropic Hardening

For the isotropic hardening the yield condition is written $F(\sigma_{ij}, \chi) : f(\sigma_{ij}) - H(\chi)$, and in a specific form $f(II_{\sigma'}, III_{\sigma'}) = H(\chi)$.

It is interesting to follow the way in which the work-hardening parameter is determined. It can be the irreversible stress work per unit volume (Fig. 1.5.1), i.e.,

$$\chi(t) = W^P(t) = \int_0^t \sigma_{ij}(X, s)\dot{\varepsilon}_{ij}^P(X, s)\, ds$$

or

$$\chi(t) = \int_0^t \sqrt{\frac{4}{3}II_{\dot{\varepsilon}^P}}\, ds = \int_0^t \sqrt{\frac{2}{3}d\varepsilon_{ij}^P d\varepsilon_{ij}^P}\, ds \left(= \int_0^t \bar{\dot{\varepsilon}}^P(s)\, ds\right),$$

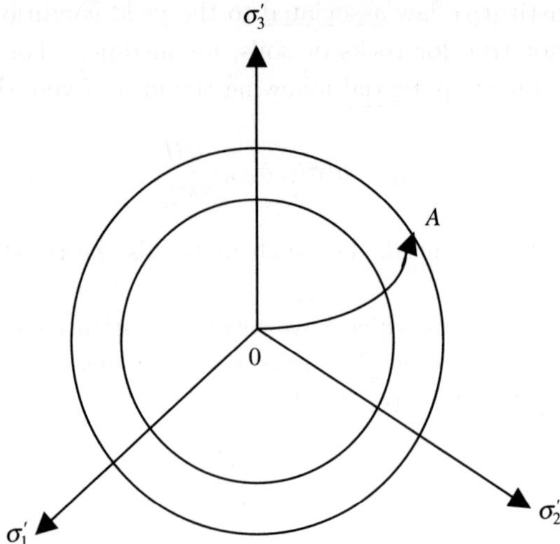

Fig. 1.5.1   The isotropic hardening.

where $\bar{\varepsilon}^P$ is not a derivative. In this case $\chi(t)$ is the history of the effective plastic strain increment. However we can still consider by definition:

$$\chi(t) = \bar{\varepsilon}^P(t) = \sqrt{\frac{2}{3}\varepsilon_{ij}^P(t)\varepsilon_{ij}^P(t)}$$

and the work-hardening parameter is equivalent to the plastic strain.

For the von Mises yield condition written as:

$$\bar{\sigma} = \sqrt{3}k$$

with $\bar{\sigma}$ the equivalent stress:

$$\bar{\sigma} = \sqrt{\frac{3}{2}}(\sigma_1'^2 + \sigma_2'^2 + \sigma_3'^2)^{1/2} = \sqrt{3II_{\sigma'}}$$

$$= \left\{\frac{1}{2}[(\sigma_{11} - \sigma_{22})^2 + (\sigma_{22} - \sigma_{33})^2 + \cdots] + 3(\sigma_{23}^2 + \cdots)\right\}^{1/2}.$$

If only one component, say, $\sigma_{11} \neq 0$, and all the others $\sigma_{ij} = 0$, then $\bar{\sigma} = |\sigma_{11}|$. For such a component we write:

$$\bar{\sigma}(t) = F(W^P(t)),$$

$$\bar{\sigma}(t) = H\left(\int_0^t \bar{\varepsilon}^P(s)\,ds\right),$$

$$\bar{\sigma}(t) = G(\bar{\varepsilon}^P(t)),$$

which describe various possible variants of the isotropic work-hardening.

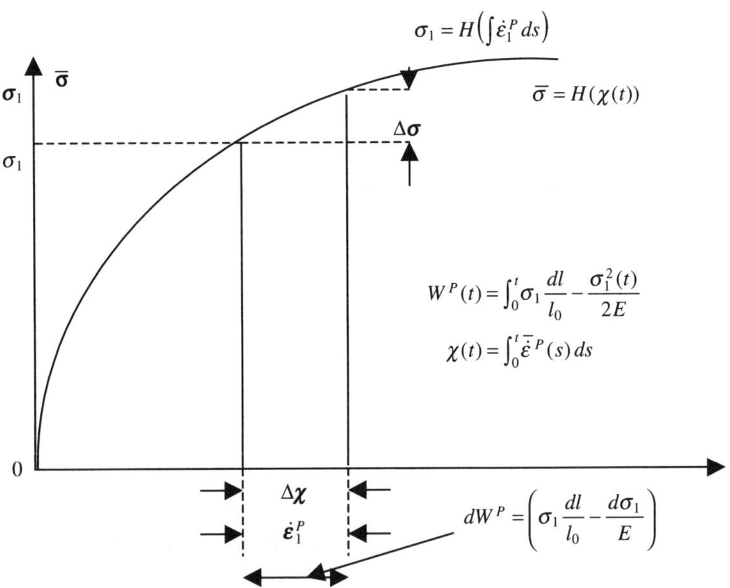

Fig. 1.6.1   The universal stress–strain curve.

1.6   The Universal Stress–Strain Curve

If we do a single test we have $\bar{\sigma} = \sigma_{11} = \sigma_1$ (assume that σ_{11} is positive). If the specimen is cylindrical, $\varepsilon_2^P = \varepsilon_3^P$, and if we assume plastic incompressibility $\varepsilon_1^P = -2\varepsilon_2^P$. Thus if we take all these into account we have (Fig. 1.6.1):

$$\int \sqrt{\frac{2}{3} d\varepsilon_{ij}^P d\varepsilon_{ij}^P} = \int \sqrt{\frac{2}{3}\left(d\varepsilon_1^{P2} + \frac{1}{2}d\varepsilon_1^{P2}\right)} = \int \dot{\varepsilon}_1^P \, ds \, .$$

Thus if we are doing a single test we discover a three-dimensional law to be applied generaly. We have to apply the general law since the function H is already known.

If the constitutive equation is

$$\dot{\varepsilon}_{ij}^P(t) = \lambda(t)\sigma_{ij}'(t)$$

with $\lambda \geq 0$, i.e., the principal directions of $\dot{\varepsilon}_{ij}^P$ and of σ_{ij}' <u>coincide</u>, then we have

$$\dot{W}^P(t) = \bar{\sigma}(t)\bar{\dot{\varepsilon}}^P(t) \, .$$

That is easy to show since

$$\sigma_{ij}'\lambda\sigma_{ij}' = \sqrt{\frac{3}{2}\sigma_{ij}'\sigma_{ij}'}\sqrt{\frac{2}{3}\dot{\varepsilon}_{kl}^P\dot{\varepsilon}_{kl}^P} = \bar{\sigma}\bar{\dot{\varepsilon}}^P \, .$$

Thus the work-hardening law is

$$\bar{\sigma}(t) = F\left(\int_0^t \bar{\sigma}(s)\bar{\dot{\varepsilon}}(s)\, ds\right) \, .$$

We have the suggestion to write this law

$$\bar{\sigma}(t) = H\left(\int_0^t \bar{\dot{\varepsilon}}^P(s)\,ds\right)$$

and it is easy to pass from one to the other.

If the variation of the plastic strains is proportional, i.e.,

$$\varepsilon_{ij}^P(t, X) = h(t)\varepsilon_{ij_0}^P(X)$$

at every particle X. Then

$$\frac{d\varepsilon_{ij}^P}{dt} = \varepsilon_{ij_0}^P \frac{dh}{dt}$$

i.e.,

$$\bar{\dot{\varepsilon}}^P = \bar{\varepsilon}_0^P \frac{dh}{dt}\,.$$

By integration with respect to time at a fixed particle X with $h(0) = 0$, we get

$$\int_0^t \bar{\dot{\varepsilon}}^P(s)\,ds = \bar{\varepsilon}_0^P h(t)\,.$$

Generally a proportional variation of ε_{ij}^P in every point of the body is not possible, but in very special designed laboratory test. The equality

$$\int_0^t \bar{\dot{\varepsilon}}^P(s)\,ds = \bar{\varepsilon}^P(t)$$

do not hold for all practical cases. Thus the work-hardening condition

$$\bar{\sigma}(t) = G(\bar{\varepsilon}^P(t))$$

is generally not correct, though extensively used. It disregards the history dependence of the plastic deformation. This is a closed relation to nonlinear elasticity.

1.7   Constitutive Equation for Isotropic Work-Hardening Materials

From the work-hardening condition

$$f(\sigma_{ij}') = F(\chi)\,,$$

we get the consistency condition

$$\frac{\partial f}{\partial \sigma_{ij}'}\dot{\sigma}_{ij}' = \frac{\partial F}{\partial \chi}\dot{\chi} = \frac{\partial F}{\partial \chi}\sigma_{kl}'\frac{\lambda}{2G}\sigma_{kl}'\,.$$

If we solve with respect to λ and we introduce in the constitutive equation

$$\dot{\varepsilon}_{ij}^P = \frac{\lambda}{2G}\sigma_{ij}'\,.$$

We get

$$\dot{\varepsilon}_{ij}^{P} = \frac{1}{2II_{\sigma_{ij}}(\partial F/\partial \chi)} \left(\frac{\partial f}{\partial \sigma'_{kl}} \dot{\sigma}'_{kl} \right) \sigma'_{ij} \,.$$

Thus the general constitutive equation is

$$\dot{\varepsilon}_{ij} = \frac{\dot{\sigma}'_{ij}}{2G} + \frac{1}{2II_{\sigma'}(\partial F/\partial \chi)} \left\langle \frac{\partial f}{\partial \sigma'_{kl}} \dot{\sigma}'_{kl} \right\rangle \sigma'_{ij}$$

with the bracket $\langle \ \rangle$ defined by

$$\langle A \rangle = \begin{cases} A \quad \text{if} \quad f(\sigma'_{ij}) = F(\chi) \quad \text{and} \quad \dfrac{\partial f}{\partial \sigma'_{ij}} \dot{\sigma}'_{ij} > 0 \\[2ex] 0 \quad \text{if} \begin{cases} f(\sigma'_{ij}) = F(\chi) \quad \text{and} \quad \dfrac{\partial f}{\partial \sigma'_{ij}} \dot{\sigma}'_{ij} \leq 0 \quad \text{or} \\[2ex] f(\sigma'_{ij}) < F(\chi) \,. \end{cases} \end{cases}$$

If we write the yield condition in the form

$$F(\sigma_{ij}, \chi) := f(\sigma'_{ij}) - H(\chi) \,,$$

then the constitutive equation is

$$\dot{\varepsilon}_{ij}^{P} = \frac{\langle (\partial F/\partial \sigma'_{kl}) \dot{\sigma}'_{kl} \rangle}{(\partial F/\partial \chi)(\partial F/\partial \sigma'_{mn}) \sigma'_{mn}} \frac{\partial F}{\partial \sigma'_{ij}}$$

with a similar definition of the bracket $\langle \ \rangle$.

1.8   The Drucker's Postulate

For stable work-hardening materials Drucker [1951] has established the following postulate. Assume that a stress–strain curve is built. It is showing as in the Fig. 1.8.1. We start from a stress σ_0. We load by producing also some plastic deformation $\Delta \varepsilon_1^{P}$. Then we unload, up to the initial stress σ_0. If the material is a stable work-hardening, then two conditions are to be satisfied:

$$\dot{\varepsilon}_{ij}^{P} \dot{\sigma}_{ij} \geq 0 \,,$$

$$\dot{\varepsilon}_{ij}^{P} (\sigma_{ij} - \sigma_{ij}^{0}) \geq 0 \,.$$

Thus the plastic work done by external agency during the application of an additional stress is positive and the net total work performed by the external agency during the cycle of adding and removing stress is non-negative.

The Drucker's postulate has significant consequences for metal plasticity. Thus the yield condition must be convex, nowhere concave. That is shown in Fig. 1.8.2.

Another similar postulate was invented by Iliushin, but it is connected to the strain concept. One is starting from a certain strain ε_{ij} and one make a loading

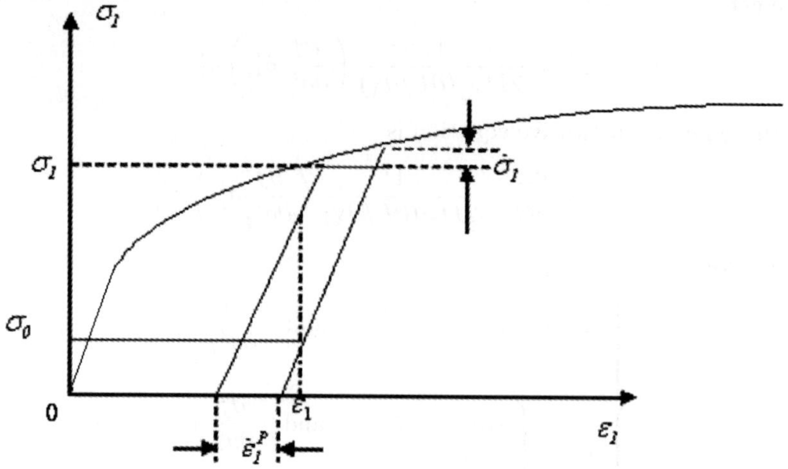

Fig. 1.8.1   The Drucker's postulate.

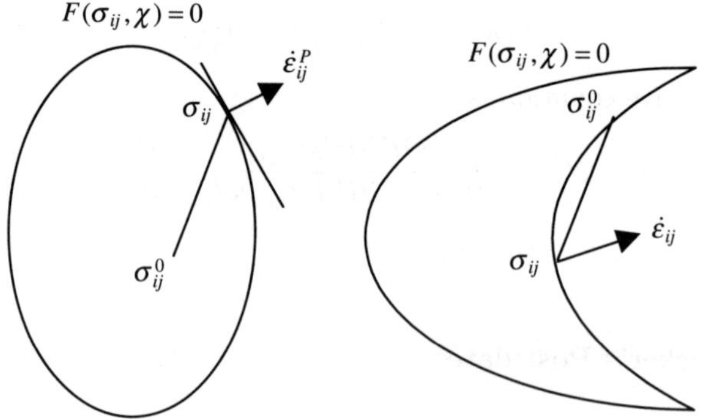

Fig. 1.8.2   Consequences of the Drucker's postulate.

and unloading until one reaches the same strain (see Fig. 1.8.1). A weakened form of Iliushin's postulate, which says that the changing rate of the stress work done along every standard strain cycle should be non-negative, whenever the incorporated plastic sub path tends to vanish, is due to Bruhns *et al.* [2005]. The Drucker's postulate has been extensively considered in the literature. It is a nonenergetic and equivalent to associated flow rule (see Stoughton and Yon [2005] for literature).

1.9   Kinematical Work-Hardening

Another work-hardening law is the kinematical work-hardening. It is shown in Fig. 1.9.1 for a plastic rigid model. The yield surface does not change shape; it translates in the stress space. The size of the elastic domain remains constant and

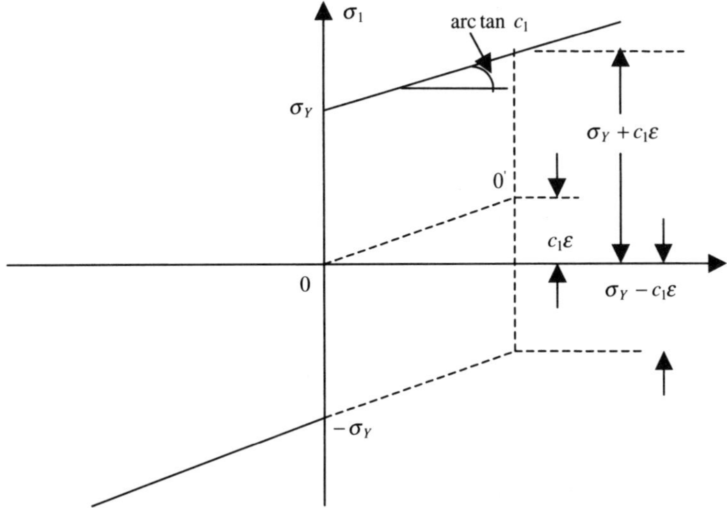

Fig. 1.9.1   The work-hardening model.

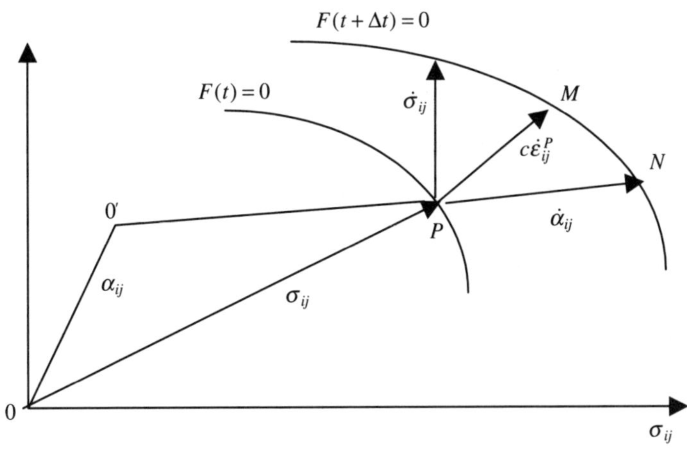

Fig. 1.9.2   The kinematical work-hardening assumptions.

equal to $2\sigma_Y$ (rigid). The origin translates with $c_1\varepsilon$ in the case of linear work-hardening/rigid model. c_1 is a constant work-hardening modulus. This model is describing an idealized Bauschinger effect.

For the general three-dimensional case the initial surface is $F(\sigma_{ij}) = 0$. The translated surface is $F(\sigma_{ij} - \alpha_{ij}) = 0$, where the coordinates of the new origin are α_{ij}. Some additional postulates are to be satisfied by α_{ij} in order to satisfy the history of the plastic deformation. It must satisfy the initial condition $\boldsymbol{\alpha} = 0$ if $\varepsilon^P = 0$. It is called sometimes "back stresses".

Prager [1959] postulated that the yield surface translates in the direction normal $\partial F/\partial\sigma_{ij}$ at the current stress state σ_{ij} in the nine-dimensional space, i.e., in the direction of PM. That is

$$\dot{\alpha}_{ij} = c\dot{\varepsilon}_{ij}^P$$

if we assume $\dot{\varepsilon}_{ij}^P \sim \partial F/\partial\sigma_{ij}$. Consideration in spaces with smaller dimensions than nine, may lead to mistakes since the normal may not remain normal.

The constitutive equation is (Fig. 1.9.2)

$$\dot{\varepsilon}(\sigma, \varepsilon^P, \dot{\sigma}) = \eta(\sigma, \varepsilon^P, \dot{\sigma})\frac{\partial F(\sigma - \alpha)}{\partial \sigma}\,.$$

The scalar η is obtained from the consistency condition

$$\frac{\partial F(\sigma - \alpha)}{\partial \sigma} : (\dot{\sigma} - \dot{\alpha}) = 0\,,$$

where, if one takes into account the above relations we have

$$\dot{\varepsilon}_{ij}^P = \frac{(\partial F(\sigma_{kl} - \alpha_{kl})/\partial\sigma_{mn})\dot{\sigma}_{mn}}{c(\partial F(\sigma_{kl} - \alpha_{kl})/\partial\sigma_{gh})(\partial F(\sigma_{kl} - \alpha_{kl})/\partial\sigma_{gh})}\frac{\partial F(\sigma_{kl} - \alpha_{kl})}{\partial\sigma_{ij}}\,.$$

If we make the assumption that the yield surface translates in the direction PN, i.e.,

$$\dot{\alpha} = (\sigma - \alpha)\dot{\mu}\,,$$

where $\dot{\mu} > 0$ is an additional unknown. It is a difficult problem with this additional unknown.

1.10   Further Developments

Combined Isotropic and Kinematical Hardening. Some authors have combined the two models to get a combination in the form

$$f(\sigma - \alpha, \chi) = 0\,,$$

where α and χ are the two work-hardening coefficients, the kinematical and work-hardening.

As an example one can consider yield surfaces of the form

$$f = \frac{1}{2}(\sigma - \alpha)(\sigma - \alpha) - \frac{1}{3}\sigma_Y^2 = 0\,.$$

Mróz [1969] has introduced the multi-surface model to describe the cyclic effect and the nonlinearity of stress–strain loops. There are several such surfaces

$$f_i(\sigma - \alpha, \chi_i) = 0\,, \quad i = 1, 2, \ldots, n\,.$$

Dafalias and Popov [1975] have reduced the number of surfaces at two, one corresponding to the conventional yield surface and the other is a bounding surface.

Thus the two surfaces are

$$f = (\boldsymbol{\sigma} - \boldsymbol{\alpha})^2 - \sigma_Y^2 = 0\,, \quad f^* = (\boldsymbol{\sigma}^* - \boldsymbol{\alpha}^*)^2 - \sigma_Y^{*2} = 0\,,$$

with $\boldsymbol{\alpha}$ and $\boldsymbol{\alpha}^*$ the two centers σ_Y and σ^* their respective sizes.

Valanis [1971] has proposed a new theory in which the present state of the material depends on the present values and the past history of observable variables. The theory is called endochronic and is expressed by

$$\boldsymbol{\sigma}' = 2G \int_0^z \rho(z - z') \frac{d\mathbf{e}^P}{dz'}\, dz'\,,$$

where the yield surface is a derivable result of the theory. $\rho(z)$ is a material function and z is the intrinsic time defined by

$$dz = \frac{d\zeta}{f(\zeta)}\,,$$

where $f(\zeta)$ is the non-negative intrinsic time scale, with $f(0) = 1$ and ζ defined by

$$d\zeta = \sqrt{d\varepsilon^P d\varepsilon^P}\,.$$

Some other further developments will no more be cited.

1.11   Experimental Tests

Some authors have tested the yield surfaces in plasticity or the work-hardening surfaces. Thus Lode [1926] has tested the yield surfaces of Tresca and Mises. Some of his results are shown in Fig. 1.11.1. One can see that he found mostly the Mises yield condition to be satisfied. He has tried steel, copper and nickel, and he find all to better satisfy the Mises condition.

Also from the old results are to be mentioned the Taylor and Quinney [1931] results. They have tested copper, aluminum and mild steel. From their results we

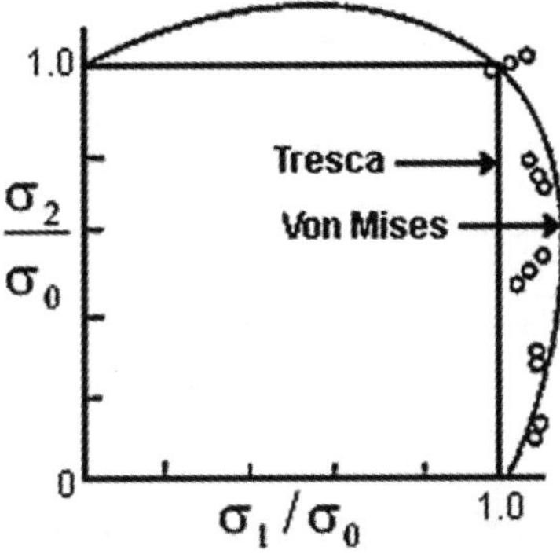

Fig. 1.11.1   The Lode's tests results.

give here Fig. 1.11.2. As one can see according to Taylor and Quinney the Mises yield condition is in better agreement with the data, than the Tresca.

From the old papers devoted to tests is also the paper of Roš and Eichinger [1929]. They have also compared the Tresca and Mises yield condition. Generally they find the Mises yield condition in better agreement with experimental data. That is shown on Fig. 1.11.3 given below.

From more recent papers in which the authors tried to compare the initial and subsequent yield surfaces in plasticity is the paper of Naghdi *et al.* [1958]. The initial surface is an ellipse, while the next ones, when only an shearing stress is applied, are deformed, trying to form an pointed corner at a thin walled tube (Fig. 1.11.4).

Outer tests are reported by Phillips [1986]. Figure 1.11.5 is shoving some of his results, obtained at several temperatures shown. No lateral influence is shown but a significant Bauschinger effect is there. The surfaces are translated and deformed.

The last authors are Williams and Svensson [1970] [1971]. In the first paper they try to apply a tensile plastic strain to 1100-F aluminum. They found again a

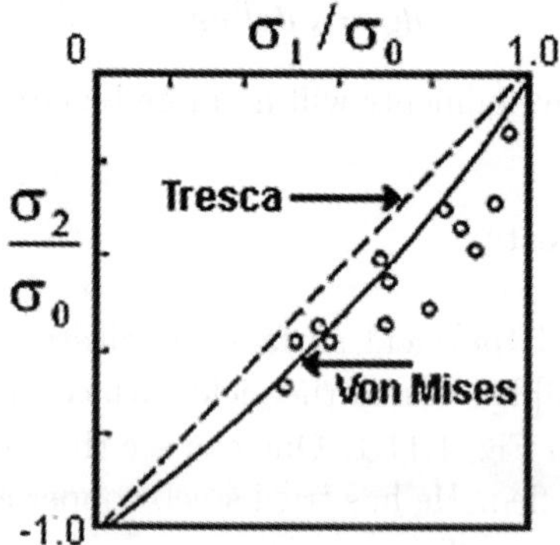

Fig. 1.11.2   Experimental results of Taylor and Quinney [1931].

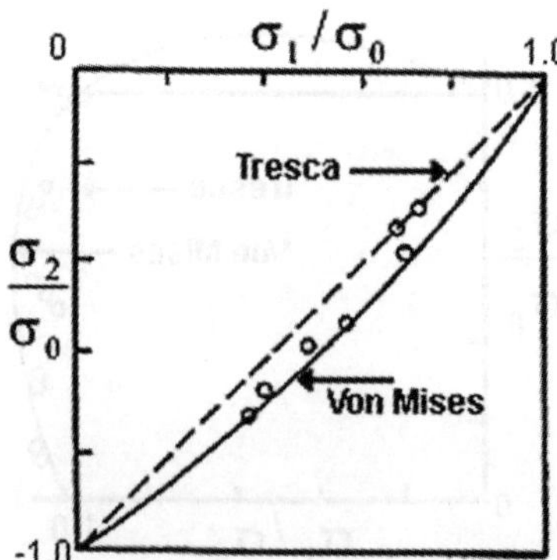

Fig. 1.11.3   Experiments of Roš and Eichinger [1929].

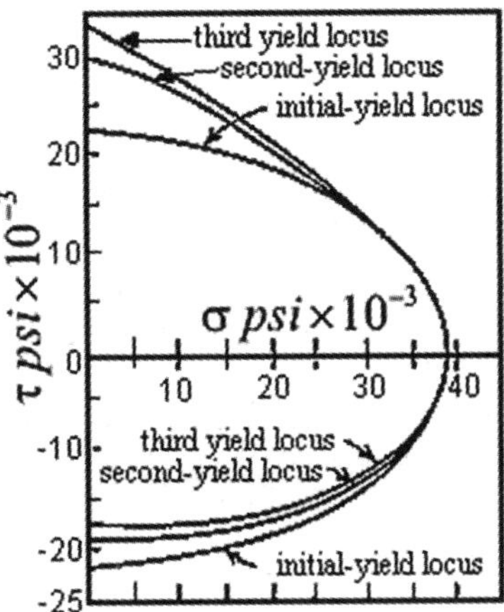

Fig. 1.11.4   The tests of Naghdi *et al.* [1958] for initial and subsequent yield loci.

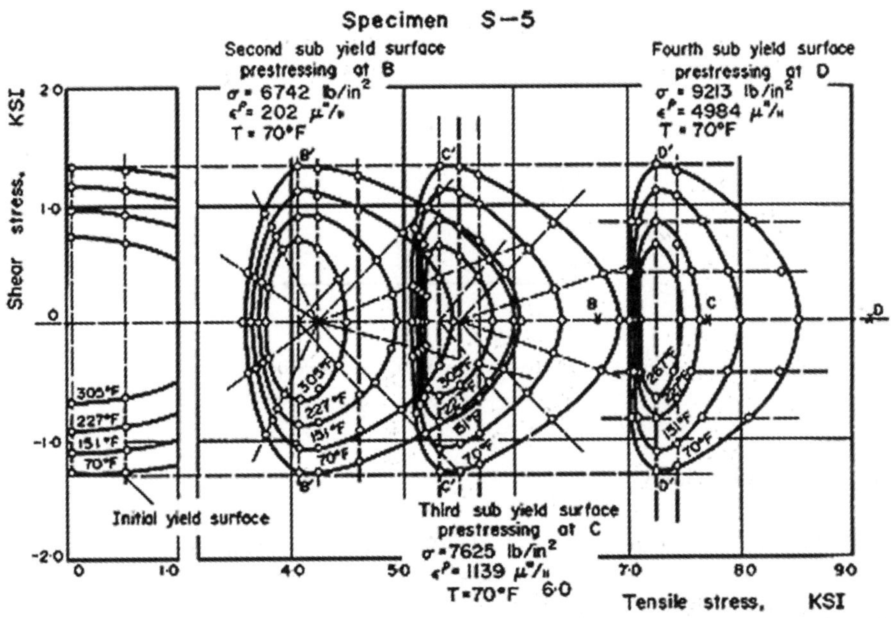

Fig. 1.11.5   Second, third and fourth subsequent yield surfaces (Phillips [1986]).

considerable Bauschinger effect, but they found also a cross effect. They found also corners in tension, as shown in Fig. 1.11.6. In another paper they apply a torsion plastic prestrain. This time they do not find corners, but strong Bauschinger effect is still there.

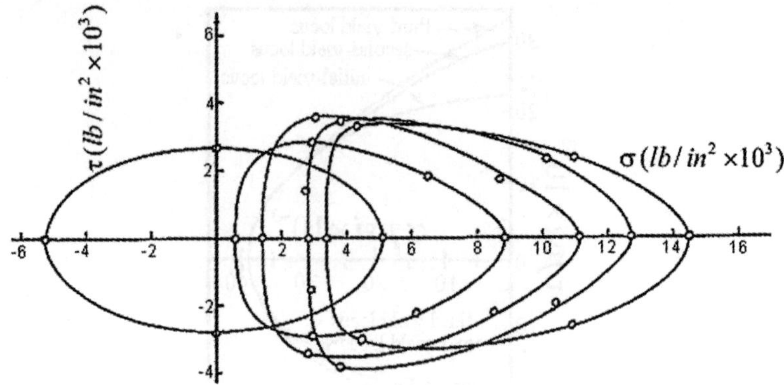

Fig. 1.11.6   Comparison of initial and subsequent yield loci (Williams and Svensson [1970]).

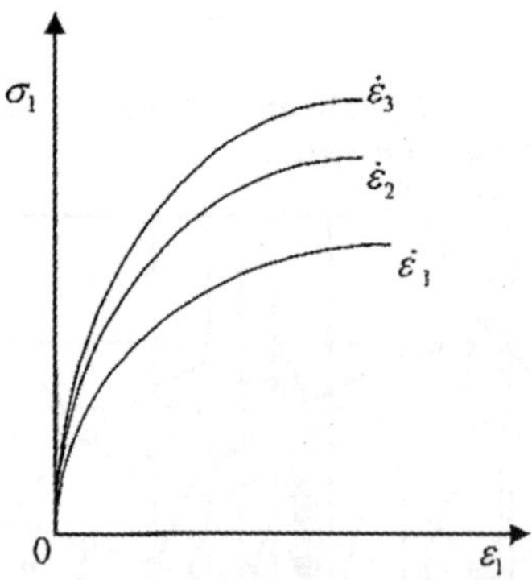

Fig. 1.12.1   Tests done with various velocities.

1.12   Viscoplasticity

In all the constitutive equations of plasticity, they are time-independent. Thus several phenomena cannot be represented. For instance, if one is doing the same experiment with various speeds, one is obtaining distinct curves as a result (see Fig. 1.12.1). That cannot be described with a time-independent constitutive equation. Also, if we stop the test at a certain stress level, and we quip the stress constant, we observe a phenomena which is creep (Fig. 1.12.2). Also, if we quip the strain constant we observe that the stress relaxes. Any other behavior intermediate is possible.

There are a number of materials which have such mechanical propriety which cannot be described: waxes, heavy oils, pastes, butter, creams, tooth pastes, etc. For such materials a new type of model was developed. It is called Bingham body.

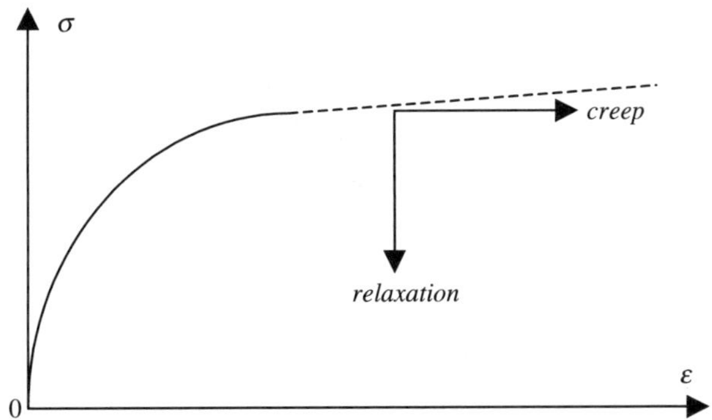

Fig. 1.12.2 Curves for creep or stress relaxation.

The assumptions for Bingham bodies are:
Incompressibility, i.e.,

$$\operatorname{tr} \dot{\boldsymbol{\varepsilon}} = 0. \tag{1.12.1}$$

A viscoplastic function is defined as

$$g(\boldsymbol{\sigma}_{ij}) := \sqrt{II_{\sigma'}} - k. \tag{1.12.2}$$

The constitutive equation is:

$$\dot{\varepsilon}'_{ij} = \begin{cases} 0 & \text{if } II_{\sigma'} \leq k^2 \\[2mm] \dfrac{1}{2\eta}\left(1 - \dfrac{k}{\sqrt{II_{\sigma'}}}\right)\dot{\sigma}'_{ij} & \text{if } II_{\sigma'} > k^2. \end{cases} \tag{1.12.3}$$

Here k and η are two constants which characterize the model. They are absolute constants or relative constants, i.e., they may depend on the particle X.

From (1.12.3) we get:

$$\boldsymbol{\sigma}' = \mathbf{K} + \mathbf{G}, \tag{1.12.4}$$

$\mathbf{G}$ is due to a Newtonian viscosity

$$G_{ij} = 2\eta\dot{\varepsilon}'_{ij} \tag{1.12.5}$$

with η a viscosity coefficient.

$\mathbf{K}$ is a plastic deviatory stress in the sense of Saint-Venant:

$$f(K) := II_K - k^2 \leq 0 \text{ yield condition} \tag{1.12.6}$$

$$\dot{\varepsilon}'_{ij} = 2\lambda K_{ij}. \tag{1.12.7}$$

We assume $\eta > 0$ and $\lambda \geq 0$ with

$$\lambda \begin{cases} = 0 & \text{if } f(K_{ij}) < 0 \\[1mm] > 0 & \text{if } f(K_{ij}) = 0. \end{cases} \tag{1.12.8}$$

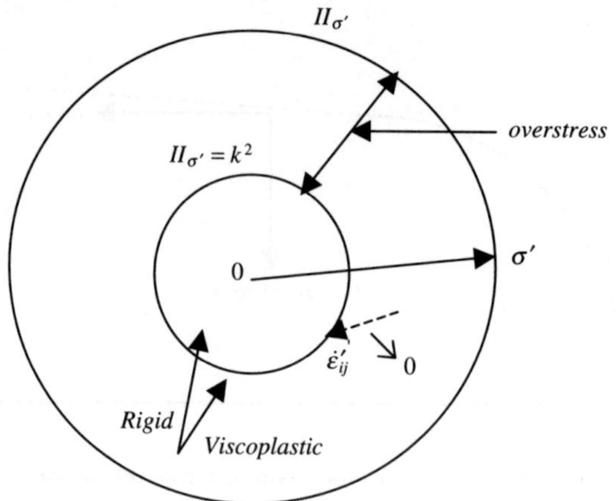

Fig. 1.12.3   The Bingham model.

From (1.12.7), (1.12.5) and (1.12.4) we get

$$\sigma'_{ij} = (1 + 4\eta\lambda)K_{ij} \tag{1.12.9}$$

and for the invariants

$$II_{\sigma'} = (1 + 4\eta\lambda)^2 II_K \, . \tag{1.12.10}$$

If $II_K < k^2$ from these formulae follows $\lambda = 0$, and from (1.12.10) we have $II_{\sigma'} = II_K < k^2$. If $II_K = k^2$ from (1.12.8)$_2$ and (1.12.10) follows

$$\lambda = \frac{1}{4\eta}\left(\frac{\sqrt{II_{\sigma'}}}{k} - 1\right) > 0 \, . \tag{1.12.11}$$

From here we have $II_{\sigma'} > k^2$. Thus the constitutive equation is

$$\dot{\varepsilon}'_{ij} = \begin{cases} 0 & \text{if } II_{\sigma'} < k^2 \\[2mm] \dfrac{1}{2\eta}\left(1 - \dfrac{k}{\sqrt{II_{\sigma'}}}\right)\sigma' & \text{if } II_{\sigma'} > k^2 \, . \end{cases} \tag{1.12.12}$$

The states $II_{\sigma'} = k^2$ are not in the constitutive equation, so that strain rates for those states are not defined. However, when $II_{\sigma'} \searrow k^2$, from (1.12.12) it follows.

From here follows that if $II_{\sigma'}$ is increased the strain rate is also increasing. If $II_{\sigma'}$ is decreased, the strain rate is decreasing. If $II_{\sigma'}$ is decreased towards k^2 the strain rates decreases towards zero $\dot{\varepsilon}'_{ij} \to 0$. Thus we obtain the constitutive equation (1.12.3) (Fig. 1.12.3).

If $II_{\sigma'} > k^2$ from (1.12.12)$_2$ we get

$$(2\eta)^2 II_{\dot{\varepsilon}} = \left(1 - \frac{k}{\sqrt{II_{\sigma'}}}\right)^2 II_{\sigma'} \, ,$$

and from here

$$2\eta\sqrt{II_{\dot{\varepsilon}}} + k = \sqrt{II_{\sigma'}}\,. \tag{1.12.13}$$

If $II_{\sigma'} < k^2$ from $(1.12.12)_1$ we have $\dot{\varepsilon}'_{ij} = 0$ or $II_{\dot{\varepsilon}'} = 0$.

$(1.10.12)$ can be used to inverse the constitutive equation. We obtain

$$\sigma'_{ij} = \begin{cases} \text{undetermined with } \sqrt{II_{\sigma'}} \le k \quad \text{for } \dot{\varepsilon}'_{ij} = 0 \\[2mm] \left(2\eta + \dfrac{k}{\sqrt{II_{\dot{\varepsilon}}}}\right)\dot{\varepsilon}'_{ij} \qquad\qquad \text{for } \dot{\varepsilon}'_{ij} \ne 0\,. \end{cases} \tag{1.12.14}$$

Since in some of the problems one is giving the velocity (experimentally suggested) it is useful to get also the <u>stress power</u>. For $\dot{\varepsilon}'_{ij} \ne 0$ from $(1.10.14)$ we obtain

$$\sigma'_{ij}\dot{\varepsilon}'_{ij} = 2k\sqrt{II_{\dot{\varepsilon}'}} + 4\eta II_{\dot{\varepsilon}'} \tag{1.12.15}$$

and it is expressed in terms of kinematics variables <u>only</u>.

Concerning the reference configuration it is the actual one, since strain is not involved. If the model is used with variable k and η (nonhomogeneous) then the reference configuration is the initial one.

The generalization of this constitutive equation is presented in the next chapters. There are very many generalizations based on the concept of "overstress" due especially to Krempl and his collaborators (Krempl [1979]). For instance in the paper of Ho and Krempl [2000], the constitutive equation for small strain, for volume preserving inelastic deformation is

$$\dot{\mathbf{e}} = \dot{\mathbf{e}}^{\text{el}} + \dot{\mathbf{e}}^{\text{in}} = \frac{1+v}{E}\dot{\mathbf{s}} + \frac{3}{2}\frac{\mathbf{s}-\mathbf{g}}{Ek[\Gamma]} = \frac{1+v}{E}\dot{\mathbf{s}} + \frac{3}{2}F[\Gamma]\frac{\mathbf{s}-\mathbf{g}}{\Gamma}\,,$$

where $\mathbf{s}$ and $\mathbf{e}$ are the deviator stress and strain tensors, respectively. The inelastic part of the flow law can either be written using the decreasing, positive viscosity function $k[\Gamma]$ with $k[0] \ne 0$ having the dimension of time, or in terms of the increasing, positive flow function $F[\Gamma]$ with $F[0] = 0$ having the dimension of 1/time. The overstress invariant is

$$\Gamma = \sqrt{\frac{3}{2}\text{tr}\left((\mathbf{s}-\mathbf{g})(\mathbf{s}-\mathbf{g})\right)}$$

and $\mathbf{g}$ is the deviator of the equilibrium stress. The elastic volumetric relation augments the deviator law by

$$\text{tr}\,\dot{\boldsymbol{\varepsilon}} = \frac{1-2v}{E}\text{tr}\,\dot{\boldsymbol{\sigma}}\,.$$

The new growth law for the deviator state variable equilibrium stress $\mathbf{g}$ is

$$\dot{\mathbf{g}} = \frac{\psi[\Gamma]}{E}\left(\dot{\mathbf{s}} + \frac{\mathbf{s}-\mathbf{g}}{k[\Gamma]} - E\dot{p}\frac{\mathbf{g}-\mathbf{f}}{A+\beta\Gamma}\right) + \left(1 - \frac{\psi[\Gamma]}{E}\right)\dot{\mathbf{f}}\,.$$

For the deviator kinematics stress $\mathbf{f}$ we have

$$\dot{\mathbf{f}} = \frac{\hat{E}_t}{E}\frac{\mathbf{s}-\mathbf{g}}{k[\Gamma]}$$

and finally a simple growth law for the secular isotropic stress is

$$\dot{A} = A_c(A_f - A)\dot{p}\,.$$

The positive, decreasing shape function $\psi[\Gamma]$ controls the transition from the initial quasi linear behavior to filly established flow and is bounded by $E > \psi[\Gamma] > E_t$ where E_t is the tangent modulus at the maximum strain of interest. It is related to the tangent modulus based on inelastic strain $\hat{E}_t$ by $\hat{E}_t = E_t(1 - (E_t/E))$. The effective inelastic strain rate is given by $\dot{p} = \Gamma/(Ek[\Gamma]) = F[\Gamma]$. The quantities A_c and A_f are constants with no dimension and of the dimension of stress, respectively. A_c controls the speed with witch the final value of A, i.e., A_f, is reached.

The model is slightly changed to describe various effects, as: relaxation, the cyclic Swift effect, the pre-necking and post-necking relaxation, creep, free-end torsion, rate-dependent deformation behavior of metals and solid polymers, high temperature applications, etc.

1.13   Rate Type Constitutive Equations

From now on, we will assume that always exists an instantaneous elastic response. It always exists for one-dimensional rate-type models, i.e., for any given state. If a jump in strain is produced, the stress also jumps and its jump is uniquely determined by the strain alone.

For several components of stress it is no more so evident. We will assume however that the instantaneous response is path independent. From physical point of view one can accept this assumption, and from experimentally point of view one is not able to imagine tests which will contradict.

The purpose of this section is (Suliciu [1989]), to examine the consistency of a simplified form of the instantaneous modules with the requirement of existence of instantaneous elastic response.

The constitutive equation we deal with is of the form

$$\dot{\mathbf{T}} = \varepsilon(\mathbf{E}, \mathbf{T})\dot{\mathbf{E}} + \mathbf{G}(\mathbf{E}, \mathbf{T}, k)\,, \quad \dot{k} = \hat{k}(\mathbf{E}, \mathbf{T}, k)\,, \tag{1.13.1}$$

where $\mathbf{E}$ and $\mathbf{T}$ are symmetric and objective strain and stress tensors conjugated with respect to the stress power w:

$$w = \frac{1}{\rho_0}\mathbf{T} \cdot \dot{\mathbf{E}}\,, \tag{1.13.2}$$

where ρ_0 is the mass density in the reference configuration, and k is a isotropic work-hardening parameter. $\mathbf{T}$ is the Cauchy stress tensor and $\mathbf{E}$ is the strain tensor.

For a body modeled by constitutive equation (1.13.1), a process starting at a state $(\mathbf{E}_0, \mathbf{T}_0, k_0) \in D$ is a curve $(\mathbf{E}(t), \mathbf{T}(t), k(t))$ in D for $t \in [0, t_0)$, $t_0 \geq 0$ with $\mathbf{E}(0) = \mathbf{E}_0$, $\mathbf{T}(0) = \mathbf{T}_0$, $k(0) = k_0$ which verifies (1.13.1).

One may term the set D_e from D where both $\mathbf{G}$ and $\hat{k}$ vanish the equilibrium set of the model since a process starting at such a state, and generated by the constant strain $\mathbf{E}(t) = \mathbf{E}_0$ for all $t \geq 0$ will be the constant process $(\mathbf{E}(t) = \mathbf{E}_0$, $\mathbf{T}(t) = \mathbf{T}_0$, $k(t) = k_0)$ for all $t \geq 0$. Depending on the structure of the equilibrium set D_e we may classify the models described by the constitutive equation of the form (1.13.1) as viscoelastic or viscoplastic models. If the function $\mathbf{G}$ in (1.12.1) is independent of k and if there is a function of strain $\mathbf{T}_R(\mathbf{E})$ for $\mathbf{E}$ in some strain domain $D_E \subset S$ such that $(\mathbf{E}, \mathbf{T}_R(\mathbf{E})) \in D$ for all $\mathbf{E} \in D_E$ and $\mathbf{G}(\mathbf{E}, \mathbf{T}) = 0$ if and only if $\mathbf{T} = \mathbf{T}_R(\mathbf{E})$ then the model may be called *viscoelastic* (the evolution equation for k may be disregarded). In other terms, the model is called *viscoelastic* if the equilibrium state of stress is determined by the state of strain alone. The model (1.13.1) may be called *viscoplastic* if the equilibrium set D_e contains an open set in D and $D_e \subset D$ is a strict inclusion.

The influence of the temperature in one-dimensional stress–strain curves are described by many authors. For stainless steel it is given in Xue *et al.* [2004] which wave used an explosion technique to study the thick-walled cylinder under high-strain deformation of $\sim 10^4$ s^{-1}. The shear-band initiation and propagation were examined. Several grain size were considered.

1.14   General Principles

For the general principles, see Malvern [1969], Mase and Mase [1999] and Fung and Tong [2001].

<u>Conservation of Mass.</u> Let us consider a volume V bounded by a fixed surface S. The mass is $M = \int_V \rho \, dV$. The rate of increase of mass is (Fig. 1.14.1):

$$\frac{\partial M}{\partial t} = \int_V \frac{\partial \rho}{\partial t} \, dV \tag{1.14.1}$$

if no mass is created or destroyed inside. That must be equal to the rate of inflow trough the surface which is $\int_S -\rho v_n \, dS = -\int_S \rho \mathbf{v} \cdot \hat{\mathbf{n}} \, dS = -\int_V \Delta \cdot (\rho \mathbf{v}) \, dV$, the last one is coming from the divergence theorem. Combining with (1.14.1) we have

$$\int_V \left[\frac{\partial \rho}{\partial t} + \Delta \cdot (\rho v) \right] dV = 0 \, .$$

Thus must be true for arbitrary V, thus

$$\frac{\partial \rho}{\partial t} + \Delta \cdot (\rho \mathbf{v}) = 0 \tag{1.14.2}$$

or, in rectangular Cartesian coordinates

$$\frac{\partial \rho}{\partial t} + \frac{\partial (\rho v_i)}{\partial x_i} = 0 \, . \tag{1.14.2a}$$

Since

$$\frac{\partial (\rho v_i)}{\partial x_i} = \frac{\partial \rho}{\partial x_i} v_i + \rho \frac{\partial v_i}{\partial x_i} \, ,$$

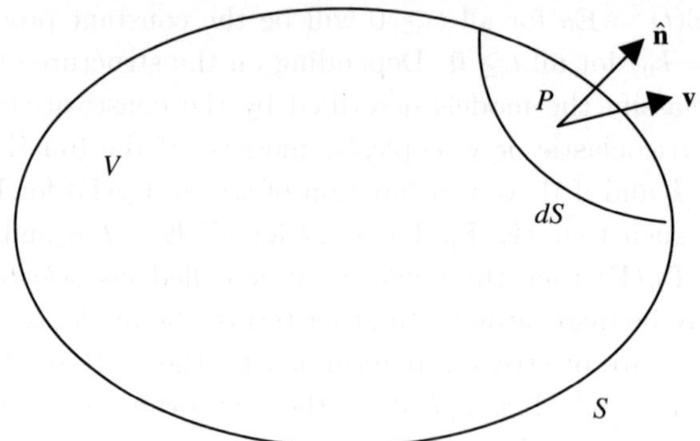

Fig. 1.14.1   The reference volume in space.

and from the material derivative

$$\frac{d\rho}{dt} = \frac{\partial\rho}{\partial t} + v_i\frac{\partial\rho}{\partial x_i}\,,$$

the above law is

$$\frac{d\rho}{dt} + \rho\frac{\partial v_i}{\partial x_i} = 0 \tag{1.14.3}$$

or in vector form

$$\frac{d\rho}{dt} + \rho\,\operatorname{div}\mathbf{v} = 0\,. \tag{1.14.3a}$$

That is the <u>continuity equation</u> due to Euler in 1757. If the density of all the particles is constant, then

$$\operatorname{div}\mathbf{v} = 0\,, \tag{1.14.4}$$

which is the condition of <u>incompressibility</u>.

The continuity condition can be written in material form. If V is the volume of material at time t, and V_0 the volume at the same material at time t_0, we have (Fig. 1.14.2)

$$\int_{V_0} \rho(\mathbf{X}, t_0)\, dV_0 = \int_V \rho(\mathbf{x}, t)\, dV = \int_{V_0} \rho(\mathbf{x}(\mathbf{X}, t), t)|J|\, dV_0$$

with $|J|$ the absolute value of the Jacobian determinant

$$J \equiv \begin{vmatrix} \dfrac{\partial x}{\partial X} & \dfrac{\partial x}{\partial Y} & \dfrac{\partial x}{\partial Z} \\[2mm] \dfrac{\partial y}{\partial X} & \dfrac{\partial y}{\partial Y} & \dfrac{\partial y}{\partial Z} \\[2mm] \dfrac{\partial z}{\partial X} & \dfrac{\partial z}{\partial Y} & \dfrac{\partial z}{\partial Z} \end{vmatrix}\,.$$

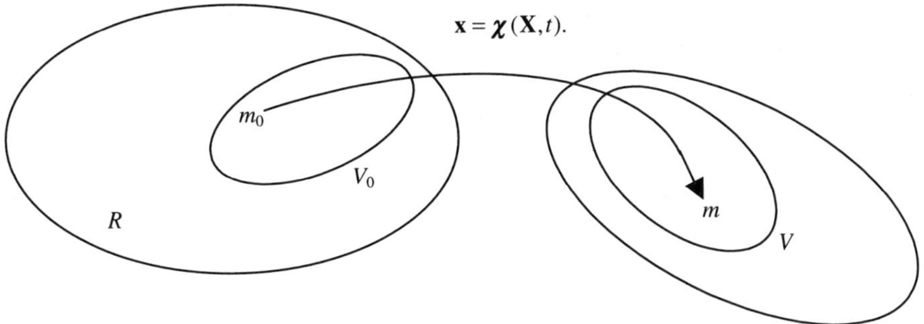

Fig. 1.14.2   The material form of continuity condition.

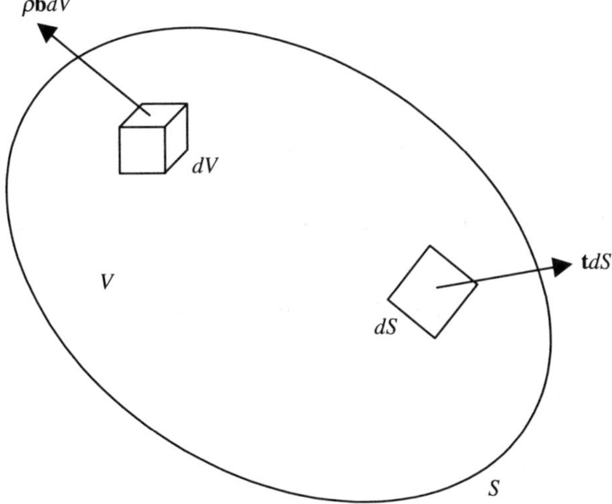

Fig. 1.14.3   The momentum balance.

We have from conservation of mass $\int_{V_0} [\rho_0 - \rho|J|]\, dV_0 = 0$. But for arbitrary volume V_0 we have $\rho|J| = \rho_0$, and since at $t = t_0$ we have no empty spaces, $\rho = \rho_0 > 0$ it follows $J = 1$ and therefore $J > 0$ for $t > t_0$. Therefore we have

$$\rho J = \rho_0 \, . \tag{1.14.5}$$

This is also due to Euler in 1762.

Equation of Motion. Let us consider a body V (see Fig. 1.14.3), which is a closed thermodynamic system, which is not exchanging matter.

We postulate that

$$\int_S \mathbf{t}\, dS + \int_V \rho \mathbf{b}\, dV = \frac{d}{dt} \int_V \rho \mathbf{v}\, dV \, ,$$

where $(d/dt)\int$ is the material derivative of the integral. If we replace here $\mathbf{t} = \hat{\mathbf{n}} \cdot \mathbf{T}$, we have in Cartesian coordinates

$$\int_S T_{ji} n_j \, dS + \int_V \rho b_i \, dV = \frac{d}{dt} \int_v \rho v_i \, dV \,.$$

Using the divergence theorem to pass to a volume integral, and since $\rho \, dV = const.$, we can write

$$\int_V \left[\frac{\partial T_{ji}}{\partial x_j} + \rho b_i - \rho \frac{dv_i}{dt} \right] dV = 0 \,.$$

For an arbitrary V,

$$\frac{\partial T_{ji}}{\partial x_j} + \rho b_i = \rho \frac{dv_i}{dt} \,, \tag{1.14.6}$$

which are the <u>equations of motion</u> of Cauchy's. In vector form, they can be written

$$\Delta \cdot \mathbf{T} + \rho \mathbf{b} = \rho \frac{d\mathbf{v}}{dt} \tag{1.14.6a}$$

independent of the system of coordinates. For the equilibrium problems we have $\Delta \cdot \mathbf{T} + \rho \mathbf{b} = 0$.

<u>Moment of Momentum Principle</u>. The total moment of momentum is

$$\int_S (\mathbf{r} \times \mathbf{t}) \, dS + \int_V (\mathbf{r} \times \rho \mathbf{b}) \, dV = \frac{d}{dt} \int_V (\mathbf{r} \times \rho \mathbf{v}) \, dV$$

and in Cartesian components

$$\int_S e_{rmn} x_m t_n \, dS + \int_V e_{rmn} x_m b_n \rho \, dV = \frac{d}{dt} \int_V e_{rmn} x_m v_n \rho \, dV \,.$$

Here we take care that $t_n = T_{jn} n_j$, we use the divergence theorem

$$\int_S x_m T_{jn} n_j \, dS = \int_V \frac{\partial (x_m T_{jn})}{\partial x_j} \, dV \,,$$

and $\rho \, dV = const.$, to obtain

$$\int_V e_{rmn} \left[\frac{\partial (x_m T_{jn})}{\partial x_j} + x_m b_n \rho \right] dV = \int_V e_{rmn} \frac{d(x_m v_n)}{dt} \rho \, dV \,.$$

If we develop the derivatives

$$\int_V e_{rmn} \left[x_m \frac{\partial T_{jn}}{\partial x_j} + \delta_{mj} T_{jn} + \rho x_m b_n \right] dV = \int_V e_{rmn} \left(\frac{dx_m}{dt} v_n + x_m \frac{dv_n}{dt} \right) dV \,.$$

Grouping now the terms we have

$$\int_V e_{rmn} \left[x_m \left(\frac{\partial T_{jn}}{\partial x_j} + \rho b_n - \frac{dv_n}{dt} \right) + \delta_{mj} T_{jn} \right] dV = \int_V e_{rmn} v_m v_n \, dV \,.$$

We take into account that $e_{rmn} v_m v_n = 0$ since $v_m v_n$ is symmetric in mn, so that $\int_V e_{rmn} T_{mn} \, dV = 0$, or of any V, $e_{rmn} T_{mn} = 0$. Thus for $r = 1$ we have $T_{23} - T_{32} = 0$, and so on. Thus the stress tensor is symmetric.

<u>Energy Balance</u>. We chose again a closed thermodynamic system not exchanging mater with the surroundings. The work is on the system, not by the system.

Power input is

$$P_{\text{input}} = \int_S \mathbf{t} \cdot \mathbf{v} \, dS + \int_V \rho \mathbf{b} \cdot \mathbf{v} \, dV \,.$$

In Cartesian coordinates

$$\begin{aligned}
P_{\text{input}} &= \int_S T_{ji} n_j v_i \, dS + \int_V \rho b_i v_i \, dV \\
&= \int_V \left[v_i \left(\frac{\partial T_{ji}}{\partial x_j} + \rho b_i \right) + T_{ji} \frac{\partial v_i}{\partial x_j} \right] dV \\
&= \int_V \left\{ v_i \left(\rho \frac{dv_i}{dt} \right) + T_{ji} [D_{ij} + W_{ij}] \right\} dV \\
&= \int_V \frac{1}{2} \frac{d \, (v_i v_i)}{dt} \rho \, dV + \int_V T_{ji} D_{ij} \, dV \\
&= \frac{d}{dt} \int_V \left(\frac{1}{2} \rho v_i v_i \right) dV + \int_V T_{ij} D_{ij} \, dV \,.
\end{aligned}$$

In vector form we have

$$P_{\text{input}} = \frac{d}{dt} \int_V \left(\frac{1}{2} \rho \mathbf{v} \cdot \mathbf{v} \right) dV + \int_V \mathbf{T} : \mathbf{D} \, dV \,.$$

Thus the rate of work which is the power input is equal to the variation of the kinetic energy plus the total stress power.

The heat input rate is

$$Q_{\text{input}} = - \int_S \mathbf{q} \cdot \hat{\mathbf{n}} \, dS + \int_V \rho r \, dV \,,$$

where the first integral is the heat flux through the surface of contact S ($\mathbf{q}$ is the heat flux vector), while the second is the distributed internal heat source of strength r per unit mass. The total energy E is

$$\dot{E}_{\text{total}} = \dot{K} + \dot{U} = P_{\text{input}} + Q_{\text{input}} \,,$$

where K is the kinetic energy and U the internal energy (elastic stored energy, and all other nonspecified energies). Thus we have

$$\frac{dK}{dt} + \frac{d}{dt}\int_V \rho u\, dV = \left[\frac{dK}{dt} + \int_V \mathbf{T}:\mathbf{D}\, dV\right]$$
$$+ \left[-\int_S \mathbf{q}\cdot\hat{\mathbf{n}}\, dS + \int_V \rho r\, dV\right].$$

Simplifying and transforming the surface integral in a volume integral, we have for arbitrary V in vector form and coordinate form:

$$\rho\frac{du}{dt} = \mathbf{T}:\mathbf{D} + \rho r - \Delta\cdot\mathbf{q}\,,$$

$$\rho\frac{du}{dt} = T_{ij}D_{ij} + \rho r - q_{j,j}\,,$$

(1.14.7)

which is the <u>energy equation</u> due to Kirchhoff in 1894. Here $\rho\, du/dt$ is the rate of increase of the internal energy ρu per unit volume, ρr is the internal supply of heat per unit volume, and $-q_{i,i}$ the inflow per unit volume of heat through the boundaries of the element.

In elasticity the heat transfer is insignificant. But in plasticity we have a heating due to the plastic deformation, a heating at the tectonic plates, volcanoes, etc.

<u>The Second Law of Thermodynamics.</u> If we introduce the Piola–Kirchhoff stress tensor

$$\mathbf{S} = \rho^{-1}\mathbf{T}(\mathbf{F}^T)^{-1} = \frac{1}{\rho_0}\tilde{\mathbf{S}}\,,$$

(1.14.8)

the energy balance can be written

$$\rho\dot{e} - \rho S_{ij}\dot{F}_{ij} + \frac{\partial q_i}{\partial x_i} = \rho r\,.$$

(1.14.9)

The entropy rate γ is defined as

$$\rho\gamma = \rho\dot{\eta} - \frac{\rho r}{\theta} + \operatorname{div}\frac{q}{\theta}\,,$$

(1.14.10)

where $\eta = \eta(\mathbf{X}, t)$ is the entropy per unit mass while $\theta = \theta(\mathbf{X}, t) > 0$ is the absolute temperature.

The second law of thermodynamics, or the Clausius–Duhem inequality (not used here), states that (Truesdell and Toupin [1960])

$$\gamma \geq 0\,.$$

(1.14.11)

From (1.14.9), (1.14.10) and (1.14.11) follows

$$\gamma = \dot{\eta} - \frac{\dot{e}}{\theta} + \theta^{-1}S_{ij}\dot{F}_{ij} - \frac{1}{\rho\theta^2}q_i\frac{\partial\theta}{\partial x_i} \geq 0\,.$$

(1.14.12)

If one introduces the free energy

$$\psi = e - \theta\eta\,,$$

then (1.14.12) can be expressed as

$$\theta\gamma = -\dot{\psi} - \eta\dot{\theta} + S_{ij}\dot{F}_{ij} - \frac{1}{\rho\theta}q_i\frac{\partial\theta}{\partial x_i} \geq 0\,. \tag{1.14.13}$$

Using the tensor $\tilde{\mathbf{S}}$ given by (1.14.8), we can write the balance equations in initial coordinates

$$\rho_0\frac{\partial v_i}{\partial t} - \frac{\partial \tilde{S}_{ij}}{\partial X_j} = \rho_0 b_i\,,$$

$$\rho_0\frac{\partial e}{\partial t} - \tilde{S}_{ij}\frac{\partial F_{ij}}{\partial t} + \frac{\partial \tilde{q}_i}{\partial X_i} = \rho_0 r\,, \tag{1.14.14}$$

with $\tilde{\mathbf{q}} = J\mathbf{F}^{-1}\mathbf{q}$ (see Green and Rivlin [1964a,b]).

A regular surface Σ with equation $\varphi(\mathbf{X}, t) = 0$, is called an *acceleration wave* through the body if φ is continuous and their derivatives may have jump discontinuities across Σ, while being continuous at all points.

The quantities

$$U = -\frac{\partial\varphi/\partial t}{|\mathrm{Grad}\,\varphi|}\,, \qquad n_i = \frac{\partial\varphi/\partial X_i}{|\mathrm{Grad}\,\varphi|}\,, \tag{1.14.15}$$

are called the *propagation speed* and the *direction of propagation* of the acceleration wave, respectively.

The jumps of the derivatives of $\mathbf{v}, \mathbf{F}, \theta, \tilde{\mathbf{S}}, \psi, \ldots$ cannot be independent; they have to satisfy three types of conditions. The *geometric* and *kinematics compatibility conditions* give those relations between the jumps of $\partial\mathbf{v}/\partial t$, $\partial\mathbf{v}/\partial\mathbf{X}$, etc., that are due to the continuity of $\mathbf{v}$ or follow from the fact that $\mathbf{v}$ and $\mathbf{F}$ are the partial derivatives of the same vector function χ. These compatibility conditions can be written as

$$\left[\frac{\partial v_k}{\partial t}\right] = U^2 a_k\,, \quad \left[\frac{\partial F_{kl}}{\partial X_j}\right] = a_k n_l n_j\,, \quad \left[\frac{\partial v_k}{\partial X_l}\right] = \left[\frac{\partial F_{kl}}{\partial t}\right] = -U a_k n_l\,,$$

$$\left[\frac{\partial\theta}{\partial t}\right] = -Uv\,, \quad \left[\frac{\partial\theta}{\partial X_j}\right] = v n_j\,, \tag{1.14.16}$$

$$\left[\frac{\partial\tilde{S}_{ij}}{\partial t}\right] = -U s_{ij}\,, \quad \left[\frac{\partial\tilde{S}_{ij}}{\partial X_k}\right] = s_{ij} n_k\,, \text{ etc.}$$

Here $[f] = f^+ - f^-$, f^+ being the limit value of f at a point on the surface Σ, reached from the positive side of the normal $(\mathbf{n}, -U)$ to Σ at that point, and f^- being the limit value of f at the same point but reached from the other side of the surface. The vector $\mathbf{a} = (a_1, a_2, a_3)$ is called the *mechanical amplitude of the wave*, the scalar v is called the *thermal amplitude of the wave* and s_{ij} may be called the *stress amplitude*, etc.

The second group of restrictions imposed on the jumps comes from the balance equations. They are called *dynamic compatibility equations* and they are obtained from (1.14.14) as follows: one considers (1.14.14) on each side of the surface Σ and then, after taking limit values at a point on Σ, one calculates the difference between the two obtained relations. Since $\mathbf{b}$ and r are assumed continuous, one obtains the following homogeneous dynamic compatibility conditions.

$$\rho_0 \left[\frac{\partial v_i}{\partial t} \right] - \left[\frac{\partial \tilde{S}_{ij}}{\partial X_j} \right] = 0 \,,$$

$$\rho_0 \left[\frac{\partial e}{\partial t} \right] - \tilde{S}_{ij} \left[\frac{\partial F_{ij}}{\partial t} \right] + \left[\frac{\partial \tilde{q}_j}{\partial X_j} \right] = 0 \,. \qquad (1.14.17)$$

The third group of restrictions is imposed by the constitutive equations, i.e., those relations that must exist between $\mathbf{F}$, θ, $\mathrm{Grad}\,\theta$, $\tilde{\mathbf{S}}$, η, ψ and $\mathbf{q}$. These restrictions will be discussed in the other chapters.

In case of an isolated discontinuity surface, the geometric-kinematic and dynamic compatibility conditions have been presented for the first time by Hadamard [1903] for the general three-dimensional case.

Bibliography

Bruhns O. T., Xiao H. and Meyers A., 2005, A weakened form of Ilyushin's postulate and the structure of self-consistent Eulerian finite elastoplasticity, *Int. J. Plasticity* **21**, 199–219.

Dafalias Y. F. and Popov E. P., 1976, Plastic internal variables formalism of cyclic plasticity, *J. Appl. Mech.* **43**, 645–651.

Drucker D. C., 1951, A more fundamental approach to plastic stress–strain relations, *Proc. 1st U.S. Natl. Cogr. Appl. Mech.*, 487–491.

Fung Y. C. and Tong P., 2001, *Classical and Computational Solid Mechanics*, World Scientific, 930 pp.

Green A. E. and Rivlin R. S., 1964a, Simple force and stress multipoles, *Arch. Rational Mech. Anal.* **16**, 5, 325–353.

Green A. E. and Rivlin R. S., 1964b, Multipolar continuum mechanics, *Arch. Rational Mech. Anal.* **17**, 2, 113–147.

Hadamard J., 1903, Leçons sur la Propagation des Ondes et les Equations de l'Hydrodynamique, *Herman*, Paris.

Hencky H., 1924, Zur Theorie plastischer Deformationen und der hierdurch hervorgerufenen Nachspannungen, *Zeits. Angew. Math. U. Mech.* **4**, 323–334.

Ho K. and Krempl E., 2000, Modeling of positive, negative and zero rate sensitivity by using the viscoplasticity theory based on overstress (VBO), *Mech. Time-Dep. Materials* **4**, 21–42.

Iliushin A. A., 1961, On the postulate of plasticity, *Prikl. Mat. Mech.* **25**, 503.

Koiter W. T., 1953, Stress–strain relations, uniqueness and variational theorems for elastic, plastic materials with a singular yield surface, *Q. Appl. Math.* **11**, 350–354.

Krempl E., 1979, Viscoplasticity based on total strain: The modeling of creep with special considerations of initial strain and aging, *J. Eng. Mat. Tech.* **101**, 380–386.

Lode W., 1926, Versuche uber den Einfluss der mittleren Hauptspannung auf das Fliessen der Metalle Eisen, Kupfer, und Nickel, *Z. Physik* **36**, 913–939.

Malvern L. E., 1969, *Introduction to the Mechanics of a Continuous Medium*, Prentice-Hall, Inc., 713 pp.

Mase G. Th. and Mase G. E., 1999, *Continuum Mechanics for Engineers*, CRC Press, 377 pp.

Mises R. von, 1913, Mechanik der festen Körper in plastisch deformable Zustand, *Nachr. Akad. Wiss. Göttingen Math.-Phis. K1* **H.4**, 582–592.

Mróz Z., 1969, An attempt to describe behaviour of metals under cyclic loads using a more general workhardening model, *Acta. Mech.* **7**, 199–212.

Nadai A., 1923, Der Beginn des Fliessvorganges in einem tortierten Stab, *Zeits. Angew. Math. Mech.* **3**, 442–456.

Naghdi P. M., Essenburg F. and Koff. W., 1958, An experimental study of initial and subsequent yield surfaces in plasticity, *J. Appl. Mech.* **25**, 201–209.

Phillips A., 1986, A review of quasistatic experimental plasticity and viscoplasticity, *Int. J. Plasticity* **2**, 315–328.

Prager W., 1959, *An Introduction to Plasticity*, Addison-Westley Publishing Co.

Prandtl L., 1924, Spannungsgsverteilung in plastischen Koerpern, *Proc. 1st Int. Congr. Appl. Mech.* (Delft), 43–54.

Reuss E., 1930, Beruecksichtigung der elastischen Formaenderungen in der Plasizitaets theorie, *Zeits. Angew. Math. U. Mech.* **10**, 266–274.

Ros M. and Eichinger A., 1926, *Proc. 2nd Int. Congr. Appl. Mech.* (Zürich), 315.

Ros M. and Eichinger A., 1929, Metalle Diskussionsbericht No. 34 der Eidgenossen Materialprufungsanstalt, Zurich, 315.

Saint-Venant, Barré de., 1870, Mémoire sur l'établissement des équations différentielles des mouvements intérieure opérés das les corps solides ductiles au delà des limite où l'élasticité pourrait les ramener à leur premier état, *C. R. Ac. Sci.* (Paris) **70**, 473–480.

Stoughton T. B. and Yon J. W., 2005, Review of Drucker's postulate and the issue of plastic stability in metal forming, *Int. J. Plasticity* (in print).

Suliciu I., 1989, Some remarks on the instantaneous response in rate-type viscoplasticity, *Int. J. Plasticity* **5**, 173–181.

Taylor G. I. and Quinney H., 1931, The plastic distortion of metals, *Philos. Trans. Roy. Sos. London A* **230**, 323–362.

Tresca H., 1868, Memoire sur l'ecoulment des corps solides, *Mém. Sav. Acad. Sci. Paris*, 733–799.

Truesdell C. and Toupin E. R., 1960, *The Classical Field Theories*, Handbuch der Physik, Vol. III/1, Springer Verlag, Berlin.

Valanis K. C., 1980, Fundamental consequences of a new intrinsic time measure plasticity as a limit of the endoshronic theory, *Arch. Mech.* **32**, 171.

Williams J. F. and Svensson N. L., 1970, Effect of tensile prestrain on the yield locus of 1100-F Aluminium, *J. Strain Analysis* **5**, 2, 128–139.

Williams J. F. and Svensson N. L., 1971, Effect of torsional prestrain on the yield locus of 1100-F Aluminium, *J. Strain Analysis* **6**, 4, 263–272.

Xue Q., Meyers M. A. and Nesterenco V. F., 2004, Self organization of shear bands in stainless steel, *Mater. Sci. Eng. A* **384**, 35–46.

Chapter 2

Rocks and Soils

2.1   Introduction

The constitutive equation for rocks or soils is much more complicated. The reason
is that the experimental results are much more involved. Even the elastic constants
are no more what they were for metals, and the methods of finding them from tests
are different from what we know from metals. The yield conditions are different.
While for metals a cylindrical constitutive yield condition, independent of the mean
stress is quite normal, such a yield condition for rocks is not acceptable. One has
tried to introduce a yield condition depending also on mean stress and on the third
deviator stresses

$$F(I_\sigma, \sqrt{II_{\sigma'}}, \theta) = 0\,,$$

where

$$\theta = \frac{1}{3}\sin^{-1}\frac{3\sqrt{3}III_{\sigma'}}{2\sqrt{II_{\sigma'}^3}}$$

is the Lode angle. The most familiar are the Mohr–Coulomb and Drucker–Prager
yield conditions.

The first condition is written

$$\tau = c - \sigma\tan\phi$$

with

$$c = \frac{1}{2}\sqrt{\sigma_c\sigma_t} \quad\text{and}\quad \phi = \sin^{-1}\left(\frac{\sigma_c - \sigma_t}{\sigma_c + \sigma_t}\right) \quad\text{with } 0 \le \phi < \frac{\pi}{2}\,.$$

The Drucker–Prager yield condition is

$$\sqrt{II_{\sigma'}} + aI_\sigma - k = 0\,,$$

with a and k constants. For a fixed pressure the two surfaces are represented in
Fig. 2.1.1. They are conical in three-dimensional space, but for a constant pressure
they show as in figure.

53

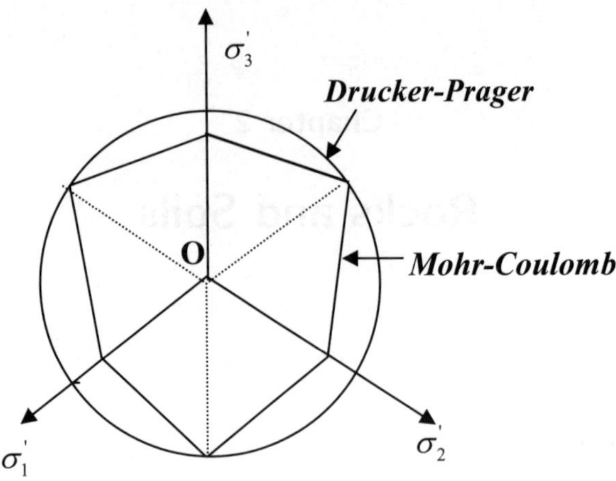

Fig. 2.1.1   The Mohr–Coulomb and Drucker–Prager yield conditions, for a constant pressure.

Generally the criteria can be written (Li and Aubertin [2003], Aubertin and Li [2004], Li *et al.* [2005]):

$$F = \sqrt{II} - F_0 F_\pi = 0 \,, \quad F_0 = [\alpha^2(I_\sigma^2 - 2a_1 I_\sigma) + a_2^2 - a_3\langle I_\sigma - I_c\rangle^2]^{1/2} \,,$$

where α, a_1, a_2 and I_c are material parameters. α is expressed as function of the friction angle $\alpha = 2\sin\phi/\sqrt{3}(3 - \sin\phi)$ and $II = J_2$, while

$$a_1 = \left(\frac{\sigma_c - \sigma_t}{2}\right) - \frac{\sigma_c^2 - (\sigma_t/b)^2}{6\alpha^2(\sigma_c + \sigma_t)} \,, \quad a_2 = \left\{\left(\frac{\sigma_c + (\sigma_t/b)^2}{3(\sigma_c + \sigma_t)} - \alpha^2\right)\sigma_c\sigma_t\right\}^{1/2} \,,$$

where b is linked to the shape of the surface in the π plane. The authors have investigated the relationship between porosity and uniaxial strength of various materials (in compression and tension). This leads to the development of a general nonlinear relationship, which is used here to define the uniaxial strength as a function of porosity:

$$\sigma_{un} = \left\{\sigma_{u0}\left(1 - \sin^{x_1}\left(\frac{\pi}{2}\frac{n}{n_C}\right)\right) + \langle\sigma_{u0}\rangle\cos^{x_2}\left(\frac{\pi}{2}\frac{n}{n_C}\right)\right\}\left\{1 - \frac{\langle\sigma_{u0}\rangle}{2\sigma_{u0}}\right\} \,,$$

where σ_{un} may be used for compression ($\sigma_{un} = \sigma_{cn}$) or tension ($\sigma_{un} = \sigma_{tn}$); the subscript n indicates the magnitude of the parameter at a porosity n. In this equation, n_C is the critical porosity for which σ_{un} becomes negligible, in tension ($n_C = n_{Ct}$) and in compression ($n_C = n_{Cc}$). For a given material, it is expected that $n_{Ct} \cong n_{Cc} \leq 1$. Parameter σ_{u0} represents the theoretical (extrapolated) value of σ_{un} for $n = 0$. Exponents x_1 and x_2 are material parameters. A graphical representation of the proposed general equation is given in Fig. 2.1.2. The index n identifies porosity-dependent parameters. The parameters a_3 and I_c control the behavior of the porous material under high hydrostatic compression, when the surface

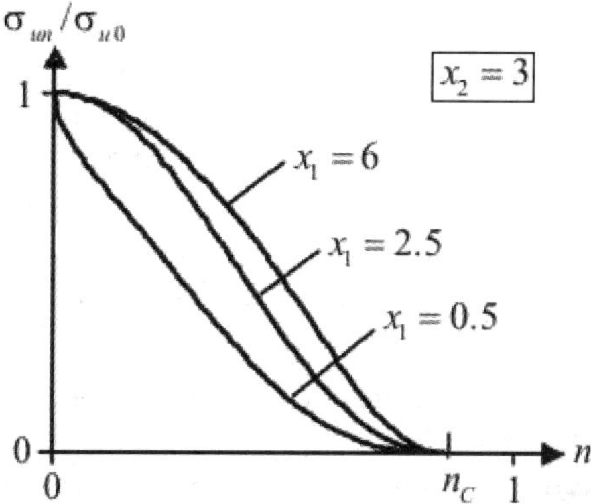

Fig. 2.1.2 Graphical representation of the general equation for uniaxial strength.

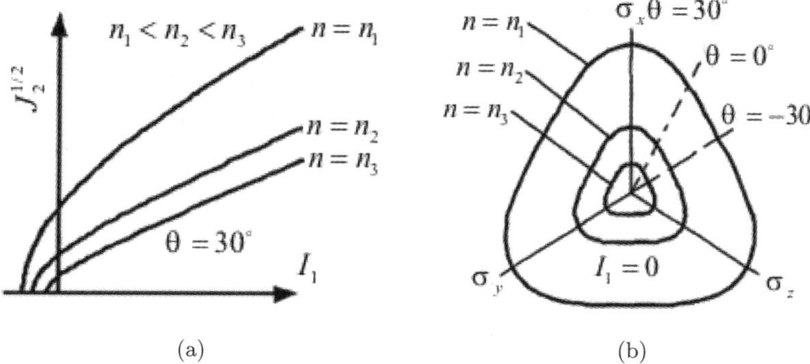

(a)   (b)

Fig. 2.1.3 Schematic representation of the criterion for low porosity materials ($I_1 = I_\sigma < I_c$ and $v = 1$).

closes with a "Cap" on the positive side of I_σ. The surface in the octahedral (π) plane is represented by the following function of the Lode angle:

$$F_\pi = \left(\frac{b}{[b^2 + (1 - b^2)\sin^2(45° - 1.5\theta)]^{1/2}} \right)^v,$$

with $v = \exp(-v_1 I_\sigma)$, in which exponent v reflects the influence of hydrostatic pressure on the evolution of the surface shape in the π plane and v_1 is a material parameter.

In Fig. 2.1.3 is given some examples. In Fig. 2.1.4 are given other examples corresponding to high values of I_σ when v tends towards 0 or F_π tends towards 1. A most recent version of SUVIC, a viscoplastic model with internal state variables, is presented by Aubertin and Gill [1993], Aubertin *et al.* [1999a, b].

They are then used to represent the behavior of polycrystalline sodium chloride submitted to conventional triaxial compression and reduced triaxial extension tests, using results that highlight mixed, kinematics and isotropic, hardening of the material, hence showing a type of Bauschinger effect. In another paper

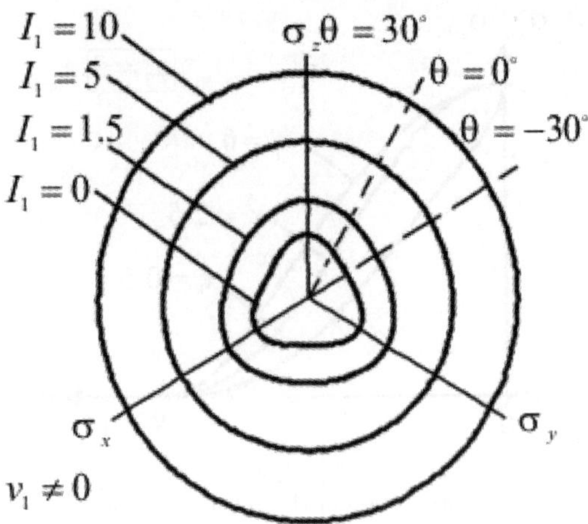

Fig. 2.1.4   Illustration of the evolution of the surface in the π plane when $I_\sigma = I_1$ increases.

Aubertin [1996] discusses a paper on triaxial stress relaxation tests. Another model is given by Jeremić *et al.* [1999] for a nonassociative yielding, both deviatoric and volumetric, and hardening/softening characteristics. Wathugala and Pal [1999] define the yield surface F, in terms of first stress invariant J_1, the second invariant of the stress deviator J_{2D}, and the third invariant of the deviatoric stress tensor J_{3D}, as

$$F \equiv \left(\frac{J_{2D}}{p_a^1} \right) - \left[-\alpha_{ps} \left(\frac{J_1}{p_a} \right) + \left(\frac{J_1}{p_a} \right)^2 \right] (1 - \beta S_r)^{-0.5} = 0 \,,$$

where p_a is the atmospheric pressure, and α_{ps} is the hardening or growth function space, γ, β and n are material parameters. S_r is defined as a stress ratio, and given by

$$S_r \equiv \frac{\sqrt{27}}{2} J_{3D} J_{2D}^{-3/2} \,.$$

For a general literature about the rocks and special the time effects, see the books by Cristescu [1989] and Cristescu and Hunsche [1998]. In these books is given not only constitutive equations, but also many mining problems: failure, vertical galleries, horizontal circular tunnels, rooms rectangular and the problem of tunnel linings. Everything is described taking into account that the surrounding rocks are creeping, either by dilatancy or compressibility. The propagation of pressure waves in bubble-field liquids, considered to be locking media, is due to Dresner [1973]. The study of impact of a piston on a strain-dependence dynamic compaction of a locking media is due to Lundberg [1974]. For an early presentation of plastic waves propagation in soils see the book edited by Cizek [1985]. The time-dependent tunnel convergence and various problems associated were considered by Pan and Dong [1991a] [1991b]; they have considered the rheological properties, the advancement of the tunnel and the tunnel-support interaction. Also, a significant review paper of Jing [2003] is a comprehensive literature for numerical modeling in rock mechanics and rock engineering.

We assume that the rock has a small initial porosity. If it not so, i.e., if the rock has a significant porosity, the things may change. Thus Maranini and Brignoli [1999] have tested a limestone of initial porosity of 38% and find that at first the rock is dilatant and after a certain stress it is compressible.

A detailed description of the so-called composite model for transient and steady state creep which is based on micro mechanisms is given by Hunsche and Hampel [1999]. This model is not only able to model deformation in a wide range of stresses and temperatures, but also after a stress decrease, as well as the large differences in creep caused by variations in the distribution of impurities. As a result of this sound physical and experimental basis, the modeling results can be reliably extrapolated. Dilatancy, healing, damage, failure, and deformation of rock salt are described by an elasto-viscoplastic constitutive equation. An important feature is the stress dependent dilatancy boundary which separates the dilatant domain from the compressible one and forms a kind of safety boundary. The relation between deformation, dilatancy, and permeability is also addressed.

A nonassociated elasto-viscoplastic general model for rock salt is given by Nicolae [1999]. The constitutive functions and parameters are determined using a significant number of laboratory tests performed either with classical, or with original devices. The matching of data and the theoretical prediction is good from both a qualitative and quantitative point of view. The paper presents the analysis of the stress distribution in the neighborhood of a horizontally mining excavation, with either a circular or noncircular transverse cross-section (a square with rounded corners). The computations are compared with *in situ* displacement field measurements.

An elastic/viscoplastic model for transient creep of rock salt is due to Jin and Cristescu [1998a]. Both yield function and viscoplastic potential are determined from experimental data. Singularity and asymptotic properties of the yield surfaces and viscoplastic potential are considered. The model is matching quite well the experimental data and can be incorporated into a finite element program. As an example the stress distribution around a vertical cylindrical cavity is analyzed.

The split-Hopkinson pressure bar used for rocks is studied by Shan *et al.* [2000]. Complete stress–strain curves for rock (marble and granite) are established.

The general geomechanical stability and integrity of waste disposal mines in salt structures is due to Langer and Heusermann [2001]. Thousands of salt caverns (100 in France alone) are being used to store hydrocarbons. This is the safest way to store large quantities of hydrocarbons: salt formations are almost perfectly impermeable, and fire or explosion is impossible underground. However, a small number of accidents (blow-out, product seepage, cavern instability) have occurred in the past. Cavern abandonment is also a concern in some cases. These accidents have been described and the lessons that have been drawn from them, leading to considerable improvements in storage design and operation is due to Bérest and Brouard [2003]. The dynamic loading with viscoplasticity and temperature effects on the evolution of damage in metal forming processes is due to Gelin [1992]. One considers materials containing microscopic voids and cracks. The deformation can be decomposed in an elastic part, in a pure dilatant part, and in a plastic one that preserves the volume, the classical multiplicative decomposition is accepted. The problem of ultra-deep mines with the increased potential for squeezing conditions is mentioned in Malan and Basson [1998]. The generation and development of cracks in rock salt under the influence of mining processes was examined by Silberschmidt and Silberschmidt [2000]. They show that not all the discontinuities are closed in the course of the creep deformation of rocks, and that the parameters of cracking are determined for various conditions: age of pillars, mining technology, etc. The parameter identification for lined tunnels in a viscoplastic medium is due to Lecampion *et al.* [2002]. The paper is dedicated to the identification of constitutive parameters of elasto-viscoplastic constitutive law from measurements performed on deep underground tunnels. The method is presented for lined or unlined structures and is applied for an elasto-viscoplastic law. The interacting of two cracks is considered by Miura *et al.* [2003]. The creep failure following tertiary creep is represented as unstable extension of the interacting cracks. The time to failure is calculated for different values of axial stress, confining pressure, and environmental conditions such as temperature and presence of water. A review on creep and creep fracture/damage of engineering materials is due to Mackerle [2004]. In another paper Bérest *et al.* [2004] describe the creep of the salt at extremely very small rates of strain $\dot{\varepsilon} = 10^{-13}$ s^{-1}. The tests have been done at 160 m under the surface. One is applying the Norton–Hoff constitutive equation for steady state creep

$$\dot{\varepsilon} = A \exp\left[-\frac{Q}{RT}\right] \sigma^n \,,$$

where σ is the applied stress, T is the absolute temperature, n is between 3 and 5, $A = 0.64$ MPa^{-n} year^{-1} and $Q/R = 4100$ K. A hybrid intelligent method optimization of a soft rock replacement scheme for a large cavern excavated in alternate hard and soft rock strata is due to Feng and An [2004]; it is an integration of an evolutionary neural network and finite element analysis using a genetic algorithm.

Experimental strength results for dense Santa Monica Beach sand was given by Abelev and Lade [2004]. Also by Lade [1977], a failure surface, as a combination of the two stress invariants I_1 and I_3 with

$$\left(\frac{I_1^3}{I_3} - 27\right)\left(\frac{I_1}{p_a}\right)^m = \eta_1$$

and $I_1 = \sigma_1 + \sigma_2 + \sigma_3$, $I_3 = \sigma_1\sigma_2\sigma_3$ and the parameters η_1 and m are constant dimensionless numbers and p_a is the atmospheric pressure.

Experiments and simulations of penetration and perforation of concrete targets by steel projectiles have been performed by Unisson and Nilsson [2005]. It was possible to track the velocity history of the projectile in free flight and also in the deceleration of the projectile as it penetrated the target. The result was shown to be greatly influenced by the erosion criterion.

Similar results have been established for uniaxial compressive response of polymeric structural foams (Subhash *et al.* [2005]) investigated under quasistatic and high strain rate conditions. It is shown that Young's modulus, yields strengths, the maximum stress, and the strain to failure increased with increasing initial foam density under quasistatic loading. Under dynamic loading, the failure strength increased with strain rate but the strain to failure decreased.

There are also some additional data on strain rate effects for shock-mitingating foams (Tedesco *et al.* [1993]). They are depending on density.

A recently developed model for simulating the dynamic behavior of silicate materials is applied to the loading and unloading properties of granite by Boettger *et al.* [1995]. Four time-resolved wave profile measurements on granite are presented and used to supplement Hugoniot data to constrain the model.

A series of controlled impact experiments has been performed by Hall *et al.* [1999] to determine the shock loading and relies behavior of two types of concrete. Results indicate that the average loading and relies behavior are comparable for the three types of concrete discussed in the paper. Residual strain is also indicated from these measurements.

The compressive behavior of a polystyrene foam was investigated by Song *et al.* [2005] at strain rates from 0.001 to 950/s. The collapse stress of the foam is found to increase nearly linearly with the logarithm of the strain rate, and the elastic modulus is seen to increase with strain rate.

Dynamic compression testing of soft materials is due to Chen *et al.* [2002]. Experimental results show that homogeneous deformations at nearly constant strain rates can be achieved in materials with very low impedances, such as silicone rubber and a polyurethane foam, with the experimental modifications presented in this study.

Forrestal *et al.* [2003] conducted two sets of penetration experiments with concrete targets that had average compressive strengths of 23 and 39 MPa. They recorded acceleration during launch and deceleration during penetration.

Three diameters of concrete targets that had an average compressive strength of 23 MPa, were impacted by Frew *et al.* [2006]. Acceleration was recorded during launch and deceleration during penetration. The 13 kg projectiles had a striking velocity between 160 and 340 m/s.

2.2 Experimental Foundation

Rocks are tested in the so called three-axial testing devices. These are the devices in which a cylindrical specimen is subjected to an axial loading and to a lateral pressure loading. First both loadings are increased together, but in the second part of the test the lateral pressure is maintained constant and only the axial loading is increased. One is doing the same experiments uniaxially. The results are shown in Fig. 2.2.1 for schist. The three curves are obtained with three loading rates. One can see that the loading rates are changing the whole stress–strain curve, from the beginning. Also the failure is time-dependent since the star at the end of each curve is showing failure. The figure is also showing a creep curve. The constant stress is maintained constant 20 days, 20 days, 21 days, etc. Failure is at a much lower stress, and the strain is different than in the other stress–strain curves. Thus the whole stress–strain curve is time-dependent. That is a uniaxial stress–strain curve.

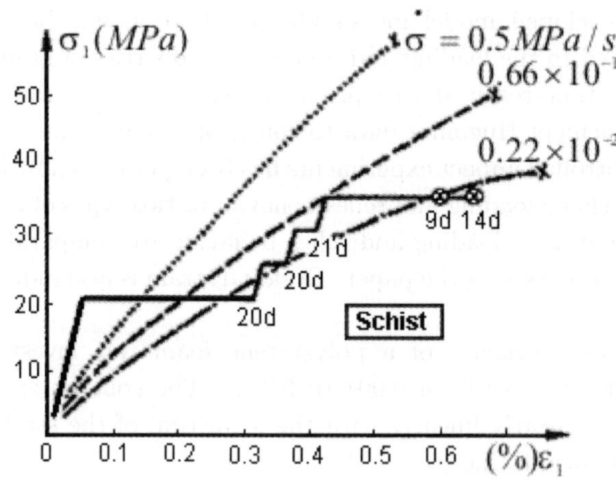

Fig. 2.2.1   Uniaxial stress–strain curves for schist for various loading rates, showing time influence on the entire stress–strain curves, including failure (Cristescu [1986]).

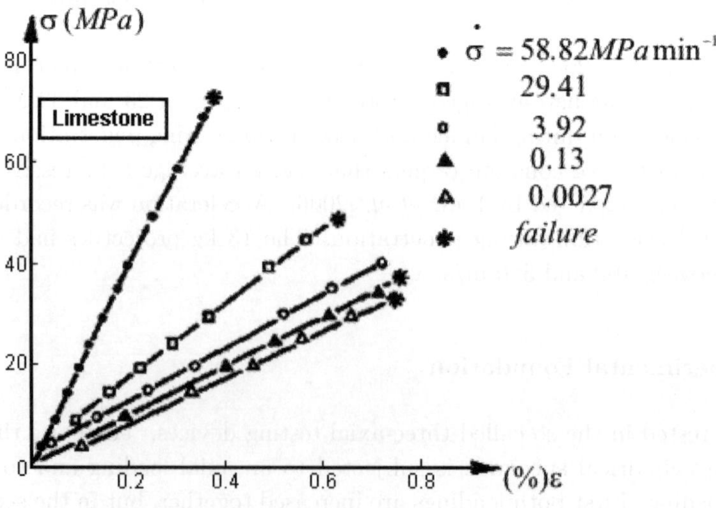

Fig. 2.2.2   Stress–strain curves for limestone for various loading rates.

Another uniaxially curve is for limestone and given on Fig. 2.2.2. The behavior is nearly perfectly linear, but obviously it is not elasticity. The star at the end of the curve is showing again failure. Thus failure is strongly time-dependent and depending on the loading rates. For the three-axial curves one obtains the results shown in Fig. 2.2.3 for granite (Maranini and Yamaguchi [2001]).

These are the

$$\bar{\sigma} - \varepsilon_1, \ \bar{\sigma} - \varepsilon_2 \quad \text{and} \quad \bar{\sigma} - \varepsilon_v$$

curves obtained for 100 kN/min. In order to find the correct values of the elastic strains the tests have been stopped at various levels of stresses for a few minutes.

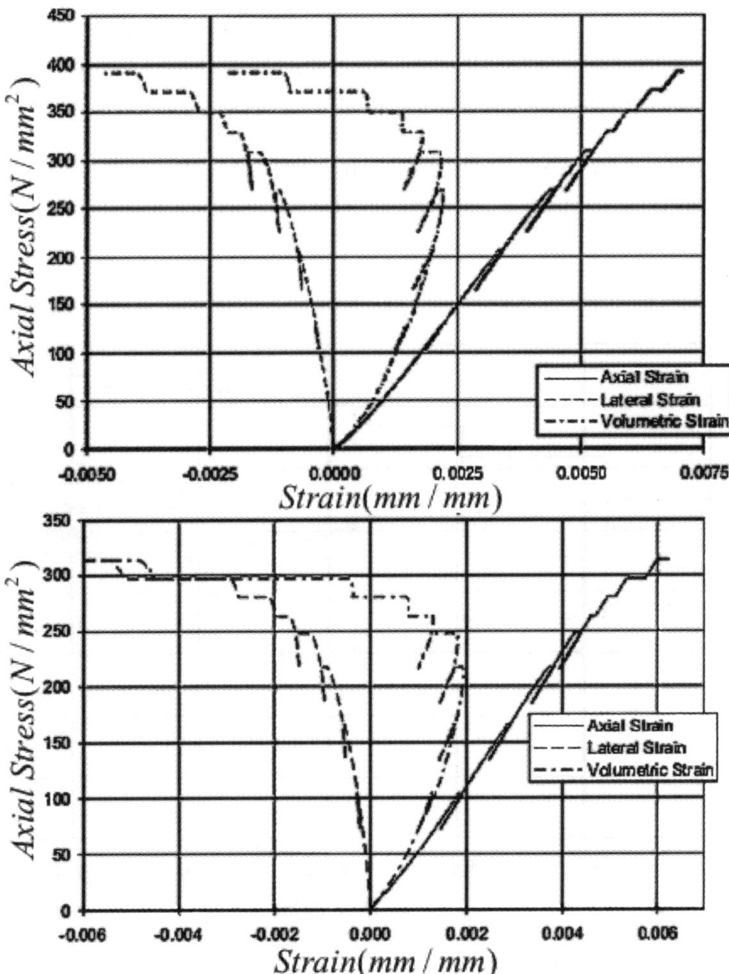

Fig. 2.2.3 The stress–strain curves for granite obtained in three axial tests. The upper curve is obtained for a confining pressure of 20 MPa, while the lover one for 10 MPa.

A significant creep is observed. After this period, a small unloading and reloading have given the correct elastic parameters.

The elastic parameters are also not really constant. They are increasing with the confining pressure. From the last curve one can see that initially the volume is compacting and afterwards dilating. The various horizontal plateaus correspond to a period of several minutes when the stress is kept constant. That was done in order to see that the time effects are present throughout the test, and that the elastic parameters are to be measured at the end of these periods. If one is measuring the elastic parameters before one is mixing the time effects (creep and relaxation) with unloading.

Also, during the period in which the stresses are constant a fast creep effect is taking place. One can plot the curves $\varepsilon_1 - t$ and $\varepsilon_2 - t$ and from here the $\varepsilon_V - t$ curves (the middle curve). That is shown on the Fig. 2.2.4. These curves stabilize very fast and that very fast they approach the horizontal line. If at the end of these periods one is producing a small unloading, this unloading is quite linear. In this way one is determining the elastic parameters of a rock or of a porous material. These parameters are not constants throughout. In the first part of the test, the hydrostatic one, all the elastic parameters are increasing. In the deviator part of the test, they continue to increase so long as the compressibility/dilatancy boundary was not reached.

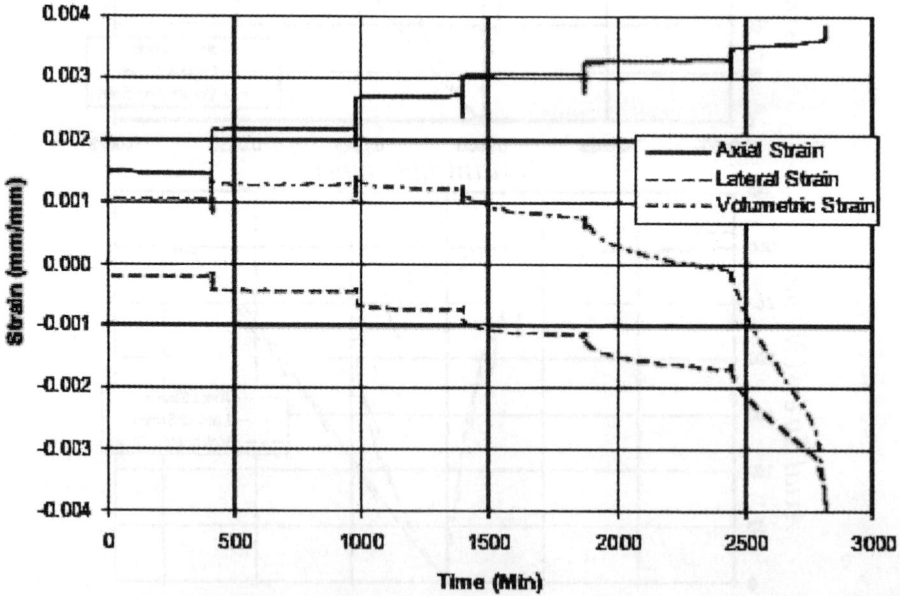

Fig. 2.2.4 Static procedure to determine the elastic parameters in unloading processes following short creep periods. The volumetric deformation is showing first compaction and afterwards dilatation.

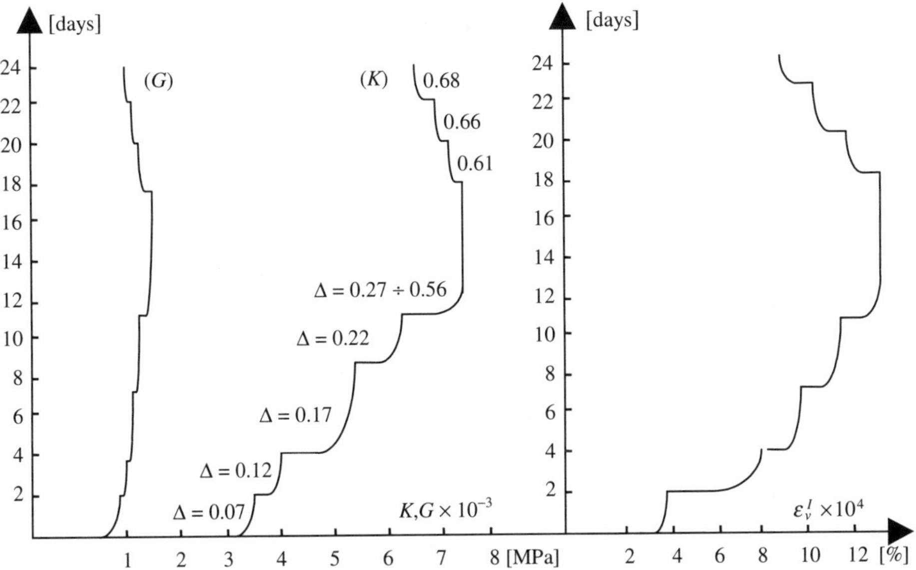

Fig. 2.2.5 Variation of elastic parameters during tests.

Afterwards the elastic parameters are decreasing until failure. That is shown for instance in Fig. 2.2.5 for rock salt (Matei and Cristescu [2000]). One can see that both the shearing modulus and the bulk modulus are behaving in this way. Also, when passing through the compressibility/dilatancy boundary these parameters are constants, but afterwards both they decrease.

That is also shown in Fig. 2.2.6. It is question of the variation of the two waves v_p and v_s during the test. They are related to the elastic parameters by

$$K = \rho \left(v_p^2 - \frac{4}{3} v_s^2 \right), \quad G = \rho v_s^2 .$$

The variation of elastic parameters due to repeated loadings is showing a decrease. This has been shown for concrete loaded repeated times (Taliercio and Gobbi [1997]).

If the rock has a significant initial porosity things may change. The elastic constants are either constant or slightly increasing with the octahedric shearing stress. The yield stress due to pore collapse is decreasing, both in hydrostatic and triaxial configuration (Maranini and Brignoli [1999]).

The same method is used for powders of various sorts. In Fig. 2.2.7 is given the stress–strain curve for microcrystalline cellulose obtained in a triaxial test. The axial stress–strain curve is practically a straight line. The diameter strain is no more linear. The small unloading following a short period of creep, are for the measurement of the elastic parameters. The curves depart from some strains which correspond to the values at the end of the hydrostatic part of the test. The curve for

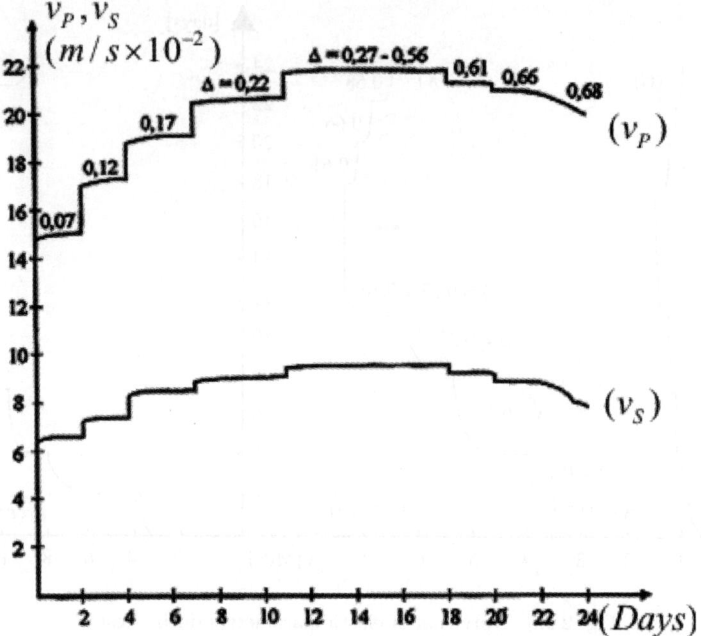

Fig. 2.2.6   Variation of the velocities of propagation of the two waves.

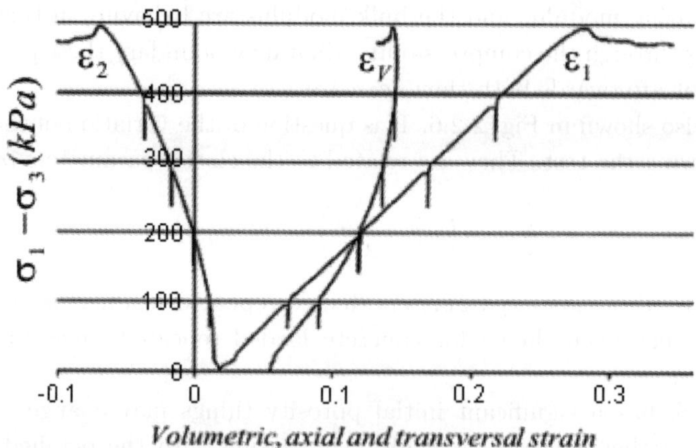

Fig. 2.2.7   Stress–strain curves for microcrystalline cellulose.

the volume is showing only vary little dilatancy, only compressibility is shown. But generally, for a general powder, the volume is compressible and afterwards dilatant.

In Fig. 2.2.8 is shown several curves for alumina powder, all for the volumetric strains, obtained in three-axial tests. The various confining pressure when these curves are changing are the behavior, are shown.

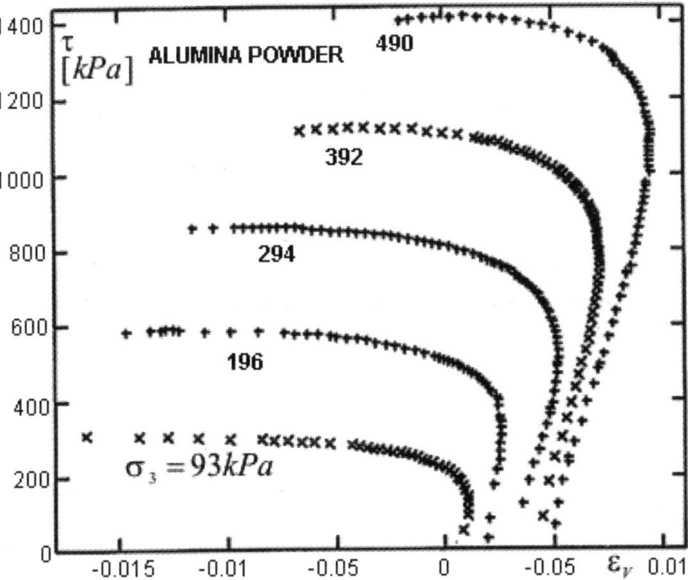

Fig. 2.2.8   The volumetric curves obtained in three axial tests shoving compressibility followed by dilatancy.

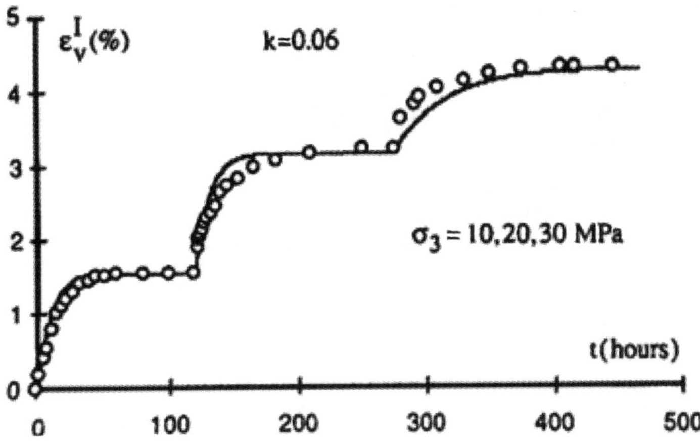

Fig. 2.2.9   Comparison between experimental and theoretical lines for hydrostatic creep.

One can see that the passing from compressibility to dilatancy depends on the value of the confining pressure. Thus all these curves are depending on the confining pressure. The coordinates of all the points where the curves are changing curvature, are on the compressibility/dilatancy boundary.

In Figs. 2.2.9 and 2.2.10 are given two creep figures for porous chalk, obtained by Dahu *et al.* [1995]. The first figure is given the result of hydrostatic creep, while

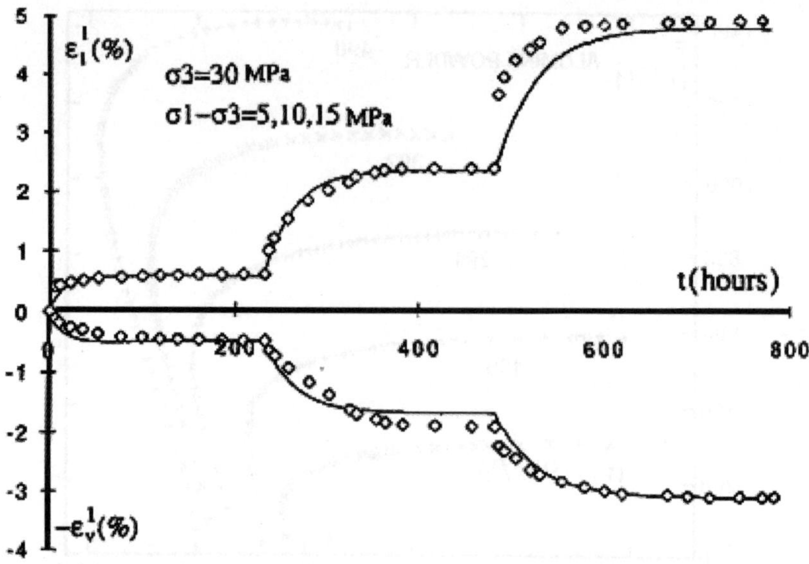

Fig. 2.2.10   Comparison between model prediction and experimental data in triaxial creep test on porous chalk.

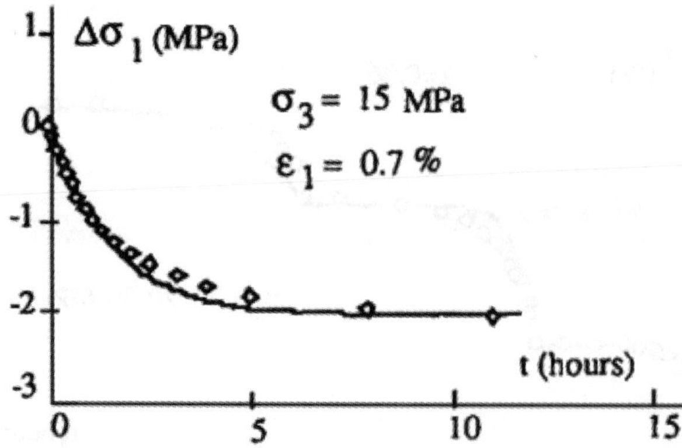

Fig. 2.2.11   Comparison between the model prediction (continuous line) and the experimental data for a relaxation test.

the second one, gives the triaxial creep tests. Both are comparison of the model with the experimental data.

Dahu *et al.* [1995] has made various comparisons of the theory with the tests. For instance, he has compared the model prediction with the experiments for a relaxation test. The results are given in the Fig. 2.2.11.

The procedure to be applied during test in order to measure correct elastic parameters is shown in Fig. 2.2.12. After each loading one is keeping the stress constant for 10 to 20 minutes. During this time a fast creep is taking place. When

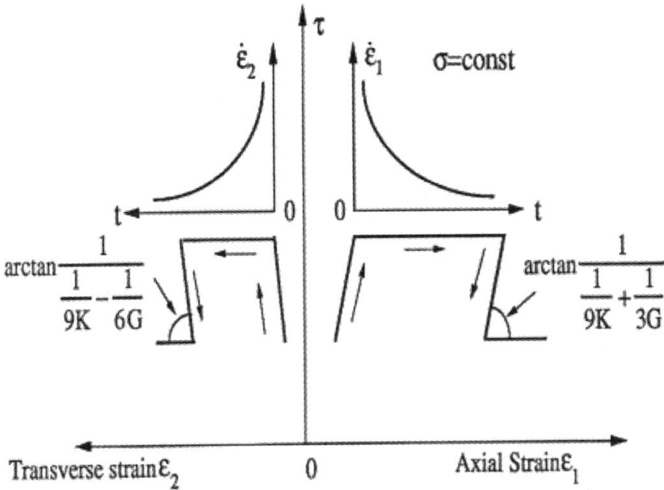

Fig. 2.2.12 Static procedure to determine the elastic parameters in unloading processes following short creep periods (Cristescu [1989]).

the strain rate are reasonable small so that during a fast unloading and reloading the rheological effect is no more influencing the unloading, one is measuring the value of the elastic parameters. If during the unloading a hysteresis loop is still observed, then one have to stay more time for creep. In other words one is not staying long enough to have creep developed. This method is now applied by other authors as well (Niandou *et al.* [1997], Cazacu [2002], Nawrochi *et al.* [1999], Maranini and Brignoli [1999]). That can be seen in Fig. 2.2.13 taken from Niandou; one can see the difference between a three-axial test done without a relaxation phase and a test done correctly. The Tournemire shale is an anisotropic rock. A creep theory for such kind of rocks is given by Pietruszczak *et al.* [2004]. Another creep theory for sedimentary rock as argillites is due to Shao *et al.* [2003] and for porous chalk by Shao *et al.* [1994]. The time-dependent deformation is described in terms of evolution of microstructure, leading to progressive degradation of elastic modulus and failure strength of material. The proposed model is applied to predict material responses in creep and relaxation tests. Again the creep of rock salt studied as differed behavior by means of multi-step creep tests with changes in deviator and temperature is due to Hamami [2000]. The study of the rock transfer function showing P-wave attenuation in the direction of loading as a function of stress and time during the test was done by Moustachi and Thimus [1997]. They show that the attenuation becomes more significant at the stage at which the rock dilates.

The results shown until now where obtained in three axial tests with cylindrical specimens, i.e., $\sigma_2 = \sigma_3$. However tests have been done with true three-axial tests. In these tests the specimen is cubical and is loaded independently on all feces. In these tests one is able to follow exactly either σ-constant or $\bar{\sigma}$-constant. In

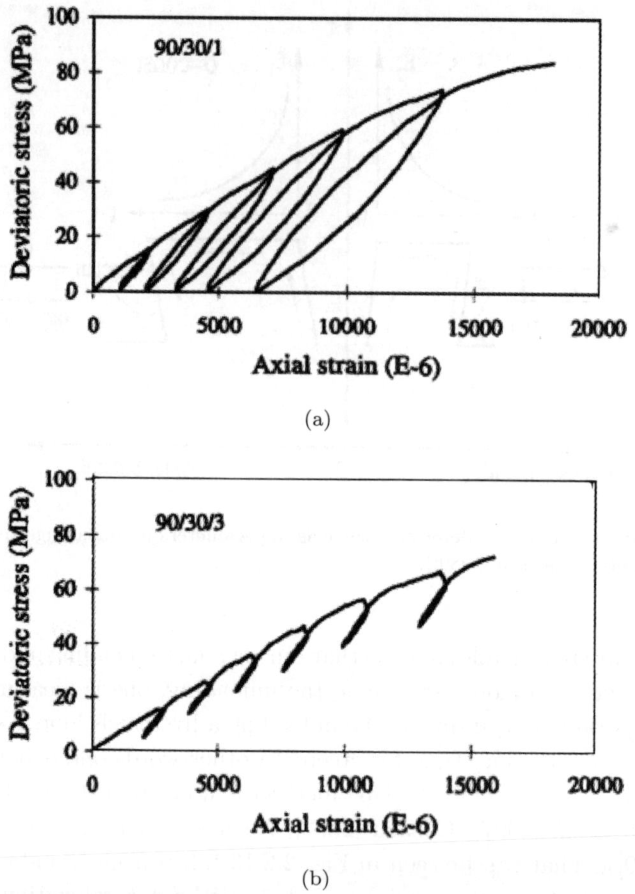

(a)

(b)

Fig. 2.2.13   (a) Stress–strain curves including unloading cycles during triaxial test without relaxation phase. (b) Stress–strain curves including unloading cycles during triaxial test with relaxation phase.

Fig. 2.2.14 is given a figure for rock salt (Hunsche), obtained in Germany at BGR. The compressibility/dilatancy boundary is the dotted line denoted by C. The mean stress is held constant while the octahedrical shearing stresses is increased and decreased above and below this line. One can see the behavior of the volume. It is first compressible for smaller values of τ and then dilatant, when τ is greater than the value corresponding to the compressibility/dilatancy boundary. The correctness of the compressibility/dilatancy boundary was tested by various authors. One has done various tests in order to see that. For instance Schultze *et al.* [2001] have tested the permeability of rock salt. This permeability is very low (less than 10^{-20} m^2). By combining measurements of ultrasonic wave velocities and permeability are used to determine the state of stresses at the compressibility/dilatancy boundary. The results confirm the boundary. The Opalinus clay (Switzerland) was also tested by

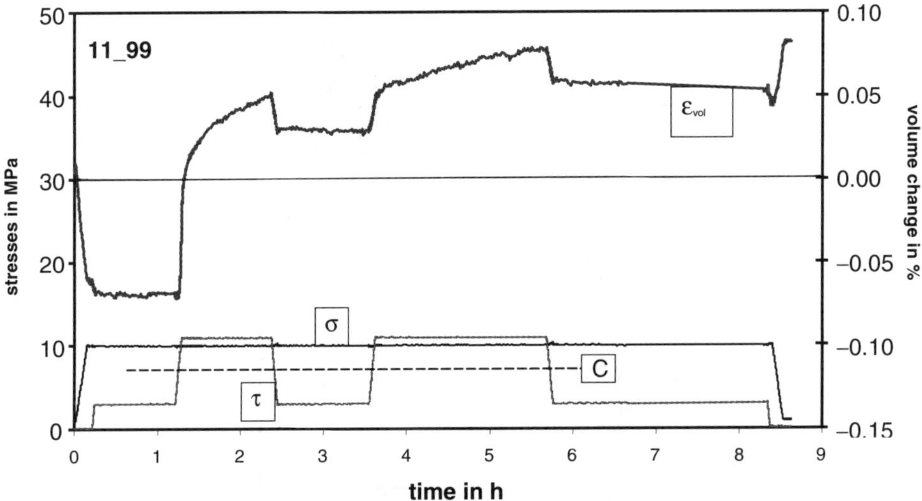

Fig. 2.2.14   Volume change in a true triaxial compressional deformation test with stepwise change of the octahedral shear stress τ below and above the dilatancy boundary Ⓒ at $\tau = 7.3$ MPa for constant mean stress $\sigma = 10$ MPa.

Hunsche *et al.* [2004] at BGR. Undrained clay specimens were compressed to study the temperature effects on their strength. Post-failure stresses were observed to drop to 40% of the failure stresses. At 600C, the failure stresses were 25% lower than at room temperature. The test data are used to develop a material law based on Burgers rheological model. The modeled strains fit reasonably well to the measured strains.

Some aspects of computational strategies for predictive geology with particular reference to the field of salt mechanics are due to Perić and Crook [2004]. The computational approach is based on the Lagrangian methodology incorporating: (i) large deformations of inelastic solids at finite strain, (ii) constitutive models for generic inelastic materials suitable for description of simultaneously active elasto-plastic, viscoplastic and viscoelastic behavior, (iii) an adaptive strategy for modeling of large deformations of inelastic solids at finite strains. A number of numerical simulations include the formation of salt diapers due to (i) compression and folding, (ii) thin skinned extension, and (iii) simulation of salt diapirism due to propagation on a basin scale.

In Fig. 2.2.15 is shown a triaxial testing device used to determine the mechanical properties of powders. For rocks it is similar, but more powerful. Some of the figures shown previously are obtained with this device.

Thus, the instantaneous response of a porous material is

$$\dot{\varepsilon}^E = \frac{\dot{\sigma}}{2G} + \left(\frac{1}{3K} - \frac{1}{2G} \right) \dot{\sigma} \mathbf{1} \tag{2.2.1}$$

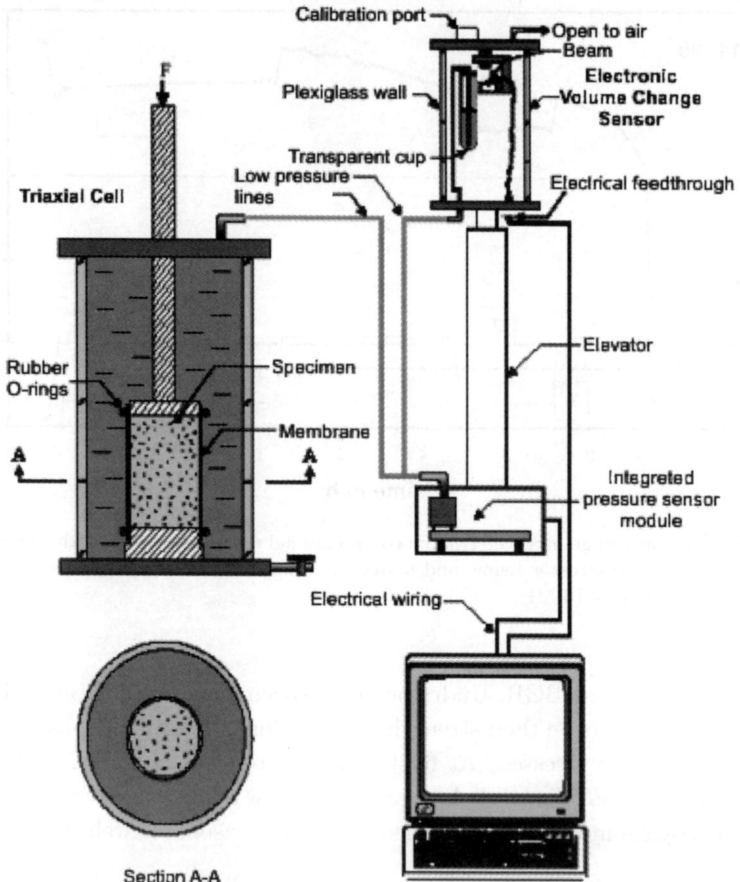

Fig. 2.2.15   A triaxial testing device to be used for powders (Abdel-Hadi and Cristescu [2004]).

with the elastic parameters variables and $\mathbf{1}$ is the unit tensor. From dynamic points of view, the two velocities of propagation, the longitudinal and transverse wave, are propagating with the velocities

$$v_P^2 = \frac{3K}{\rho}\frac{1-v}{1+v}, \quad v_S^2 = \frac{G}{\rho}, \quad v_B^2 = \frac{E}{\rho}, \tag{2.2.2}$$

where ρ is the density of the material, and v is the Poisson's ratio. These are used in order to determine dynamically the two elastic parameters, from

$$K = \rho\left(v_P^2 - \frac{4}{3}v_S^2\right), \quad G = \rho v_S^2, \quad E = \rho\frac{v_S^2(3v_P^2 - 4v_S^2)}{v_P^2 - v_S^2} = \rho v_B^2. \tag{2.2.3}$$

In three-axial tests $\varepsilon_1 = \bar{\varepsilon} + (\varepsilon_V/3)$, with $\bar{\varepsilon}$, the equivalent strain given by

$$\bar{\varepsilon} = \left(\frac{2}{3}\dot{\varepsilon}\cdot\dot{\varepsilon}\right)^{1/2}. \tag{2.2.4}$$

Besides that, one is using the true three-axial tests, where one is no more using such formulae.

2.3 The Constitutive Equation

In order to develop the constitutive equation, we have to describe both the **steady state creep** and the **transient creep**. For this purpose we use the positive part of a function

$$\langle A \rangle = \frac{1}{2}(A + |A|)\,. \tag{2.3.1}$$

For the irreversible part of the rate of deformation due to transient creep, we can use the formula

$$\dot{\varepsilon}_T^I = k_T \left\langle 1 - \frac{W(t)}{H(\boldsymbol{\sigma})} \right\rangle \frac{\partial F}{\partial \boldsymbol{\sigma}} \tag{2.3.2}$$

or if one is not able to determined the viscoplastic potential $F(\boldsymbol{\sigma})$ one can use

$$\dot{\varepsilon}_T^I = k_T \left\langle 1 - \frac{W(t)}{H(\boldsymbol{\sigma})} \right\rangle \boldsymbol{N}(\boldsymbol{\sigma})\,, \tag{2.3.3}$$

where $H(\boldsymbol{\sigma})$ is the yield function, with

$$H(\boldsymbol{\sigma}(t)) = W(t)\,, \tag{2.3.4}$$

the equation of the stabilization boundary (which is the locus of the stress states at the end of transient creep when stabilization takes place, i.e., when $\dot{\varepsilon}_T^I = \boldsymbol{0}$, $\dot{\boldsymbol{\sigma}} = \boldsymbol{0}$). This boundary depends on the loading history, with

$$\begin{aligned}
W(T) &= \int_0^T \boldsymbol{\sigma}(t) \cdot \dot{\boldsymbol{\varepsilon}}(t)\, dt \\
&= \int_0^T \sigma(t)\dot{\varepsilon}_v^I(t)\, dt + \int_0^T \boldsymbol{\sigma}'(t) \cdot \dot{\boldsymbol{\varepsilon}}^{I'}(t)\, dt \\
&= W_V(T) + W_D(T)\,,
\end{aligned} \tag{2.3.5}$$

the irreversible stress power per unit volume at time T, used as a work-hardening parameter or an internal state variable. Therefore the history is involved in $W(T)$. If F coincides with H we say that the constitutive equation is "associated" to a prescribed yield function H. Otherwise the constitutive equation is said to be "nonassociated", as is the case for most porous materials.

The function $\boldsymbol{N}(\boldsymbol{\sigma})$ is used when one is not able to determine the viscoplastic potential, and one can determine the "viscoplatic strain rate orientation tensor" (Cleja-Tigoiu [1991], Cazacu and Cristescu [1995], Cazacu *et al.* [1997]). K_T is some kind of "viscosity coefficient"; it may depend slightly on stress and strain invariants, and maybe on a damage parameter describing the history of micro cracking and/or history of pore collapse.

In order to describe the **steady-state creep** on can adapt accordingly either the function H, or one can add to (2.3.2) an additional term as for instance

$$\dot{\varepsilon}_S^I = k_S \frac{\partial S}{\partial \boldsymbol{\sigma}}\,, \tag{2.3.6}$$

where $S(\boldsymbol{\sigma})$ is a viscoplastic potential for steady-state creep and k_S is a viscosity coefficient for steady-state creep, who may possibly depend on stress invariants and on damage if necessary. This creep described by this term will last so long as stress is applied. Transient and steady-state creep is quite often difficult to distinguish. One can describe them by a single term (a better procedure) or by two additive terms. By describing them by two terms is easiest from the point of view of describing the volumetric behavior, with either compressibility or dilatancy.

Let as make the following remarks. The bracket $\langle\ \rangle$ is involved in linear form. But one can use some nonlinear forms also. For instance

$$\dot{\varepsilon}_T^I = \frac{k_T}{E}[1 - \exp(\lambda\langle H(\boldsymbol{\sigma}) - W(t)\rangle)]\frac{\partial F}{\partial\boldsymbol{\sigma}}$$

or

$$\dot{\varepsilon}_T^I = \frac{k_T}{E}\left\langle\frac{H(\boldsymbol{\sigma}) - W(t)}{a}\right\rangle^n \frac{\partial F}{\partial\boldsymbol{\sigma}}$$

if necessary (see Cristescu and Suliciu [1982]). Here $\lambda > 0$, $a > 0$ and $n > 0$ are material constants. We can use also other parameter to describe the irreversible isotropic hardening. For instance we ca use the irreversible equivalent strain

$$\bar{\varepsilon}^I(t) = \sqrt{\frac{2}{3}}(\varepsilon^I(t) \cdot \varepsilon^I(t))^{1/2}$$

or the irreversible equivalent integral of the rate of deformation tensor

$$\bar{\varepsilon}^I(T) = \sqrt{\frac{2}{3}}\int_0^T \sqrt{\dot{\varepsilon}^I(t) \cdot \dot{\varepsilon}^I(t)}\, dt\,.$$

However, both these last expressions cannot distinguish between irreversibly produced by compressibility and irreversibility produced by dilatancy. These expressions can be used only if a certain material is either compressible only or only dilatant.

We consider here only homogeneous and isotropic materials. Thus the constitutive functions will depend on stress and strain invariants only, and maybe, on an isotropic damage parameter. We will denote by

$$\sigma = \frac{1}{3}(\sigma_1 + \sigma_2 + \sigma_3) \tag{2.3.7}$$

the mean stress and by

$$\bar{\sigma}^2 = \sigma_1^2 + \sigma_2^2 + \sigma_3^2 - \sigma_1\sigma_2 - \sigma_2\sigma_3 - \sigma_3\sigma_1 \tag{2.3.8}$$

the equivalent stress, or by

$$\tau = \frac{\sqrt{2}}{3}\bar{\sigma} = \left(\frac{2}{3}II_{\sigma'}\right)^{1/2} \tag{2.3.9}$$

the octahedral shear stress τ, with $II_{\sigma'} = (1/2)\boldsymbol{\sigma'} \cdot \boldsymbol{\sigma'}$ the second invariant of the stress deviator $\boldsymbol{\sigma'} = \boldsymbol{\sigma} - \sigma\mathbf{1}$.

- We assume that the material displacements and rotations are small so that the rate of deformation components are additive

$$\dot{\varepsilon} = \dot{\varepsilon}^E + \dot{\varepsilon}^I \tag{2.3.10}$$

- The elastic rate of deformation component $\dot{\varepsilon}^E$ is given by (2.2.1).
- The irreversible rate of deformation component $\dot{\varepsilon}^I$ satisfies (2.3.2) or (2.3.6), or is the sum of the two terms, if transient and steady-state creeps are to be considered.
- The initial yield stress of the particular material can be assumed to be zero, or very close to it.
- The constitutive equation is valid in a certain constitutive domain bounded by a short-term failure surface, which are also time-dependent and will be included in the constitutive equation.

The **constitutive equation** will be written in the form

$$\dot{\varepsilon} = \frac{\dot{\sigma}}{2G} + \left(\frac{1}{3K} - \frac{1}{2G}\right)\dot{\sigma}\mathbf{1} + k_T\left\langle 1 - \frac{W(t)}{H(\sigma)}\right\rangle\frac{\partial F}{\partial \sigma} \tag{2.3.11}$$

if only transient creep is considered. The volumetric irreversible rate of deformation component is

$$(\dot{\varepsilon}_V^I)_T = k_T\left\langle 1 - \frac{W(t)}{H(\sigma)}\right\rangle\frac{\partial F}{\partial \sigma}\cdot\mathbf{1}. \tag{2.3.12}$$

A stress variation from $\sigma(t_0)$ to $\sigma(t) \neq \sigma(t_0)$ with $t > t_0$, will be called **loading** if

$$H(\sigma(t)) > H(\sigma(t_0)) \tag{2.3.13}$$

and three cases are possible depending on which of the following inequalities is satisfied by the new stress state:

$$\frac{\partial F}{\partial \sigma}\cdot\mathbf{1} > 0 \quad \text{or} \quad \frac{\partial F}{\partial \sigma} > 0 \text{ compressibility} \tag{2.3.14}$$

$$\frac{\partial F}{\partial \sigma}\cdot\mathbf{1} = 0 \quad \text{or} \quad \frac{\partial F}{\partial \sigma} = 0 \text{ compressibility/dilatancy boundary} \tag{2.3.15}$$

$$\frac{\partial F}{\partial \sigma}\cdot\mathbf{1} < 0 \quad \text{or} \quad \frac{\partial F}{\partial \sigma} < 0 \text{ dilatancy}. \tag{2.3.16}$$

Let us observe that $(\partial F/\partial \sigma)\cdot\mathbf{1} = \partial F/\partial \sigma$ if F depends on the stress invariants only. Therefore the behavior of the volume is governed by the orientation of the normal to the surface $F(\sigma) = constant$ at the point representing the actual stress state (see Fig. 2.3.1). If the projection of this normal $(\partial F/\partial \sigma)\cdot\mathbf{1}$ on the σ-*axis* is pointing towards the positive orientation of this axis, that stress state produces irreversible compressibility, otherwise dilatancy. There where this normal is orthogonal to the σ-*axis*, there are no irreversible volumetric changes, and the volume change is elastic (Cristescu [1994]).

If instead of (2.3.13) the new stress state satisfies

$$H(\boldsymbol{\sigma}(t)) < W(t_0) \qquad\qquad (2.3.17)$$

then an **unloading** takes place and the response of the material is elastic, according to (2.2.1).

The viscoplastic potential is determined from triaxial tests where $\sigma_2 = \sigma_3$, from the formulas

$$\frac{\partial F}{\partial \sigma} = \frac{\dot{\varepsilon}_V^I}{k\langle 1 - (W(t)/H(\sigma))\rangle}, \qquad \frac{\partial F}{\partial \bar{\sigma}} = \frac{2}{3}\frac{\dot{\varepsilon}_1^I - \dot{\varepsilon}_2^I}{k\langle 1 - (W(t)/H(\sigma))\rangle}. \qquad (2.3.18)$$

In these expressions the brackets part are already known. The derivative $\partial F/\partial \sigma$ is determined in the hydrostatic part and afterwards in the deviator part. The last part is determined by determining first the equation of the compressibility/dilatancy boundary and by writing the derivative $\partial F/\partial \sigma$ to be zero on this boundary and to satisfy some other conditions (failure and variation with respect to σ). Thus is determined the viscoplastic potential.

If we make a creep test and after the time interval $t - t_0$ the axial stress is increased in successive steps, and during this short time interval the strain increase by creep, according to

$$\varepsilon_1^R = \left(\frac{1}{3G} + \frac{1}{9K}\right)\sigma_1^R + \frac{\langle 1 - (W(t_0)/H(\sigma))\rangle \partial F/\partial \sigma_1}{1/H((\partial F/\partial \sigma)\,\sigma + (\partial F/\partial \bar{\sigma})\,\bar{\sigma})}$$

$$\times \left\{1 - \exp\left[\frac{k}{H}\left(\frac{\partial F}{\partial \sigma}\sigma + \frac{\partial F}{\partial \bar{\sigma}}\bar{\sigma}\right)(t_0 - t)\right]\right\}$$

$$\varepsilon_2^R = \left(-\frac{1}{6G} + \frac{1}{9K}\right)\sigma_1^R + \frac{\langle 1 - (W(t_0)/H(\sigma))\rangle \partial F/\partial \sigma_2}{1/H((\partial F/\partial \sigma)\,\sigma + (\partial F/\partial \bar{\sigma})\,\bar{\sigma})}$$

$$\times \left\{1 - \exp\left[\frac{k}{H}\left(\frac{\partial F}{\partial \sigma}\sigma + \frac{\partial F}{\partial \bar{\sigma}}\bar{\sigma}\right)(t_0 - t)\right]\right\}$$

in order to determine the strains after each stress increase occurring at time t_0. Similar formulae are used for the next loading step, and so forth. Here the upper script "R" means "relative".

The picture Fig. 2.3.1 is real, but a little complicated. If we need a simpler model we have to choose a picture which is simpler. In Fig. 2.3.2 is shown a possible example. We choose a compressibility/dilatancy boundary which is a simple increasing curve, tending towards horizontal.

The behavior of Berea sandstone under confining pressure was studied by Khan *et al.* [1991, 1992]. An elastic-plastic associated model to deal with the irreversible deformation is given. The Berea sandstone is compressible/dilatant only for small confining pressures.

The constitutive equation for rock salt by Cristescu and Hunsche [1998] was used by Mahnken and Kohlmeier [2001] to describe the steady state creep in rock

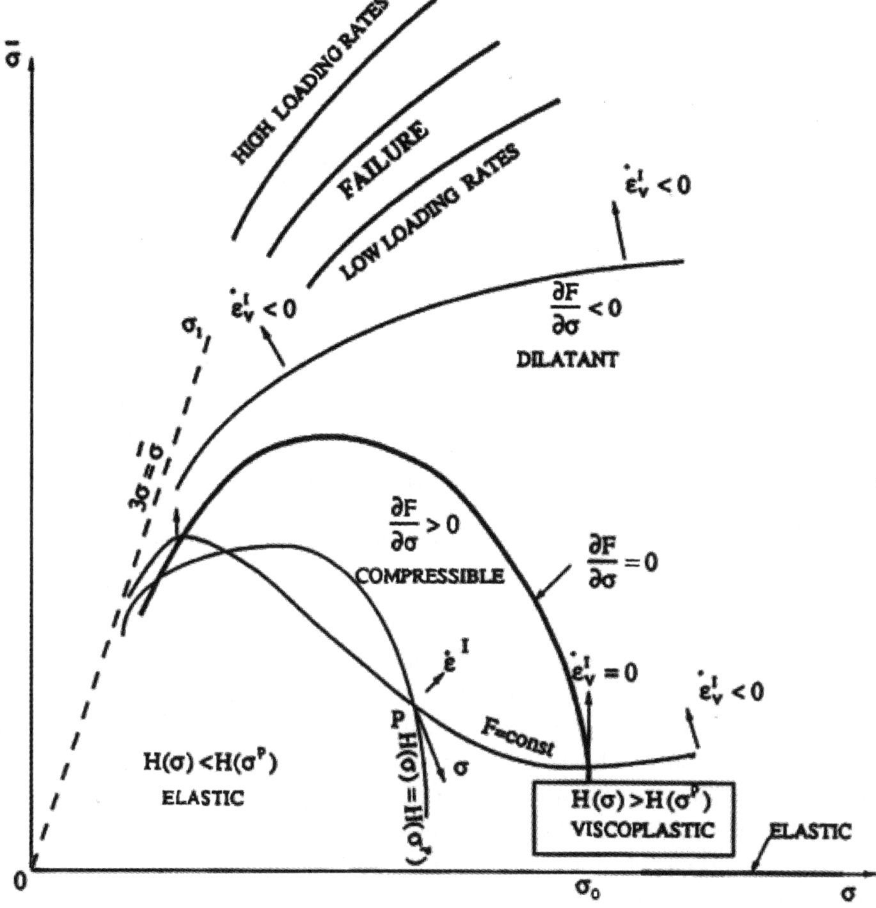

Fig. 2.3.1 Domains of compressibility, dilatancy, and elasticity in the constitutive domain; thick line is compressibility/dilatancy boundary $\partial F/\partial\sigma = 0$; failure depends on the loading rate.

salt. Not only the long-term stress-induced deformation but also the simultaneous fluid permeation of rock salt was considered. A coupled finite-element strategy is presented, where the problem is formulated in the context of the theory of fluid saturated porous media, considering two phases, i.e., the rock salt as a solid phase in which the remaining pores are filled with brine as a fluid phase. The constitutive model for the rock salt is based on true triaxial experimental tests thus taking into account a distinct boundary between the dilatancy and the compression domains in the octahedral stress space.

In another paper Hatzor and Heyman [1997] have tested the salt from Mount Sendom diaper at strain rates of 10^{-5} s^{-1}. The strengthening effect of confining pressure is observed up to 4.5 MPa. They are confirming the compressibility-dilatancy boundary of Cristescu and Hunsche [1992].

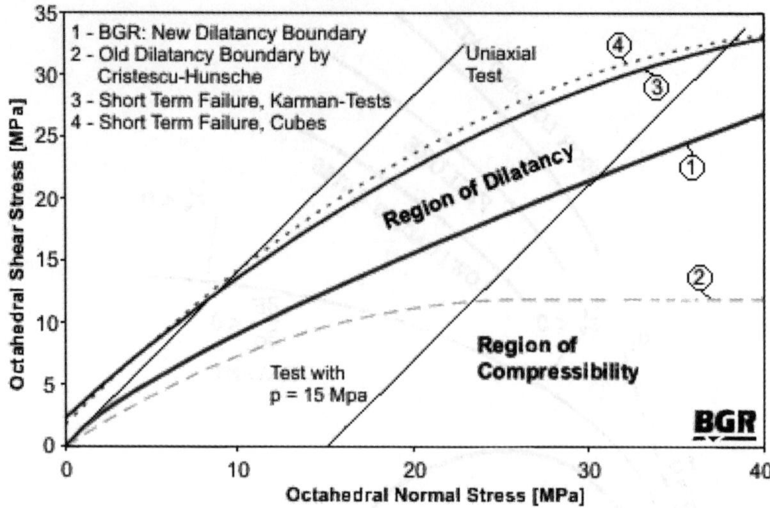

Fig. 2.3.2   A simple figure showing a particular compressible/dilatant boundary, and a single failure surface.

Another constitutive equation for rock salt is due to Aubertin *et al.* [1999]. A general format for internal state variable modeling to describe the rate-dependent behavior of rock salt in the ductile regime is given. The model is representing fairly well the fully plastic behavior of rock salt submitted to triaxial tests at relatively high confining pressures. Then Yahya *et al.* [2000] have investigated the development and applications of elaborate constitutive equations for ductile, fully plastics, behavior of rock salt. The unified model relies on the use of internal state variables attached to a specific phenomena, including isotropic and kinematics hardening. These concepts are illustrated by results obtained on rock salt samples submitted to different loading conditions, including constant strain rate tests, creep tests, and relaxation tests.

A theory for pressure sensitive inelastic flow and damage evolution in crystalline solids was evaluated against triaxial creep experiments on rock salt, is due to Chan *et al.* [1994]. The model was then utilized to obtain the creep response and damage evolution in rock salt as function of confining pressure and stress difference.

Some nonlinear one-dimensional models used to describe creep are the model of Wang *et al.* [2004]. Another model to describe the nonlinear viscoplastic creep of mudstone around an underground excavation is due to Song [1993].

2.4   Failure

It has been shown by several examples given previously **that failure is strongly a time-dependent phenomenon**, while ultimate failure depends also on confining pressure, temperature, humidity, and geological factors (Cristescu [1993]). Let us

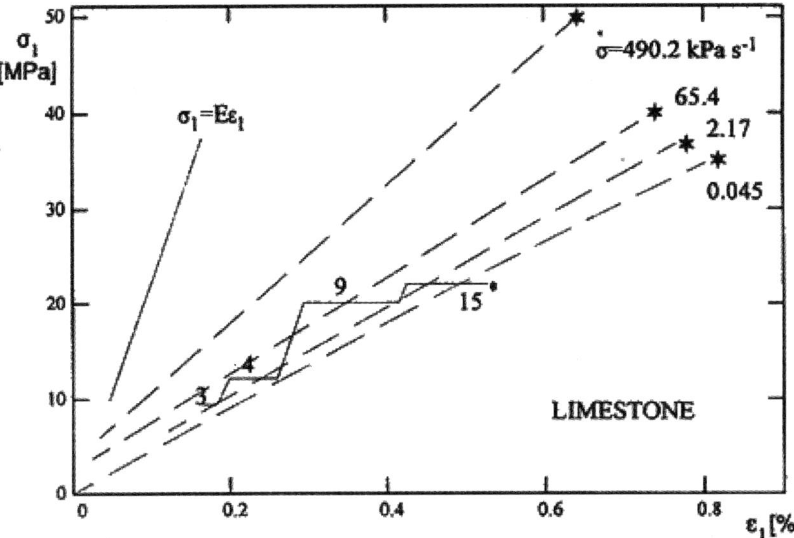

Fig. 2.4.1   Stress–strain curves for limestone showing a strong influence of loading history on failure (stars).

give here additional examples. In Fig. 2.4.1 are given several uniaxial stress–strain curves for limestone (Cristescu [1989b]) obtained with four distinct loading rates shown. Failure is marked by a star. Thus failure depends on the loading rate. In the very long time intervals involved in mining or petroleum applications failure may occur at a much lower stress state, but after a certain long time interval. That is shown in Fig. 2.4.1 by the full line: the test is a creep test in which stress was increased in steps, and after each increase it was held constant 3 days, then 4, 9, and finally 15 days. Thus failure occurring after 31 days is taking place at a much smaller stress state than in conventional tests performed with constant loading rate. The strain at failure is also **loading-history dependent**. Figure 2.4.1 is also showing the elastic slope, but that has been obtained not from the initial slope of the stress–strain curve but using the method described.

It is also important to mention that as the strain rate is increased, the dilatancy and compressibility at failure diminishes. Since dilatancy is related to micro cracking and pore formation, one can conclude that not only ultimate failure but also damage **evolution depends on the loading history**. Also, in vary slow loading rates the rock can sustain much more damage before ultimate failure.

It has been shown by Kranz *et al.* [1982] that the time to failure depends on the magnitude of stress difference (or octahedral shear stress). Figure 2.4.2 shows (Kranz *et al.* [1982]) the time to failure for dray and wet granite. The time to failure increases significantly with decrease of applied stress. This time increase seems to be asymptotic, i.e., increasing very much when the decreasing stress approaches a certain level — the compressibility/dilatancy boundary.

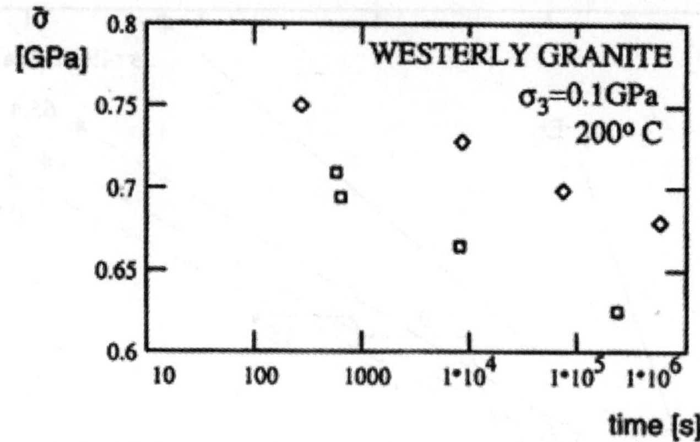

Fig. 2.4.2   Time to failure for various levels of applied stress for wet (squares) and dray (diamonds) granite.

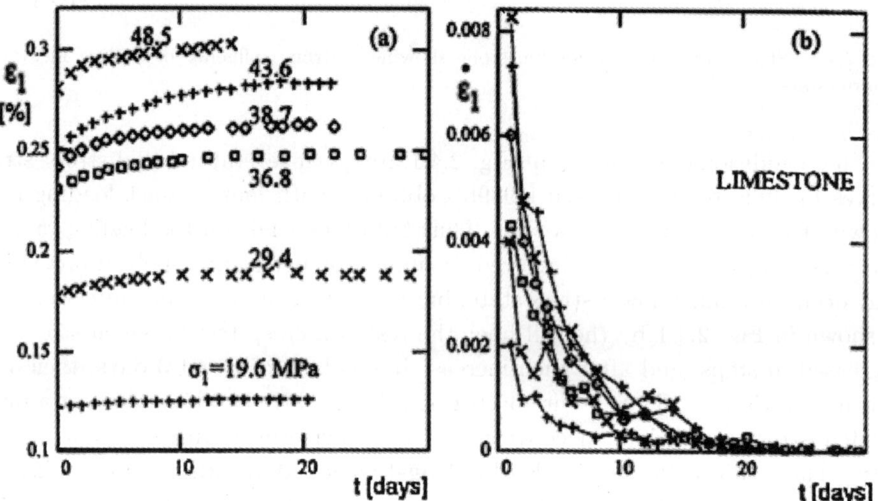

Fig. 2.4.3   Creep curves for limestone in uniaxial compression test showing that creep failure is possible for $\sigma_1 > 0.6\sigma_c$ only.

Failure after a long-term loading depends obviously on the stress level. Figure 2.4.3 shows several creep curves (Cristescu [1975]) obtained in uniaxial creep tests on limestone from Palazu Mare (initial porosity 4 to 6%). All curves have been obtained with a single specimen loaded successively with increasing stress level. The short term uniaxial compression strength of this rock is $\sigma_c = 72.4$ MPa.

For smaller loading stress the transient creep ends by stabilization after about 20 days, as shown in figure. If the loading stress is less than a certain limit, which depends on the short term compression strength of the rock, practically only transient creep is apparently exhibited by the rock, i.e., stabilization is considered to be realized after 5–10 additional days under the same constant stress no strain increase is exhibited.

For this particular limestone axial loading stresses $\sigma_1 < 0.6\sigma_c$ will result more or less in transient creep only. For $\sigma_1 = 0.672\sigma_c$ the creep becomes steady-state and failure is obtained 14 days after the last reloading. No tertiary creep was exhibited, but for other specimens of the same rock tertiary creep was also observed several days before failure. For the test shown in Fig. 2.4.3 the total duration up to failure of specimen deformation was 119 days [upper curve in Fig. 2.4.3(a)]. Figure 2.3.3(b) shows the variation of strain rate, since from such figures one can distinguish if in a certain test the creep is transient (decreasing $\dot{\varepsilon}$), steady state (constant $\dot{\varepsilon}$) or tertiary creep occurs (increasing $\dot{\varepsilon}$). The limit stress up to where only transient creep is observed is considerably distinct from one rock to the other one. For several rocks (such as rock salt) this limit stress is quite small or even not existent, but for most other rocks this limit stress is a certain fraction of the short-term compression strength, but decreases with increasing temperature. Generally the higher the applied stresses difference the shorter the time to failure.

Kranz [1980] has shown that "higher pressure require more volumetric strain to accumulate prior to onset of instability". By testing small circular openings in samples of jointed coal, Kaiser *et al.* [1982] have found that the rupture process is time-dependent and more readily detectable by observing creep deformation than from the instantaneous response to loading.

A quite large literature is devoted to Acoustic Emission. With AE, one is usually recording the total number N of AE, or their rate $\dot{N}$ (number of events per unit time). For more advanced evaluation one can also measure the location of the events, the amplitude of AE, the energy of the AE (square of the amplitude) or the total energy of AE (sum of squares of all amplitudes). Let us mention here only the data by Fota [1983] on andesitic rock (for which $K = 13.3$ MPa, $G = 18.2$ MPa,

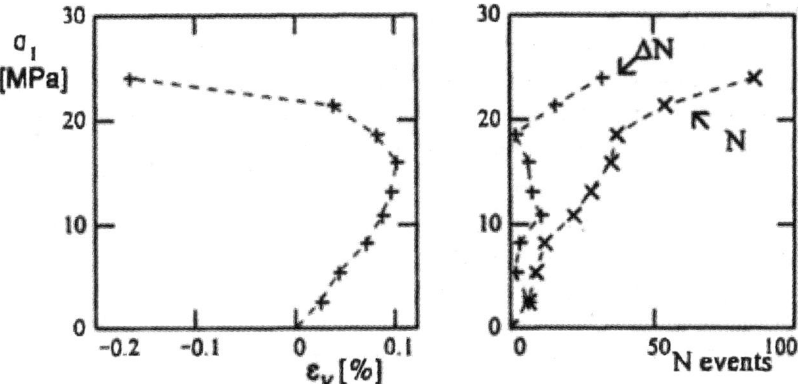

Fig. 2.4.4 Variation of volumetric strain, of total events N and of the total events per each loading level with axial stress, for andesite.

$E = 37.5$ MPa, $v = 0.031$). The axial stress σ is increased in successive steps and after each increase is held constant for several minutes. Figure 2.4.4 shows for that rock the successive final values of ε_v and those of $\dot{N}$ at the end of each time interval in which stress was held constant; the successive loading stress increment is $\Delta\sigma_1 = 24.5$ MPa and after each increment the stress is held constant for 15 minutes. During this time interval (Cristescu [1989b]) creep takes place and a number of events are recorded. The initial porosity of the rock is $n = 1.84\%$. So long as the rock is in the compressibility state, the total number of events ΔN as recorded at the end of each of the successive loading steps (i.e., ΔN is the total number of events recorded during a single loading stress) is somewhat higher than that recorded during those loading steps which correspond to the stress interval during which the rock is passing from compressibility to dilatancy. Finally, during dilatancy the number of events per loading step increases quite fast from one step to the next one. Figure 2.4.4 shows also the total number of events N recorded during the test. Similar results have been obtained by other researchers. The smallest values of $dN/d\sigma_1$ always correspond to the passage from compressibility to dilatancy and the larges one to the advanced stage of dilatancy just preceding failure.

The damage was also estimated by dynamic procedures. If we measure the **seismic velocities** v_P, for the longitudinal wave, and v_S for the shearing waves, as well as the **bar velocity** v_B, we can determine the elastic constants from

$$E = \rho\frac{v_S^2(3v_P^2 - 4v_S^2)}{v_P^2 - v_S^2} = \rho v_B^2\,, \quad K = \rho\left(v_P^2 - \frac{4}{3}v_S^2\right),$$

$$G = \rho v_S^2\,, \qquad\qquad v = \frac{v_P^2 - 2v_S^2}{2(v_P^2 - v_S^2)} = \frac{v_B^2}{2v_S^2} - 1\,.$$

As a general trend obtained from tests, all the velocities of propagation generally increase with increasing pressure. To illustrate this statement, Fig. 2.4.5 shows the variation of the velocities of propagation v_P and v_S with pressure for several granite specimens according to the experimental data by Bayuk [1966]. The densities and initial porosities are also given. The solid lines represent the mean of all cases shown. The increase of the velocities with pressure is more significant for relatively small pressures, while at high pressure this increase is very slow and tending towards a limit value, when all microcracks and pores are closed. On the other hand, the increase of v_P with pressure is more significant than that of v_S. The variation of velocities with pressure is also more important for rocks with high initial porosities, while for rocks with very small initial porosity this variation is less pronounced and sometimes even negligible. At high pressures the rocks seems isotropic. For wet and dry amphibolite and amphibolite/gneiss Popp and Kern [1994] (Fig. 2.4.6) have shown that both v_P and v_S increase with increasing pressure.

Yanagidani *et al.* [1985] have shown that in granite at earlier stage before the onset of dilatancy the v_P velocities increase: "This corresponds to the closure of pre-existing cracks regardless of their directions. After the onset of dilatancy, $v_{P\perp}$

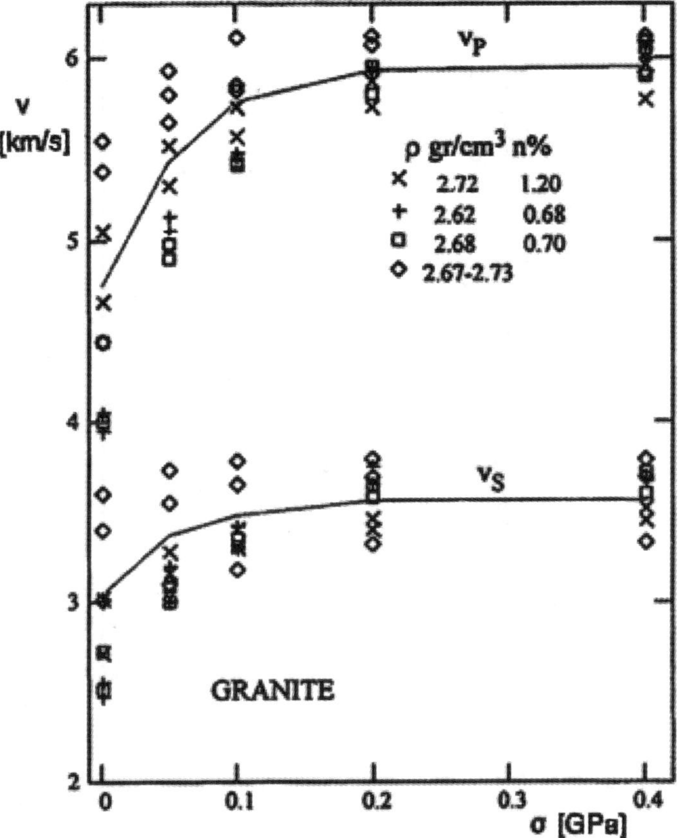

Fig. 2.4.5 Variation of longitudinal velocity of propagation v_P and of the shear velocity of propagation v_S with pressure, for granite.

(in planes perpendicular to the loading axis) began to decrease gradually while $v_{P\parallel}$ (in planes parallel to it) hardly changed, which is explained by the opening of axially induced cracks". These authors also show that "during the primary creep, the changes in $v_{P\perp}$ tracked well with those of the average circumferential straining on the other hand, there was no change in $v_{P\parallel}$. However, after the primary creep terminated, $v_{P\perp}$ and $v_{P\parallel}$ hardly changed". A rapid decrease in both $v_{P\perp}$ and $v_{P\parallel}$ take place just before faulting.

Similar results are reported by other authors. Velocity-pressure curves for the Haast schist, New Zealand, have been published by Okaya and McEvilly [2003]. They have tested the anisotropy of various rocks in various directions. From all these follows that the damage or microcracks and pores existing in a rock can be estimated by the measurements of the travel time of seismic waves. At high pressures the cracks of the rock are closed and the rock elasticity is in fact the elasticity of the constituent minerals (without cracks and pores).

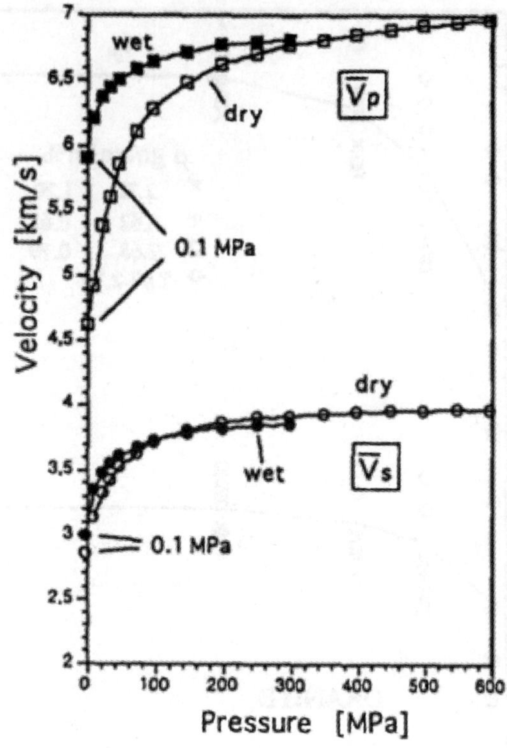

Fig. 2.4.6 Effect of pressure on averaged P- and S-wave velocities.

From the above one can conclude that damage and failure of rock are progressive and **related to the same mechanisms which produce dilatancy**. We recall that work-hardening by dilatancy and/or compressibility is described by the irreversible stress work per unit volume

$$W(T) = \int_0^T \boldsymbol{\sigma}(t) \cdot \dot{\boldsymbol{\varepsilon}}^I(t)\, dt \qquad (2.4.1)$$

and the compressibility/dilatacy boundary is defined by

$$\frac{\partial F}{\partial \boldsymbol{\sigma}} \cdot \mathbf{1} = 0 \quad \text{or} \quad N_1 = 0. \qquad (2.4.2)$$

We introduce an energetic damage parameter, which can describe the evolution in time of the damage of a geomaterial (Cristescu [1986, 1989a]). We recall that the irreversible stress work per unit volume (2.4.1) can be decomposed into two parts:

$$W(T) = \int_0^T \sigma(t)\dot{\varepsilon}_V^I(t)\, dt + \int_0^T \boldsymbol{\sigma}'(t) \cdot \dot{\boldsymbol{\varepsilon}}^{I'}(t)\, dt \qquad (2.4.3)$$

corresponding to the volumetric deformation and to the change in shape (deviatoric).

Let us consider now a typical low porosity compressible/dilatant isotropic rock and its mechanical behavior during triaxial tests. Figure 2.4.7(a) shows two possible stress trajectories followed in such tests. In this figure the C/D boundary is shown as a "transition" zone of incompressibility, between the compressibility domain and the dilatant one. During the hydrostatic portion of the test it is only W_V which increases while $W_D = 0$. A schematic representation of the W_V increase during hydrostatic test is shown in Fig. 2.4.7(b). If this test is carried out up to very high pressures, W_V reaches a constant maximum value $W_{V(\max)}$, at point E corresponding to the value σ_0 of σ, the one that closes all pores and microcracks. However, in standard true triaxial test starting from an intermediate value of W_V [see point A on the Fig. 2.4.7(b), corresponding to the point A in Fig. 2.4.7(a)], mean stress is held constant and it is τ which is increased only. This is shown in Fig. 2.4.7(c), where the point A corresponds to point A in Figs. 2.4.7(a) and (b). W_V continues to increase until it reaches the C/D boundary [Fig. 2.4.7(a)], stays constant in the incompressible transition zone between point B and C shown on both Figs. 2.4.7(a) and (c), and afterwards decreases, becoming ultimately negative [see Fig. 2.4.7(c)]. The point A, corresponding to $\tau = 0$, in Fig. 2.4.7(c), shows the magnitude of W_V reached at the end of the hydrostatic stage of the test [point A in Fig. 2.4.7(b)]. The starting value for W_V or ε_V is dependent on the time spent at point A (creep taking place when passing from hydrostatic to deviatoric test). Ultimately the decrease of W_V is abrupt.

The mechanical properties shown in Fig. 2.4.7 correspond to rocks with small initial porosity, with particle size relative big, over 20 microns. Some other rocks may possess different properties. For dilatant rocks it is natural to consider the total decrease of W_V starting from its maximal value $W_{V(\max)}$ to be the measure of the damage of the rock produced by the loading along portions of the segments CD or C'D' shown in Fig. 2.4.7(a). Let us introduce the notation

$$d(t) = W_{V(\max)} - W_V(t) \tag{2.4.4}$$

for the damage parameter of the rock at time t. Here t is a moment of time during the period when dilatancy takes place and therefore $t > t_{(\max)}$, where $t_{(\max)}$ is the time when W_V reaches its maximum.

Thus the damage parameter is a measure of the energy release due to microcracking when dilatancy takes place. The damage rate is defined by the evolution law

$$\dot{d}(t) = -\dot{W}_V(t) = -\sigma(t)\dot{\varepsilon}_V^I(t) = -k_T \left\langle 1 - \frac{W(t)}{H(\sigma)} \right\rangle \frac{\partial F}{\partial \sigma}\sigma - k_S \frac{\partial S}{\partial \sigma}\sigma \tag{2.4.5}$$

$$\text{if} \quad \frac{\partial F}{\partial \sigma} < 0 \quad \text{and} \quad \frac{\partial S}{\partial \sigma} < 0,$$

and thus by the constitutive equation itself.

Thus the energetic damage parameter is $d_f = 0.71$ MPa for rock salt, $d_f = 0.167$ MPa for granite, $d_f = 4 \times 10^{-3}$ MPa for some coals, and $d_f = 0.5$ MPa for sandstone. For more details see Cristescu [1989a] and Cristescu and Hunsche [1998].

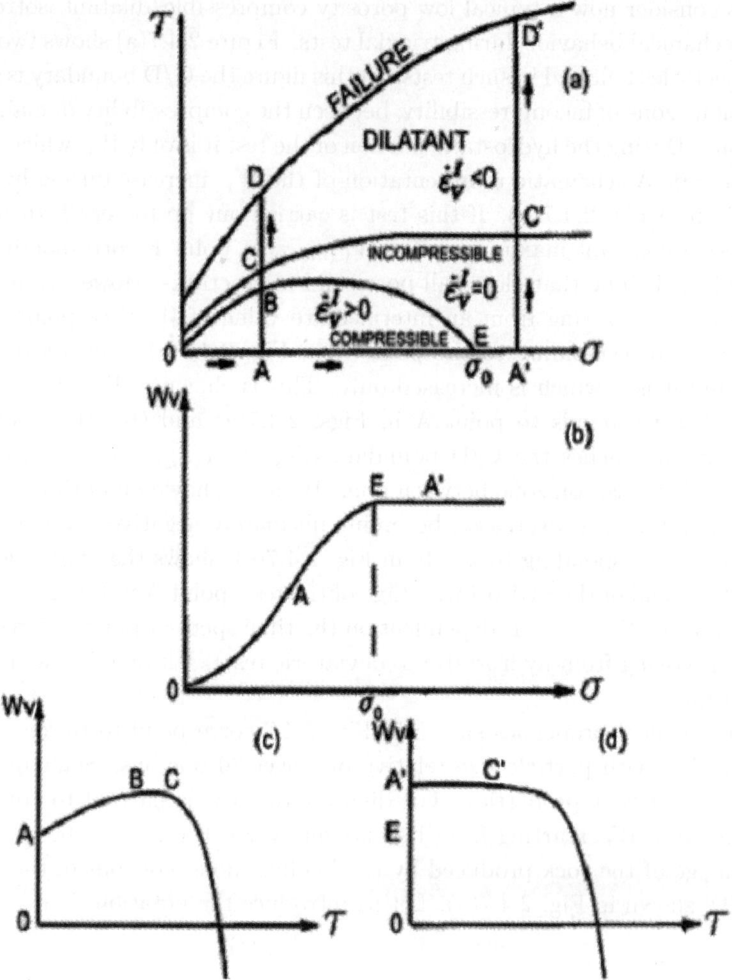

Fig. 2.4.7   Schematic variation of W_V during true triaxial test: (a) stress trajectories shown in constitutive plane; (b) during hydrostatic test; (c) during deviatoric test starting from small or moderate pressure; (d) during deviatoric test starting from very high pressures above σ_0.

The above concepts can be used to describe the damage during creep tests. In creep tests one can integrate the constitutive equation to get

$$W_V(t) = \sigma \varepsilon_V^I(t) = \frac{\langle 1 - (W_T(t)/H(\boldsymbol{\sigma}))\rangle (\partial F/\partial \sigma)\,\sigma}{(1/H)\,(\partial F/\partial \sigma)\cdot \boldsymbol{\sigma}}\left\{1 - \exp\left[\frac{k_T}{H}\frac{\partial F}{\partial \boldsymbol{\sigma}}\cdot \boldsymbol{\sigma}(t_0 - t)\right]\right\}$$

$$+ k_S \frac{\partial S}{\partial \sigma}\sigma(t_0 - t) + W_V^P$$

where t_0 is the time of the beginning of the test and W_V^P stands for the initial "primary" value of W_V at time t_0.

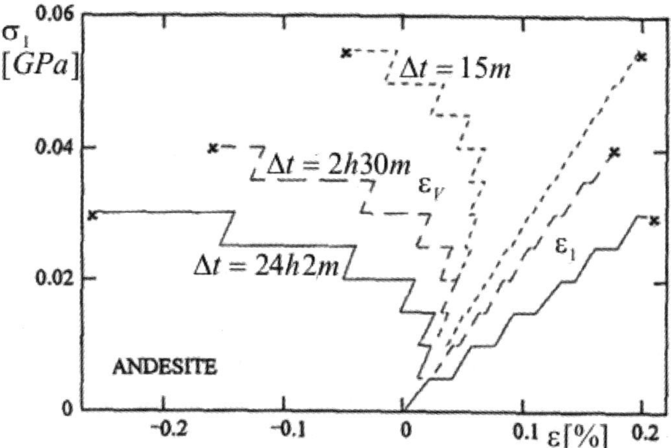

Fig. 2.4.8   Incremental creep tests for andesite: after each stress increase, the stress is kept constant for the time interval shown.

As an example Fig. 2.4.8 shows several stress–strain curves obtained numerically in uniaxial creep tests for andesite (Cristescu [1986]), where an associate model was used. The average loading rates are $\dot{\sigma} = 6.06 \times 10^{-8}$ GPa s^{-1}, 6.17×10^{-7} GPa s^{-1} and 5.7×10^{-6} GPa s^{-1}, respectively. After each stress increase, stress was held constant for the time interval shown. Following the last loading, failure occurred after the time intervals 12 h 30 m (bottom curve), 30 m and 8 m 20 s (top curve), respectively. The star is showing where failure is taking place. Another theory of fracture for oil shale is due to Grady and Kipp [1980]. The damage D is a scalar parameter satisfying $0 \leq D \leq 1$ so that $D = 0$ corresponds to the intact rock and $D = 1$ to full fragmentation. Further the elastic energy is

$$E = \frac{1}{2}K(1 - D)\varepsilon^2$$

where ε is a one-dimensional tensile strain.

Theories of failure for anisotropic rocks are due to Cazacu *et al.* [1998], Mróz and Maciejewski [2002] and Dubey and Gariola [2000]. Kachanov [1993] has developed a theory in which the displacement jump is decomposed in a normal component and a shear component.

Similar theories have been developed for a variety of materials. An associated and a nonassociated theory to be applied to bituminous concrete, was developed by Florea [1994a, b], and Tashman *et al.* [2005]. Lemiszki *et al.* [1994] consider a range of geologically reasonable boundary conditions to show that at one extreme, fracturing can occur as a result of only minor shortening by folding to the other extreme where a tight can form with no associated extension fracturing. For folds formed at shallow depths, where the confining stress on the system is less than the bending stresses in the layer and where the confining stress has greatly increased the fracture toughness of the rock, hinge-parallel extension fractures can grow under hydrostatic fluid pressure conditions.

A continuum damage model developed to describe the rate-dependent dynamic response of laminated polymer composites is investigated by Nemes and Spéciel [1995], for its behavior during strain-softening. It is assumed that the damage enters in the elastic coefficients. Eftis and Nemes [1996] have developed a thesis in which the damage of a porous material is increased by damage; the elastic parameters are increasing by damage.

A paper which deals with propagation of plane wave fronts in a solid with a nonlinear relation between stress and deformation is due to Osimov [1998]. The objective is to calculate the distance that a wave front covers. The formulae derived for a general quasi-linear system of two equations are applied to the propagation of plane compression waves in dry and partially saturated granular bodies. Numerical calculations relevant to soil mechanics are presented.

Another theory for deformability of brittle rock-like materials in the presence of an oriented damage of their internal structure is formulated by Litewka and Debinski [2003]. It was assumed that a material response, represented by the strain tensor, is a function of two tensorial variables: the stress tensor and the damage effect tensor that is responsible for the current state of internal structure of the material.

Also extensive study devoted to failure of porous metallic materials saturated with liquid is due to Martin *et al.* [1997], Martin *et al.* [1999]. Their study is related to tensile behavior of visoplastic porous metallic materials saturated with liquid. Simple tensile experiments together with ring extension tests are carried out to study the fracture behavior of this class of material. Ring tests consist in applying an internal pressure on a specimen with a ring shape. A model is presented. A simple modification of the model allows the treatment of the strong asymmetry between tension and compression which is exhibited by these materials.

In a review paper Gioda and Swoboda [1999] discuss and present various aspects related to rock mechanics, mainly the problems related to tunnels. The mentioned literature is also impressive.

A computational method is proposed by Yamada [1999] for simulating rheological phenomena of a body composed of a material such as soft soil, powder, and granular material under large excitations. These phenomena are considered to be elastic-viscoplastic flows with moving boundaries. The constitutive law is of the form

$$\gamma' = \frac{\sigma'}{\mu} + \theta\frac{\sigma}{\eta}, \quad \theta = \frac{1}{2}\left(\frac{|\sigma|}{\beta} - 1 + \left|\frac{|\sigma|}{\beta} - 1\right|\right).$$

Here γ is the strain, μ the elastic coefficient, η the viscoplastic coefficient, β is the yield stress and $(')$ is derivative with respect to time. Fundamental equations for the dynamics are constructed from the Eulerian viewpoint. The constitutive laws for pressure volume and deviatoric stress strain are elastic-viscoplastic and are expressed by differential equations. The pressure and deviatoric stress are combined to satisfy yield conditions. The domain containing an elastic-viscoplastic body is divided into cells by applying a finite difference method to the partial differential equation system. The collapse of an elastic-viscoplastic square body is shown as an application example. In another paper by Yamada [2000] a computational method is introduced for simulating seismic wave propagation in elastic-viscoplastic shear layers. The fundamental dynamics are expressed by two partial differential equations for shear stress and velocity, the balance of momentum, and the elastic-viscoplastic constitutive law. Seismic wave propagation is computed (finite difference method), considering the reflection and refraction rule at the contact boundary of the two layers, as well as the boundary conditions at the bottom and top.

Popp and Kern [2000] consider that rock salt formations are prime candidates for underground cavities or radioactive waste disposal sates, primarily because of their extremely low permeabilities. Combined gas-permeability and P- and S-wave velocity measurements were carried out on natural rock salt samples in order to investigate the transport properties of rock salt under mechanical stresses. Experiments were done at temperatures up to 60°C under conditions of hydrostatic compaction and triaxial compressive and extensional strain. The crack-sensitivity of P- and S-wave velocities is used for monitoring the *in situ* state of the microcracking during deformation. Triaxial deformation of the compacted rock salt samples is accompanied by the onset of dilatancy, that is, the opening of microcracks. The orientation of cracks is controlled by the symmetry of the applied stress field. Cracks are mostly oriented parallel to the maximum principal stress direction leading to an anisotropic crack array within the samples. A marked permeability increase is observed under compressive strain because in this case an interconnecting permeability network is generated parallel to the deformation and measuring axis. The inversions of P- and S-wave velocities are used to define the boundary between the dilatant and compressive domains. The results confirm the equation for the dilatancy boundary given by Cristescu and Hunsche [1998].

A review paper on research work related to the micromechanical modeling of behavior of hard rock under compression is due to Okui and Horii [2000]. First are provided the observed macroscopic behavior of compressive rock for both time-independent and time-dependent cases. Crack growth laws in rock for both short and long term loading are next reviewed.

Various theories of viscoplasticity applied to geomaterials are due to Cristescu and Cazacu [2000] (see also Rocchi *et al.* [2003]).

A constitutive model for crushed salt is presented by Olivella and Gens [2002]. A creep constitutive model is developed first and compared with test results. The constitutive model presented here concentrates on creep deformation because saline media behave basically in a ductile and time-dependent way. The model is able to predict strain rates that compare well with results from laboratory tests under isotropic and oedometric conditions. Macroscopic laws are written using a nonlinear viscous approach.

The stress state around various tunnels or other excavation, or stress state around lined tunnels, are presented in Cristescu [1989a], and Cristescu and Hunsche [1998] and will not be discussed here. I would like to present here a parameter identification for lined tunnels in viscoplastic medium by Lecampion *et al.* [2002], from measurements performed on deep underground tunnels. This inverse problem is solved by the minimization of a cost functional of least-squares type. The method is presented for lined or unlined structures and is applied for an elastoviscoplastic constitutive law. Several identification problems are presented in one and two dimensions for different tunnel geometries.

It has been determined through laboratory tests by Hou [2003] that the development of the excavation disturbed zone is closely related to the corresponding damage boundary and the associated damage. This leads to prediction of the time-dependent development of the excavation disturbed zone and, together with a permeability model, calculation of the permeability of rock salt.

A theory of piezocone penetration is due to Voyiadjis and Kim [2003]. It has been conducted using the elastoplastic-viscoplastic bounded surface model.

2.5   Examples

The above constitutive equation was applied to a variety of materials. First to rocks as granite, schist, shale, coal, rock salt, porous chalk (Dahou *et al.* [1995]), etc., to wet and dry sand, to a lot of powders as alumina of various powder magnitude (Jin and Cristescu [1998a], Cazacu *et al.* [1996] , Cristescu *et al.* [1997], Jin and Cristescu [1998b]), etc.

Let us give the details for microcrystalline cellulose powder PH-105 (Abdel-Hadi *et al.* [2002], Zhupanscka *et al.* [2002]). We start from the short term failure surface which is

$$\tau^* = 2.03\sigma^* + 21$$

where

$$\sigma^* = \frac{\sigma}{1\ \text{kPa}}, \quad \tau^* = \frac{\tau}{1\ \text{kPa}}.$$

We determine first the elastic parameters from the tests mentioned and using the following formulas

$$K = \frac{\Delta\sigma_1}{3\Delta\varepsilon_v^E}, \quad G = \frac{3K(\Delta\sigma_1/\Delta\varepsilon_1^E)}{9K - (\Delta\sigma_1/\Delta\varepsilon_1^E)}, \quad v = \frac{1 - (2G/3K)}{2(1 + (G/3K))}.$$

These are not constant, as already mentioned. Since the pressures are small, these are increasing parameters. This increase is shown in the Fig. 2.5.1. Both elastic parameters are increasing. To make a simple analysis we have chosen the following average values $K = 35$ MPa and $G = 30$ MPa. With that, the elastic response is determined.

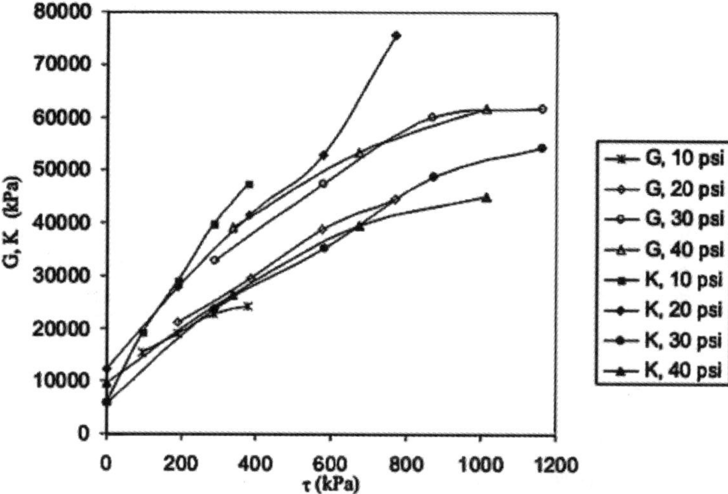

Fig. 2.5.1   The increase of the elastic parameters at all confining pressures.

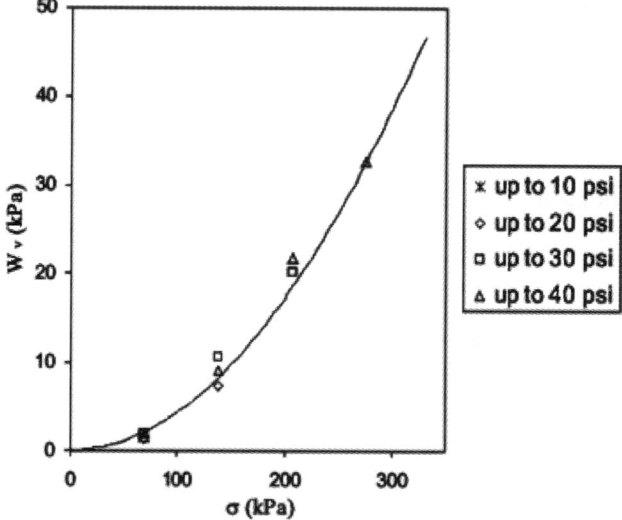

Fig. 2.5.2   The variation of the hydrostatic work-hardening.

We are now testing the irreversible stress work. From the formula (2.3.5) we determine W_v, shown in the figure. This expression is approximated very easy with the parabola

$$H_H(\sigma) = a(\sigma^*)^2, \quad \text{where } a = 0.00043 \text{ kPa}.$$

Then we calculate the deviator work-hardening. We start from the figure.

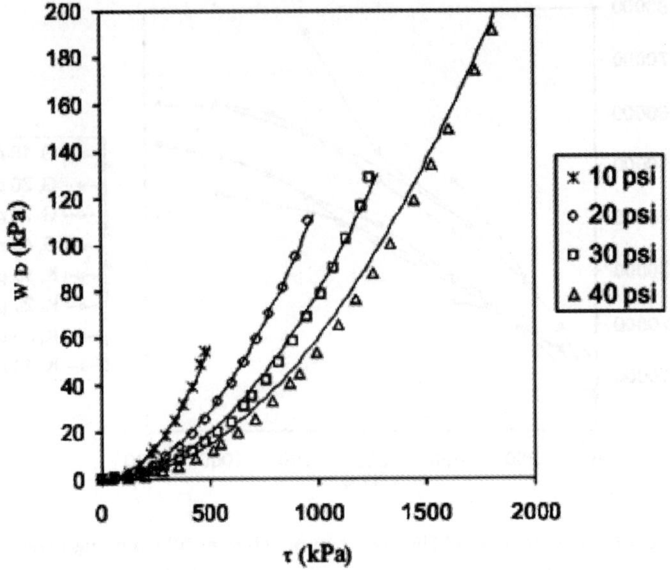

Fig. 2.5.3   The irreversible stress work and the aproximation (full line).

Now we approximate the figure with

$$H_D(\sigma,\tau) = b\frac{(\tau^*)^2}{\sigma^* - (\tau^*/3) + 7}$$

and $b = 0.017$ kPa. We can now write the whole yield function

$$H(\sigma,\tau) = a(\sigma^*)^2 + b\frac{(\tau^*)^2}{\sigma^* - (\tau^*/3) + 7}\,.$$

The compressibility/dilatancy boundary was obtained from the experimental data and can be described by equation

$$\tau^* = 1.99\sigma^* + 16\,.$$

The experimental data shows that microcrystalline cellulose PH-105, 20 μm is practically only compressible. Therefore in our model we do not take into account the small possible dilatancy deformation, which may occur just before the failure. We construct the model that describes only the compressible behavior of the given material. We have determined directly from the experimental data that

$$k_T\frac{\partial F}{\partial \sigma} = 0.000003(\sigma^*)^{0.29} - 0.0000005\frac{(\tau^*)^{1.2}}{(\sigma^*)^{0.8}}\,,$$

$$k_T\frac{\partial F}{\partial \tau} = 0.000003(\sigma^*\tau^*)^{0.2} + 0.000013(\tau^*)^{0.2}\,.$$

Note that $\partial F/\partial\sigma$ is correctly determined in both deviator and hydrostatic tests. Integrating these equations, we finally have the expression for the viscoplastic potential

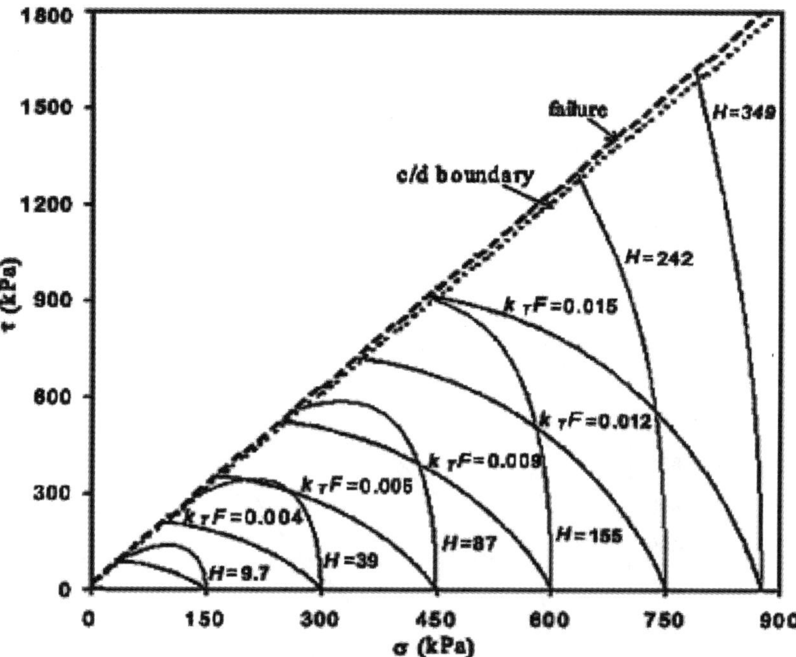

Fig. 2.5.4   Stabilization boundaries (gray full lines), viscoplastic potential surfaces (black full lines), compressibility/dilatancy boundary (dot line), failure surface (dash line).

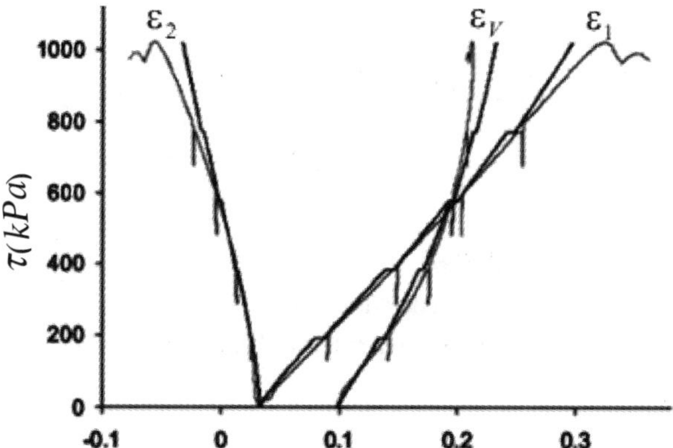

Fig. 2.5.5   Experimentally obtained (gray lines) and theoretically predicted (black lines) stress–strain curves in the deviator test under confining pressure 20 psi for microcrystalline cellulose PH-105, 20 μm.

92   *Dynamic Plasticity*

$$k_T F(\sigma, \tau) = -0.0000025(\sigma^*)^{0.2}(\tau^*)^{1.2} + \frac{0.000003}{1.29}(\sigma^*)^{1.29} + \frac{0.000013}{1.2}(\tau^*)^{1.2}.$$

To show what we have found we give on Fig. 2.5.4 the characteristic plane. One can see that the yield surfaces and the viscoplastic surfaces are quite distinct. The compressibility/dilatancy boundary is very close to failure. Thus the dilatancy domain is disregarded. Fig. 2.5.7 is showing the influence of air passing through the specimen. At the beginning, there is a lot of instabilities in the curve a. Then,

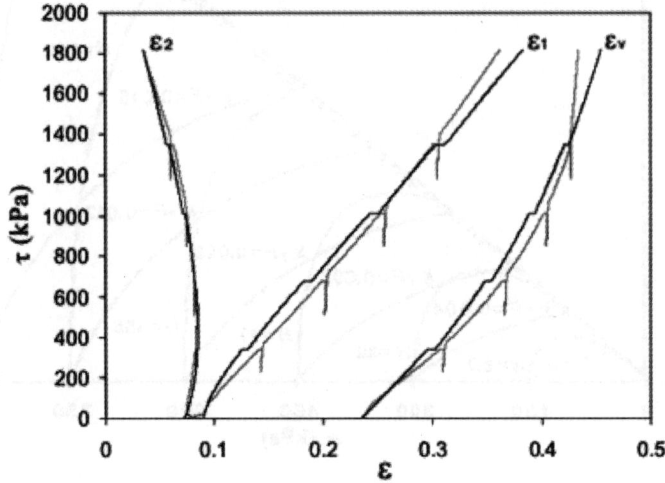

Fig. 2.5.6   Experimentally obtained (gray lines) and theoretically predicted (black lines) stress–strain curves for deviatoric tests for confining pressure 40 psi.

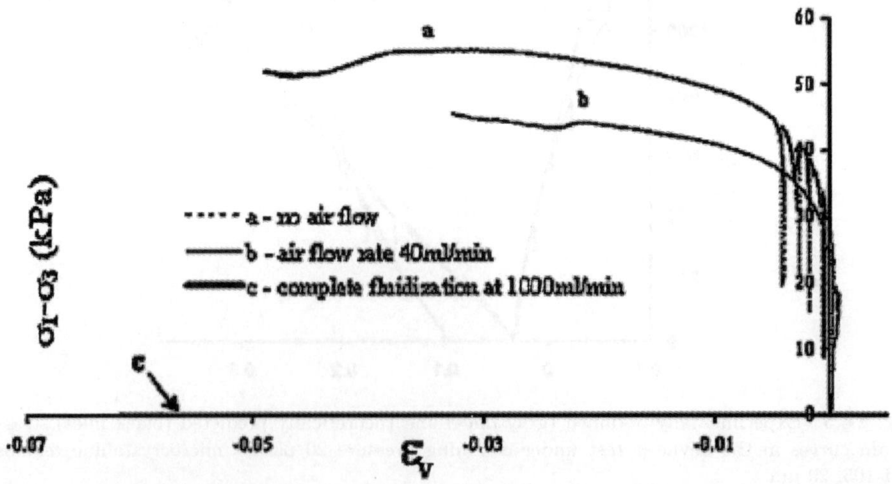

Fig. 2.5.7   Effect of air flow on the volumetric response of Alumina powder (100 μm), low confining pressure of 2.38 psi (16.4 kPa).

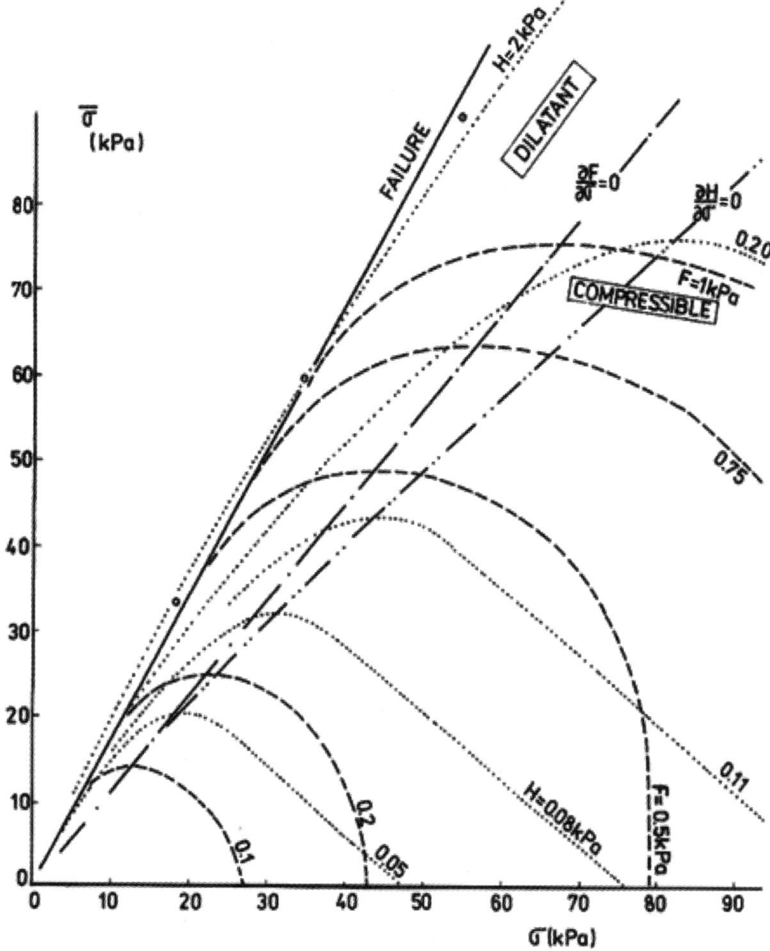

Fig. 2.5.8   The characteristic plane for saturated sand. The surfaces H = const. and F = const. are quite distinct.

with an air flow of 40 ml/min, the curve b drops quite a lot, with quite smaller instabilities. Finale, the curve c is showing that the specimen collapses without any load (Abdel-Hadi and Cristescu [2004]).

The model was applied to many materials. We give here the application for saturated sand (Cristescu [1991]). In Fig. 2.5.8 is given the characteristic plane for saturated sand. We will give no further details concerning this model. As one can see the stress–strain curves are close to the experimental.

Another viscoplastic constitutive model including damage for time-dependent behavior of rock is due to Pellet *et al.* [2005]. The inclusion of the Drucker–Prager yield surface in the Lemaitre viscoplastic constitutive equation showed the ability of the model to exhibit dilatancy and compressibility associated with viscoplastic

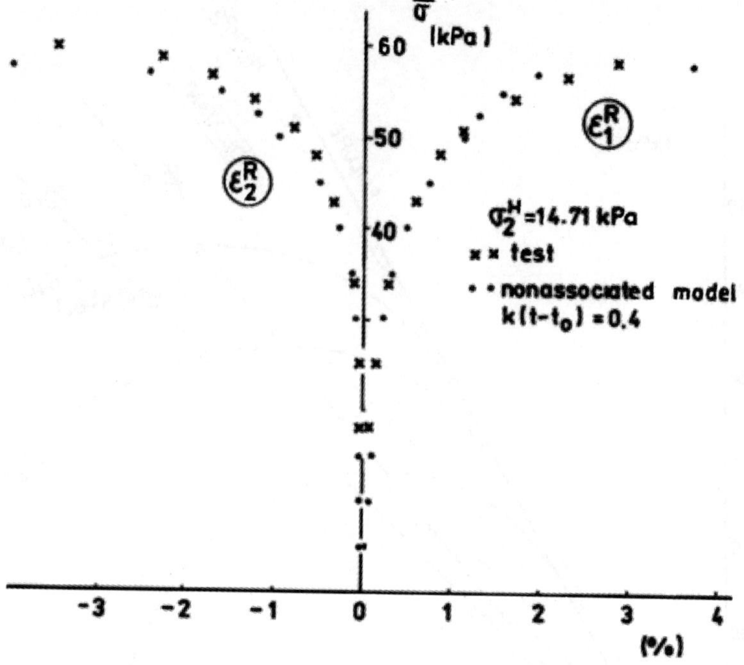

Fig. 2.5.9   Stress–strain curves for saturated sand for confining pressure of 14.71 kPa.

strain. The model requires ten parameters. The model was compared with creep tests and with strain-controlled compression tests.

2.6   Viscoelastic Model

A general form of linear viscoelastic model of rate-type for homogeneous and isotropic bodies (Cristescu [1989a]) is

$$\dot{\sigma} = 3K\dot{\varepsilon} - \eta_v(\sigma - 3K_0\varepsilon)$$
$$\dot{\sigma}' = 2G\dot{\varepsilon}' - \eta(\sigma' - 2G_0\varepsilon'),$$

(2.6.1)

or

$$\dot{\varepsilon} = -k_v\left(\varepsilon - \frac{\sigma}{3K_0}\right) + \frac{1}{3K}\dot{\sigma}$$
$$\dot{\varepsilon}' = -k\left(\varepsilon' - \frac{\sigma'}{2G_0}\right) + \frac{1}{2G}\dot{\sigma}',$$

(2.6.2)

with

$$k_v = \frac{K_0}{K}\eta_v\,, \quad k = \frac{G_0}{G}\eta\,.$$

(2.6.3)

The constants K and G are the dynamic (elastic) moduli, and are the static moduli, while k_v and k (or η_v and η) are the constant viscosity coefficients.

Let us consider the *velocity of propagation* of the dilatational and shear waves, as

$$v_p^2 = \frac{1}{\rho}\left(K + \frac{4}{3}G\right), \quad v_s^2 = \frac{G}{\rho}. \tag{2.6.4}$$

From the assumption that v_s is real it follows that

$$G > 0. \tag{2.6.5}$$

Since the natural order of propagation is $v_p > v_s$, from the condition that the bar velocity $(E/\rho)^{1/2}$ is real, it follows that

$$\frac{E}{\rho} = \frac{v_s^2(3v_p^2 - 4v_s^2)}{v_p^2 - v_s^2} > 0, \tag{2.6.6}$$

and therefore $v_p^2 > 4/3\, v_s^2$. Thus

$$K = \rho\left(v_p^2 - \frac{4}{3}v_s^2\right) > 0. \tag{2.6.7}$$

Let as consider now the *creep tests*, i.e., the tests done under constant stress. We assume that at the moment t_0 we have $\varepsilon(t_0) = \varepsilon_0$, $\sigma(t_0) = \sigma_0$, and for $t \geq t_0$ the stress is kept constant: $\sigma(t) = \sigma(t_0) = \sigma_0$. From (2.6.2) we have

$$\varepsilon'(t) = \left(\varepsilon_0' - \frac{1}{2G_0}\sigma_0'\right)\exp[-k(t - t_0)] + \frac{1}{2G_0}\sigma_0'$$

$$\varepsilon(t) = \left(\varepsilon_0 - \frac{1}{3K_0}\sigma_0\right)\exp[-k_v(t - t_0)] + \frac{1}{3K_0}\sigma_0. \tag{2.6.8}$$

Therefore, from the condition of stabilization of the deformation in creep tests we must have, when $t \to \infty$

$$k > 0, \quad k_v > 0. \tag{2.6.9}$$

In a similar way from the condition of stabilization of the *stress relaxation* under constant strain, when at $t = t_0$, $\varepsilon(t_0) = \varepsilon_0$, $\sigma(t_0) = \sigma_0$, and for $t \geq t_0$ the strain is kept constant, $\varepsilon(t) = \varepsilon(t_0) = \varepsilon_0$. It follows that

$$K_0 > 0, \quad G_0 > 0, \tag{2.6.10}$$

if (2.6.5), (2.6.7), and (2.6.9) are also taken into account. From (2.6.5), (2.6.7), (2.6.9), and (2.6.10) we further get

$$K > K_0, \quad G > G_0. \tag{2.6.11}$$

If inequalities (2.6.9) hold, for $t \to \infty$ the stabilization state satisfies

$$t \to \infty \begin{cases} \left.\varepsilon'(t)\right|_{t\to\infty} = \dfrac{1}{2G_0}\sigma_0' & \text{if} \quad k > 0 \\[3mm] \left.\varepsilon(t)\right|_{t\to\infty} = \dfrac{1}{3K_0}\sigma_0 & \text{if} \quad k_v > 0. \end{cases} \tag{2.6.12}$$

If the inequalities (2.6.10) and (2.5.11) are satisfied, the energy is positive-definite.

Let us consider a *uniaxial compressive creep test* in which at $t = t_0 = 0$ a sudden loading is applied followed by a very long time interval in which the stress is kept constant,

$$\sigma_1 - \sigma_1^0 = const., \quad \sigma_2 = \sigma_3 = 0. \tag{2.6.13}$$

According to the model the instantaneous response is

$$\varepsilon_1^0 - \varepsilon_2^0 = \frac{1}{2G}\sigma_1^0, \quad \varepsilon_1^0 + 2\varepsilon_2^0 = \frac{1}{3K}\sigma_1^0. \tag{2.6.14}$$

These strain values are the initial data for the deformation by creep which follows, according to

$$\dot{\varepsilon}_1 - \dot{\varepsilon}_2 = -k\left[\varepsilon_1 - \varepsilon_2 - \frac{1}{2G_0}\sigma_1^0\right]$$

$$\dot{\varepsilon}_1 + 2\dot{\varepsilon}_2 = -k_v\left[\varepsilon_1 + 2\varepsilon_2 - \frac{1}{3K_0}\sigma_1^0\right]. \tag{2.6.15}$$

After integration we get

$$3\varepsilon_1 = \left(\frac{1}{G} - \frac{1}{G_0}\right)\sigma_1^0\exp(-kt) + \left(\frac{1}{3K} - \frac{1}{3K_0}\right)\sigma_1^0\exp(-k_vt)$$

$$+ \left(\frac{1}{G_0} + \frac{1}{3K_0}\right)\sigma_1^0$$

$$3\varepsilon_2 = -\left(\frac{1}{2G} - \frac{1}{2G_0}\right)\sigma_1^0\exp(-kt) + \left(\frac{1}{3K} - \frac{1}{3K_0}\right)\sigma_1^0\exp(-k_vt)$$

$$+ \left(-\frac{1}{2G_0} + \frac{1}{3K_0}\right)\sigma_1^0. \tag{2.6.16}$$

Since during the uniaxial (unconfined) compressive creep test $\dot{\varepsilon}_1 > 0$ and $\dot{\varepsilon}_2 < 0$, it follows from the second condition [with (2.6.11)] that

$$k\left(\frac{1}{2G} - \frac{1}{2G_0}\right)\exp(-kt) < k_v\left(\frac{1}{3K} - \frac{1}{3K_0}\right)\exp(-k_vt) < 0 \tag{2.6.17}$$

for any t. Therefore for $t \to \infty$, the inequality (2.6.17) becomes

$$k\frac{G - G_0}{2GG_0} \geq k_v\frac{K - K_0}{3KK_0} > 0 \tag{2.6.18}$$

this is an additional inequality to be satisfied by the constitutive constants involved in the model.

Further, from (2.6.17) it follows also for $t \to \infty$ that

$$k_v \geq k. \tag{2.6.19}$$

Furthermore, from (2.6.16) it follows that for $t \to \infty$ the stable states must satisfy the relationships

$$\varepsilon_1 = \frac{1}{3}\left(\frac{1}{G_0} + \frac{1}{3K_0}\right)\sigma_1^0, \quad \varepsilon_2 = \frac{1}{3}\left(-\frac{1}{2G_0} + \frac{1}{3K_0}\right)\sigma_1^0. \tag{2.6.20}$$

Since the slope of $(2.6.20)_2$ is negative we get

$$2G_0 < 3K_0 \,. \qquad (2.6.21)$$

In a similar way from (2.6.14) it follows that the instantaneous curve

$$2\varepsilon_2^0 = \left(-\frac{1}{2G} + \frac{1}{3K} \right) \sigma_1^0$$

also has a negative slope, i.e.,

$$2G < 3K \,. \qquad (2.6.22)$$

Let as consider now the *uniaxial tests with constant loading rate* $\dot\sigma_1^0 > 0$:

$$\sigma_1(t) = \dot\sigma_1^0 t \,, \quad \sigma_2 = \sigma_3 = 0 \,. \qquad (2.6.23)$$

Introducing (2.6.23) in the model, we get for this test

$$\dot\varepsilon_1 - \dot\varepsilon_2 = -k \left(\varepsilon_1 - \varepsilon_2 - \frac{1}{2G_0}\dot\sigma_1^0 t \right) + \frac{1}{2G}\dot\sigma_1^0$$

$$\dot\varepsilon_1 + 2\dot\varepsilon_2 = -k_v \left(\varepsilon_1 + 2\varepsilon_2 - \frac{1}{3K_0}\dot\sigma_1^0 t \right) + \frac{1}{3K}\dot\sigma_1^0 \,.$$

By integrating with zero initial data, we obtain

$$\varepsilon_1 - \varepsilon_2 = \frac{\dot\sigma_1^0}{k} \left[\frac{k}{2G_0}t + \left(\frac{1}{2G} - \frac{1}{2G_0} \right) [1 - \exp(-kt)] \right]$$

$$\varepsilon_1 + 2\varepsilon_2 = \frac{\dot\sigma_1^0}{k_v} \left[\frac{k_v}{3K_0}t + \left(\frac{1}{3K} - \frac{1}{3K_0} \right) [1 - \exp(-k_v t)] \right] \,. \qquad (2.6.24)$$

Taking into account (2.6.23), i.e., by replacing $\dot\sigma_1^0$, we express the equation of the stress–strain curves as

$$3\varepsilon_1 = \left(\frac{1}{3K_0} + \frac{1}{G_0} \right) \sigma_1 + \left(\frac{1}{3K} - \frac{1}{3K_0} \right) \frac{\dot\sigma_1^0}{k_v} \left[1 - \exp\left(-k_v \frac{\sigma_1}{\dot\sigma_1^0} \right) \right]$$

$$+ \left(\frac{1}{G} - \frac{1}{G_0} \right) \frac{\dot\sigma_1^0}{k} \left[1 - \exp\left(-k \frac{\sigma_1}{\dot\sigma_1^0} \right) \right]$$

$$3\varepsilon_2 = \left(\frac{1}{3K_0} - \frac{1}{2G_0} \right) \sigma_1 + \left(\frac{1}{3K} - \frac{1}{3K_0} \right) \frac{\dot\sigma_1^0}{k_v} \left[1 - \exp\left(-k_v \frac{\sigma_1}{\dot\sigma_1^0} \right) \right]$$

$$- \left(\frac{1}{2G} - \frac{1}{2G_0} \right) \frac{\dot\sigma_1^0}{k} \left[1 - \exp\left(-k \frac{\sigma_1}{\dot\sigma_1^0} \right) \right] \,. \qquad (2.6.25)$$

The slopes of these stress–strain curves at the origin coincide with the slope of the instantaneous response strait angles, i.e.,

$$\frac{d\varepsilon_1}{d\sigma_1}\bigg|_{\sigma_1=0} = \frac{1}{3} \left(\frac{1}{3K} + \frac{1}{G} \right), \quad \frac{d\varepsilon_2}{d\sigma_1}\bigg|_{\sigma_1=0} = \frac{1}{3} \left(\frac{1}{3K} - \frac{1}{2G} \right) \,. \qquad (2.6.26)$$

The same "instantaneous response" slopes are obtained if anywhere along the stress–strain curves we make $\dot\sigma_1^0 \to \infty$. Also, if in (2.6.25) me make $\dot\sigma_1^0 \to 0$, that is we are making very slow tests, we obtain the slopes

$$\varepsilon_1|_{\dot\sigma_1^0 \to 0} = \frac{1}{3}\left(\frac{1}{3K_0} + \frac{1}{G_0}\right)\sigma_1, \quad \varepsilon_2|_{\dot\sigma_1^0 \to 0} = \frac{1}{3}\left(\frac{1}{3K_0} - \frac{1}{2G_0}\right). \qquad (2.6.27)$$

The curves (2.6.25) may have some oblique asymptotes $\varepsilon_1 = A_1\sigma_1 + B_1$ with

$$A_1 = \lim_{\sigma_1 \to \infty} \frac{\varepsilon_1(\sigma_1)}{\sigma_1}, \quad B_1 = \lim_{\sigma_1 \to \infty} [\varepsilon_1(\sigma_1) - A\sigma_1].$$

One find easy

$$A_1 = \frac{1}{3}\left(\frac{1}{3K_0} + \frac{1}{G_0}\right), \quad B_1 = \left(\frac{1}{3K} - \frac{1}{3K_0}\right)\frac{\dot\sigma_1^0}{3k_v} + \left(\frac{1}{G} - \frac{1}{G_0}\right)\frac{\dot\sigma_1^0}{3k}.$$

Though the two asymptotes are

$$\varepsilon_1 = \frac{1}{3}\left(\frac{1}{3K_0} + \frac{1}{G_0}\right)\sigma_1 + \frac{1}{3}\left(\frac{1}{3K} - \frac{1}{3K_0}\right)\frac{\dot\sigma_1^0}{k_v} + \frac{1}{3}\left(\frac{1}{G} - \frac{1}{G_0}\right)\frac{\dot\sigma_1^0}{k},$$

$$\varepsilon_2 = \frac{1}{3}\left(\frac{1}{3K_0} - \frac{1}{2G_0}\right)\sigma_1 + \frac{1}{3}\left(\frac{1}{3K} - \frac{1}{3K_0}\right)\frac{\dot\sigma_1^0}{k_v} - \frac{1}{3}\left(\frac{1}{2G} - \frac{1}{2G_0}\right)\frac{\dot\sigma_1^0}{k}.$$

The asymptotes are represented in principle in Fig. 2.6.1 given below. The constants can be determined from: the equilibrium two straight lines one determine K_0 and G_0, from the instantaneous straight lines one can determine K and G, while from the two asymptotes one can determine k and k_v.

Examples. Such models have been used for *coal*. For the coal from Baraolt one has: $G = 210$ MPa, $K = 1954$ MPa, $E = 608$ MPa, $G_0 = 30.6$ MPa, $K_0 = 32.6$ MPa,

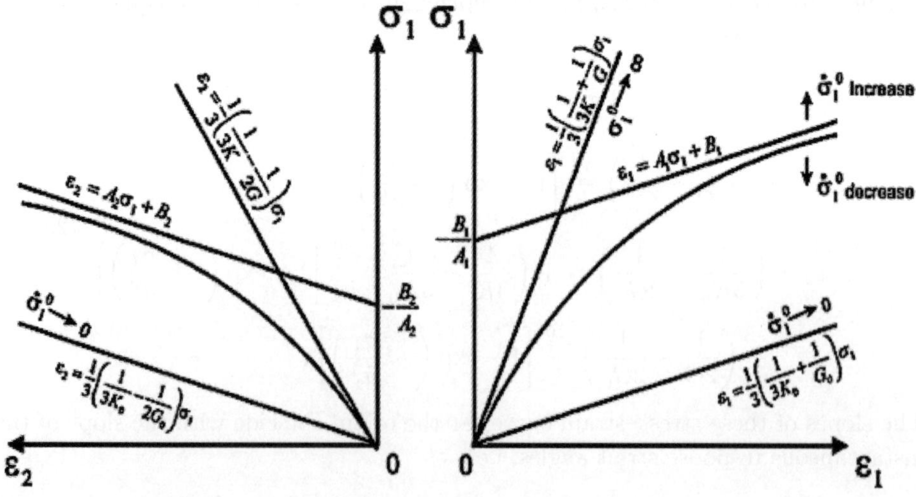

Fig. 2.6.1   Schematic diagram for the determination of constants.

$k = 1.4 \times 10^{-6}$ s^{-1}, $k_v = 1.9 \times 10^{-6}$ s^{-1}, and the short-term tensile strength as determined in a Brazilian test ranges between 0.53 MPa and 1.24 MPa. For another coal from Racoş, we have: $G = 107.4$ MPa, $K = 1923.4$ MPa, $E = 316$ MPa, $G_0 = 39.2$ MPa, $K_0 = 53.9$ MPa, $k = 1 \times 10^{-6}$ s^{-1}, $k_v = 1.23 \times 10^{-6}$ s^{-1}, and the short-term tensile strength (Brazilian) ranges between 0.56 MPa and 0.65 MPa.

In a paper by Pan *et al.* [1997] presents a boundary element formulation for 3D linear and viscoelastic bodies subjected to the body force of gravity. The Laplace transformation is first used to suppress the time variable, and solutions of displacements and stresses are found in the transformed domain. The time domain solutions are then found by an accurate and efficient numerical inversion method which requires only real calculations for all quantities. Several numerical examples involving 3D viscoelastic bodies are presented.

Bibliography

Abdel-Hadi A. I. and Cristescu N. D., 2002, A new experimental setup for the characterization of bulk mechanical properties of aerated particulate systems, *Part. Sci. and Tech.* **20**, 197–207.

Abdel-Hadi A. I., Zhupanska O. I. and Cristescu N. D., 2002, Mechanical properties of microcrystalline cellulose, Part I. Experimental results, *Mech. of Mater.* **34**, 373–390.

Abdel-Hadi A. I. and Cristescu N. D., 2004, unpublished.

Abelev A. V. and Lade P. V., 2004, Characterization of failure in cross-anisotropic solids, *J. Eng. Mech.* **130**, 5, 599–606.

Aubertin M., 1996, Triaxial stress relaxation tests on Saskatchewan potash: Discussion, *Can. Geotech. J.* **33**, 375–377.

Aubertin M. and Gill D. E., 1993, Elastoplastic modeling for the ductile behavior of polycrystalline sodium chloride with SUVIC, *Int. J. Plasticity* **9**, 4, 479–505.

Aubertin M., Julien M. R., Servant S. and Gill D. E., 1999a, A rate-dependent model for the ductile behavior of salt rocks, *Can. Geotech. J.* **36**, 660–674.

Aubertin M. and Li L., 2004, A porosity-dependent inelastic criterion for engineering materials, *Int. J. Plasticity* **20**, 2179–2208.

Aubertin M., Yahya O. M. L. and Julien M., 1999b, Modeling mixed hardening of alkali halides with a modified version of an internal state variables model, *Int. J. Plasticity* **15**, 1067–1088.

Bayuk E. I., 1966, Velocities of waves in specimens of eruptive and metamorphic rocks at pressures up to 4000 kg/cm^2, *Trudy Inst, Fiz. Zemli.* No. **37** (204), 16 36, (in Russian).

Bérest P. and Brouard B., 2003, Safety of salt caverns used for underground storage, blow out; mechanical instability; seepage; cavern abandonment, *Oil & Gas Science and Technology-Rev. IFP* **58**, 3, 361–384.

Bérest P., Blum P. A., Charpentier J. P., Gharbi H. and Valès F., 2004, Fluage du sel gemme sous très faibles charges, *C. R. Géoscience* **336**, 1337–1344.

Boettger J. C., Furnish M. D., Dey T. N. and Grady D. E., 1995, Time-resolved shock-wave experiments on granite and numerical simulations using dynamic phase mixing, *J. Appl. Phys.* **78**, 8, 5155–5165.

Cazacu O., 2002, A new hyperelastic model for transversely isotropic solids, *ZAMP* **53**, pp. 901–911.

Cazacu O. and Cristescu N. D. 1995, Failure of an anisotropic compressive shale, *Joint Applied Mechanics and Materials Summer Meeting*, ASME, AMD-Vol. **205**, pp. 1–8.

Cazacu O., Shao J. F., Henry J. P. and Cristescu N. D., 1996, Elastic/viscoplastic constitutive equation for anisotropic shale, in: *North American Rock Mechanics Symposium, Rock Mechanics, Tool and Techniques* Aubertin M., Hassani F. and Mitri H. (eds.), Balkema, Rotterdam, pp. 1683–1690.

Cazacu O., Cristescu N. D., Shao J. F. and Henry J. P., 1998, A new anisotropic failure criterion for transversely isotropic solids, *Mech. Choesive-Frictiona Materials* **3**, 89–103.

Cazacu O., Jin J. and Cristescu N. D., 1997, A new constitutive model for alumina powder compaction, *KONA* **15**, 103–112.

Chan K. S., Brodsky N. S., Fossum A. F., Bodner S. R. and Munson D. E., 1994, Damage-induced nonassociated inelastic flow in rock salt, *Int. J. Plasticity* **10**, 6, 623–642.

Chen W., Lu F., Frew D. J. and Forrestal M. J., 2002, Dynamic compression testing of soft materials, *J. Appl. Mech. Trans. ASME* **69**, 214–223.

Cizek J. C. (ed.), 1985, *Response of Geologic Materials to Blast Loading and Impact*, ASME, 181 pp.

Cleja-Tigoiu S., 1991, Elasto-viscoplastic constitutive equations for rock-type materials (finite deformation), *Int. J. Engng. Sci.* **29**, 1531–1544.

Cristescu N., 1975, *Rheological Properties of Some Deep Located Rocks*. University of Bucharest Report.

Cristescu N., 1986, Damage and failure of viscoplastic rock-like materials, *Int. J. Plasticity* **2**, 2, 189–204.

Cristescu N., 1989a, *Rock Rheology*, Kluver Academic Pub., 336 pp.

Cristescu N. 1989b, *Proc. 4th Conf. on AcusticEmission/Microseismic Activity in Geologic Structures and Materials*, Pennsylvania State University, Oct. 1985, Trans. Tech. Publ., Clausthal-Zellerfeld, 559–567.

Cristescu N., 1991, Nonassociated elastic/viscoplastic constitutive equations for sand, *Int. J. Plasticity* **6**, 41–64.

Cristescu N. D., 1993, Failure and creep failure around an underground opening, *Asses. Preven. Failure Phen. Rock Engng.* Pasamehmetoglu *et al.* (eds.), Balkema, Rotterdam, 205–210.

Cristescu N. 1994, A procedure to determine nonassociated constitutive equations for geomaterials, *Int. J. Plasticity* **10**, 2, 103–131.

Cristescu N. and Suliciu I., 1982, *Viscoplasticity*, Martinus Nijhoff Publishers, The Hague. 307 pp.

Cristescu N. D., Cazacu O. and Jin J., 1997, Constitutive equations for compaction of ceramic powders, *IUTAM Symposium on Mechanics of Granular and Porous Materials*, Fleck N. A. and Cocks A. C. F. (eds.), 117–128.

Cristescu N. D. und Hunsche U. E., 1992, Determination of a nonassociated constitutive equation for rock salt from experiments, in Finite Inelastic Deformation — Theory and Applications, *Proceedings of IUTAM Symposium*, Besdo D. and Stein E. (eds.), 511–523, Springer Verlag, New York.

Cristescu N. D., and Hunsche U., 1998, *Time Effects in Rock Mechanics*, Wiley, 342 pp.

Cristescu N. D. and Cazacu O., 2000, Viscoplasticity of geomaterials, in *Modeling in Geomechanics*, Zaman M., Gioda G. and Booker J. (eds.), Chichester, John Wiley & Sons, 129–153.

Dahou A., Shao J. F. and Bederiat M., 1995, Experimental and numerical investigations on transient creep of porous chalk, *Mech. of Mat.* **21**, 147–158.

Dresner L., 1973, Propagation of pressure waves in bubble-filled liquids, *J. Appl. Phys.* **44**, 2, 680–686.

Dubey R. K. and Gariola V. K., 2000, Influence of structural anisotropy on the uniaxial compressive strength of pre-fatigued rocksalt from Himachal Pradesh, India, *Int. J. Rock Mech. Min. Sci.* **37**, 993–999.

Eftis J. and Nemes J. A., 1996, On the propagation of elastic-viscoplastic waves in damage-softening polycrystalline materials, *Int. J. Plasticity* **12**, 8, 1005–1022.

Feng X.-T. and An H., 2004, Hybrid intelligent method optimization of a soft rock replacement scheme for a large cavern excavated in alternate hard and soft rock strata, *Int. J. Rock Mech. Min. Sci.* **41**, 4, 655–667.

Florea D., 1994a, Associated elastic/viscoplastic model for bituminous concrete, *Int. J. Engng. Sci.* **32**, 1, 79–86.

Florea D., 1994b, Nonassociated elastic/viscoplastic model for bituminous concrete, *Int. J. Engng. Sci.* **32**, 1, 87–93.

Forrestal M. J., Frew D. J., Hickerson J. P. and Rohwer T. A., 2003, Penetration of concrete targets with deceleration-time measurements, *Int. J. Impact Engng.* **28**, 5, 479–497.

Fota D., 1983 (unpublished results).

Fragaszy R. J. and Voss M. E., 1986, Undrained compression behaviour of sand, *J. Geotech. Eng. — ASCE* **112**, 3, 334–345.

Frew D. J., Forrestal M. J., and Cargile J. D., 2006, The effect of concrete target diameter on projectile deceleration and penetration depth, *Int. J. Impact Engng.* (in print).

Gelin J. C., 1992, Dynamic loading, viscoplasticity and temperature effects on the evolution of damage in metal forming processes. *J. of Material Processing Tech.* **32**, 169–178.

Gioda G. and Swoboda G., 1999, Developments and applications of the numerical analysis of tunnels in continuous media, *Int. J. Numer. Anal. Meth. Geomech.* **23**, 1393–1405.

Grady D. E. and Kipp M. E., 1980, Continuum modeling of explosive fracture in oil shale, *Int. J. Rock Mech. Min. Sci. & Geomch. Abstr.* **17**, 147–157.

Hall C. A., Chhabildas L. C. and Reinhart W. D., 1999, Shock Hugoniot and release in concrete with different aggregate sizes from 3 to 23 GPa, *Int. J. Impact Engng.* **23**, 341–351.

Hamami M., 2000, Proposition d'une loi "hybride" modélisat l'écruissage du sel gemme, *Can. Geotech.* **37**, 898–908.

Hatzor Y. H. and Heyman E. P., 1997, Dilatation of anisotropic rock salt: Evidence from Mount Sendom diaper, *J. Geophys. Res.* **102**, B7, 14853–14868.

Hou Z., 2003, Mechanical and hydraulic behavior of rock salt in the excavation distributed zone around underground facilities, *Int. J. Rock Mech. Min. Sci.* **40**, 725–738.

Hunsche U. and Hampel A., 1999, Rock salt-the mechanical properties of the host rock material for a radioactive waste repository, *Engineering Geology* **52**, 271–291.

Hunsche U., Walter F. and Schnier H., 2004, Evolution and failure of the Opalinus clay: relationship between deformation and damage, experimental results and constitutive equation, *Applied Clay Sci.* **26**, 403–411.

Jeremić B., Runesson K. and Sture S., 1999, A model for elastic-plastic pressure sensitive materials subjected to large deformations, *Jnt. J. Solids Structures* **36**, 4901–4918.

Jin J. and Cristescu N. D., 1998a, A constitutive model for powder materials, *J. Enge. Mat. And Technology* **120**, 97–104

Jin J. and Cristescu N. D., 1998b, An elastic/viscoplastic model for transient creep of rock salt, *Int. J. Plasticity* **14**, 1–3, 85–107.

Jing L., 2003, A review of techniques, advances and outstanding issues in numerical modeling for rock mechanics and rock engineering, *Int. J. Rock Mech. Min. Sci.* **40**, 283–353.

Kachanov M., 1993, Elastic solids with many cracks and related problems, in *Advances in Applied Mechanics*, Vol. **30**, Hutchinson J. and Wu T. (eds.), Academic Press, N. Y., 259–445.

Kaiser P. K., Guenot A. and Morgenstern N. R., 1985, Reformation of small tunnels_IV — Behavior during Failure, *Int. J. Rock Mech. Min. Sci. Geomech. Abstr.* **22**, 3, 141–152.

Khan A. S., Xiang Y. and Huang S., 1991, Behavior of Berea sandstone under confining pressure, Part 1: yield and failure surfaces, and nonlinear elastic response, *Int. J. Plasticity* **7**, 607–624.

Khan A. S., Xiang Y. and Huang S., 1992, Behavior of Berea sandstone under confining pressure Part II: elastic-plastic response, *Int. J. Plasticity* **8**, 209–220.

Kranz R. J., 1980, The effects of confining pressure and stress difference on static fatigue of granite, *J. Geophys. Res.* **85**, B4, 1854–1866.

Kranz R. L., Harris W. J. and Carter N. L., 1982, Static fatigue of granite at 200°C, *Geophis. Res. Lett.* **9**, 1, 1–4.

Lade P. V., 1977, Elasto-plastic stress–strain theory for cohesionless soil with curved yield surfaces, *Int. J. Solids Structures* **13**, 1019–1035.

Langer M. and Heusermann S., 2001, Geomechanical stability and integrity of waste disposal mines in salt structures, *Engineering Geology* **61**, 2–3, 155–161.

Lecampion B., Constantinescu A. and Nguyen Minh D., 2002, Parameter identification for lined tunnels in a viscoplastic medium, *Int. J. Numer. Anal. Meth. Geomech.* **26**, 12, 1191–1211.

Lemiszki P. J., Landes J. D. and Hatcher Jr. R. D., 1994, Controls on hinge-parallel extension fracturing in single-layer tangential-longitudinal strain folds, *J. Geoph. Res.* **99**, No. B11, 22,027–22,041.

Li L. and Aubertin M., 2003, A general relationship between porosity and uniaxial strength of engineering materials, *Can. J. Civ. Eng.* **30**, 644–658.

Li L., Aubetin M., Simon R. and Bussiere B., 2005, Formulation and application of a general inelastic locus for geomaterials with variable porosity, *Can. Geotech. J.* **42**, 601–623.

Litewka A. and Debinski J., 2003, Load-induced oriented damage and anisotropy of rock-like materials, *Int. J. Plasticity* **19**, 2171–2191.

Lundberg B., 1974, Impact energy absorption by a rate-dependent locking target, *Appl. Sci. Res.* **29**, 4, 305–318.

Mackerle J., 2004, Creep and creep fracture/damage finite element modeling of engineering materials and structures: an addendum, *Int. J. Pressure Vessels and Piping* **81**, 5, 381–392.

Malan D. F. and Basson F. R. P., 1998, Ultra-deep mining: The increased potential for squeezing conditions, *The J. of The South African Institute of Mining and Metallurgy*, Nov/Dec. 353–363.

Mahnken R. and Kohlmeier M., 2001, Finite element simulation for rock salt with dilatancy boundary coupled to fluid permeation, *Comput. Methods Appl. Mech. Engrg.* **190**, 4259–4278.

Maranini E. and Brignoli M., 1999, Creep behaviour of a week rock: experimental characterization, *Int. J. Rock. Mech. Min. Sci.* **36**, 1, 127–138.

Maranini E. and Yamaguchi T., 2001, A non-associated viscoplastic model for the behavior of granite in triaxial compression, *Mech. of Mater.* **33**, 5, 283–293.

Martin C. L., Favier D. and Suery M., 1997, Viscoplastic behaviour of porous metallic materials saturated with liquid. Part I: constitutive equations, *Int. J. Plasticity* **13**, 3, 215–235.

Martin C. L., Favier D. and Sury M., 1999, Fracture behaviour in tension of viscoplastic porous metallic materials saturated with liquid, *Int. J. Plasticity* **15**, 981–1008.

Matei A. and Cristescu N. D., 2000, The effect of volumetric strain on elastic parameters for rock salt, *Mech. Cohes.-Frict. Mater.* **5**, 113–124.

Miura K., Okui Y. and Horii H., 2003, Micromechanics-based prediction of creep failure of hard rock for long-term safety of high-level radioactive waste disposal system, *Mech. Mater.* **35**, 587–601.

Moustachi O. and Thimus J. F., 1997, P-wave attenuation in creeping rock and system identification, *Rock Mech. Rock. Engng.* **30**, 4, 169–180.

Mróz Z. and Maciejewski J., 2002, Failure criteria of anisotropically damaged materials based on the critical plane concept, *Int. J. Numer. Anal. Meth. Geomech.* **26**, 407–431.

Nawrochi P. A., Cristescu N. D., Dusseault M. B. and Bratli R. K., 1999, Experimental methods for determining constitutive parameters for nonlinear rock modeling, *Comut. Meth. Advanc. Geomech.* **36**, 5, 659–672.

Nemes J. A. and Spéciel E., 1995, Use of a rate-dependent continuum damage model to describe strain-softening in laminated composites, *Comp. Struct.* **58**, 6, 1083–1092.

Niandou H., Shao J. F., Henry J. P. and Fourmaintraux D., 1997, Laboratory investigation of the mechanical behaviour of Tournemiere Shale, *Int. J. Rock Mech. Min. Sci.* **34**, 1, 3–16.

Nicolae M., 1999, Non-associated elasto-viscoplastic models for rock salt, *Int. J. Engn. Sci.* **37**, 269–297.

Okaya D. A. and McEvilly T. V., 2003, Elastic wave propagation in anisotropic crystal material possessing arbitrary internal tilt, *Geophys. J. Int.* **153**, 344–358.

Okui Y. and Horii H., 2000, Application of micromechanics to modeling compressive failure of rock, *Mat. Sci. Res. Int.* **6**, 2, 65–73.

Olivella S. and Gens A., 2002, A constitutive model for crushed salt, *Int. J. Numer. Anal. Meth. Geomech.* **26**, 719–746.

Osimov V. A., 1998, On the formation of discontinuities of wave fronts in a saturated granular body, *Continuum Mech. Thermodyn.* **10**, 253–268.

Pan Y.-W. and Dong J.-J., 1991a, Time-dependent tunnel convergence — I. Formulation of the model, *Int. J. Rock Mech. Min. Sci. & Geomech. Abstr.* **28**, 6, 469–475.

Pan Y.-W., and Dong J.-J., 1991b, Time-dependent tunnel convergence — II. Advances rate and tunnel-support interaction, *Int. J. Rock Mech. Min. Sci. & Geomech. Abstr.* **28**, 6, 477–488.

Pan E., Sassolas C., Amadei B. and Pfeffer W. T., 1997, A 3-D boundary element formulation of viscoplastic media with gravity, *Computational Mechanics* **19**, 308–316.

Pellet F., Hajdu A., Deleruyelle F. and Besnus F., 2005, A viscoplastic model including anisotropic damage for the time dependent behaviour of rock, *Int. J. Num. Anal. Meth. Geomech.* **29**, 941–970.

Perić D. and Crook A. J. L., 2004, Computational strategies for predictive geology with reference to salt tectonics, *Comput. Methods Appl. Mech. Engrg.* **193**, 5195–5222.

Pietruszczak S., Lydzba D. and Shao J. F., 2004, Description of creep in inherently anisotropic frictional materials, *J. Eng. Mech.* **130**, 6, 681–690.

Popp T. and Kern H., 1994, The influence of dry and water saturated cracks on seismic velocities of crystal rocks- A comparison of experimental data with theoretical model, *Surveys in Geophysics* **15**, 443–465.

Popp T. and Kern H., 2000, Monitoring the state of microfracturing in rock salt during deformation by combined measurements of permeability and P- and S-wave velocities, *Phys. Chem. Earth (A)* **25**, 2, 149–154.

Rocchi G., Fontana M. and Da Prat M., 2003, Modeling of natural soft clay destruction processes using viscoplasticity theory, Géotehnique **53**, 8, 729–745.

Schultze O., Popp T., and Kern H., 2001, Development of damage and permeability in deforming rock salt, *Engineering Geology* **61**, 2–3, 163–180.

Shan R., Jiang Y. and Li B., 2000, Obtaining dynamic complete stress–strain curves for rock using the Split Hopkinson Pressure Bar technique, *Int. J. Rock Mech. Min. Sci.* **37**, 983–992.

Shao J. F., Bederiat M. and Schroeder Ch., 1994, Elasto-viscoplastic modeling of a porous chalk, *Mec. Res. Com.* **21**, 1, 63–75.

Shao J. F., Zhu Q. Z. and Su K., 2003, Modeling of creep in rock materials in terms of material degradation, *Computers and Geotechnics* **30**, 549–555.

Silberschmidt V. G. and Silberschmidt V. V., 2000, Analysis of cracking in rock salt, *Rock Mech. Rock. Engng.* **33**, 1, 53–70.

Song B., Chen W. W., Dou S., Winfree N. A., and Kang J. H., 2005, Strain-rate effects on elastic and early cell-collapse responses of a polystyrene foam, *Int. J. Impact Engineering* **31**, 509–521.

Song D., 1993, Non-linear viscoplastic creep of rock surrounding an underground excavation, *Int. J. Rock. Mech. Sci. & Geomech. Abstr.* **39**, 6, 653–658.

Subhash G., Liu Q. and Gao X. L., 2005, Quasistatic and high strain rate uniaxial compressive response of polymeric structural foams, *Int. J. Impact Engineering* (in print).

Taliercio A. L. F. and Gobbi E., 1997, Effect of elevated triaxial cyclic and constant loads on the mechanical properties of plain concrete, *Mag. of Concrete Res.* **49**, 181, 353–365.

Tashman L., Masad E., Little D. and Zbib H., 2005, A microstructure-based viscoplastic model for asphalt concrete, *Int. J. Plasticity* **21**, 1659–1685.

Tedesco J. W., Ross C. A., and Kuennen S. T., Strainrate effects on the compressive strength of shock-mitingating foams, *Jour. Sound Vibr.* **165**, 2, 374–384.

Unosson M. and Nilsson L., 2005, Projectile penetration and perforation of high performance concrete: experimental results and macroscopic modeling, *Int. J. Impact Engineering* (in print).

Voyiadjis G. Z. and Kim D., 2003, Finite element analysis of the piezocone test in cohesive soils using an elastoplastic-viscoplastic model and updated Lagrangian formulation, *Int. J. Plasticity* **19**, 253–280.

Wang C. G., Song Z. Q., Chen W. Z., Liu Q. S., and Yang C. H., 2004, Study of thermo-rheology characteristics of rock under the uni-axial compression, *Key Engineering Materials*, 261–263, 639–644.

Wathugala G. W. and Pal S., 1999, comparison of different implementation algorithms for HiSS constitutive models in FEM, *Int. J. Solids Struct.* **36**, 4941–4962.

Yahya O. M. L., Aubertin M. and Julien M. R., 2000, A unified representation of the plasticity, creep and relaxation behavior of rocksalt, *Int. J. Rock Mech. Min. Sci.* **37**, 787–800.

Yamada K., 1999, Computation for Rheological phenomena of elastic-viscoplastic body, *J. Engn. Mechanics* **125**, 1, 11–18.

Yamada K., 2000, Seismic wave propagation in elastic-viscoplastic shear layers, *J. Geotechnical and Geoenvironmental Engn.* **126**, 3, 218–226.

Yanagidani T., Ehara S., Nishizawa O., Kusunose K. and Terada M., 1985, Localization of dilatancy in Ohshima Granite under constant uniaxial-stress, *J. Geophys. Res.* **90**, No. B8, 6840–6858.

Zhupanska O. I., Abdel-Hadi A. I. and Cristescu N. D., 2002, Mechanical properties of microcrystalline cellulose, Part II. Constitutive model, *Mech. of Mater.* **34**, 391–399.

Chapter 3

The Propagation of Longitudinal Stress Waves in Thin Bars

3.1　The Equation of Motion

The propagation of longitudinal elastic-plastic waves in thin rods or wires was the first problem to be considered in dynamic plasticity, since it is also the simplest. It is the only possible one-dimensional problem, both because only a single stress component and a single strain component arise, and because a single spatial coordinate occurs in the problem. The bar will be assumed to be semi infinite and thin, the influence of the shape of the section of the rod will be disregarded. Only the area of the cross section will be taken into account. The bar is considered "thin" in the sense that lateral inertia can be neglected, the material particles can move freely in the direction transverse to the generatrices of the bar. That is a proper one-dimensional problem since a single space coordinate is involved x and a single variable t, besides one stress σ, one strain ε and one particle velocity v. In the presentation of the problem generally we follow the book by Cristescu [1967]. For the older literature see Huffington [1965], Miklowitz [1969], Cristescu [1968], Kinslow [1970], Chou and Hopkins [1973], Varley [1976], Nowacki [1978], Lee [1973] [1974], Kawata and Shioiri [1978]. They will be mentioned no more.

The coordinate axis will be chosen with the origin at the end of the bar and the Ox positive axis directed along the bar. One supposes that for $t < 0$ the bar is at rest, and for $t = 0$ the end of the bar is struck by a rigid body, so that for $t > 0$ the bar is no more at rest. Dynamic buckling will be disregarded.

We assume that the cross section of the rod, which is plane before the impact, remains plane after it too. Thus the displacements of the particles parallel to the axis of the bar are equal in a given transverse plane, for the whole cross section.

If one consider the initial conditions:

$$\left.\begin{array}{c} t = 0 \\ x \geq 0 \end{array}\right\} \ \sigma(x,0), \ \varepsilon(x,0), \ v(x,0) \text{ all prescribed}$$

and the boundary conditions:

$$\left.\begin{array}{c} t > 0 \\ x = 0 \end{array}\right\} \ \sigma(0,t) \text{ or } \varepsilon(0,t) \text{ or } v(0,t) \text{ prescribed (only one!)}$$

105

As notations, we denote by X the material coordinate and by $x(X,t) = X + u(X,t)$ the space coordinate and u the displacement. From here

$$\frac{\partial x}{\partial X} = 1 + \frac{\partial u}{\partial X} = 1 + \varepsilon$$

with

$$\varepsilon = \frac{ds - dS}{dS}$$

with $ds > dS$ a finite measure of strain.

We recall the Green definition of strain

$$C_{IJ} = \frac{\partial x_k}{\partial X_I} \frac{\partial x_k}{\partial X_J}.$$

Which, for thin bars, is written

$$C_{11} = \left(1 + \frac{\partial u}{\partial X}\right)^2.$$

Also the Lagrange definition is:

$$E_{IJ} = \frac{1}{2}\left(\frac{\partial x_k}{\partial X_I} \frac{\partial x_k}{\partial X_J} - \delta_{IJ}\right)$$

which for thin bars is

$$E_{11} = \frac{1}{2}\left[\left(1 + \frac{\partial u}{\partial X}\right)^2 - 1\right].$$

Thus $\varepsilon = \partial u/\partial X = \sqrt{1 + 2E_{11}} - 1 = \sqrt{C_{11}} - 1$ is a finite measure. As a convention we consider tension to be positive.

From conservation of mass

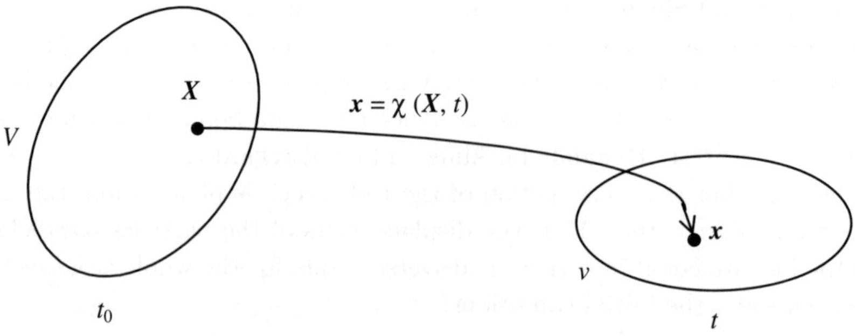

Fig. 3.1.1   Conservation of mass.

we have:

$$\int_V \rho_0(\mathbf{X}, t_0)\, dX = \int_v \rho(\mathbf{x}, t)\, dx$$

which for thin bars is written:

$$\int_V \rho_0(X,t) A_0(X,t)\, dX = \int_V \rho(x(X,t),t) A(x(X,t),t) \frac{\partial x}{\partial X}\, dX\,.$$

When all functions are sufficiently smooth:

$$\rho_0 A_0 = \rho A \frac{\partial x}{\partial X} \quad \text{or} \quad \rho A = \frac{\rho_0 A_0}{1+\varepsilon}\,.$$

From the balance of momentum we have:

$$\int_S \boldsymbol{t}\, dS + \int_V \rho \boldsymbol{b}\, dV = \frac{d}{dt}\int_V \rho \boldsymbol{v}\, dV$$

or

$$\int_V \left(\frac{\partial T_{ji}}{\partial x_j} + \rho b_i - \rho \frac{dv_i}{dt} \right) dV = 0\,.$$

For bars we have:

$$\frac{\partial T}{\partial X} \frac{\partial X}{\partial x} = \rho \frac{dv}{dt}$$

or

$$\frac{\partial (F/A)}{\partial X} \frac{\rho A}{\rho_0 A_0} = \rho \frac{\partial v}{\partial t}$$

by neglecting the term $v_j(\partial v_i/\partial x_j)$. For cylindrical bars $\partial A/\partial X = 0$. Thus, we have

$$\frac{\partial \sigma}{\partial X} = \rho_0 \frac{\partial v}{\partial t}$$

with $v = \partial u/\partial t$ for the particle velocity. From now on, we change the notation and we denote by x the material coordinate, and by ρ the initial density.
Thus we obtain

$$\frac{\partial^2 u}{\partial t^2} = \frac{1}{\rho}\frac{\partial \sigma}{\partial x}\,. \tag{3.1.1}$$

Equation (3.1.1) is the equation of motion that will be examined below. To this equation one must add the constitutive equation which establishes a relation between the strain ε and the stress σ and which describes the mechanical properties of the material considered. The strain will always be taken as the sum of two components, the elastic and the plastic:

$$\varepsilon = \varepsilon^E + \varepsilon^P\,. \tag{3.1.2}$$

The elastic one is defined by $\varepsilon = \sigma/E$, where σ is the initial stress and E is the Young's modulus.

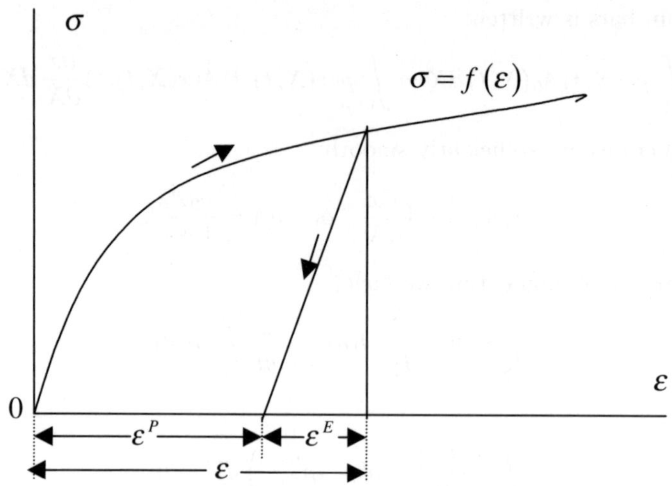

Fig. 3.2.1   Stress–strain curve for a work hardening material.

3.2   The Finite Constitutive Equation

In the present chapter we shall consider the constitutive equation of the form

$$\sigma = f(\varepsilon) \qquad (3.2.1)$$

where the function f is a monotonically increasing function of ε. First we assume that it is a monotonically decreasing function as shown in the Fig. 3.2.1. Introducing (3.2.1) in (3.1.1) the equation of motion becomes

$$\frac{\partial^2 u}{\partial t^2} = \frac{1}{\rho} \frac{d\sigma}{d\varepsilon} \frac{\partial^2 u}{\partial x^2} . \qquad (3.2.2)$$

This is a quasi linear equation of the second order of the type of wave equation. We can consider an equivalent system of equation

$$\frac{\partial v}{\partial t} = c^2(\varepsilon) \frac{\partial \varepsilon}{\partial x}$$
$$\frac{\partial v}{\partial x} = \frac{\partial \varepsilon}{\partial t} \qquad (3.2.3)$$

where $v = u_t = \partial u/\partial t$ is the particle velocity and

$$c^2(\varepsilon) = \frac{1}{\rho} \frac{d\sigma}{d\varepsilon} \qquad (3.2.4)$$

is the velocity of propagation of the wave. For all kinds of relations (3.2.1) we have $c(\varepsilon) \geq 0$.

If $d\sigma/d\varepsilon = E = const.$, we obtain $c_0^2 = E/\rho$ the constant bar velocity. The equation of motion becomes

$$\frac{\partial^2 u}{\partial t^2} = c_0^2 \frac{\partial^2 u}{\partial x^2} .$$

Let us introduce the new variables

$$\alpha = x - c_0 t, \quad \beta = x + c_0 t$$

i.e.,

$$\frac{\partial u}{\partial t} = \frac{\partial u}{\partial \alpha}\frac{\partial \alpha}{\partial t} + \frac{\partial u}{\partial \beta}\frac{\partial \beta}{\partial t} = -c_0\frac{\partial u}{\partial \alpha} + c_0\frac{\partial u}{\partial \beta},$$

$$\frac{\partial^2 u}{\partial t^2} = c_0^2\left(\frac{\partial^2 u}{\partial \alpha^2} + \frac{\partial^2 u}{\partial \beta^2}\right), \quad \text{etc.}$$

Now the equation of motion is written

$$\frac{\partial^2 u}{\partial \alpha \partial \beta} = 0.$$

The solution of this equation (since $\partial u/\partial \alpha = f'(\alpha)$, etc.) is:

$$u = f(x - c_0 t) + g(x + c_0 t).$$

$u_1 = f(x - c_0 t)$ are the waves propagating towards right, and are the solution of the equation $(\partial/\partial t + c_0\,\partial/\partial x)u = 0$. The waves $u_2 = g(x + c_0 t)$ are propagating towards the left and are the solution of the equation $(\partial/\partial t - c_0\,\partial/\partial x)u = 0$. The second order equation can be written as

$$\left(\frac{\partial}{\partial t} + c_0\frac{\partial}{\partial x}\right)\left(\frac{\partial}{\partial t} - c_0\frac{\partial}{\partial x}\right)u = 0$$

f and g are determined from initial data prescribed along $t = 0$ or from boundary conditions prescribed along $x = const$.

<u>Exercise.</u> *Car traffic on a highway.*

We introduce the notation: $\rho(x,t)$ for the car density on a unit length of highway, $q(x,t)$ for the number of cars passing through the place x in unit time. q obviously depends on ρ by the local constitutive equation $q = Q(\rho)$. The flow speed is $v = q/\rho = v(x,t)$. From the conservation of cars on the portion $x_1 \le x \le x_2$ of the highway, if there are no exits on this portion;

$$\frac{d}{dt}\int_{x_1}^{x_2} \rho(x,t)\,dx + q(x_1,t) - q(x_2,t) = 0,$$

and if $\rho(x,t)$ is smooth (continuous derivative) and $x_1 \to x_2$, we have

$$\frac{\partial \rho}{\partial t} + \frac{\partial q}{\partial x} = 0 \quad \text{or} \quad \frac{\partial \rho}{\partial t} + Q'(\rho)\frac{\partial \rho}{\partial x} = 0$$

or

$$\rho_t + c(\rho)\rho_x = 0$$

this is quite similar to the above equation.

By introducing the notation

$$U = \left\{\begin{array}{c} \varepsilon \\ v \end{array}\right\}, \quad A = \left\{\begin{array}{cc} 0 & -1 \\ -c^2 & 0 \end{array}\right\}$$

  Dynamic Plasticity

the system (3.2.3) can be written in matrix notation:

$$U_t + AU_x = 0 \,. \tag{3.2.5}$$

Let us assume that the system (3.2.2) has in a certain domain D in the characteristic plane x, t, a smooth solution. Consider an arbitrary point $(x_0, t_0) \in D$ and a curve Γ passing through x_0, t_0. A small displacement along Γ has the coordinates dx and dt. The problem is that assuming that U is known along Γ, can we find U_t and U_x along Γ?

In order to find the characteristics of these equations we are looking for the curves in the plane xOt at the crossing of which v and ε are continuous but their derivatives may be discontinuous. We add to (3.2.3) the relations

$$
\begin{aligned}
\frac{\partial v}{\partial s} &= \frac{\partial v}{\partial x}\frac{dx}{ds} + \frac{\partial v}{\partial t}\frac{dt}{ds}\,, \\[2mm]
\frac{\partial \varepsilon}{\partial s} &= \frac{\partial \varepsilon}{\partial x}\frac{dx}{ds} + \frac{\partial \varepsilon}{\partial t}\frac{dt}{ds}
\end{aligned}
\tag{3.2.6}
$$

where the derivatives dx/ds and dt/ds are computed along one of the characteristic directions so that $\partial v/\partial s$ and $\partial \varepsilon/\partial s$ are the interior derivatives. In order to get some simplifications we can write (3.2.6) in the form

$$
\begin{aligned}
dv &= \frac{\partial v}{\partial x}\,dx + \frac{\partial v}{\partial t}\,dt \\[2mm]
d\varepsilon &= \frac{\partial \varepsilon}{\partial x}\,dx + \frac{\partial \varepsilon}{\partial t}\,dt \,.
\end{aligned}
\tag{3.2.7}
$$

From (3.2.3) and (3.2.7) we get

$$
\frac{\partial v}{\partial t} = \frac{c^2(dv\,dt - d\varepsilon\,dx)}{-dx^2 + c^2 dt^2}\,, \qquad
\frac{\partial v}{\partial x} = \frac{c^2 d\varepsilon\,dt - dv\,dx}{-dx^2 + c^2 dt^2}\,,
$$

$$
\frac{\partial \varepsilon}{\partial x} = -\frac{d\varepsilon\,dx - dv\,dt}{-dx^2 + c^2 dt^2}\,,
$$

if

$$
\Delta \equiv \det \begin{vmatrix} I & A \\ I\,dt & I\,dx \end{vmatrix}
= \begin{vmatrix} 1 & 0 & 0 & -1 \\ 0 & -c^2 & 1 & 0 \\ dt & dx & 0 & 0 \\ 0 & 0 & dt & dx \end{vmatrix}
= -(dx)^2 + c^2(dt)^2 \neq 0\,,
$$

giving the definition of the characteristic lines as

$$\frac{dx}{dt} = \pm c(\varepsilon)\,. \tag{3.2.8}$$

Let us assume that Γ is characteristic. Since we have assumed that in D the system (3.2.2) has a solution, with prescribed values along Γ, we must have:

$$
rang \left\{ \begin{matrix} I & A & 0 \\ I\,dt & I\,dx & dU \end{matrix} \right\} = rang\Delta
$$

i.e., any fourth order determinant chosen from

$$\left\{ \begin{array}{ccccc} 1 & 0 & 0 & -1 & 0 \\ 0 & -c^2 & 1 & 0 & 0 \\ dt & dx & 0 & 0 & d\varepsilon \\ 0 & 0 & dt & dx & dv \end{array} \right\}$$

must be zero along Γ. Thus are obtained the differential relations satisfied along these lines which are

$$dv = \pm c(\varepsilon)\, d\varepsilon \qquad (3.2.9)$$

or

$$\frac{\partial v}{\partial s_1} = +c(\varepsilon)\frac{\partial \varepsilon}{\partial s_1}, \quad +\frac{\partial v}{\partial s_2} = -c(\varepsilon)\frac{\partial \varepsilon}{\partial s_2},$$

along the characteristic lines $s_2 = const.$ and $s_1 = const.$ respectively. Since from constitutive relation follows $d\varepsilon = \pm(d\sigma/c^2(\sigma)\rho)$, (3.2.9) can be written

$$dv = \pm\frac{d\sigma}{\rho c(\sigma)}. \qquad (3.2.9)'$$

Thus the considered system is of hyperbolic type. The slopes of the characteristic lines are variable and depend on the unknown function ε. Thus the characteristic lines are usually two families of curved lines, unknown before finding the solution of the system. For totally hyperbolic systems the number of differential relations satisfied along the characteristic lines is equal to the number of characteristic lines. Let us mention that the lower and upper sign corresponds to each other. If the sign convention for σ and ε is different, the sign in (3.2.8) and (3.2.9) change.

A wave will be defined as a solution $u(x,t)$ of equation (3.2.2), determined within a certain range of variation of the variables x and t, and possessing continuous first and second order derivatives within this domain. The geometrical locus of the points which separate two waves and move along the bar in time, will be called wave front. Across a wave front the velocity v and the strain ε are continuous, but their first derivatives are discontinuous. Thus the wave fronts coincide with the characteristics of the equation of motion; more precisely in a certain dynamic problems the characteristics of the equation of motion can be considered to be wave fronts, while sometimes else some of the characteristics have no mechanical meaning. If the first derivatives of ε and v are discontinuous across the wave front, the corresponding wave is called a weak wave, smooth wave or finally acceleration wave. Occasionally, one encounters strong discontinuity waves or shock waves, the fronts of which are surfaces of discontinuity even for ε and v.

In the next sections we shall consider only smooth waves, which will be simple called waves. If we write the relations (3.2.7) on both sides of a wave front and the symbol $[f] = f^+ - f^-$ is used to denote the magnitude of the jump of a function f across the wave front, we obtain, since the bar remains continuous,

$$\left[\frac{\partial v}{\partial x}\right] dx + \left[\frac{\partial v}{\partial t}\right] dt = 0, \quad \left[\frac{\partial \varepsilon}{\partial x}\right] dx + \left[\frac{\partial \varepsilon}{\partial t}\right] dt = 0, \qquad (3.2.10)$$

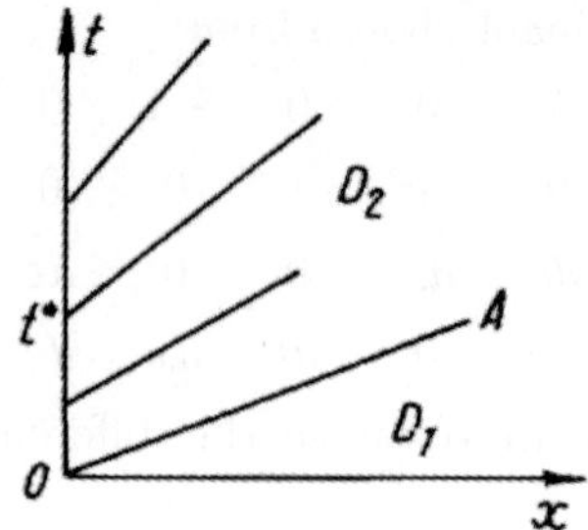

Fig. 3.2.2   Characteristic field showing a divergent family of characteristic lines.

where dx and dt are connected by one of the two relations (3.2.8). The relations (3.2.10) are the so-called kinematics compatibility conditions. Another condition which must be satisfied by the same jumps can be obtained directly from the equation of motion (3.2.2)

$$\left[\frac{\partial v}{\partial t}\right] = c^2(\varepsilon)\left[\frac{\partial \varepsilon}{\partial x}\right]. \tag{3.2.11}$$

This is the "dynamic compatibility condition", which is not independent on the relations (3.2.10). In the theory of wave propagation the jump conditions play an important role because they show which derivatives are discontinuous across a certain wave front and consequently which functions are affected by the wave considered. However, it is not possible to determine the magnitude of these jumps, but only the ratio between them.

From equation (3.2.2) it follows that the variation of the velocity of propagation, as a function of ε, is governed by the slope of the stress–strain curve. But the constitutive equation used for various plastic materials is vary diverse, thus the laws describing the variation of c as a function of ε also vary considerably. For instance, for most materials the stress–strain curve takes the form illustrated in Fig. 3.2.1, i.e., where $\sigma d^2\sigma/d\varepsilon^2 < 0$ for any ε. For such stress–strain curves the velocity of propagation $c(\varepsilon)$ decreases when the stress increases: $dc/d\varepsilon < 0$ for any $d\varepsilon > 0$. Considering that the stress at the end of the bar is increasing continuously, the waves generated successively at the end of the bar will propagate with continually decreasing velocities. Furthermore, the corresponding wave fronts will be represented in a characteristic plane by a divergent family of curves, whose slopes will increase with the stress (Fig. 3.2.2). This means that for such materials the distance between the wave fronts will increase during propagation, i.e., the waves will spread.

There are however materials for which the diagram representing the constitutive equation takes the form shown in Fig. 3.2.3, i.e., for which the slope increases continuously ($\sigma d^2\sigma/d\varepsilon^2 > 0$ for any ε). Such constitutive equations apply to some rubbers, soils, and even to certain metals. Because the slope of the stress–strain curve increases continuously, the velocity of propagation will increase when the stress increases ($dc/d\varepsilon > 0$ for any $d\varepsilon > 0$). A representation of the wave fronts

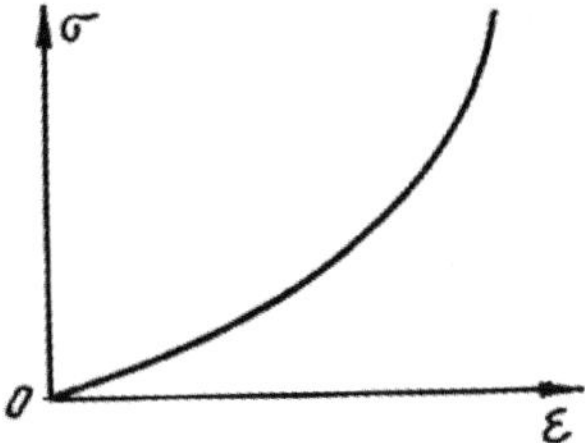

Fig. 3.2.3   A stress–strain curve concave towards the stress axis.

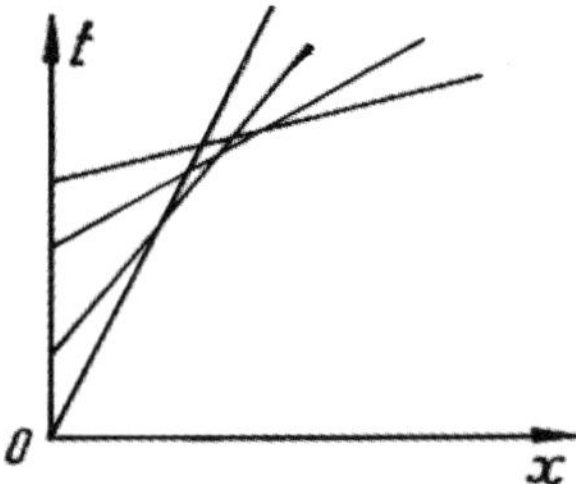

Fig. 3.2.4   Characteristic field showing a convergent bundle of characteristics lines.

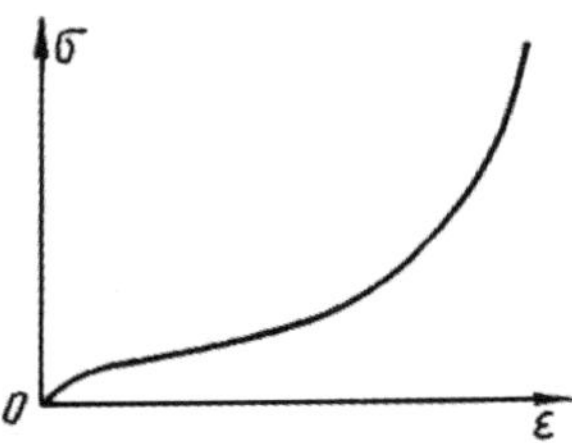

Fig. 3.2.5   Stress–strain curve of variable concavity.

near the end of the bar is given in Fig. 3.2.4. In this case the distance between the wave fronts decreases during propagation and here is a tendency to form shock waves.

Finally it is possible that the curvature of the stress–strain curve changes at a certain moment (Fig. 3.2.5). In this case the wave fronts will first diverge ($dc/d\varepsilon <$ 0) and then converge ($dc/d\varepsilon > 0$) (Fig. 3.2.6). In the following sections only the case when $dc/d\varepsilon \leq 0$ for any $d\varepsilon > 0$ will be considered.

If the impact at the end of the bar is not sufficiently strong, the yield stress is not reached and the waves generated at the end of he bar are pure elastic ones. In this case the constitutive equation is the Hook's law (Fig. 3.2.7)

$$\sigma = E\varepsilon \tag{3.2.12}$$

where E is Young's modulus. The velocity of propagation is now constant

$$c_0^2 = E/\rho . \tag{3.2.13}$$

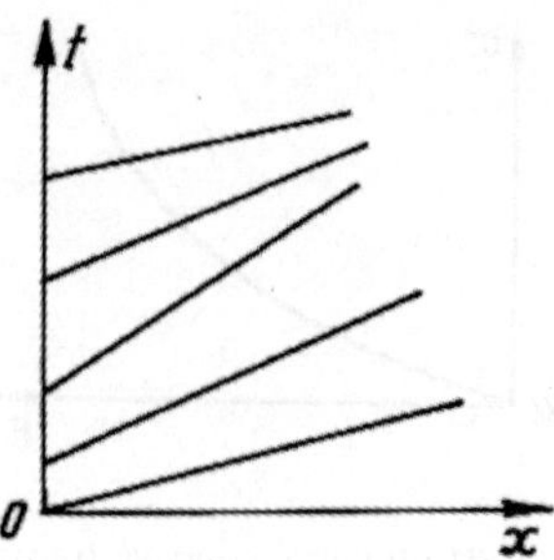

Fig. 3.2.6 Characteristic field for a stress–strain curve of variable concavity; first part of the bundle is divergent, and the other convergent.

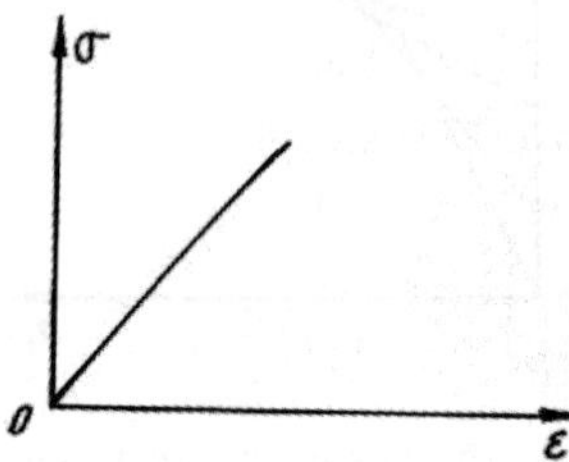

Fig. 3.2.7 A linear stress–strain curve.

In this case the characteristics take the form of parallel strain lines.

A similar situation arises when the material is no longer elastic, but the stress–strain curve is linear on certain portions. For instance, linear work hardening materials give the following stress–strain relationship

$$\sigma = \sigma_Y + E_1(\varepsilon - \varepsilon_Y) \tag{3.2.14}$$

where E_1 is the constant work-hardening modulus, while σ_Y and ε_Y are the stress and strain at the yield point. Here again the velocity of propagation is constant

$$c^2 = E_1/\rho. \tag{3.2.15}$$

In most of the cases this velocity is much smaller than the velocity of propagation of elastic waves c_0.

It is possible that in certain portions of the stress–strain curve $d\sigma/d\varepsilon \to 0$, as for instance for perfectly plastic materials. In such cases $c(\varepsilon) \to 0$, and the wave considered can thus no longer be propagated. In some other cases, as for instance for locking materials, when $d\sigma/d\varepsilon \to \infty$ the problem becomes parabolic, and it is an infinite velocity of propagation.

It is interested to consider now the initially undeformed semi-infinite bar at rest. This is the case considered by Rakhmatulin [1945]. In this case, the initial conditions (the Cauchy data) are

$$t = 0 \quad \text{and} \quad x > 0: \ \varepsilon(x,0) = v(x,0) = \sigma(x,0) = 0, \tag{3.2.16}$$

while the boundary conditions are

$$x = 0 \quad \text{and} \quad t \geq 0 : \text{ given } \varepsilon(0,t) \text{ or } \sigma(0,t) \text{ or } v(0,t) \,. \tag{3.2.17}$$

The conditions (3.2.16) are satisfied in the domain D_1 in Fig. 3.2.2.OA is the first wave front propagated along the bar. A solution can be obtained by integrating the relations (3.2.9) along the corresponding characteristic lines. We obtain

$$v = \int c(\varepsilon)\, d\varepsilon + k_1 = \psi(\varepsilon) + k_1(s_2) \,,$$

$$\tag{3.2.18}$$

$$v = - \int c(\varepsilon)\, d\varepsilon + k_2 = -\psi(\varepsilon) + k_2(s_1) \,,$$

where the Riemann invariant parameters $k_1(s_2)$ and $k_2(s_1)$ have different constant values on different characteristic lines. However, all the characteristics of negative slope intersect the line OA and therefore start from the undisturbed domain D_1. It follows that all the constants $k_2 = 0$ and thus throughout the domain D_2 the relation between the velocity and the strain is

$$v = -\psi(\varepsilon) \,. \tag{3.2.19}$$

Introducing (3.2.19) into the first relation (3.2.18), we conclude that both v and ε are constants along the characteristics of positive slope. The corresponding constants are determined from the boundary conditions. Thus if the bar is initially unperturbed, the characteristics of positive slope are straight lines

$$x = c(\varepsilon(t^*))(t - t^*) \,, \tag{3.2.20}$$

where t^* is the time when the straight line (3.2.20) intersects the Ot axis.

Other methods to find the characteristic lines. Let us consider the equation

$$\frac{\partial^2 u}{\partial t^2} = c^2(\varepsilon)\frac{\partial^2 u}{\partial x^2} \,. \tag{3.2.21}$$

Now if we change the variables

$$\alpha = \alpha(x,t) \,, \quad \beta = \beta(x,t) \,, \quad \text{with} \quad \frac{D(\alpha,\beta)}{D(x,t)} \neq 0 \,,$$

since

$$\frac{\partial u}{\partial x} = \frac{\partial u}{\partial \alpha}\frac{\partial \alpha}{\partial x} + \frac{\partial u}{\partial \beta}\frac{\partial \beta}{\partial x} \,, \dots \dots$$

$$\frac{\partial^2 u}{\partial x^2} = \frac{\partial^2 u}{\partial \alpha^2}\left(\frac{\partial \alpha}{\partial x}\right)^2 + \frac{\partial u}{\partial \alpha}\frac{\partial^2 \alpha}{\partial x^2} + \frac{\partial^2 u}{\partial \beta^2}\left(\frac{\partial \beta}{\partial x}\right)^2 + \frac{\partial u}{\partial \beta}\frac{\partial^2 \beta}{\partial x^2} \,,$$

$$\frac{\partial^2 u}{\partial t^2} = \frac{\partial^2 u}{\partial \alpha^2}\left(\frac{\partial \alpha}{\partial t}\right)^2 + \frac{\partial u}{\partial \alpha}\frac{\partial^2 \alpha}{\partial t^2} + \frac{\partial^2 u}{\partial \beta^2}\left(\frac{\partial \beta}{\partial t}\right)^2 + \frac{\partial u}{\partial \beta}\frac{\partial^2 \beta}{\partial t^2} \,,$$

and plugging into (3.2.21) we obtain

$$\left[\left(\frac{\partial \alpha}{\partial t}\right)^2 - c^2(\varepsilon)\left(\frac{\partial \alpha}{\partial x}\right)^2\right]\frac{\partial^2 u}{\partial \alpha^2} + \left[\left(\frac{\partial \beta}{\partial t}\right)^2 - c^2(\varepsilon)\left(\frac{\partial \beta}{\partial x}\right)^2\right]\frac{\partial^2 u}{\partial \beta^2}$$

$$+ \left[\frac{\partial^2 \alpha}{\partial t^2} - c^2(\varepsilon)\frac{\partial^2 \varepsilon}{\partial x^2}\right]\frac{\partial u}{\partial \alpha} + \left[\frac{\partial^2 \beta}{\partial x^2} - c^2(\varepsilon)\frac{\partial^2 \beta}{\partial t^2}\right]\frac{\partial u}{\partial \beta} = 0 \,.$$

Let us assume that along a curve $\alpha = \alpha_0 = const.$: u and $\partial u/\partial \alpha$ are known functions of β, and by derivation with respect to β we have also $\partial u/\partial \beta$, $\partial^2 u/\partial \alpha \partial \beta$, $\partial^2 u/\partial \beta^2$. Then if

$$\left(\frac{\partial \alpha}{\partial t}\right)^2 - c^2(\varepsilon)\left(\frac{\partial \alpha}{\partial x}\right)^2 \neq 0$$

one can obtain $\partial^2 u/\partial \alpha^2$ too. The curves $\alpha = const.$ with $grad\,\alpha \neq 0$, satisfying

$$\left(\frac{\partial \alpha}{\partial t} - c(\varepsilon)\frac{\partial \alpha}{\partial x}\right)\left(\frac{\partial \alpha}{\partial t} + c(\varepsilon)\frac{\partial \alpha}{\partial x}\right) = 0$$

are the characteristic lines of (3.2.21). Along these lines, the solution u must satisfy

$$\left[\left(\frac{\partial \beta}{\partial t}\right)^2 - c^2(\varepsilon)\left(\frac{\partial \beta}{\partial x}\right)^2\right]\frac{\partial^2 u}{\partial \beta^2} + \left[\frac{\partial^2 \alpha}{\partial t^2} - c^2(\varepsilon)\frac{\partial^2 \alpha}{\partial x^2}\right]\frac{\partial u}{\partial \alpha}$$

$$+ \left[\frac{\partial^2 \beta}{\partial x^2} - c^2(\varepsilon)\frac{\partial^2 \beta}{\partial t^2}\right]\frac{\partial u}{\partial \beta} = 0$$

which are the relations to be satisfied by u and $\partial u/\partial \alpha$ along the characteristic lines. Thus u and $\partial u/\partial \alpha$ can not be prescribed arbitrarily along a characteristic line and the formulation of the Cauchy problem along a characteristic line is not possible.

Formulation of the initial conditions for

$$\rho_t + c(\rho)\rho_x = 0\,. \tag{a}$$

Let us find the curves (with $q = Q(\rho)$, $c(\rho) = Q'(\rho)$)

$$\frac{dx}{dt} = c(\varepsilon) \tag{b}$$

for which the expression $\rho_t + c(\rho)\rho_x$ is a total differential in that direction, i.e.,

$$\frac{d\rho}{dt} = \frac{\partial \rho}{\partial t} + \frac{\partial \rho}{\partial x}\frac{dx}{dt}\,. \tag{c}$$

Thus along the curves (b) introducing (a) in (c) we have

$$\frac{d\rho}{dt} = 0\,, \quad \frac{dx}{dt} = c(\rho) \tag{d}$$

i.e., $\rho = const.$ on each curve (b) and the slope of this curve is constant (straight line). Generally the characteristic slope is not constant and the solution is not constant along a characteristic line. However for (a) that is true.

The Cauchy problem (initial data):

$$\left.\begin{array}{c} t = 0 \\ -\infty < x < +\infty \end{array}\right\} \rho = f(x)\,. \tag{e}$$

If (b) intersects x-axis in ξ then $\rho = f(\xi)$ along the whole straight line $t = 0$. The family of characteristic lines when ξ varies is

$$x = \xi + \bar{c}(\xi)t \quad \text{with} \quad \bar{c}(\xi) = c(f(\xi))\,. \tag{f}$$

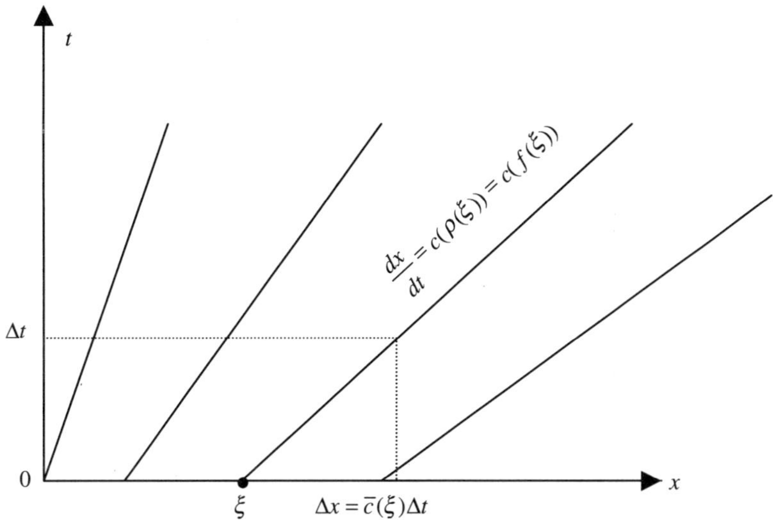

Fig. 3.2.8   The Cauchy problem.

Along each $\rho = f(\xi) = const.$, and thus we get the solution of the problem. We say that $\rho(\xi)$ propagates with the velocity $\bar{c}(\xi)$: after the time Δt, $\rho(\xi)$ propagates towards right up to the position $\bar{c}(\xi)\Delta t$.

Let us check if $\rho = f(\xi)$ is a solution of (a), if ξ is defined implicitly by (f). We have

$$\rho_t = f'(\xi)\xi_t\,, \quad \rho_x = f'(\xi)\xi_x\,. \tag{g}$$

From (f), by derivation with respect to t and x:

$$0 = \xi_t + \bar{c}(\xi) + \bar{c}'(\xi)t\xi_t\,, \quad 1 = \xi_x + \bar{c}(\xi)t\xi_x\,. \tag{h}$$

Combining (g) and (h):

$$\rho_t = -\frac{f'(\xi)\bar{c}(\xi)}{1 + \bar{c}'(\xi)t}\,, \quad \rho_x = -\frac{f'(\xi)}{1 + \bar{c}'(\xi)t}\,, \tag{i}$$

which satisfies (a).

Let us consider a constitutive equation of the form

$$q = Q(\rho)\,, \quad c(\rho) = Q'(\rho)$$

with $Q'(\rho) > 0$ and $c'(\rho) = Q''(\rho) > 0$ (see Fig. 3.2.9). The initial condition is shown in Fig. 3.2.10 at $t = 0$. The first part of the curve corresponds to expansion, while the second one to compression. In this case the wave breaks. We have a shock formation and non-unique solution.

The characteristic lines can have an envelope which can be obtained from

$$x = \xi + \bar{c}(\xi)t\,, \quad 0 = 1 + \bar{c}'(\xi)t\,. \tag{j}$$

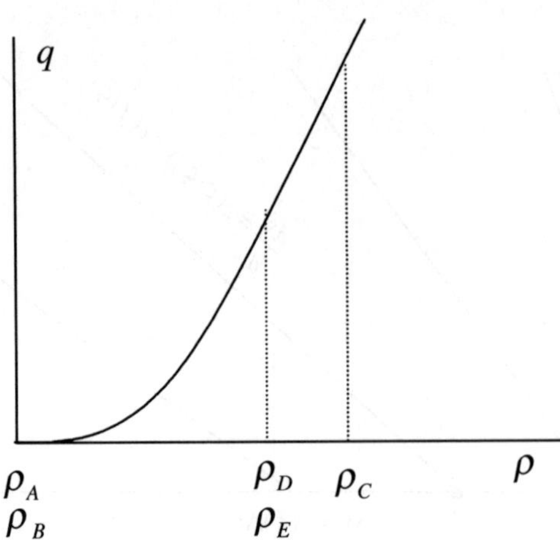

Fig. 3.2.9   The constitutive equation considered.

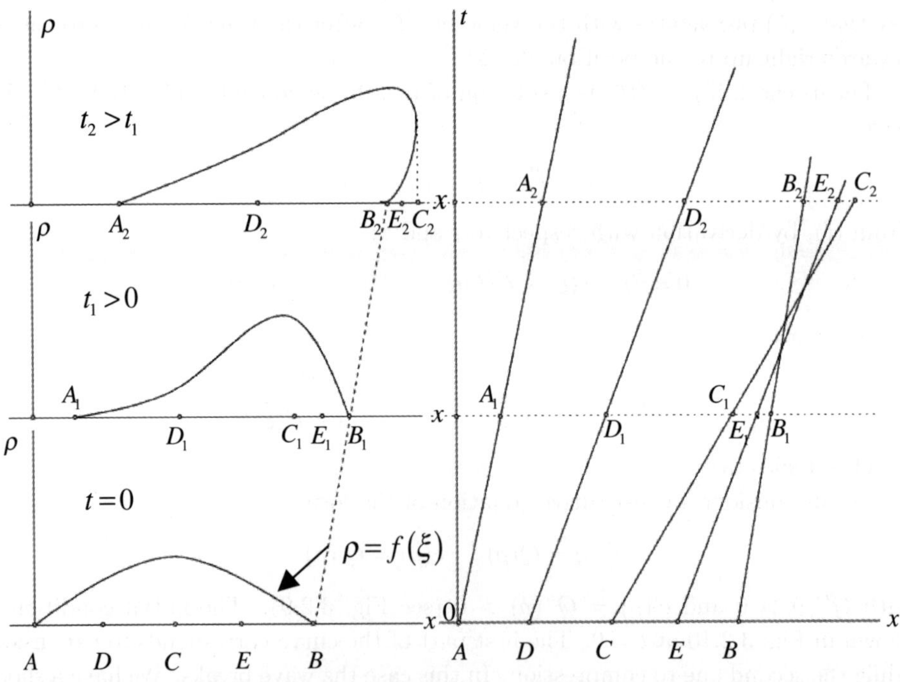

Fig. 3.2.10   Distortion of a wave in the non linear case for $c'(\rho) > 0$.

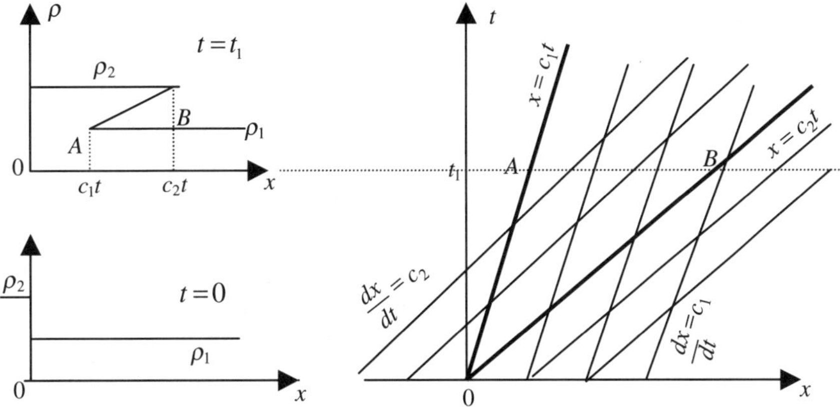

Fig. 3.2.11   The first case when we do not have uniqueness.

The first point on this envelope when t increases and $t > 0$, corresponds to the maximum of $\bar{c}'(\varepsilon)$ or $\rho_x \to \infty$ at time $t = -1/\bar{c}'(\xi)$ at position $x = \xi - \bar{c}(\xi)/\bar{c}'(\xi)$. Particular case when function $f(x)$ from (e) is discontinuous:

$$f(x) = \begin{cases} \rho_1 & \text{for } x > 0 \\ \rho_2 & \text{for } x < 0 \end{cases} \quad \text{and} \quad \bar{c}(x) = \begin{cases} \bar{c}(\rho_1) = c_1 = const. \ x > 0 \\ \bar{c}(\rho_2) = c_2 = const. \ x < 0 \end{cases}$$

<u>1st case:</u> $\rho_2 > \rho_1$ and $c_2 > c_1$ $(c'(\rho) > 0)$ we have a non unique solution.

The problem must be reformulated. In AB $(c_1 < x < c_2 t)$ the solution is not unique. The problem must be reformulated from the physical point of view in order to get uniqueness (shocks).

<u>2nd case:</u> $\rho_2 < \rho_1$ and $c_2 < c_1$ $(c'(\rho) > 0)$. On $AB : \rho \in [\rho_2, \rho_1]$ and the characteristics have constant slope $x = c(\rho)t$ from where $c = x/t$. Thus for an arbitrary t:

$$x \leq c_2 t \Rightarrow c = c_2$$

$$c_2 t \leq x \leq c_1 t \Rightarrow c(\rho) = \frac{x}{t}$$

$$c_1 t \leq x \Rightarrow c = c_1 .$$

Formulation of initial and boundary value problem for a quasilinear system of hyperbolic type

$$I\frac{\partial U}{\partial t} + A(U, x, t)\frac{\partial U}{\partial x} = 0 .$$

We make the following assumption:

- The system is totally hyperbolic system i.e., all eigenvalues are distinct finite real numbers in the strip $a \leq x \leq b$, $t \geq 0$. Let λ_k $(k = 1, \ldots, n)$ be the eigenvectors. For

$$\left.\begin{array}{c} t = 0 \\ a \leq x \leq b \end{array}\right\} \text{ is prescribed } U(x, 0) = U_0(x)$$

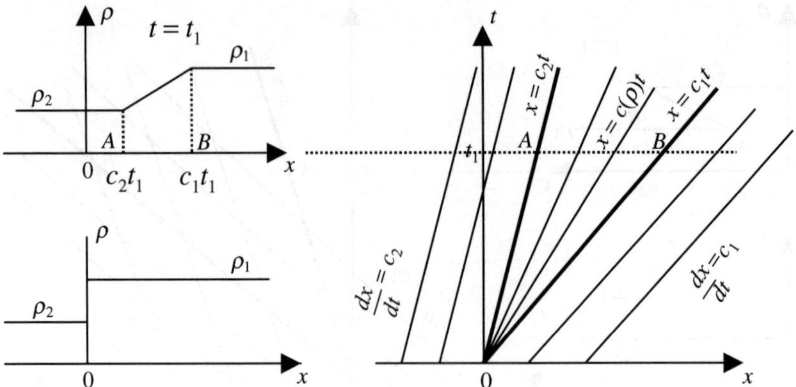

Fig. 3.2.12   Centered expansion wave.

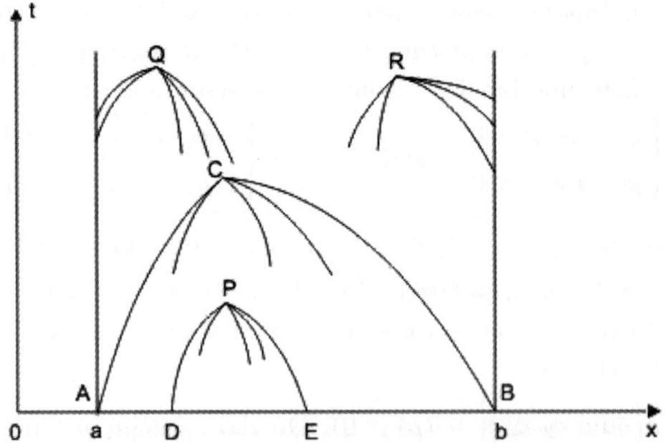

Fig. 3.2.13   Formulation of the problem.

while at

$$x = a, \ t > 0 : U_i(a,t) = U_i^a(t), \ i = 1,\ldots,r$$

$$x = b, \ t > 0 : U_j(b,t) = U_j^b(t), \ j = 1,\ldots,s\,.$$

If these data are compatible with the system, then r and s are not arbitrary.

- The initial and boundary data are smooth enough and that at $t = 0$, $x = a$ and at $t = 0$, $x = b$ they coincide. If not, shocks may be generated.

Through a generic point P (Fig. 3.2.13) one drives backwards all the characteristics. Using the differential relations along characteristics one can determine all required functions in P. DE is the domain of dependence of P.

In a similar way we draw the characteristics backwards through C. AC and BC are the "outer" characteristics through C. In the domain ABC the Cauchy problem can be solved. ABC is the domain of determinacy of the segment AB.

Above AC and CB one has to solve a <u>mixed boundary value problem</u>. If we consider a generic point Q and if $n = 5$ (say) and $p = 2$, $q = 3$, in order to determine the solution in Q by the characteristic method one must have $r = p = 2$. Similarly from R: $s = q = 3$.

Since $\lambda_k(U)$ (quasi linear system) the characteristics can be drawn only step by step together with the finding of the solution. This is a typical quasilinear system when compared with semi linear or linear systems, for which the characteristics mesh can be drawn a priori. Shocks may occur if the characteristics of some family (positive or negative) are convergent when t increases.

<u>Particular cases of characteristic lines.</u> Let us consider some particular cases of characteristic lines which may be considered in applications. We assume that the stress increase is continues. Thus for

$$\left.\begin{array}{c} \text{any } \varepsilon \\ \sigma\dfrac{d^2\sigma}{d\varepsilon^2} < 0 \end{array}\right\} \Rightarrow \left\{\begin{array}{c} d\varepsilon > 0 \\ \dfrac{dc}{d\varepsilon} < 0 \end{array}\right.$$

we have the wave starting from the end of the bar are a straight diverging lines:

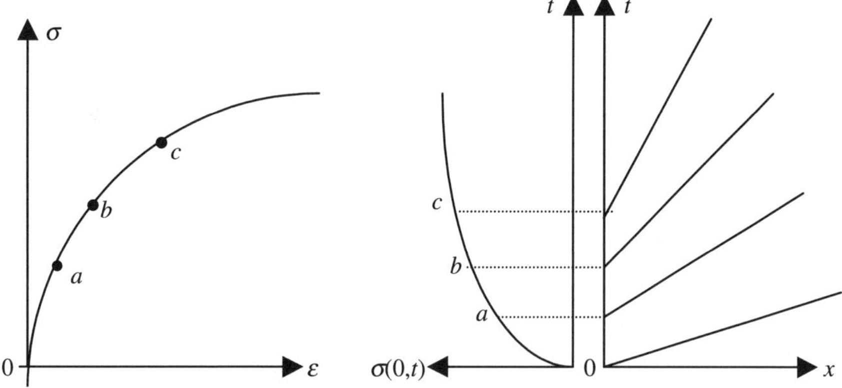

Fig. 3.2.14   Standard case for formulation of the problem.

Let us assume that the bar is power work-hardening i.e., $\sigma = E\varepsilon^{1/n}$, $n > 1$. In this case we have the differential relations as

$$-v = \varphi(\sigma) = \int_0^\sigma \frac{d\sigma}{\rho c(\sigma)} = \int_0^\sigma \sqrt{\frac{n}{\rho E}}\left(\frac{\sigma}{E}\right)^{(n-1)/2} d\sigma$$

$$= \frac{2}{n+1}\sqrt{\frac{n}{\rho E^n}}\,\sigma^{(n+1)/2}.$$

Another case is the elastic case. In this case $\sigma = E\varepsilon$ while the velocity is a constant $c_0 = E/\rho$. The stress and the first waves are:

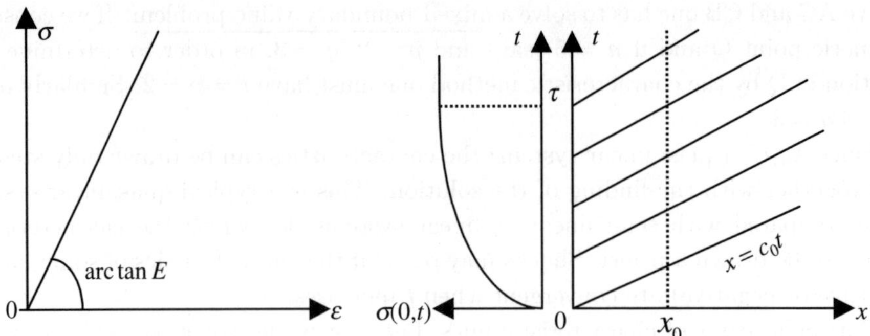

Fig. 3.2.15   First waves in an elastic bar.

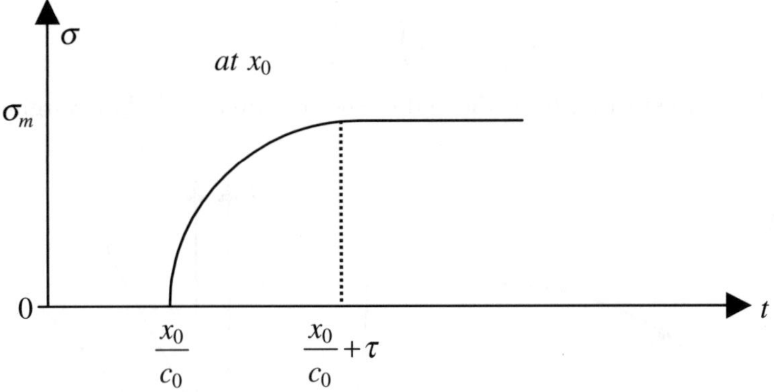

Fig. 3.2.16   What an observer at x_0 sees; the same variation of the stress.

The differential relations along the characteristics are

$$v = -c_0\varepsilon \quad \text{and} \quad \sigma = -\rho c_0 v.$$

An observer at x_0 will see a variation of the stress as shown in Fig. 3.2.16.
Another important case is the case of linear work-hardening. In this case we have:

$$\sigma = E\varepsilon \qquad\qquad \text{if } \sigma \leq \sigma_Y \Rightarrow c_0^2 = \frac{E}{\rho}$$

$$\sigma = \sigma_Y + E_1(\varepsilon - \varepsilon_Y) \quad \text{if } \sigma \geq \sigma_Y \Rightarrow c_1^2 = \frac{E_1}{\rho}$$

and $c_1 \ll c_0$, since $E_1 \ll E$.
The differential relations are

$$-v = \int_0^{\sigma_Y} \frac{d\sigma}{\rho c_0} + \int_{\sigma_Y}^{\sigma} \frac{d\sigma}{\rho c_1} = \frac{\sigma_Y}{\rho c_0} + \frac{\sigma - \sigma_Y}{\rho c_1}$$

An observer placed at x_0 will see the variation in Fig. 3.2.18:

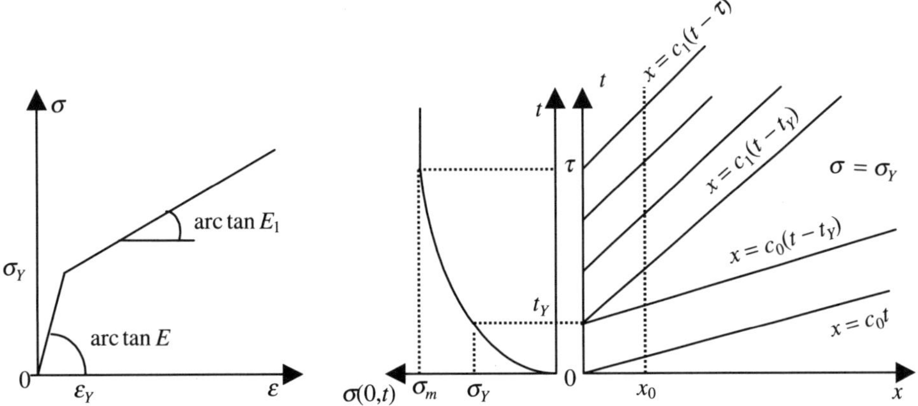

Fig. 3.2.17   The first waves in the linear work-hardening material.

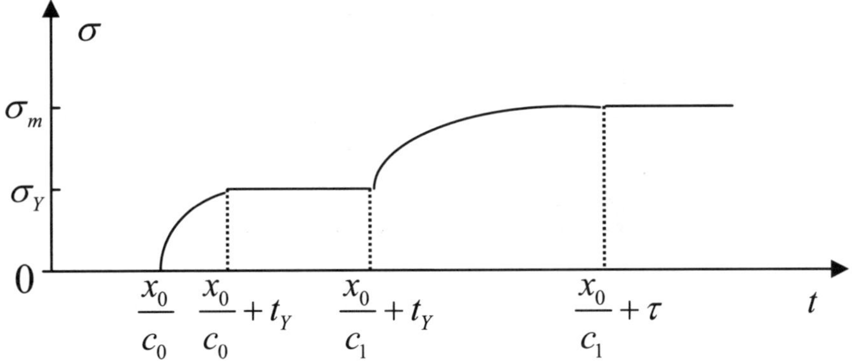

Fig. 3.2.18   What an observer at x_0 will see.

The characteristics are two families of straight lines shown in Fig. 3.2.17. Due to the change in slope at σ_Y we obtain a region in the characteristic plane where the stress is constant and equal to σ_Y. The various significant lines are written on the figure. What an observer placed at x_0 will see is shown in Fig. 3.2.18.

There are also several other cases which are important, though they are not so often met. So is the case when the stress–strain curve is changing concavity. Then we have a possibility of shock formation.

There is the case of perfect plasticity i.e., when at least on a part of the stress–strain curve we have $\sigma = \sigma_Y = const$. Now, on this part of the curve we have $d\sigma/d\varepsilon = 0$ that is $c = 0$. Though no waves are propagating in the bar, and the system is only hyperbolic, not totally hyperbolic.

Another case is the locking case, when the stress–strain curve becomes vertical. Now $d\sigma/d\varepsilon \to \infty$ and also $c \to \infty$. In this case we have an evolution system the

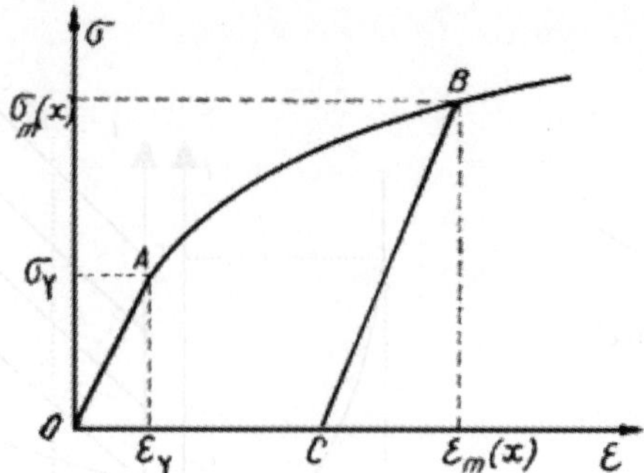

Fig. 3.2.19   Stress–strain curve illustrating perfectly elastic unloading.

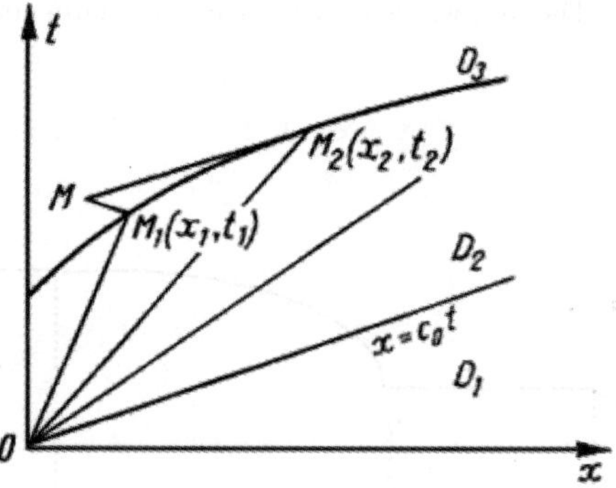

Fig. 3.2.20   Characteristic field which loading and unloading domains.

system is changing from hyperbolic to parabolic. The propagation is instantaneous and the characteristic lines superpose. The differential relation is reduced only to one $d\varepsilon = 0$.

3.3   The Unloading Problem

We have considered until now the loading problem, when the stresses are increasing or staying constant. Let us consider now the unloading when the stress after increasing is decreasing. During unloading, especially for metals, the unloading is a perfectly elastic one. For instance, in Fig. 3.2.19 the unloading begins at point B and takes place along the straight line BC, parallel to the segment OA, which is the initial elastic segment. Thus, the following constitutive equation should be used during unloading

$$\sigma = \sigma_m(x) + E[\varepsilon - \varepsilon_m(x)], \tag{3.3.1}$$

where $\sigma_m(x)$ and $\varepsilon_m(x)$ are the stress and strain corresponding to the point B (Fig. 3.2.19). For each section x, $\sigma_m(x)$ and $\varepsilon_m(x)$ are the maximum stress and maximum strain, respectively. The difficulty of the unloading problem consists in the fact that in each section of the bar, the unloading process begins at a different maximum stress and maximum strain. That is to say, that different constitutive equation (3.3.1) must be used for each section of the bar.

In the characteristic plane xOt (Fig. 3.2.20) there will be two domains in which stress, strain and velocity are either increasing or decreasing. In the domain D_2 the plastic strain increases, while in the domain D_3 the plastic part of the strain decreases. These two domains are separated by a curve, called loading/unloading boundary. The shape of this boundary depends on the mechanical properties of the material, but also on the boundary and initial conditions. By definition the loading/unloading boundary is the geometrical locus of points in the characteristic plane xOt, in which the maximum strain has been reached in each section of the bar (it may contain also an area). This gives the following for the loading/unloading condition

$$\frac{\partial \sigma}{\partial t} > 0 \text{ for loading, i.e. } \frac{\partial \varepsilon^P}{\partial t} > 0 \,,$$

$$\frac{\partial \sigma}{\partial t} \leq 0 \text{ for unloading, i.e. } \frac{\partial \varepsilon^P}{\partial t} = 0 \,. \tag{3.3.2}$$

The equation of motion in the unloading domain is obtained by introducing (3.3.1) in (3.2.2):

$$\frac{\partial^2 u}{\partial t^2} = c_0^2 \frac{\partial^2 u}{\partial x^2} + \frac{1}{\rho} \frac{d\sigma_m(x)}{dx} - c_0^2 \frac{d\varepsilon_m(x)}{dx} \,, \tag{3.3.3}$$

where c_0 the constant velocity of propagation is expressed by (3.2.13), while σ_m and ε_m depend only on x and are unknown functions. The general solution of equation (3.3.3) is

$$u = F_1(c_0 t + x) + F_2(c_0 t - x) - \frac{1}{E} \int_0^x (\sigma_m - E\varepsilon_m)\, dx \,, \tag{3.3.4}$$

where F_1 and F_2 are arbitrary functions to be determined by the boundary conditions formulated for the end of the bar and the loading/unloading boundary.

The characteristics of equation (3.3.3) are

$$dx/dt = \pm c_0 \,, \tag{3.3.5}$$

while the differential relations satisfied along them are

$$dv = \pm \frac{1}{\rho c_0}\, d\sigma \,. \tag{3.3.6}$$

The lower and upper signs correspond to each other. With the aid of (3.3.1) the last equation can also be written in the form

$$dv = \pm c_0 d\varepsilon \pm \frac{1}{\rho c_0}\, d\sigma \mp c_0 d\varepsilon \,. \tag{3.3.7}$$

Let us assume that the loading/unloading boundary is known. Take two points on the boundary, $M_1(x_1, t_1)$ and $M_2(x_2, t_2)$ (Fig. 3.2.20). Through M_1 is drawn the characteristic of negative slope in the unloading domain. Similarly the characteristic of positive slope is drawn through M_2. These two characteristics intersect at a point $M(x, t)$. The equations of these characteristics and the differential relations along them are

$$x - x_2 = c_0(t - t_2), \quad x - x_1 = -c_0(t - t_1),$$

$$v - v_2 = c_0(\varepsilon - \varepsilon_2) + \frac{1}{\rho c_0}[\sigma_m(x) - \sigma_m(x_2)] - c_0[\varepsilon_m(x) - \varepsilon_m(x_2)],$$

$$v - v_1 = -c_0(\varepsilon - \varepsilon_1) - \frac{1}{\rho c_0}[\sigma_m(x) - \sigma_m(x_1)] + c_0[\varepsilon_m(x) - \varepsilon_m(x_1)].$$

From these relations, and taking into account that on the loading/unloading boundary

$$\varepsilon_1 = \varepsilon_m(x_1), \quad \varepsilon_2 = \varepsilon_m(x_2), \quad v_1 = -\psi(\varepsilon_m(x_1)), \quad v_2 = -\psi(\varepsilon_m(x_2))$$

we obtain

$$v = -\frac{1}{2}\{\psi(\varepsilon_m(x_1)) + \psi(\varepsilon_m(x_2))\} + \frac{1}{2\rho c_0}[\sigma_m(x_1) - \sigma_m(x_2)]$$

$$\varepsilon = \frac{-\psi(\varepsilon_m(x_1)) + \psi(\varepsilon_m(x_2))}{2c_0} - \frac{1}{2E}[2\sigma_m(x) - \sigma_m(x_1) - \sigma_m(x_2)] + \varepsilon_m(x).$$

$$(3.3.8)$$

These formulae yield the velocity and the strain at any point M in the unloading domain, if the loading/unloading boundary is known. In particular these may coincide with the prescribed conditions at the end of the bar.

In order to determine the loading/unloading boundary it is sometimes useful to associate to the characteristic plane xOt, a hodografic plane $vO\sigma$. In this last plane we represent the differential relations satisfied along the characteristics. The loading/unloading boundary will be represented on the hodographic plane by the curve (3.2.19), in which the strain is replaced by the stress

$$v = -\varphi(\sigma). \tag{3.3.9}$$

If the bar is initially unperturbed, there is a one-to-one correspondence between the points of this curve and the characteristics of positive slope in the loading domain of the characteristic plan. Because these characteristics intersect the loading/unloading boundary, it follows that there is a one-to-one correspondence between the points of the loading/unloading boundary and the points of the curve (3.3.9).

If the constitutive equation is given, one can find easily the hodographic plane (3.3.9). For instance if the material is linear work-hardening so that $\sigma = \sigma_Y + E_1(\varepsilon - \varepsilon_Y)$, (3.3.9) becomes

$$-v = \int_0^{\sigma_Y} \frac{d\sigma}{\rho c_0} + \int_{\sigma_Y}^{\sigma} \frac{d\sigma}{\rho c} = \frac{\sigma_Y}{\rho c_0} + \frac{\sigma - \sigma_Y}{\rho c_1}, \tag{3.3.10}$$

where $c_1 = (E_1/\rho)^{1/2}$. Thus in the case of linear work-hardening material, the image of the loading/unloading boundary is a straight line.

If the material is power work-hardening, so that

$$\sigma = E\varepsilon^{1/n} \qquad (3.3.11)$$

where E and n are constants $(n > 1)$, we obtain

$$-v = \varphi(\sigma) = \int_0^{\sigma} \sqrt{\frac{n}{\rho E}} \left(\frac{\sigma}{E}\right)^{(1/2)(n-1)} d\sigma = \frac{2}{n+1} \sqrt{\frac{n}{\rho E^n}} \sigma^{(1/2)(n+1)}, \qquad (3.3.12)$$

i.e., the curve (3.3.9) is a generalized parabola. Similar considerations can be made for any other constitutive equation.

3.4 Determination of the Loading/Unloading Boundary

The grapho-analytical method of Shapiro–Biderman. This method can be applied to sufficiently long, thin, bars (no reflections), initially at rest and undeformed.

The shape of the loading/unloading boundary is determined in two stages. First the slope of this boundary near the end of the bar must be found, and secondly the remainder of the boundary must be determined.

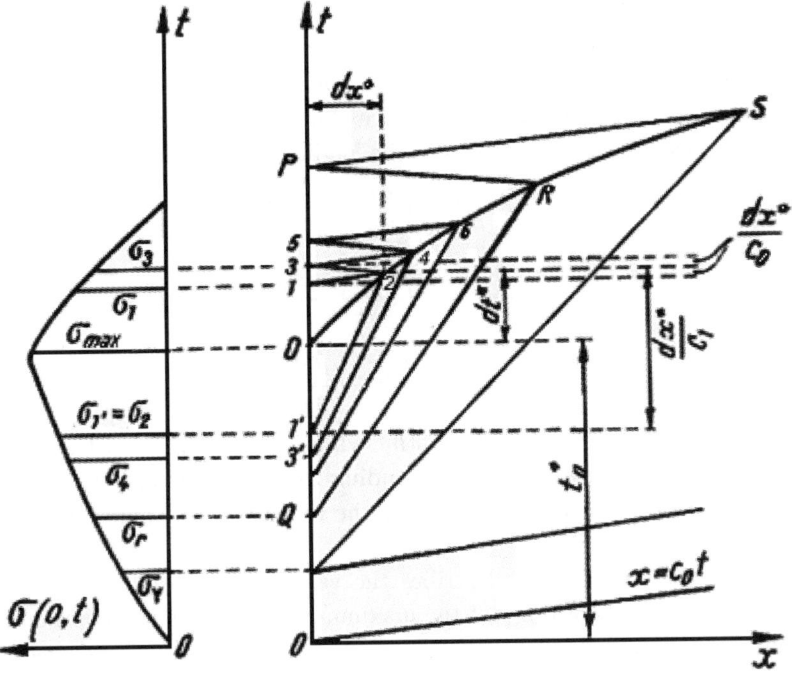

Fig. 3.4.1   Characteristic fields and boundary condition for the determination of the initial slope of graphic-analytical method.

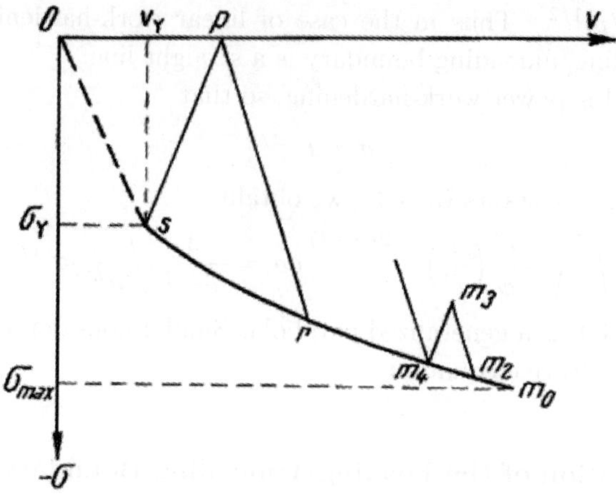

Fig. 3.4.2   Hodographic plane associated to the characteristic field of Fig. 3.4.1.

It will be assumed that the stress at the end of the bar increases in accordance with a certain law up to a value σ_{max}, and then decreases to zero (Fig. 3.4.1). We consider first a law of variation of the stress at the end of the bar, according to which the slope is discontinuous at the maximum. On the loading/unloading boundary we consider a point M_2 (denoted simply by 2 in Fig. 3.4.1) of coordinates $x_2^* = dx^*$ and $t_2^* = t_0^* + dt^*$, where the asterisks are used to mean that the coordinates of the point indicated lie on the loading/unloading boundary. Through M_2 one draws the two characteristics of positive and negative slope into the unloading domain and the characteristics of positive slope into the loading domain. These lead trivially to

$$t_{1'} = t_0^* - \left(\frac{dx^*}{c_1} - dt^* \right),$$

$$t_1 = t_0^* + \left(dt^* - \frac{dx^*}{c_0} \right), \tag{3.4.1}$$

$$t_3 = t_0^* + \left(dt^* + \frac{dx^*}{c_0} \right).$$

Here t_0^* is the ordinate of the point M_0, while $c_1 = c(\sigma_{max})$ is the velocity of propagation of the plastic wave corresponding to the maximum stress. Therefore we can assume that in the interval $M_{1'} - M_0$ the velocity $c(\sigma)$ remains approximately constant and equal to $c(\sigma_{max})$.

If the slope of curve which describes the variation of the stress at the end of the bar possesses a discontinuity at the maximum, then in the neighborhood of this point the stress is given by:

$$\sigma^-(0,t) = \sigma_{max} + k_1(t - t_0) \quad \text{for } t < t_0,$$

$$\sigma^+(0,t) = \sigma_{max} + k_2(t - t_0) \quad \text{for } t > t_0,$$

where k_1 and k_2 are the two values of the derivative $d\sigma(0,t)/dt$ (higher order terms are neglected).

The stress at point $M_{1'}$ is

$$\sigma_{1'} = \sigma_{\max} - k_1(t_0 - t_{1'}) = \sigma_{\max} - k_1\left(\frac{dx^*}{c_1} - dt^*\right), \tag{3.4.2}$$

and the velocity follows from (4.4.2) and (4.3.9):

$$v_{1'} = -\int_0^{\sigma_{1'}} \frac{d\sigma}{\rho c(\sigma)} = -\int_0^{\sigma_{\max}} \frac{d\sigma}{\rho c(\sigma)} + \frac{\sigma_{\max} - \sigma_{1'}}{\rho c_1}$$

$$= v_{\max} + \frac{k_1}{\rho c_1}\left(\frac{dx^*}{c_1} - dt^*\right), \tag{3.4.3}$$

where $v_{\max}$, is the velocity at the point M_0. Clearly, $\sigma_2 = \sigma_{1'}$ and $v_2 = v_{1'}$.

For the points M_1 and M_3 we obtain

$$\sigma_1 = \sigma_{\max} + k_2(t_1 - t_0) = \sigma_{\max} + k_2\left(dt^* - \frac{dx^*}{c_0}\right),$$

$$\sigma_3 = \sigma_{\max} + k_2(t_3 - t_0) = \sigma_{\max} + k_2\left(dt^* + \frac{dx^*}{c_0}\right),$$

$$\tag{3.4.4}$$

$$v_1 = v_{\max} + j(t_1 - t_0) = v_{\max} + j\left(dt^* - \frac{dx^*}{c_0}\right),$$

$$v_3 = v_{\max} + j(t_3 - t_0) = v_{\max} + j\left(dt^* + \frac{dx^*}{c_0}\right),$$

where $j = dv/dt$ is the acceleration of the end of the bar for $t \geq t_0$.

Along the characteristics $M_1 M_2$ and $M_2 M_3$ it follows from (3.3.6) that:

$$v_1 - \frac{\sigma_1}{\rho c_0} = v_2 - \frac{\sigma_2}{\rho c_0}, \quad v_3 + \frac{\sigma_3}{\rho c_0} = v_2 + \frac{\sigma_2}{\rho c_0}.$$

Combining these expressions with (3.4.2), (3.4.3) and (3.4.4), we obtain:

$$j\left(\frac{1}{b^*} - \frac{1}{c_0}\right) - \frac{k_1}{\rho c_1}\left(\frac{1}{c_1} - \frac{1}{b^*}\right) = \frac{1}{\rho c_0}\left[k_1\left(\frac{1}{c_1} - \frac{1}{b^*}\right) + k_2\left(\frac{1}{b^*} - \frac{1}{c_0}\right)\right],$$

$$j\left(\frac{1}{b^*} + \frac{1}{c_0}\right) - \frac{k_1}{\rho c_1}\left(\frac{1}{c_1} - \frac{1}{b^*}\right) = -\frac{1}{\rho c_0}\left[k_1\left(\frac{1}{c_1} - \frac{1}{b^*}\right) + k_2\left(\frac{1}{b^*} + \frac{1}{c_0}\right)\right],$$

where $b^* = dx^*/dt^*$ is the initial velocity of propagation of the loading/unloading boundary. Eliminating j from these two formulae, we obtain the formula for the initial slope (the slope at point M_0) of the loading/unloading boundary as

$$b^* = \sqrt{\frac{c_1^2 c_0^2 (k_1 - k_2)}{c_0^2 k_1 - c_1^2 k_2}}. \tag{3.4.5}$$

This formula is due to Biderman (see Ponomarev *et al.* [1959] Ch. X).

Many special cases of the above can be considered. If $k_1 = 0$ and $k_2 \neq 0$ [Fig. 3.4.3(a)], or if k_1 is finite while k_2 is infinite [Fig. 3.4.3(b)], then $b^* = c_0$ and

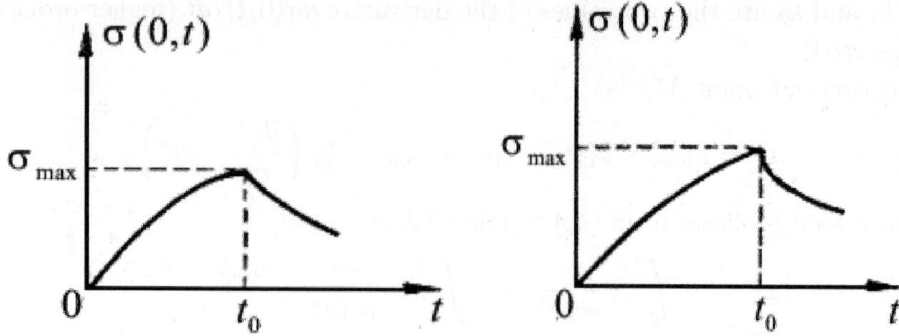

Fig. 3.4.3   Particular boundary conditions for which the initial slope of the loading/unloading boundary coincides with the elastic wave fronts.

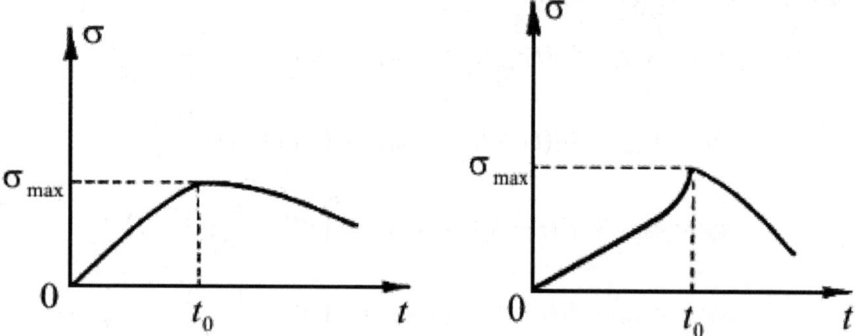

Fig. 3.4.4   Particular boundary conditions for which the initial slope of the loading/unloading boundary coincides with the slope of the plastic waves fronts, corresponding to the maximum strain.

the loading/unloading boundary propagates with the velocity of the elastic waves. If $k_1 \neq 0$ and $k_2 = 0$ [Fig. 3.4.4(a)] or if k_2 is finite while k_1 is infinite [Fig. 3.4.4(b)], then $b^* = c_1$ and initially the loading/unloading boundary propagates with the velocity of the plastic loading wave corresponding to the maximum stress.

If, at the point in which $\sigma(0,t) = \sigma_{\max}$ the curve representing the variation of the stress $\sigma(0,t)$ does not possess a discontinuity of the slope (i.e., $k_1 = k_2 = 0$), then the above formula cannot be applied. In this case in the calculation of the stress one should also take into account the higher order derivatives. With $k_1 = k_2 = 0$ and the derivative $d^2\sigma(0,t)/dt^2$ continuous for $\sigma(0,t) = \sigma_{\max}$, Biderman obtains for b^* the formula

$$b^* = c_0 \left(\sqrt{\frac{c_0^2}{c_1^2} + 3} - \frac{c_0}{c_1} \right). \tag{3.4.6}$$

The initial direction of the loading/unloading boundary may thus be calculated for any case. The remaining of this boundary can be determined by a method proposed by Shapiro [1946].

After drawing the initial direction of the loading/unloading boundary, we choose in this direction a point M_2 fairly close to M_0. The stress at M_2 is obtained from (3.4.2). We now go to the image of the loading/unloading boundary in the hodographic plane (Fig. 3.4.2), and mark on it the points m_0 and m_2, which are the correspondent in this plane of M_0 and M_2. We then draw the characteristic $m_2 m_3$ which corresponds to the characteristic $M_2 M_3$. The position of the point m_3 on this characteristic is known since we know the stress σ_3 from the boundary conditions. We then draw the characteristic $m_3 m_4$ and the corresponding $M_3 M_4$. The stress σ_4 is obtained by the intersection of the straight line $m_3 m_4$ with the image (3.3.9) of the loading/unloading boundary. Thus is located the point m_4 and since $\sigma_{3'} = \sigma_4$ we determine the position of the point $M_{3'}$. We can draw now the characteristic

$$x = c(\sigma_4)(t - t_{M_{3'}})$$

in the loading domain. Its intersection with the characteristic $M_3 M_4$ fixes the position of the point M_4 on the loading/unloading boundary. Continuing in the same manner step by step, other points of the loading/unloading boundary can be located. When a high accuracy is required one can also use intermediate points.

We must now locate the last point S (Fig. 3.4.1) of the loading/unloading boundary, on which the stress becomes σ_Y. The image s of the point S, of co-ordinates σ_Y and $v_Y = \sigma_Y / \rho c_0$ in the hodographic plane, is known. We draw the line sp up to the intersection with $\sigma = 0$, and then pr, up to the intersection with the hodographic plane. This gives σ_r and as result the position of the point Q in the characteristic plane ($\sigma_r = \sigma_q$). We then draw the characteristic QR until it intersects the already establish portion of the loading/unloading boundary at R. Then from R we draw the characteristics RP and PS. The latter intersects the first plastic loading wave front and that gives the position of the point S. The

$$x = c(\sigma_Y)(t - t_Y),$$

abscissa of this point is the length of the bar deformed plastically, due to the impact. In the other parts of the bar only elastic vibrations are present.

Several examples have been considered with the above method. We give here only one, examined by Shapiro [1946] and then by Biderman (see Ponomarev *et al.* [1959]. The bar is considered semi-infinite, linear work-hardening and initially undeformed. The law of variation of the stress at the end of the bar takes the form

$$-\sigma(0,t) = 8\sigma_Y \frac{t}{T}\left(1 - \frac{t}{T}\right), \tag{3.4.7}$$

where T is the time during which the stress acts at the end of the rod. The graphical representation of this function is given in Fig. 3.4.5. The maximum stress is reached at $t = 1/2\,T$, when it equals $2\sigma_Y$. Considering the properties of the material, we assume that between the elastic modulus E and the work-hardening modulus

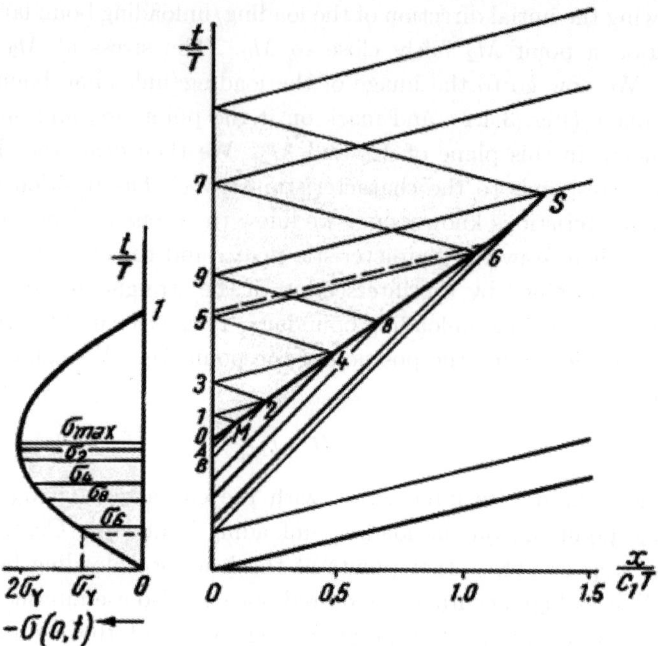

Fig. 3.4.5   Determination of the loading/unloading boundary in the example.

E_1 there is a relation $E = 16\,E_1$. Thus, the relation between the velocities of propagation is $c_0 = 4c_1$. Figure 3.4.5 gives all details of propagation. To this figure corresponds the hodographic plane from Fig. 3.4.6, with the coordinates σ and v/v_Y, and with $v_Y = \sigma/\rho c_0$. The initial shape of the loading/unloading boundary was obtained using the formula (3.4.6) $b^* = dx^*/dt^* = 0.359c_0$. A first segment OM (Fig. 3.4.5) of the loading/unloading boundary may now be drawn. By drawing the characteristics AM in the loading domain we obtain the stress σ_M at the point M. In practice, if $\partial\sigma(0,t)/\partial t = 0$ at the point of maximum stress, then the stress at the point M differs very little from the maximum stress $\sigma_{\max} = 2\sigma_Y$. In this case the two corresponding points 0 and m in the plane $vO\sigma$ practically coincide. Beyond this stage, the construction of the loading/unloading boundary is done as in the previous example. We draw the characteristic MM_1, and obtain the stress σ_1 at this point from (3.4.7). We then draw the corresponding characteristics mm_1 in the plane $vO\sigma$ and, knowing σ_1, the position of point m_1 is obtained. We then draw the characteristics m_1m_2 of negative slope in the same plane, and this gives $\sigma_2 = 1.96\sigma_Y$, namely the stress at this point. Introducing this numerical value into (3.4.7) and solving with respect to t, we obtain t_B and hence the position of point B (Fig. 3.4.5). Furthermore the point M_2 in the loading/unloading boundary is located at the intersection of the characteristics M_1M_2 and BM_2. In a similar way, we can locate the other points on the loading/unloading boundary, ending with the

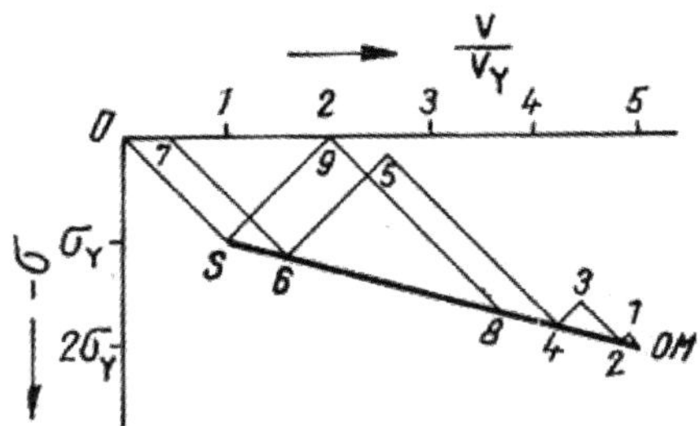

Fig. 3.4.6   Hodographic plane associated with the Fig. 3.4.5.

last one, S. Further one can give diagrams giving the variation of the stress or of the strain.

The first scientists which have considered the propagation of plastic waves in bars are also Kármàn and Duwez [1950], who have used material coordinates, and G. I. Taylor [1946, 1958a, 1958b] who has used spatial coordinates. The propagation of plastic waves in thin rods with quiescent of temperature gradients are due to Francis [1966]. He found that a thermal gradient magnifies the particle velocity at the impacted end of the bar. Later the same theory is due to Clifton and Bodner [1966] for pressure pulse of short duration, for isotropic work hardening for materials in tension is independent of any previous compression. Further, for aluminum rods at room temperature and at elevated temperatures by Bodner and Clifton [1967].

Experimental results are presented by Santosham and Ramsey [1970] on the propagation of small plastic strain waves in statically priestessed rods of soft electrolytic copper at room temperature. Strain pulses of order 500 μin./in. amplitude were generated by mechanical impact, and the average strain rate during loading was of order 1 sec^{-1}. The special feature of these experiments is that the rod specimens were long enough to permit observation of strain pulses at several positions along the roads without interference from reflected waves. Propagation velocity as a function of strain and the loading-unloading boundary in the x-t plane were observed directly. All the results obtained conform to the one-dimensional rate-independent theory of plastic wave propagation, although the dynamic stress–strain relations obtained differ considerably from the quasi-static one.

The uniaxial motion of interfaces between regions deforming elastically and regions deforming plastically is considered by Kenning [1974]. The governing constitutive stress-rate/strain-rate equations in both elastic and plastic regions are taken to be non-linear.

The influence of dynamical thermal expansion on the propagation of plane elastic-plastic stress waves was considered by Raniecki [1971]. It is shown how the plane waves caused by mechanical impulse and sudden heating at the boundary of an elastic-plastic half-space are influencing. It is shown that the effect of dynamical thermal expansion is to reduce the jump in the stress at waves of strong discontinuity. The stress and temperature fields dealt with here are assumed to be thermodynamically uncoupled.

In a paper by Misra *et al.* [1987] one is studying the distribution of stress, displacement and temperature in an elastic inhomogeneous semi-infinite rod, one end of which is heated non-uniformly for a finite interval of time. Three types of temperature functions viz. (i) linear, (ii) periodic and (iii) exponential are considered. The elastic moduli and the density are considered to be functions of position. The modified heat conduction equation has been used. The problem is treated by using Laplace transforms. Since it is not possible to obtain a closed form solution, Laplace inversion is carried out by using short time range approximations and by using the complex inversion formula.

Soft-recovery plate impact experiments have been conducted by Raiser *et al.* [1994] to study the evolution of damage in polycrystalline Al_2O_3 samples. Velocity-time profiles measured at the rear surface of the momentum trap indicate that the compressive pulse is not fully elastic even when the maximum amplitude of the pulse is significantly less than the elastic limit.

Experimental techniques are described by Nemat-Nasser *et al.* [1994] and illustrated for direct measurements of temperature, strain-rate, and strain effects on the flow stress of metals over a broad range of strains and strain rates. The approach utilizes: (1) the dynamic recovery Hopkinson bar technique, (2) direct measurements of sample temperature by high-speed infra-red detectors, and (3) ability to change the strain rate during the course of experiment at high strain rates. Taylor anvil tests are performed, accompanied by high-speed photographic recording of deformation, and the results are compared with those obtained by finite-element simulation.

Also experimental results are given by Chen *et al.* [1999]. The stress–strain responses of a half-hardened copper and an annealed tantalum as a function of temperature and strain-rate were investigated. The rate-dependent yield stress and work-hardening behavior were described. The yield and flow stresses exhibit very high strain rate sensitivities which is typical in bcc materials at low temperature or at high strain rate. The rod velocity is 178 m/s for the hhCu and 175 m/s for Ta, respectively.

The high strain rate deformation behaviors of kinetic energy penetrator materials during ballistic impact were considered by Magness [1994]. He considers velocity of impact exceeding 1000 m/s and strain rates of 10^6 s^{-1}, and study via metallographic examinations.

3.5 The Finite Bar

Conditions under which a shock wave may appear. The plastic shock waves i.e. of waves whose wave fronts are traveling discontinuities surfaces even for stress and

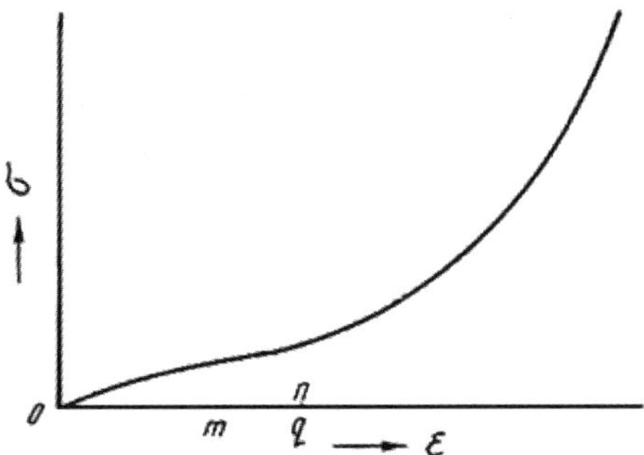

Fig. 3.5.1   An S-shaped stress–strain curve.

for the first order derivatives of displacements, appear in several circumstances. We consider here velocities of impact smaller than the velocity of propagation of waves.

First can be considered (Cristescu [1967]) stress–strain curves which have a concavity upwards. For such kind of curves

$$\sigma = E\varepsilon \quad \text{for } \varepsilon \leq \varepsilon_Y$$

$$\sigma = \sigma_Y + \frac{a(\varepsilon - \varepsilon_Y)}{\varepsilon_Y + b - \varepsilon} \quad \varepsilon \geq \varepsilon_Y$$

where a and b are material constants, the velocity of propagation is

$$c(\varepsilon) = \sqrt{\frac{ab}{\rho}} \left[\frac{1}{\varepsilon_Y + b - \varepsilon} \right].$$

Such kind of theory was applied of the behavior of the armor, due to an impact from a fragment simulating projectile (Mines [2004]). Good agreement between the results from the literature and the stress wave model, is achieved, and the development model is then used to gain insight into the effect of the variation of various parameters.

Secondly plastic shock waves can appear if the boundary conditions and the initial conditions are not equal at the impacted end, and the constitutive equation is linear (elastic, linear work-hardening, etc.). In the second case a shock wave can appear when the velocity of propagation is increasing during propagation. This case can happen for instance if the bar is heated non-uniformly and the velocity of propagation is increasing during propagation [either E is increasing during propagation, or even $c(\varepsilon)$]. A third case is the one due to a particular material property. For instance for rubber the stress–strain curve is as shown in Fig. 3.5.1. For such material the constitutive equation is (Cristescu [1957])

$$\sigma = \alpha\varepsilon + \beta\varepsilon^2 + \gamma\varepsilon^3, \tag{3.5.1}$$

with α, β, γ constant. This relation is represented in Fig. 3.5.1. The curvature of the curve is changing at a certain point m. If we compute the envelope of the

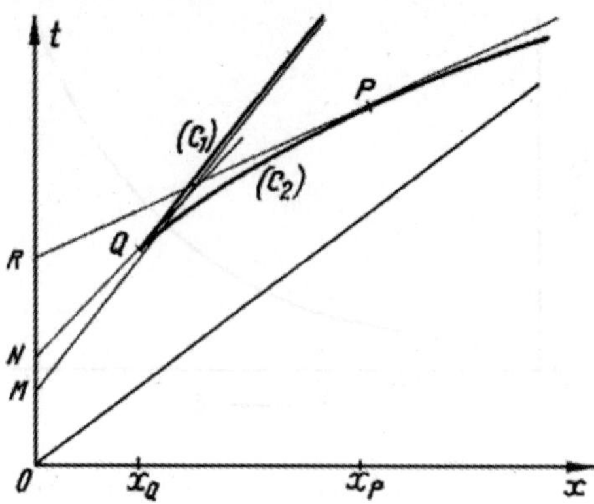

Fig. 3.5.2   Envelope of characteristics for an S-shape curve.

smooth plastic wave fronts for a linear variation of the strain at the end of the bar, we obtain a curve which possesses a cusp.

The wave fronts starting after point n, form an envelope. This envelope is a shock wave front traveling with variable velocity (Fig. 3.5.2).

Such a case is shown in the test by Kolski [1969]. He has tested the propagation of finite tensile pulses traveling along bands of highly stretched vulcanized natural rubber. A band of vulcanized natural rubber gum stock of square cross-section (1/2 inch × 1/2 inch in the unstretched state) was prestrained to five times its original length. A 13 foot length was clamped at opposite ends of an optical bench and a section 10 inches long, at one end of the stretched rubber, was then stretched even further. The additional stretch was maintained by a piece of steel piano wire. When this wire was suddenly volatilized by a heavy electric current, a tensile pulse traveled along the stretched rubber, and the velocity-time profiles of the pulse were observed on the cathode-ray oscilograph. Light wires attached to the rubber cut the lines of force of constant strength magnetic fields which were produced by permanent magnets. Small amplitude pulses (< 10 per cent) propagated along the rubber without measurable change in shape, but the front of larger pulses became sharper as they progressed along the band. Fig. 3.5.3 shows the change in shape of the velocity-time profile when the 10 inch length was extended to 11 inches, so that the strain in this section was 450 per cent (additional strain 50 per cent). When the pulse has traveled about nine feet, it may be regarded as a shock wave.

Shape memory alloys have recently been considered for dynamic loading applications for energy absorbing and vibration damping devices. Such a body considered by Lagoudas *et al.* [2003], is subjected to external dynamic loading, will experience large inelastic deformations that will propagate through the body as phase transformation and/or detwinning shock waves. Numerical solutions for various boundary conditions are presented for stress induced martensite and detwinning of martensite. The dynamic response of a rod is also studied experimentally in a split Hopkinson bar apparatus under detwinning conditions.

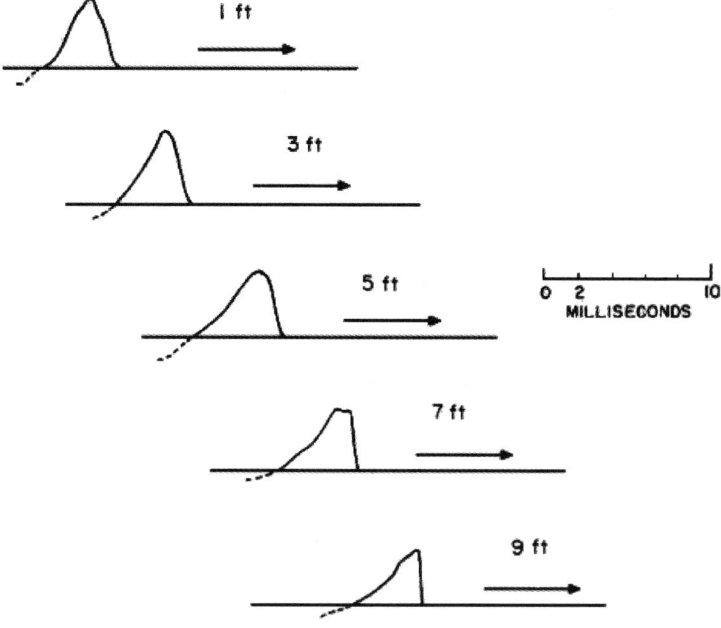

Fig. 3.5.3   Experimental formation of a shock wave (Kolski [1969]).

Elementary theory of shock waves.
We consider the process as an adiabatic one. We disregard any thermal effect. We consider a bar in tension $\varepsilon > 0$, $\sigma > 0$. We write the displacement on one side and another side and because this displacement is the same

$$(du)^- = v^- dt + \varepsilon^- dx$$

$$(du)^+ = v^+ dt + \varepsilon^+ dx$$

or

$$[v] = -c[\varepsilon]. \tag{3.5.2}$$

This is the kinematics condition.

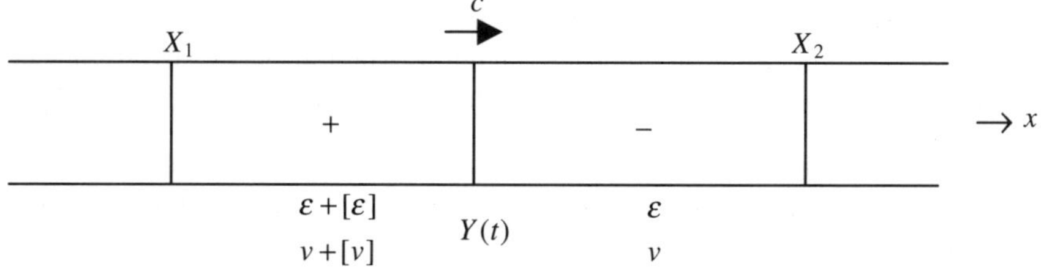

Fig. 3.5.4   Formation of shock waves.

We write now the balance of linear momentum. We have

$$\frac{d}{dt}\int_{X_1}^{Y(t)}\rho v(X,t)\,dX + \frac{d}{dt}\int_{Y(t)}^{X_2}\rho v(X,t)\,dX$$

$$= \int_{X_1}^{X_2}\rho b(X,t)\,dX + \sigma(X_2,t) - \sigma(X_1,t)$$

or making the derivatives

$$\rho v^+(Y(t),t)\dot{Y}(t) + \int_{X_1}^{Y(t)}\rho\frac{\partial v(X,t)}{\partial t}\,dX - \rho v^-(Y(t),t)\dot{Y}(t) + \int_{Y(t)}^{X_2}\rho\frac{\partial v(X,t)}{\partial t}\,dX$$

$$= \int_{X_1}^{X_2}\rho b(X,t)\,dX + \sigma(X_2,t) - \sigma(X_1,t)\,.$$

We introduce new notations when $X_1 \to Y(t)$, $X_2 \to Y(t)$ and

$$v^+(Y(t),t) = \lim v(X,t)$$

$$X \to Y(t)$$

$$X < Y(t)$$

and the velocity of propagation $c = \dot{Y}(t)$. With these notations we obtain

$$c\rho[v] = -[\sigma] \tag{3.5.3}$$

which is the dynamic jump condition. We obtain also from here

$$c = \sqrt{\frac{1}{\rho}\frac{[\sigma]}{[\varepsilon]}} \tag{3.5.4}$$

for the velocity of propagation. It is certainly not independent and does not coincide with the velocity of the acceleration waves.

We have a total of four unknown functions σ, ε, v, c but we have only three equations. Another condition is obtained from boundary conditions or some other condition which may result from the problem.

This is a very simple theory, which replaces the differential relations with algebraic linear equations. It allows an elementary study of the interaction between elastic (loading and unloading) and plastic waves. It also allows a very simple description of reflection and refraction of waves.

Reflection of waves. When waves are reflected from the end of the bar, many situations may arise, depending on whether the end of the bar is free, fixed, or whether the bar continues with another bar, with different material proprieties.

If an impact is produced at one end $x = -l$ of the bar, the first wave to reach the other end $x = 0$ is always an elastic wave and $\sigma = -\rho c_0 v$. The displacement due to this elastic wave is

$$u_1 = f_1(c_0 t - x)\,, \tag{3.5.5}$$

where f_1 is an arbitrary function to be determined by the initial conditions. The corresponding stress is

$$\sigma_1 = E(\partial u_1/\partial x) = -\rho c_0(\partial u_1/\partial t) \tag{3.5.6}$$

for the inverse waves (propagating in the opposite direction)

$$u_2 = f_2(c_0 t + x), \tag{3.5.7}$$

and

$$\sigma_2 = E(\partial u_2/\partial x) = \rho c_0(\partial u_2/\partial t). \tag{3.5.8}$$

Let us consider the case when the end $x = 0$ of the bar is free. After the reflection of the elastic wave, the total stress will be

$$\sigma = \sigma_1 + \sigma_2 = E\{f_2'(c_0 t + x) - f_1'(c_0 t - x)\}. \tag{3.5.9}$$

But at the end $x = 0$ the stress vanishes for all t

$$f_2'(c_0 t) - f_1'(c_0 t) = 0. \tag{3.5.10}$$

One can therefore conclude that $f_1(z) = f_2(z) = f(z)$, and the displacement after reflection will be

$$u = u_1 + u_2 = 2f(c_0 t). \tag{3.5.11}$$

Furthermore, due to reflection at the free end the displacement produced by the wave is doubled, as is also the velocity. With regard to stress, we have in this case $\sigma_1 = -\sigma_2$, i.e., a compressive wave is reflected as a tensile one and vice versa.

If we now assume that the second end of the bar is fixed rigid, then the displacement of this end is zero for all t

$$f_1(c_0 t) + f_2(c_0 t) = 0,$$

and therefore $f_1(z) = -f_2(z)$, i.e., the displacement produced by the reflected wave is equal to that produced by the incident wave, but in the reverse direction. This stress

$$\sigma = \sigma_1 + \sigma_2 = 2E f_2'(c_0 t) \tag{3.5.12}$$

is doubled due to reflection from the fixed end. For this reason, an elastic wave can be reflected a plastic one, if on reflection, the stress carried by the reflected wave exceeds the yield stress.

If the bar is continued with another one, possessing different mechanical properties the situation is more complicated: a reflected wave of intermediate intensity will propagate from the end of the bar, but at the same time a wave will also be reflected in the other medium (see Kolsky [1953]).

The previous arguments are generally also valid for plastic waves.

The finite bar. Many authors have considered the problem of normal impact of a finite bar against a rigid target. Let us examine the problem as done by Lenski [1949]. A bar of length l moves with velocity V from right to left (Fig. 3.5.4).

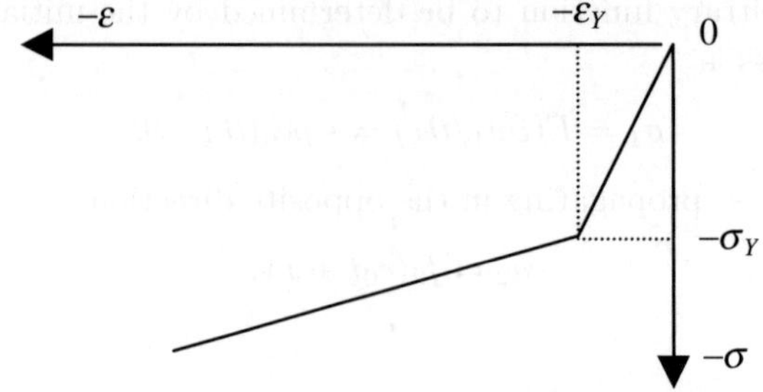

Fig. 3.5.5   Elastic linearly work-hardening plastic bar striking a rigid target.

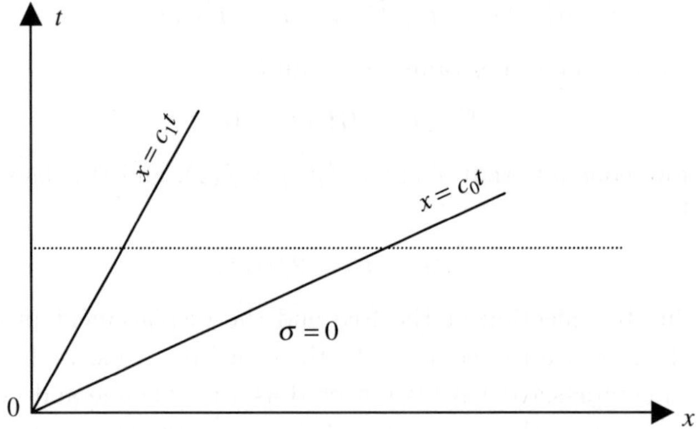

Fig. 3.5.6   The first waves to propagate in the bar.

At time $t = 0$ the bar strikes a rigid wall $x = 0$. It is assumed that the impact velocity is sufficiently high to produce elastic-plastic strains and that the material of the bar is linear work-hardening. Due to the sudden impact and linear work-hardening only shock waves are considered. Thus, after impact two shock waves, one elastic and one plastic will propagate. Their wave fronts delimitate three distinct regions in the bar.

In the first region (Fig. 3.5.6) $\varepsilon_0 = 0$, $v_0 = -V$, in the second region $\varepsilon_1 = -\varepsilon_Y$, $\sigma_1 = -E\varepsilon_Y$ and $v_1 = -V + c_0\varepsilon_Y$. In the third region $v_2 = 0$ and

$$\varepsilon_2 = (1/c_1)(c_0\varepsilon_Y - V) - \varepsilon_Y, \quad \sigma_2 = (E_1 - E)\varepsilon_Y + E_1\varepsilon_2. \tag{3.5.13}$$

Therefore, for the plastic strains to appear, the velocity of impact must satisfy the condition $V > c_0\varepsilon_Y$.

After the time $t = l/c_0$ the elastic wave front is reflected from the free end. Behind this front

$$\sigma_3 = \varepsilon_3 = 0, \quad v_3 = 2c_0\varepsilon_Y - V. \tag{3.5.14}$$

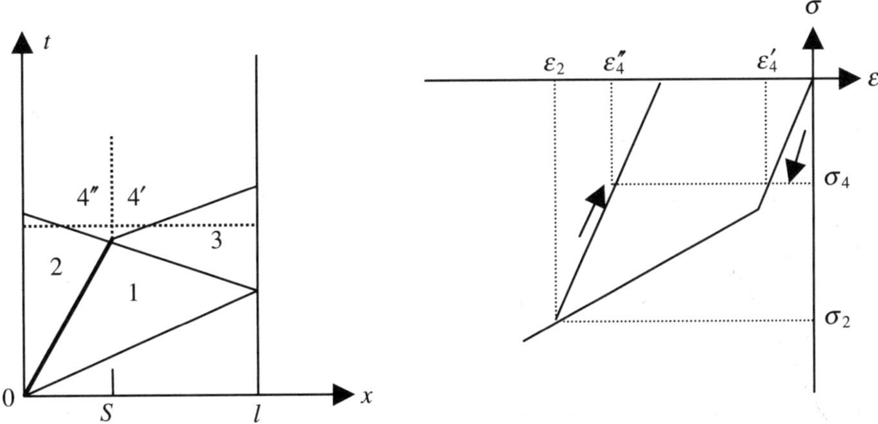

Fig. 3.5.7   Stresses and strains during unloading after the first reflexion.

When the wave fronts of the plastic and of the elastic unloading waves meet at the section S (Fig. 3.5.7) several situations are possible, depending on the magnitude of the velocity V. Let us consider first the case when, after the two wave fronts have met, the plastic wave does not propagate any more. In this case the elastic waves will spread in both directions from S (Fig. 3.5.10). The leftward wave is always an elastic one. On both sides of the section S the particle velocities are equal to $v'_4 = v''_4$, otherwise the bar will break. From the balance of the force it also follows that, $\sigma'_4 = \sigma''_4$ (the stresses are initial stresses). The strains in the two regions will be different. In the region $4'$ the material is in an elastic state, the yield point was never reached, while in the region $4''$ the material is also in an elastic state, but is being unloaded after a plastic deformation. This situation is shown in Fig. 3.5.7. Thus the stress–strain relation that acts on the two sides of the section S is

$$\sigma = E\varepsilon \quad \text{and} \quad \sigma = \sigma_2 + E(\varepsilon - \varepsilon_2). \tag{3.5.15}$$

From these relations and the conditions $\sigma'_4 = \sigma''_4$, $v'_4 = v''_4$ and the jump conditions, we obtain

$$\varepsilon'_4 = \frac{c_0 + c_1}{2c_0^2}(c_0\varepsilon_Y - V), \quad \varepsilon''_4 = \frac{2c_0^2 + c_0c_1 - c_1^2}{2c_0^2 c_1}(c_0\varepsilon_Y - V),$$

$$v'_4 = v''_4 = \frac{3c_0 - c_1}{2c_0}(c_0\varepsilon_Y - V) + V.$$

Thus the strain is discontinuous across the section S. This is a non-propagable discontinuity surface.

If the plastic wave stops at S (Fig. 3.5.10), the strain for $x > x_S$ must be bigger than the strain corresponding to the yield stress $\varepsilon'_4 > -\varepsilon_Y$. From this condition it follows that

$$V < \left(1 + \frac{2c_0}{c_0 + c_1}\right)\varepsilon_Y c_0.$$

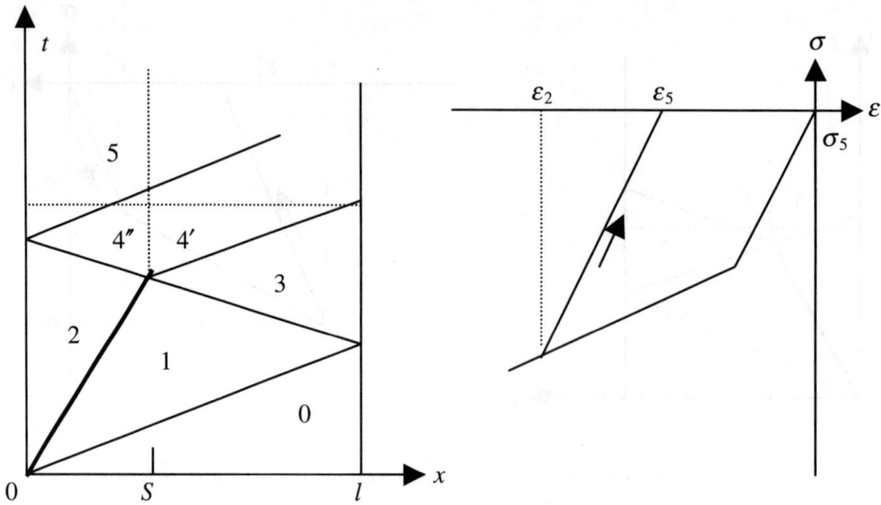

Fig. 3.5.8   Possible separation from the wall at the first reflection.

If we have also $v_4' = v_4'' > 0$ when the elastic wave propagating to the left reaches the end of the bar, this end can possibly be pulled away from the wall (Fig. 3.5.8). If the separation takes place $\sigma_5 = 0$ and

$$\varepsilon_5 = -\varepsilon_P = \frac{c_0^2 - c_1^2}{c_0^2 c_1}(c_0 \varepsilon_Y - V), \quad v_5 = 2c_0\varepsilon_Y - V,$$

where ε_P is the plastic strain. The plastic strain is constant in the part of the bar to the left of section S, whose length is

$$x_S = 2c_1 l/(c_0 + c_1).$$

From here we obtain also the separation condition $V < 2c_0\varepsilon_Y$. The time during which the bar is in contact with the wall is $2l/c_0$. After the separation the bar will have a sudden change of diameter.

If the inequality

$$2c_0\varepsilon_Y < V < \left(1 + \frac{2c_0}{c_0 + c_1}\right)c_0\varepsilon_Y$$

is satisfied, then the elastic waves will again propagate from S in both directions, but the bar will continue to touch the wall for a little longer: $4l/(c_0 + c_1)$. Now the elastic wave propagating to the right is reflected and when it arrives at the impact end separation takes place.

If finally the inequality

$$V > \left(1 + \frac{2c_0}{c_0 + c_1}\right)c_0\varepsilon_Y$$

holds, then $\varepsilon_4' < -\varepsilon_Y$ and the plastic wave continues to propagate to the right from the section S (Fig. 3.5.10). There are two possibilities. Either the elastic wave which

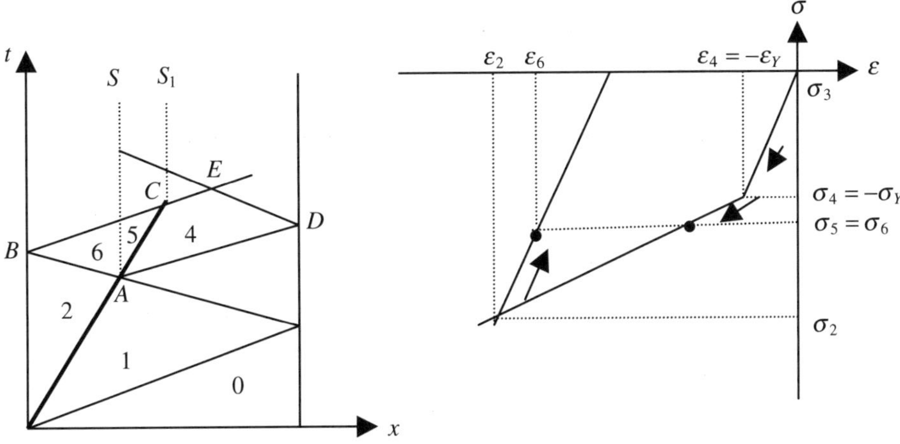

Fig. 3.5.9   Characteristic field for the case when two stationary discontinuities occur.

propagates to the left from S is reflected from the end $x = 0$ and then meets the plastic wave at S_1, which in the meantime continues to propagate. Or the elastic wave which propagates from S to the right and is reflected from the end $x = l$ meets the plastic wave first. That depends on the velocities of propagation of the elastic and plastic waves. In Fig. 3.5.9 the first of these two possible cases is represented, when $c_0/c_1 > 4.236$. Now the length of the plastically deformed portion of the bar, the abscissa of the section S_1, is

$$x_{S_1} = 2lc_1/(c_0 + c_1).$$

The magnitudes of the strain and velocity in regions 1, 2 and 3 coincide with those given in the previous example. In region 4, since $\varepsilon = -\varepsilon_Y$, it follows that $v_4 = 3c_0\varepsilon_Y - V$. The stresses in regions 5 and 6 are equal and satisfy the relations:

$$\sigma_5 + \sigma_Y = E(\varepsilon_5 + \varepsilon_Y) \quad \text{and} \quad \sigma_6 - \sigma_2 = E(\varepsilon_6 - \varepsilon_2).$$

Using the momentum equation for shock waves together with the condition $v_6 = v_5$ and the previous relation we obtain (Fig. 3.5.10)

$$\varepsilon_5 = \frac{3c_0^2 - c_1^2}{c_1(c_0 + c_1)}\varepsilon_Y - \frac{V}{c_1}, \quad \varepsilon_6 = \frac{c_1^2 + c_0^2}{c_1(c_0 + c_1)}\varepsilon_Y - \frac{V}{c_1},$$

$$v_5 = v_6 = \frac{2c_0c_1}{c_0 + c_1}\varepsilon_Y.$$

Finally the bar will possess two stationary discontinuity fronts. If necessary the analysis can be continued in a similar way for greater impact velocities.

Many such kinds of problems, described only by shock waves, can be found in the papers by De Juhasz [1949a], [1949b] and in the book by Rakhmatulin and Demianov [1961].

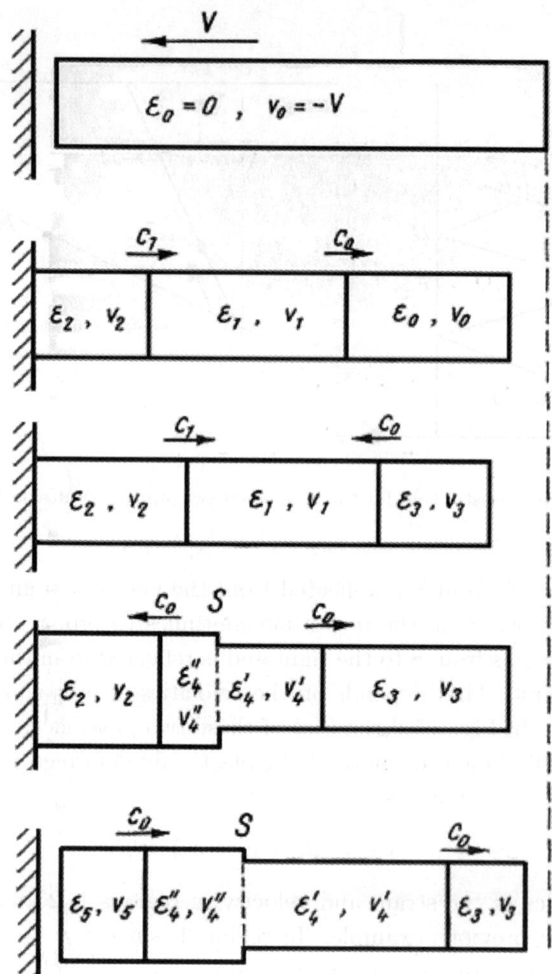

Fig. 3.5.10   Elastic linearly work-hardening plastic bar striking a rigid target.

3.6   Examples

The examples given are the impact of two identical aluminum 1100 bars (Cristescu and Bell [1970], Cristescu [1970]). The bars are 25 cm in length and one inch in diameter. The experimental results are due to J. F. Bell published in several papers and the books by Bell [1968, 1973].

Numerical integration method. The constitutive equation is written $\sigma = F(\varepsilon)$ or in an inverse $\varepsilon = F^{-1}(\sigma)$. F is a strictly increasing function. We consider σ and ε positive in compression. The density is $\rho = 0.002757$ g cm^{-4} sec^2. The plastic deformation is $\varepsilon^P = F^{-1}(\sigma) - \sigma/E$. The characteristics are

$$\frac{dx}{dt} = \pm c = \pm\sqrt{\frac{E}{\rho(1 + E\varphi)}}$$

(3.6.1)

and

$$d\sigma = \mp \rho c \, dv \, .$$

(3.6.2)

The differential relations along the characteristics are:

$$dx = 0 \quad \text{(twice)}$$

(3.6.3)

and

$$d\varepsilon^P = \varphi d\sigma$$

$$E d\varepsilon^E = d\sigma \, .$$

(3.6.4)

The constitutive equation is written as

$$\dot\varepsilon = \frac{\dot\sigma}{E} + \chi\left[(F^{-1}(\sigma))' - \frac{1}{E}\right]\dot\sigma = \frac{\dot\sigma}{E} + \varphi(\sigma)\dot\sigma$$

and

$$\chi = \begin{cases} 1 & \text{if } \sigma = \sigma_m(x) \\ 0 & \text{if } \sigma \leq \sigma_Y \text{ or } \sigma_Y < \sigma < \sigma_m(x) \, . \end{cases}$$

(3.6.5)

We introduce the dimensionless variables

$$S = \frac{\sigma}{E} \, , \quad V = \frac{v}{c_0} \, , \quad T = kt \, , \quad X = \frac{kx}{c_0} \, .$$

Now we choose the highest value of dx/dt in the whole field, i.e., $dx/dt = \pm c_0$ or $dX/dT = \pm 1$ the bar velocity, with $c_0 = 5080$ m/sec.

The principle of the method is an integration along characteristic lines. We will not give details. The method is convergent since the monotone bounded sequence

$$\sqrt{\frac{F'(\varepsilon)}{\rho}} < c^{(i)}(\sigma, \varepsilon) \leq c^{(i-1)}(\sigma, \varepsilon) \leq \cdots \leq c_0$$

is always satisfied. It is very rapid, since we do 2-3 iterations in one point.

Initial and boundary conditions. The initial conditions are for the specimen (S):

$$\left.\begin{matrix} t = 0 \\ 0 \leq x \leq l \end{matrix}\right\} : \sigma = \varepsilon = v = 0$$

and for the hitter (H):

$$\left.\begin{matrix} t = 0 \\ -l \leq x \leq 0 \end{matrix}\right\} : \sigma = \varepsilon = 0 \, , \quad v = V \, .$$

The boundary conditions are, if $v(0, t)$ is prescribed:

$$x = l \, , \quad t \geq 0 : \sigma = 0 \, ,$$

$$x = -l \, , \quad t \geq 0 : \sigma = 0 \, ,$$

$$x = 0 \begin{cases} 0 \leq t \leq t^* = \dfrac{2l}{c_0} : v_H = v_S = v_{\max} = \dfrac{V}{2} \quad \text{or} \quad \sigma_H = \sigma_S = \sigma_{\max}\,, \\[2ex] \quad t^* \leq t < T_c : v_H = -v_S \quad \text{symmetry of motion}\,, \\[2ex] \quad\quad T_c \leq t : \sigma_S = 0\,. \end{cases}$$

Here T_c is the time of contact between the two bars. The boundary conditions can be prescribed for $\sigma(0,t)$ if the stress is measured in a hard transmitter bar or with a piezocrystal, and $v = -\varphi(\sigma)$.

The initial conditions are formulated to make economy of computer time and to avoid the singular point O. The first initial conditions are named S from shock.

An initial shock wave is propagating and reflecting from the free end (see Fig. 3.6.1). Thus AR is a tensile unloading shock wave. For a constitutive equation of the form $\sigma = \beta \varepsilon^{1/2}$ we have along this line:

$$x = -c_0(t - t^*) \quad \text{and} \quad x = c(\sigma)t = \sqrt{\dfrac{1}{\rho} \dfrac{\beta^2}{2\sigma}}\, t\,.$$

From here we get:

$$\sigma_b(t) = \frac{\beta^2}{2\rho c_0^2} \frac{t^2}{(t^* - t)^2}$$

$$v_b(t) = \frac{2}{3\beta} \sqrt{\frac{2}{\rho}} \sigma^{3/2} + \left(\frac{1}{\rho c_0} - \frac{2}{3\beta} \sqrt{\frac{2}{\rho}} \sqrt{\sigma_Y} \right) \sigma_Y \tag{i}$$

where "b" stands for "before". We now use the jump condition to pass at "after" the shock.

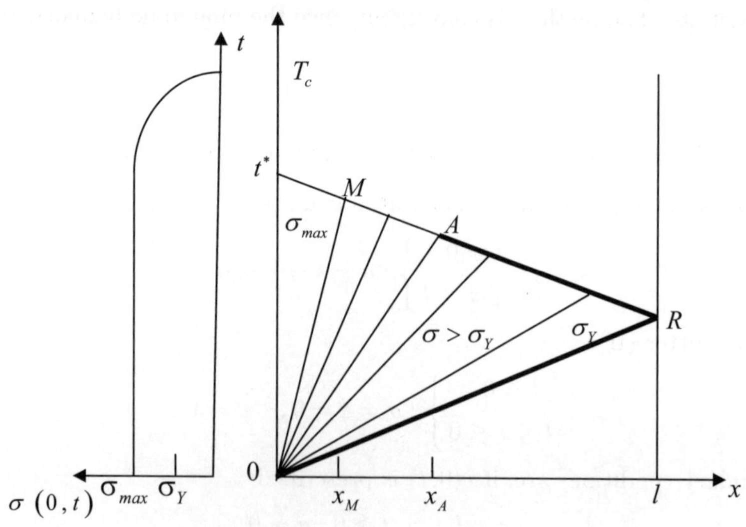

Fig. 3.6.1   The initial condition S.

We have:

$$\text{jump condition}: \sigma_a - \sigma_b = -\rho c_0(v_a - v_b),$$

$$\text{after}: -\sigma_a + \rho c_0 v_a = const.,$$

$$\text{in } R: \sigma_b = \sigma_Y = \rho c_0 v_b, \ \sigma_a = 0.$$

From here we get:

$$\sigma_a - \sigma_b = -\frac{1}{2}\{2\sigma_Y - (\rho c_0 v_b - \sigma_b)\},$$

$$v_a = v_b - \frac{1}{\rho c_0}(\sigma_a - \sigma_b), \tag{ii}$$

$$\varepsilon_a^E = \frac{\sigma_a}{E}, \quad \varepsilon_a^P = \varepsilon_b^P.$$

Though along the line $x = -c_0(t - t^*)$ for the data "after" we have:

$$0 \leq x \leq x_M : \sigma = \sigma_{\max} = const.$$

$$x_M \leq x < x_A : \text{(i)},$$

$$x_A \leq x \leq l : \text{(ii)}.$$

These are the initial data along the reflected shock wave.

We have tested also a second type of initial data called L (linear). There are no shocks now and we have the boundary conditions:

$$x = 0 \begin{cases} 0 \leq t \leq t_m = 20\ \mu \sec : \sigma = \dfrac{t}{t_m}\sigma_{\max}, \\[2mm] t_m \leq t < T_c : v_H = v_S, \\[2mm] T_c \leq t : \sigma_S = 0. \end{cases}$$

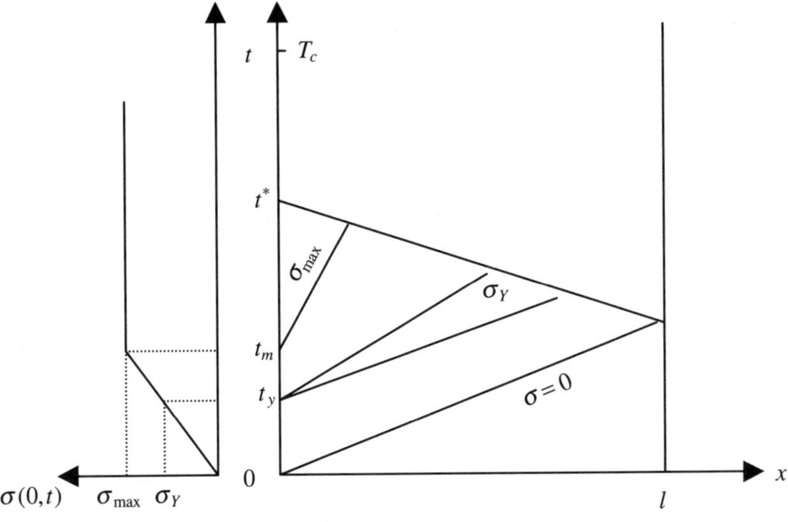

Fig. 3.6.2   The initial boundary conditions L.

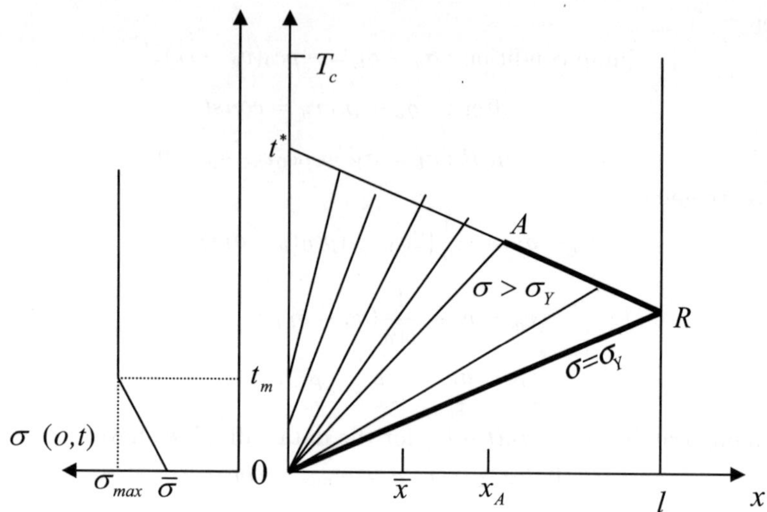

Fig. 3.6.3   The initial data of a mixed type M.

Now we get along the line:

$$x = -c_0(t - t^*) \quad \text{and} \quad x = \sqrt{\frac{1}{\rho}\frac{\beta^2}{2\sigma}}\left(t - \frac{t_m}{\sigma_m}\sigma\right).$$

From here:

$$\sqrt{\sigma_b} = \frac{-\sqrt{2\rho}c_0(t^* - t) + \sqrt{2\rho c_0^2(t^* - t)^2 + 4\beta^2(t_m/\sigma_{max})t}}{2\beta(t_m/\sigma_{max})} \tag{i}$$

and a similar formula for v_b.

A mixed initial condition called M (mixed) has also been considered. It is a one dimensional simulation of the three dimensional wave initiations at the end of the bar: a sudden shock up to $\bar{\sigma} = 0.794\sigma_{max}$ after which follows a linear variation up to σ_{max} reached at $t_m = 40\ \mu$ sec. Thus

$$x = 0 \begin{cases} 0 \leq t \leq t_m = 40\ \mu \sec : \sigma = \bar{\sigma} + \dfrac{t}{t_m}(\sigma_m - \bar{\sigma})\,, \\[2mm] \qquad t_m \leq t < T_c : v_H = v_S\,, \\[2mm] \qquad \qquad T_c \leq t : \sigma_S = 0\,. \end{cases}$$

We have to find $\bar{x}$ from

$$x = \sqrt{\frac{1}{\rho}\frac{\beta^2}{2\bar{\sigma}}}\,t \quad \text{and} \quad x = c_0(t^* - t)\,.$$

We obtain

$$\bar{x} = \frac{c_0 t^*}{1 + c_0\sqrt{2\rho\bar{\sigma}}/\beta}\,,$$

from where $\bar{x} = 2.64D < x_A$ for V $= 40$ m/sec.

For

$$0 \le x \le \bar{x} : \begin{cases} x = c_0(t^* - t) \\[2mm] x = \sqrt{\dfrac{1}{\rho}\dfrac{\beta^2}{2\bar{\sigma}}} \end{cases}$$

we have:

$$\sqrt{\sigma_b} = \sqrt{\sigma_a}$$

$$= \frac{-c_0\sqrt{2\rho}(t^* - t) + \sqrt{2\rho c_0^2(t^* - t)^2 + \{4\beta^2 t_m/(\sigma_{\max} - \bar{\sigma})[t - \bar{\sigma} t_m/(\sigma_{\max} - \bar{\sigma})]\}}}{2\beta t_m/(\sigma_{\max} - \bar{\sigma})}$$

and a similar formula for v_a. In the interval where the reflected wave is still felt:

$$\bar{x} \le x \le l : \sigma_b = \frac{\beta^2}{2\rho c_0^2}\frac{t^2}{(t^* - t)^2} \quad \text{if} \quad \begin{cases} \sigma_b > \sigma_a \to \text{(ii)} \Rightarrow \sigma_a \\[2mm] \sigma_b \le \sigma_a \to \text{(i)} \Rightarrow \sigma_a \, . \end{cases}$$

Concerning the constitutive equation we have used three. First was used the dynamic curve

$$\sigma = E\varepsilon \, , \qquad \sigma = \sigma_{Y1} = 307.5 \text{ psi}$$
$$\sigma = \beta\varepsilon^{1/2} \, , \qquad \sigma \ge \sigma_{Y1} \, .$$

with

$$\beta = 5.6 \times 10^4 \text{ psi} \, ,$$
$$E = 10.2 \times 10^6 \text{ psi} \, ,$$
$$c_0 = 5080 \text{ m/sec} \, ,$$
$$\varepsilon_{Y1} = 0.00003014 \, .$$

That was the fundamental constitutive equation. Since the yield stress was not correct, we have translated the parabola to get the correct yield stress. Thus

$$\sigma = E\varepsilon \, , \qquad\qquad \sigma \le \sigma_{Y2} = 1100 \text{ psi}$$
$$\sigma = \beta(\varepsilon + \varepsilon_0)^{1/2} \, , \qquad \sigma \ge \sigma_{Y2}$$

and $\varepsilon_{Y2} = 0.00010784$, and $\varepsilon_0 = 0.0002779988$ (the translation on horizontal).

The quasi-static stress–strain curve is $\sigma = 3.32 \times 10^4 \varepsilon^{3/8}$ from where:

$$\sigma = E\varepsilon \, , \qquad\qquad \sigma \le \sigma_{Y0} = 1068.24 \text{ psi}$$
$$\sigma = 3.32 \times 10^4 \varepsilon^{3/8} \, , \qquad \sigma \ge \sigma_{Y0} \, .$$

In the examples $\sigma_{\max}$ is given while $v_{\max}$ is computed.

First comparison was for the time of contact T_c which is obtained by observing the interruption of the longitudinal beam of light from a source on one side of the impact faces before, during, and after collision, by means of a photomultiplier tube on the opposite side (Fig. 3.6.4).

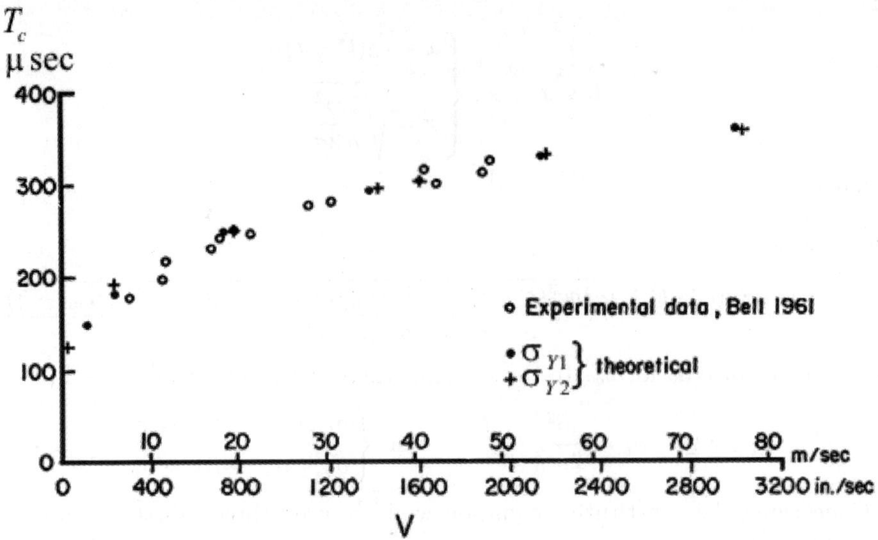

Fig. 3.6.4   Time of contact versus velocity; a comparison of theoretical and experimental data.

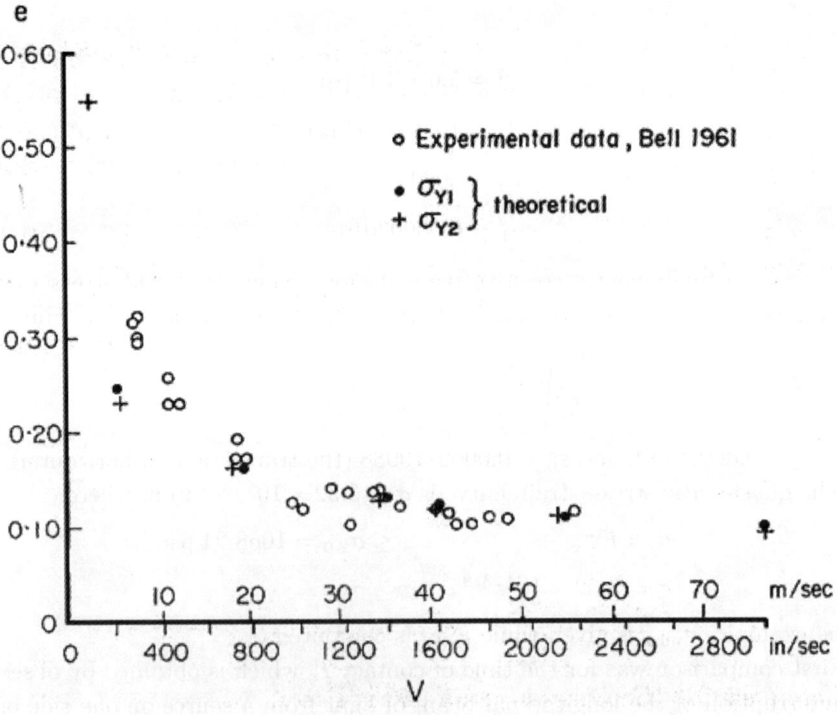

Fig. 3.6.5   Coefficient of restitution versus impact velocity.

A second comparison with the experimental data is the coefficient of restitution defined $e = (2v_f - V)/V$ with v_f the final velocity (travel time of the free end of specimen after separation) (Fig. 3.6.5).

The characteristic planes for the impact for more than 34 m/sec are given in Fig. 3.6.6. L means "loading", while U means "unloading". The time of contact coincides with the experiment.

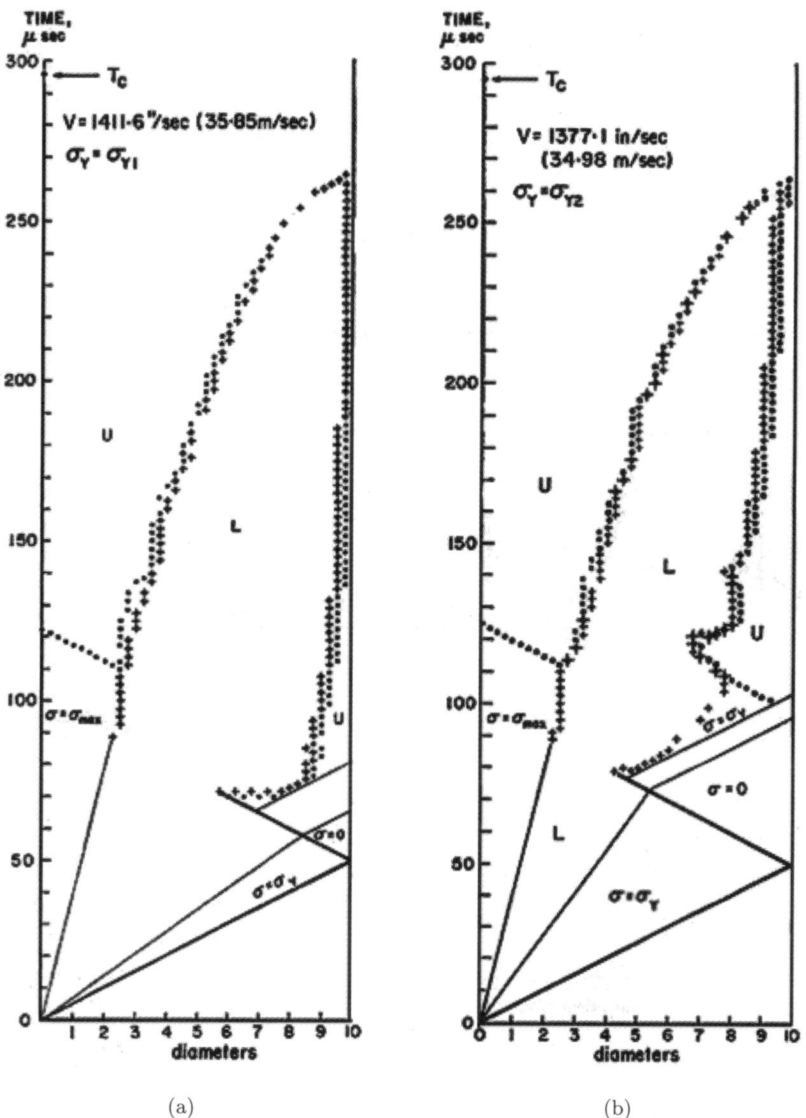

(a) (b)

Fig. 3.6.6 (a) Characteristic plane for $\sigma_Y = \sigma_{Y1}$ and $V = 35.85$ m/sec. (b) Characteristic plane for $\sigma_Y = \sigma_{Y2}$ and $V = 34.98$ m/sec.

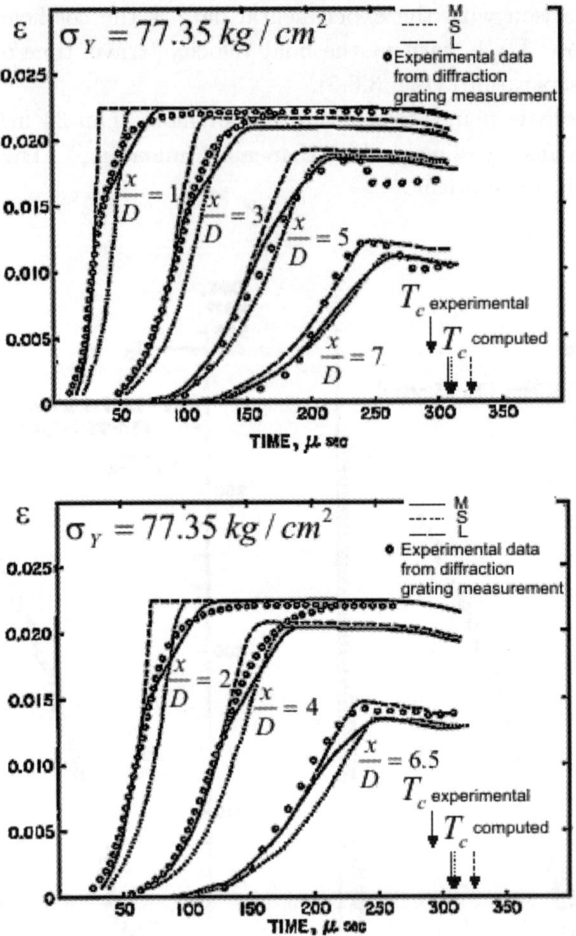

Fig. 3.6.7   A comparison of computed strain-time profiles with experimental data.

The best comparison with the experiment is on the graph of variation of strain at various cross sections. The graph given is only for the higher yield stress. All three theoretical solutions are shown. Also the time of contact is given (Fig. 3.6.7).

The displacement variation is given in Fig. 3.6.8 for the two yield stresses and the three initial boundary conditions. The diameters where the variation is measured are shown.

The variation of the stress at the impacted end is shown in Fig. 3.6.9. The initial rise of the stress at the impacted end is not given. The stress plateau is shown very clearly.

One has compared also the cross section x_A when the reflected shock wave is completely absorbed. Both theoretically and experimentally one have found x_A between 4 and 5 diameters.

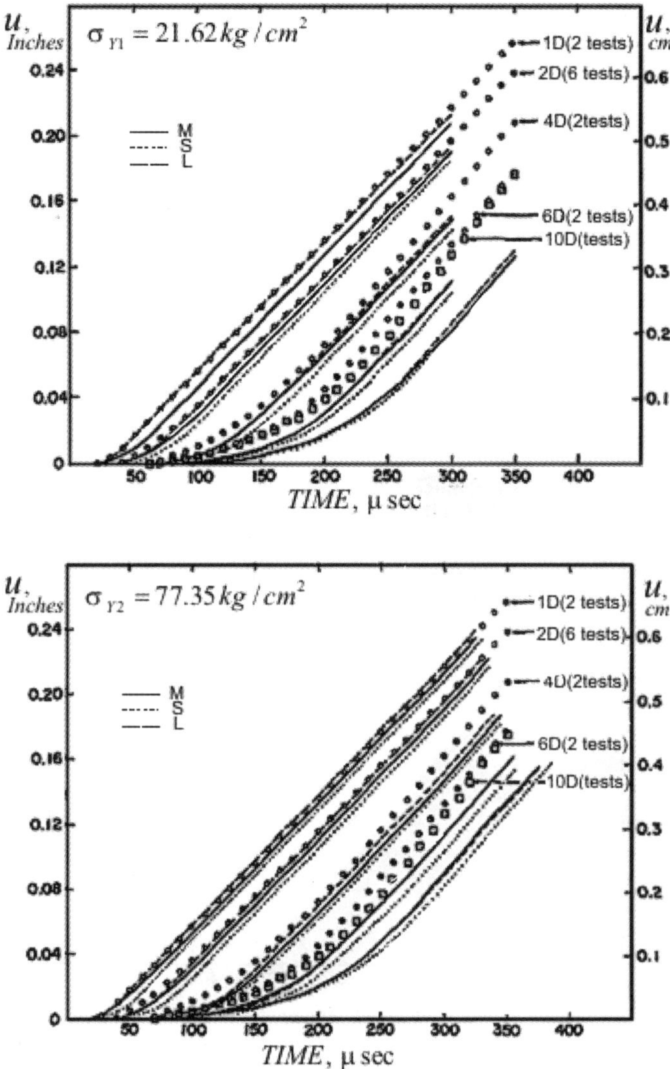

Fig. 3.6.8 Computed displacement-time curves compared with experimental values obtained with an optical displacement technique.

The velocity of propagation is checked by Swegle and Ting [1982]. They have tested if the waves travel with the velocity predicted by the one-dimensional theory, although they have larger rise-times. That was what they have found.

In a paper by Jones *et al.* [1987], a one-dimensional analysis is presented that leads to the appropriate equation of motion for the undeformed portion of a plastic, rigid rod after impact with a rigid anvil. This equation is used as a basis for deducing material properties of the rod material from post-test measurements. Data

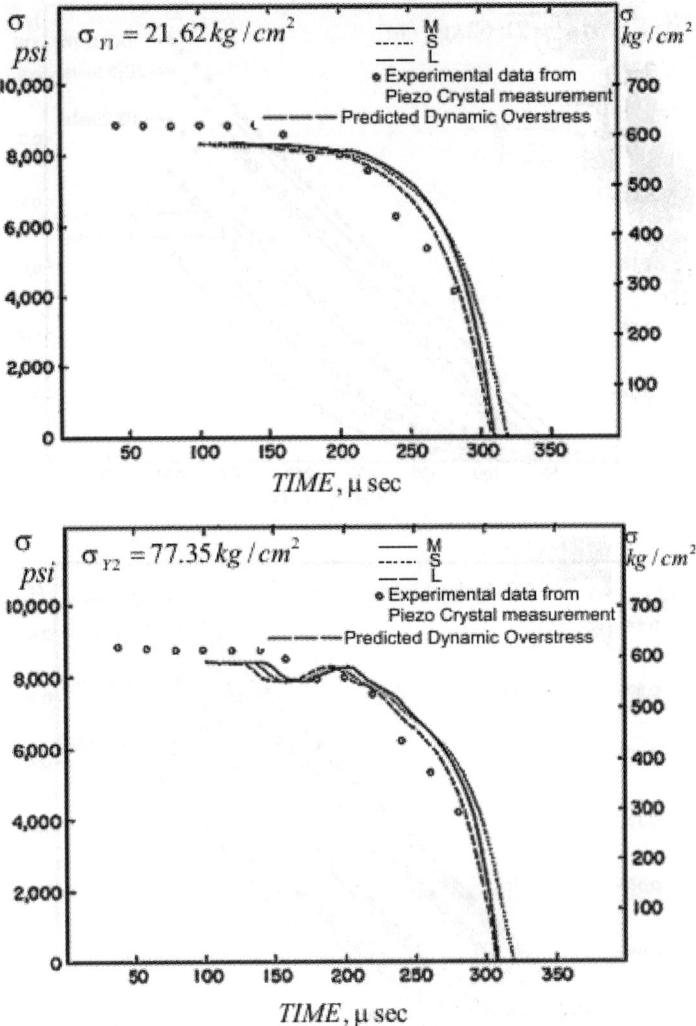

Fig. 3.6.9   Variation of stress in time at the impact end of the bar. The experiment data were obtained with piezo crystal measurements. The dynamic overstress is not included in the present discussion.

for copper, for depleted uranium, for iron and steel are given for velocity surpassing 300 m/s.

The impact of a solid cylinder (aluminum) by a compressible fluid (water) is considered by Lush [1991]. The speeds of impact are 10, 100 and 350 m s^{-1}. The main modification of the model is to incorporate the unloading of the plastic wave which is produced by the multiple reflections of the elastic release wave. In general the agreement between penetration, rate, contact pressure and depth of penetration is good.

The formation, propagation, and reflection of shock waves is studied for synthetic curved cables by Tjavaras and Triantafyllou [1996]. First, analytical solutions are derived for weightless cables with nonlinear stress–strain relation, which are then compared with numerical solutions. For very small curvature the analytical and numerical solutions show that synthetic lines can form shock waves when loaded impulsively. The tension is amplified significantly when a reflection occurs, pointing to a probable mechanism of failure. The examples are on Catenary.

In another paper by Nonaka *et al.* [1996] the integration is done again, trying to explain some damage that appeared after earthquakes. One considers finite thin elastic bars and assumes that the second end is free, then fixed, and in final assuming that at the second end the rod is miss-matched impedance. The mechanical impedance is expressed by the change of the velocity by $\alpha \rho c$ with $\alpha > 0$. For α very large one obtains various examples for steel: E $= 206$ GPa, $\rho = 7850$ kg/m^3, c $= 5120$ m/s.

Several strengths measurements are presented by several papers by Kanel [1999]. The presented experimental data for engineering metals and alloys and for metals single crystals demonstrate the effects of material grain structure and temperature on the resistance to high-rate rupture. Several figures are given, showing this influence.

A technique for rapid two-stage dynamic tensile loading of polymers, based on a tensile Hopkinson bar apparatus, was developed by Shim *et al.* [2001]. In this technique, the initial incident wave and its reflection are used to load a specimen in quick succession. Consequently, the specimen is stressed, momentarily unloaded, and then reloaded until fracture. A procedure to obtain the associated stress–strain curves for such double-stage loading is formulated. This procedure is examined experimentally and analytically to substantiate their validity. To verify the proposed approach, a relatively rate-insensitive material was subjected to two-stage dynamic tension. The stress–strain curves obtained via the procedure established were compared with results from static loading.

A flat-nosed cylinder moving at a sufficiently high impact velocity is assumed to fracture by Teng *et al.* [2005]. The fracture mechanism is investigated numerically using ABAQUS/Explicit. The impact velocity of the projectiles ranges from 240 m/s to 600 m/s. It is found that a more ductile cylinder tends to fail by petalling while a less ductile one by shear cracking.

3.7 The Elastic Solution

D'Alembert solution. For the progressive waves which translate in space we have the equation

$$\frac{\partial^2 u}{\partial t^2} = c^2 \frac{\partial^2 u}{\partial x^2} \tag{3.7.1}$$

with $c = const$. If we change the variables

$$\alpha = x - ct, \quad \beta = x + ct$$

where α, β are characteristic variables, we have

$$\frac{\partial u}{\partial x} = \frac{\partial u}{\partial \alpha} + \frac{\partial u}{\partial \beta}, \quad \frac{\partial u}{\partial t} = -c\frac{\partial u}{\partial \alpha} + c\frac{\partial u}{\partial \beta}$$

$$\frac{\partial^2 u}{\partial x^2} = \frac{\partial^2 u}{\partial \alpha^2} + 2\frac{\partial^2 u}{\partial \alpha \partial \beta} + \frac{\partial^2 u}{\partial \beta^2}, \cdots .$$

If we introduce in the equation we receive

$$\frac{\partial^2 u}{\partial \alpha \partial \beta} = 0$$

from where we get the solution

$$u = f(x - ct) + g(x + ct)$$

which is the d'Alembert solution. It gives the general solution with f and g to be determined from initial data.

$u = f(x - ct)$ are the direct waves, propagating towards the right. Let us introduce a mobile system of coordinates, for an observer which translates with the speed ct. In the new system of coordinates $\bar{x} = x - ct$, $\bar{t} = t$ we have $u(\bar{x}, \bar{t}) = f(\bar{x})$ i.e., the observer sees the same stationary wave.

The Cauchy problem for the equation (3.7.1) with initial data:

$$u(x, 0) = h(x) \quad \text{and} \quad \frac{\partial u}{\partial t}(x, 0) = k(x) \tag{3.7.2}$$

with h of class C^2 (h and first two derivatives are continuous) and k of class C^1 along $-\infty < x < \infty$ and the solution

$$u = f(x - ct) + g(x + ct) \tag{3.7.3}$$

with f and g arbitrary and of class C^2. For $t = 0$ we have

$$f(x) + g(x) = h(x). \tag{3.7.4}$$

By derivation of (3.7.3)

$$\frac{\partial u}{\partial t}(x, 0) = -cf'(x) + cg'(x) = k(x). \tag{3.7.5}$$

Integrating (3.7.5):

$$-f(x) + g(x) = \frac{1}{c}\int_{x_0}^{x} k(\xi)\, d\xi + g(x_0) - f(x_0). \tag{3.7.6}$$

From (3.7.4) and (3.7.6) we obtain

$$f(x) = \frac{1}{2}h(x) - \frac{1}{2c}\int_{x_0}^{x} k(\xi)\, d\xi - \frac{1}{2}[g(x_0) - f(x_0)]$$

$$\tag{3.7.7}$$

$$g(x) = \frac{1}{2}h(x) + \frac{1}{2c}\int_{x_0}^{x} k(\xi)\, d\xi + \frac{1}{2}[g(x_0) - f(x_0)].$$

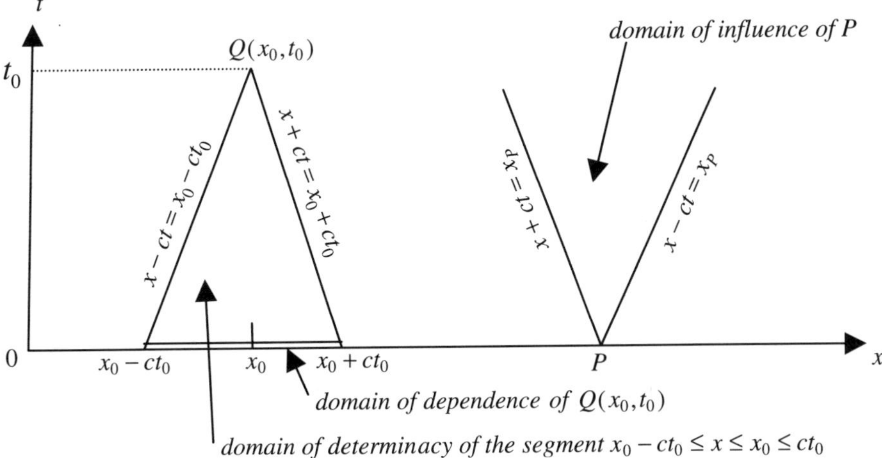

Fig. 3.7.1   The various domains existing.

We replace x by $x - ct$ in the first relation and x by $x + ct$ in the second, and by adding:

$$u(x,t) = f(x - ct) + g(x + ct)$$

$$= \frac{1}{2} \left\{ h(x - ct) + h(x + ct) - \frac{1}{c} \int_{x_0}^{x-ct} k(\xi)\, d\xi + \frac{1}{c} \int_{x_0}^{x+ct} k(\xi)\, d\xi \right\}.$$

The last two terms can be written in the form:

$$u(x,t) = \frac{h(x - ct) + h(x + ct)}{2} + \frac{1}{2c} \int_{x-ct}^{x+ct} k(\xi)\, d\xi. \tag{3.7.8}$$

The first term corresponds to the solution with initial zero velocities ($k(0) = 0$) and the last term to the solution with initial zero displacements ($h(0) = 0$).

The various domains which are involved in various problems and haw are they called are shown in Fig. 3.7.1. The prescribed data in P are not enough for the determination of the solution.

One can put the problem if the solution is *unique?* Let us assume that $u_1(x,t)$ and $u_2(x,t)$ are two distinct solutions which satisfy the initial data (3.7.2). It follows that $u = u_1 - u_2$ is also a solution of (3.7.1) with the initial values

$$u(x,0) = 0 \quad \text{and} \quad \frac{\partial u}{\partial t}(x,0) = 0$$

corresponding to $h = k = 0$. It follows from (3.7.8) that $u(x,t) = 0$ for any x and $t \geq 0$. Thus $u_1 = u_2$ and the *solution is unique.*

We have to study the *dependence of the solution on the initial data.* From (3.7.8) it follows that the solution $u(x_0, t_0)$ at $Q(x_0, t_0)$ depends on the Cauchy data along $x_0 - ct_0 \leq x \leq x_0 + ct_0$ and $t = 0$.

Let us show the continuous dependence of the solution (3.7.8) on the Cauchy data. Let us assume two sets of Cauchy data:

$$u_1(x,0) = h_1(x), \quad \frac{\partial u_1}{\partial t}(x,0) = k_1(x)$$

$$u_2(x,0) = h_2(x), \quad \frac{\partial u_2}{\partial t}(x,0) = k_2(x)$$

$$(3.7.9)$$

and let us assume that

$$|h_1(x) - h_2(x)| < \varepsilon, \quad |k_1(x) - k_2(x)| < \delta \tag{3.7.10}$$

for $-\infty < x < \infty$ and $\varepsilon > 0$, $\delta > 0$ arbitrary. From (3.7.8) we have:

$$u_1(x,t) - u_2(x,t) = \frac{h_1(x-ct) - h_2(x-ct)}{2} + \frac{h_1(x+ct) - h_2(x+ct)}{2}$$

$$+ \frac{1}{2c} \int_{x-ct}^{x+ct} [k_1(\xi) - k_2(\xi)]\, d\xi \,.$$

Taking the modulus and since $|\int_a^b f d\xi| \leq \int_a^b |f|\, d\xi$ we have

$$|u_1(x,t) - u_2(x,t)| \leq \frac{|h_1(x-ct) - h_2(x-ct)|}{2}$$

$$+ \frac{|h_1(x+ct) - h_2(x+ct)|}{2} + \frac{1}{2c} \int_{x-ct}^{x+ct} |k_1(\xi) - k_2(\xi)|\, d\xi \,.$$

Using $(3.7.10)_2$

$$\frac{1}{2c} \int_{x-ct}^{x+ct} |k_1(\xi) - k_2(\xi)|\, d\xi \leq \frac{1}{2c} \delta\{x + ct - (x - ct)\} = \delta t$$

and thus

$$|u_1(x,t) - u_2(x,t)| \leq \varepsilon + \delta t \,.$$

For t fixed if h_1 and h_2 as well as k_1 and k_2, are only "slightly" distinct, i.e., if ε and δ are "small", the solution u_2 approaches u_1 as close as desired.

In the case of non homogeneous equation

$$\frac{\partial^2 u}{\partial t^2} - c^2 \frac{\partial^2 u}{\partial x^2} = f(x,t) \tag{3.7.11}$$

we try to find a solution of the form $u = u_1 + u_2$ with u_1 a solution of homogeneous equation

$$\frac{\partial^2 u_1}{\partial t^2} - c^2 \frac{\partial^2 u_1}{\partial x^2} = 0$$

with initial data $u_1(x,0) = h(x)$, $\partial u_1/\partial t(x,0) = k(x)$ and u_2 a solution of the equation

$$\frac{\partial^2 u_2}{\partial t^2} - c^2 \frac{\partial^2 u_2}{\partial x^2} = f(x,t) \tag{3.7.12}$$

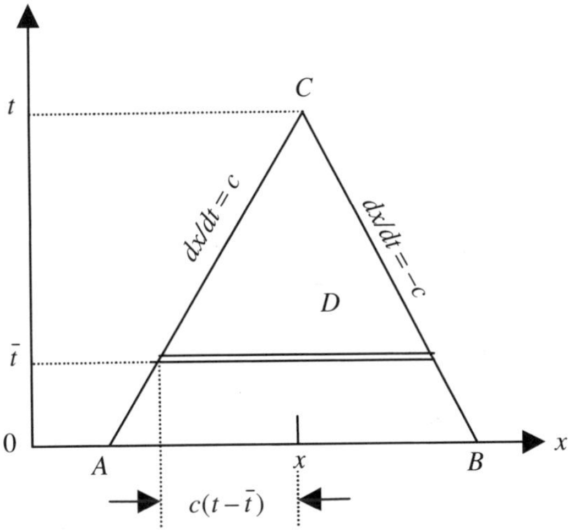

Fig. 3.7.2   Domain of integration.

with initial data

$$u_2(x,0) = 0\,, \qquad \frac{\partial u_2}{\partial t}(x,0) = 0\,. \tag{3.7.13}$$

Integrate now (3.7.12) in the domain D:

$$\iint_D \left(\frac{\partial^2 u_2}{\partial t^2} - c^2 \frac{\partial^2 u_2}{\partial x^2} \right) dx dt = \iint_D f(x,t)\, dx dt \tag{3.7.14}$$

and using the Green's formula

$$\iint_D \left(\frac{\partial P}{\partial y} - \frac{\partial Q}{\partial x} \right) dx dy = - \int_{\partial D} P dx + Q dy$$

the formula (3.7.14) becomes

$$-\int_{\partial D} \frac{\partial u_2}{\partial t}\, dx + c^2 \frac{\partial u_2}{\partial x}\, dt = \iint_D f(x,t)\, dx dt\,.$$

Since along BC we have $dx = -cdt$, along CA we have $dx = -cdt$, and along AB
we have $dt = 0$, this integral becomes

$$\int_{BC} c \left(\frac{\partial u_2}{\partial t}\, dt + \frac{\partial u_2}{\partial x}\, dx \right) - \int_{CA} c \left(\frac{\partial u_2}{\partial t}\, dt + \frac{\partial u_2}{\partial x}\, dx \right) - \int_{AB} \frac{\partial u_2}{\partial t}\, dx + c^2 \frac{\partial u_2}{\partial x}\, dt$$

$$= \iint_D f(x,t)\, dx dt\,.$$

Since in the brackets there are total differentials, the third term is zero from (3.7.13),
and the fourth is zero on AB, we obtain

$$\int_{BC} c du_2 - \int_{CA} c du_2 = \iint_D f(x,t)\, dx dt$$

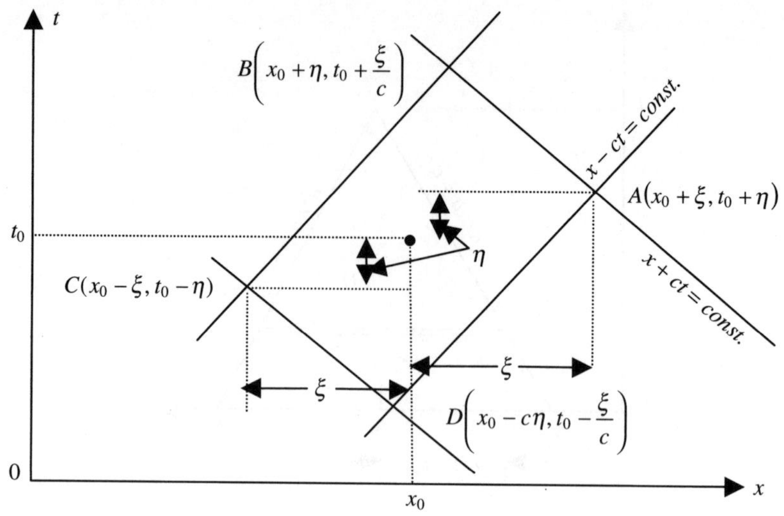

Fig. 3.7.3   Four arbitrary characteristic lines.

or by integration

$$cu_2(C) - cu_2(B) - cu_2(A) + cu_2(C) = \iint_D f(x,t)\,dxdt$$

or since the second and third terms are zero according to (3.7.13),

$$u_2(C) = u_2(x,t) = \frac{1}{2c}\int_0^t \int_{x-c(t-\bar{t})}^{x+c(t-\bar{t})} f(\bar{x},\bar{t})\,d\bar{x}d\bar{t}.$$

Thus the general solution is:

$$u(x,t) = \frac{h(x-ct)+h(x+ct)}{2} + \frac{1}{2c}\int_{x-ct}^{x+ct} k(\xi)\,d\xi + \frac{1}{2c}\int_0^t \int_{x-c(t-\bar{t})}^{x+c(t-\bar{t})} f(\bar{x},\bar{t})\,d\bar{x}d\bar{t}.$$

This solution depends on the Cauchy data on AB, and on the values of $f(x,t)$ in the internal points of the triangle D.

Let us discuss now the *boundary conditions* and *mixed boundary value problem*. The mixed boundary value problem is composed from:
boundary conditions for

$$x = a \quad \text{and} \quad t \geq 0$$

$$x = b \quad \text{and} \quad t \geq 0$$

and the initial conditions for $a \leq x \leq b$, $t = 0$.

Let us consider four arbitrary characteristic lines from $x - ct = const.$ and $x + ct = const.$ (Fig. 3.7.3). The coordinates of B are:

$$x + ct = x_0 + \xi + c(t_0 + \eta)$$

$$x - ct = x_0 - \xi - c(t_0 - \eta).$$

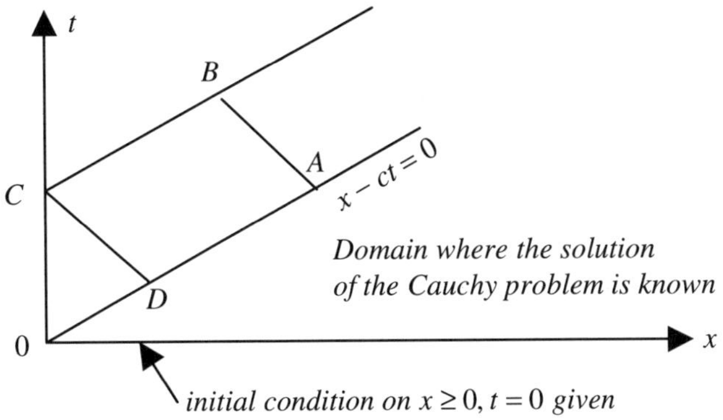

Fig. 3.7.4   Solution in B is known if given on A, D and C.

From here we have by adding: $x_B = x_0 + \eta c$, $t_B = t_0 + \xi/c$.
The coordinates from D are from:

$$x - ct = x_0 + \xi - c(t_0 + \eta)$$

$$x + ct = x_0 - \xi + c(t_0 - \eta).$$

From here we again find: $x_D = x_0 - c\eta$, $t_D = t_0 - \xi/c$.
From the general solution

$$u(x,t) = f(x - ct) + g(x + ct)$$

we obtain

$$u(A) = f(D) + g(B),$$

$$u(C) = f(B) + g(D).$$

From here by adding:

$$u(A) + u(C) = u(D) + u(B).$$

Thus if u is prescribed in A, D and C it follows in B. Thus the solution in any B can be obtained if prescribed by Cauchy data along the line $x - ct = 0$ and known in C by boundary conditions.

The solution is unique in D and the relation is also linear.

Let us consider now the reflection *from a fixed end.* We have to find out how the solution is extended by reflection. We consider the initial conditions

$$x \geq 0, \quad t = 0 : u(x,0) = h_1(x), \quad \frac{\partial u}{\partial t}(x,0) = k_1(x),$$

and the boundary condition $x = 0$, $t \geq 0 : u(x,t) = 0.$

The solution in D is

$$u(x,t) = \frac{h(x - ct) + h(x + ct)}{2} + \frac{1}{2c} \int_{x-ct}^{x+ct} k(s)\, ds.$$

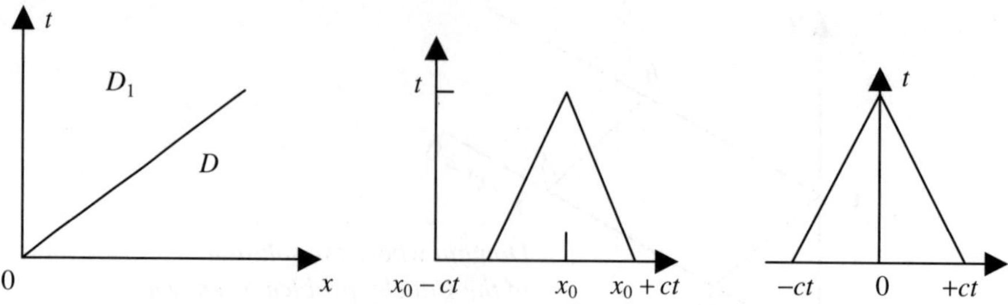

Fig. 3.7.5   Various domains for reflection of waves.

Let us find the solution in D_1. For $x = 0$ from $u(x,t) = f(x - ct) + g(x + ct)$ and $u(0,t) = 0$ we get $f(-ct) + g(ct) = 0$ or

$$f(-x) = -g(x)$$

with x arbitrary argument. From the general expressions

$$f(x) = \frac{1}{2}h(x) - \frac{1}{2c}\int_{x_0}^{x} k(\xi)\,d\xi - \frac{1}{2}[g(x_0) - f(x_0)],$$

$$g(x) = \frac{1}{2}h(x) + \frac{1}{2c}\int_{x_0}^{x} k(\xi)\,d\xi + \frac{1}{2}[g(x_0) - f(x_0)],$$

we put $x_0 = 0$ (see Fig. 3.7.5), we have:

$$f(x) = \frac{1}{2}h(x) - \frac{1}{2c}\int_{0}^{x} k(\xi)\,d\xi - \frac{1}{2}[g(0) - f(0)],$$

$$g(x) = \frac{1}{2}h(x) - \frac{1}{2c}\int_{0}^{x} k(\xi)\,d\xi + \frac{1}{2}[g(0) - f(0)].$$

We replace here x by $-x$, h by h_1, k by k_1 and we have

$$f(-x) = \frac{1}{2}h_1(-x) + \frac{1}{2c}\int_{0}^{x} k_1(-\xi)\,d\xi - \frac{1}{2}[g(0) - f(0)].$$

From the condition $f(-x) = -g(x)$ follows:

$$-\frac{1}{2}h_1(-x) + \frac{1}{2c}\int_{0}^{x} k_1(-\xi)\,d\xi - \frac{1}{2}[g(0) - f(0)]$$

$$= -\frac{1}{2}h_1(x) - \frac{1}{2c}\int_{0}^{x} k_1(\xi)\,d\xi - \frac{1}{2}[g(0) - f(0)].$$

This relation is satisfied for any x if

$$h_1(-x) = -h_1(x) \quad \text{and} \quad k_1(-x) = -k_1(x).$$

Thus the problem of a fixed end $u(0,t) = 0$ for $x = 0$ in a mixed problem, is equivalent with the Cauchy problem for the whole line $-\infty < x < +\infty$ with the <u>odd</u> functions h_1 and $k_1 \ldots$. Therefore

$$\left.\begin{array}{c} -\infty < x < +\infty \\[4pt] t = 0 \end{array}\right\} : u(x,0) = h(x), \quad \frac{\partial u}{\partial t}(x,0) = k(x)$$

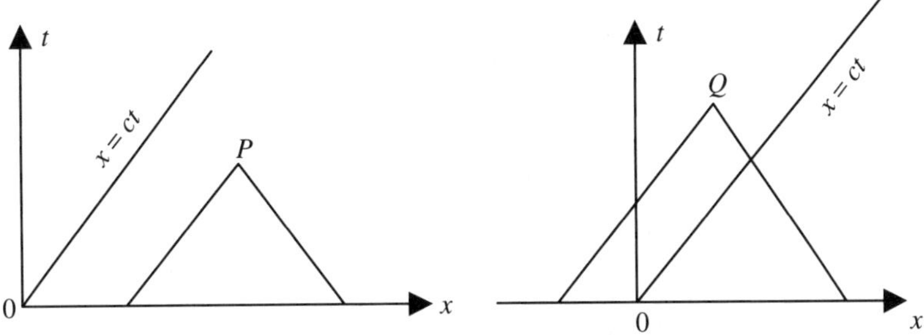

Fig. 3.7.6   The two domains where the above formula applies.

with

$$h(x) = \begin{cases} h_1(x) & \text{for } x \geq 0 \\ -h_1(x) & \text{for } x < 0 , \end{cases} \qquad k_1(x) = \begin{cases} k_1(x) & \text{for } x \geq 0 \\ -k_1(x) & \text{for } x < 0 . \end{cases}$$

We can prove also the converse: if the initial data along the line $-\infty < x < +\infty$ and $t = 0$ are $\underline{\text{odd}}$ with respect to the origin

$$h_1(x) = -h_1(x) , \quad k_1(-x) = -k_1(x) ,$$

then the solution of the wave equation is zero at that point (fixed point).

For the proof we start from the solution

$$u(0,t) = \frac{h_1(-ct) + h_1(ct)}{2} + \frac{1}{2c} \int_{-ct}^{+ct} k(s)\, ds = 0 .$$

Therefore the solution of the mixed problem in $x \geq 0$, $t \geq 0$ is

$$u(x,t) = \begin{cases} \dfrac{h(x - ct) + h(x + ct)}{2} + \dfrac{1}{2c} \displaystyle\int_{x-ct}^{x+ct} k(s)\, ds & \text{for } x > ct , \\[3ex] \dfrac{-h(ct - x) + h(x + ct)}{2} + \dfrac{1}{2c} \displaystyle\int_{-x+ct}^{x+ct} k(s)\, ds & \text{for } x < ct . \end{cases}$$

The last relation is obtained from the first by replacing $h(x - ct)$ by $-h(ct - x)$, and $k(x - ct)$ by $-k(ct - k)$. The two relations are applied in the regions shown in Fig. 3.7.6. The two formulae given above are applied in the two regions shown in this figure.

Let us study now the stress at a fixed end of an elastic bar. On the direct waves $u_1 = f_1(x - ct)$ the stress is

$$\sigma_1 = E \frac{\partial u_1}{\partial x} = -\rho c \frac{\partial u_1}{\partial t} ,$$

while on the inverse waves $u_2 = f_2(x + ct)$ the stress is

$$\sigma_2 = E \frac{\partial u_2}{\partial x} = \rho c \frac{\partial u_2}{\partial t} .$$

The total stress is

$$\sigma = \sigma_1 + \sigma_2 = \rho c \left(-\frac{\partial u_1}{\partial t} + \frac{\partial u_2}{\partial t} \right) = \rho c^2 [f_1'(x - ct) + f_2'(x + ct)] ,$$

while the total displacement is

$$u = f_1(x - ct) + f_2(x + ct)$$

from where for a fixed $x = 0$ we have $f_1(-ct) + f_2(ct) = 0$, or $f_1(-z) = -f_2(z)$. Thus at a fixed end of the bar

$$\sigma|_{x=0} = (\sigma_1 + \sigma_2)|_{x=0} = 2 f_2' E .$$

We arrive at an important conclusion: the stress is doubled due to reflection. If the incident wave reaching the fixed end is still elastic, after reflection it may be nearly $2\sigma_Y$. So that an elastic wave at reflection may become a plastic wave.

Now let us look at the conditions which exist at the free end. Let us consider the equation

$$\frac{\partial^2 u}{\partial t^2} = c^2 \frac{\partial^2 u}{\partial x^2}$$

with the initial conditions

$$\left. \begin{array}{c} t = 0 \\ 0 \leq x < +\infty \end{array} \right\} : u(x,0) = h_1(x) , \quad \frac{\partial u}{\partial t}(x,0) = k_1(x) \qquad (3.7.15)$$

and the boundary conditions at the free end $x = 0$:

$$\left. \begin{array}{c} t > 0 \\ x = 0 \end{array} \right\} : \frac{\partial u}{\partial x}(0,t) = 0 . \qquad (3.7.16)$$

Let us show that the mixed problem with a free end $\partial u/\partial x(0,t) = 0$ is equivalent with the Cauchy problem on the entire x-axis with initial conditions for displacement and velocities even.

For the proof from

$$u(x,t) = f(x - ct) + g(x + ct)$$

follows $\partial u/\partial x(0,t) = f'(-ct) + g'(ct) = 0$, i.e.,

$$f'(-x) = -g'(x) .$$

From

$$f(-x) = \frac{1}{2} h_1(-x) + \frac{1}{2c} \int_0^x k_1(-s)\, ds - \frac{1}{2}[g(0) - f(0)] ,$$

$$g(x) = \frac{1}{2} h_1(x) + \frac{1}{2c} \int_0^x k_1(s)\, ds + \frac{1}{2}[g(0) - f(0)] ,$$

follows

$$f'(-x) = \frac{1}{2} h_1'(-x) - \frac{1}{2c} k_1(-x) ,$$

$$g'(x) = \frac{1}{2} h_1'(x) + \frac{1}{2c} k_1(x) ,$$

from where

$$h_1'(-x) - \frac{1}{c}k_1(-x) = -h_1'(-x) - \frac{1}{c}k_1(x)$$

i.e.,

$$h_1'(-x) = -h_1'(x), \quad k_1(-x) = k_1(x).$$

Since the derivative of an even function $h_1(x) = h_1(-x)$ is an odd function $h_1'(x) = -h_1'(-x)$ we have

$$h_1(x) = h_1(-x).$$

Therefore the mixed problem for a *free end* is equivalent with the Cauchy problem on the entire axis with the conditions

$$\left.\begin{array}{c} -\infty < x < +\infty \\[2mm] t = 0 \end{array}\right\} \quad \begin{array}{l} u(x,0) = h(x) \\[3mm] \dfrac{\partial u}{\partial t}(x,0) = k(x) \end{array} \quad \text{with} \quad \left\{\begin{array}{l} h(x) = \left\{\begin{array}{ll} h_1(x) & \text{for } x > 0, \\ h_1(-x) & \text{for } x < 0, \end{array}\right. \\[5mm] k(x) = \left\{\begin{array}{ll} k_1(x) & \text{for } x > 0, \\ k_1(-x) & \text{for } x < 0. \end{array}\right. \end{array}\right.$$

Let us prove the converse. If the Cauchy conditions on $-\infty < x < +\infty$ are <u>even</u> $h_1(x) = h_1(-x)$, $k_1(x) = k_1(-x)$ then $\partial u/\partial x(0,t) = 0$. From the solution

$$u(x,t) = \frac{h(x-ct) + h(x+ct)}{2} + \frac{1}{2c}\int_{x-ct}^{x+ct} k(s)\,ds$$

follows for $x = 0$:

$$\frac{\partial u}{\partial x}(0,t) = \frac{h'(-ct) + h'(ct)}{2} + \frac{1}{2c}[k_1(ct) - k_1(-ct)] = 0$$

since $h_1'(-ct) = h_1'(ct)$. The solution for the mixed problem for $x > 0$, $t > 0$ is:

$$u(x,t)$$

$$= \left\{\begin{array}{ll} \dfrac{h(x-ct) + h(x+ct)}{2} + \dfrac{1}{2c}\displaystyle\int_{x-ct}^{x+ct} k(s)\,ds & \text{for } x > ct, \\[6mm] \dfrac{h(ct-x) + h(x+ct)}{2} + \dfrac{1}{2c}\left[\displaystyle\int_0^{x+ct} k(s)\,ds + \int_0^{ct-x} k(s)\,ds\right] & \text{for } x < ct, \end{array}\right.$$

which satisfies the initial conditions (3.7.15) and the boundary conditions (3.7.16).

Let us examine the *displacement at the free end* of an elastic bar. We have on the direct waves $u_1 = f_1(x - ct)$ and the stress

$$\sigma_1 = E\frac{\partial u_1}{\partial x} = -\rho c\frac{\partial u_1}{\partial t}$$

and on the inverse waves $u_2 = f_2(x + ct)$ with the stress

$$\sigma_2 = E\frac{\partial u_2}{\partial x} = \rho c\frac{\partial u_2}{\partial t}.$$

The total stress is

$$\sigma = \sigma_1 + \sigma_2 = \rho c\left(-\frac{\partial u_1}{\partial t} + \frac{\partial u_2}{\partial t}\right) = \rho c_2[f_1'(x - ct) + f_2'(x + ct)]$$

and the total displacement is $u = f_1(x - ct) + f_2(x + ct)$. If the end $x = 0$ is free, we have $\sigma = 0$ or $f_1'(-ct) + f_2'(ct) = 0$. That means

$$f_1(-ct) = f_2(ct).$$

The total displacement is

$$u|_{x=0} = (u_1 + u_2)_{x=0} = 2f_2(ct).$$

That means that at a free end the *displacement is doubled*. Because for the stress we have

$$\sigma_1 = -\sigma_2$$

a *compressive wave is reflected as a tensile wave* at the free end (or the reverse). If the mechanical proprieties are distinct in tension and in compression, that may make a difference.

Bibliography

Bell J. F., 1968, *The Physics of Large Deformation of Crystalline Solids*, Springer-Verlag, Berlin.

Bell J. F., 1973, The experimental foundations of solid mechanics, in *Handbook der Physic*, Springer-Verlag, Berlin.

Bodner S. R. and Clifton R. J., 1967, Experimental investigation of elastic-plastic pulse propagation in aluminum rods, *J. Appl. Mech. Trans. ASME* **34**, 1, 91–99.

Chen S. R., Maudlin P. J. and Gray III G. T., 1999, Constitutive behavior of model FCC, BCC, and HCP metals: experiments, modeling and validation, in *Constitutive and Damage Modeling of Inelastic Deformation and Phase Transformation*, Khan A. S. (ed.), NEAT Press, 623–626.

Chou P. C. and Hopkins A. K., (eds.) 1973, Dynamic response of materials to intense impulsive loading, *Air Force Materials Laboratory*, 555 pp.

Clifton R. J. and Bodner S. R., 1966, An analysis of longitudinal elastic-plastic pulse propagation, *J. Appl. Mech. Trans. ASME* **33**, 2, 248–255.

Cristescu N., 1957, On propagation of waves in rubber, *Prikl. Mat. Meh.* **21**, 795 800 (in Russian).

Cristescu N., 1967, *Dynamic Plasticity*, John Wiley & Sons, Inc., New York.

Cristescu N., Dynamic plasticity, *Appl. Mech. Rev.* **21**, 7, 659–668.

Cristescu N. and Bell J. F., 1970, On unloading in the symmetrical impact of two aluminum bars, In: *Inelastic Behavior of Solids*, McGraw Hill Book Co., 1970, 397–419.

Cristescu N., 1970, The unloading in symmetric longitudinal impact of two elastic-plastic bars, *Int. J. Mech. Sci.* **12**, 6, 723–738.

De Juhasz K. J., 1949a, Graphical analysis of impact of bars stresses above the elastic range. Part. I, *J. Franklin Inst.* **248**, 15–48.

De Juhasz K. J., 1949b, Graphical analysis of impact of bars stressed above the elastic range. Part. II, *J. Franklin Inst.* **248**, 113–142.

Francis P. H., 1966, Wave propagation in thin rods with quiescent temperature gradients, *J. Appl. Mech.* **33**, 3, 702–704.

Huffington N. J., (ed.) 1965, Behaviour of materials under dynamic loading, *Behavior of Materials under Dynamic Loading at the Winter Annual Meeting of the ASME*, Chicago, Illinois, 1965, 187 pp.

Jones S. E., Gillis P. P. and Foster J. C. Jr., 1987, On the equation of motion of the undeformed section of a Taylor Impact specimen, *J. Appl. Phys.* **61**, 2, 499–502.

Kanel G. I., 1999, Dynamic strength of materials, *Fatigue Fract. Engng. Mater. Struct.* **22**, 1011–1019.

Kármàn Th. and Duwez P., 1950, The propagation of plastic deformation in solids, *J. Appl. Phys.* **27**, Ser E, 107–110.

Kawata K. and Shioiri J., (eds.) 1978, *High Velocity Deformation of Solids Symposium Tokyo/Japan, 1977*, Springer Verlag, 452 pp.

Kenning M. J., 1974, Uniqueness of uniaxial elastic-plastic interface motions, *J. Mech. Phis. Solids* **22**, 437–456.

Kinslow R., (ed.) 1970, *High-Velocity Impact Phenomena*, Academic Press, 579 pp.

Kolsky H., 1953, *Stress Waves in Solids*, Clarendon Press, Oxford.

Kolsky H., 1969, Production of tensile shock waves in stretched natural rubber, *Nature* **224**, 5226, 1301.

Lagoudas D. C., Ravi-Chandar K., Sarh K. and Popov P., 2003, Dynamic loading of polycrystalline shape memory alloy rods, *Mech. Materials* **35**, 689–716.

Lee L. H. N. and Chi-Mou Ni, 1973, A minimum principle in dynamics of elastic plastic continua at finite deformation, *Archives of Mechanics* **25**, 3, 457–468.

Lee L. H. N., 1974, Dynamic plasticity, *Nuclear Eng. and Design* **27**, 386–397.

Lenski V. S., 1949a, On the elastic-plastic shock of a bar on a rigid wall, *Prikl. Mat. Meh.* **12**, 165–170 (in Russian).

Lush P. A., 1991, Comparison between analytical and numerical calculations of liquid impact on elastic-plastic solid, *J. Mech. Phys. Solids* **39**, 1, 145–155.

Magness L. S. Jr., 1994, High strain rate deformation behaviors of kinetic energy penetrator materials during ballistic impact, *Mech. of Materials* **17**, 147–154.

Miklowitz J., (ed.) 1969, Wave propagation in solids, *ASME Winter Annual Meeting*, Los Alamos, California, 1969, 183 pp.

Mines R. A. W., 2004, A one-dimensional stress wave analysis of a lightweight composite armour, *Comp. Structures* **64**, 55–62.

Misra J. C., Kar S. B. and Samanta S. C., 1987, Distribution of stresses in a heated nonhomogeneous rod, *S. M. Archives* **12/2**, 97–111.

Nemat-Nasser S., Li Y. F. and Isaacs J. B., 1994, Experimental/computational evaluation of flow stress at high strain rates with application to adiabatic shear banding, *Mec. Mater.* **17**, 111–134.

Nonaka T., Clifton R. J. and Okazaki T., 1996, Longitudinal elastic waves in columns due to earthquake motion, *Int. J. Impact. Engng.* **18**, 7–8, 889–898.

Nowacki W. K., 1978, *Problems of Wave Propagation in the Theory of Plasticity*, Mir, Moscow.

Ponomarev S. D., Biderman V. L., Likharev K. K., Makushin V. M., Malinin N. N. and Feodos'ev V. I., 1959, Resistence calculus in construction of machines, *Mashigiz*, Moscow, Tom III, 553–580 (in Russian).

Raniecki B., 1971, The influence of dynamical thermal expansion on the propagation of plane elastic-plastic stress waves, *Quart. Appl. Math.* **29**, 2, 277–290.

Raiser G. F., Wise J. L., Clifton R. J., Grady D. E. and Cox D. E., 1994, Plate impact response of ceramics and glasses, *J. Appl. Phys.* **75**, 8, 3862–3869.

Rakhmatulin H. A. and Demianov Yu. A., 1961, *Strength under Impulsive Momentary Loads*, Moscow (in Russian).

Rakhmatulin H. A., 1945, On the propagation of unloading waves, *Prikl. Mat. Meh.* **9**, 1, 91–100 (in Russian).

Shim V. P. W., Yuan J. and Lee S.-H., 2001, A technique for rapid two-stage dynamic tensile loading of polymers, *Experimental Mechanics* **41**, 1, 122–127.

Santosham T. V. and Ramsey H., 1970, Small plastic strain wave propagation in pre-stressed soft copper rods, *Int. J. Mech. Sci.* **12**, 5, 447–457.

Shapiro G. S., 1946, Longitudinal vibration of bars, *Prikl. Mat. Meh.* **10**, 597–616 (in Russian).

Swegle J. W. and Ting T. C. T., 1982, Plastic wave propagation in a circular cylindrical rod, *J. Appl. Mech. Trans. ASME* **49**, 253–256.

Taylor G. I., 1946, The testing of materials at high rates of loading, *J. Inst. Civil Engrs.* **8**, 486–519.

Taylor G. I., 1958a, The use of flat-ended projectiles for determining dynamic yield stress, *Proc. Roy. Soc. Ser. A* **194**, 289–289.

Taylor G. I., 1958b, The formation and enlargement of circular hole in a thin, plastic sheet, *Quart. J. Mech. Appl. Math.* **1**, 103–124.

Teng X., Wierzbicki T., Hiermaier S. and Rohr I., 2005, Numerical prediction of fracture in the Taylor test, *Int. J. Solids Structures* **42**, 2929–2948.

Tjavaras A. A. and Triantafyllou M. S., 1996, Shock waves in curved synthetic cables, *J. Enge. Mech.* **122**, 4, 308–315.

Varley E., (ed.) 1976, Propagation of shock waves in solids, *The Applied Mechanics Conference*, Salt Lake City, Utah, 1976.

Chapter 4

Rate Influence

4.1 Experimental Results

The rate influence seems to be the central problem of dynamic plasticity. It has been known for many years that during dynamic testing, solids behave in a more or less different way than in static tests. Many theoretical and experimental papers have been devoted to this problem. Let us consider the problem shortly.

We discuss the main conclusion resulting from the experiments that have been performed. Details concerning the experimental techniques will not be given here. We shall see however, that in some experimental papers the wave propagation phenomena were disregarded, although the rate of loading was high. This can be done only if the specimen is very short. The old tests are discussed in detail by Cristescu [1967], Cambell [1970], Cristescu and Suliciu [1982] and Goldsmith [2001], and will not be mentioned here any more. Only a few tests will be revealed. Strain rates ranging between 10^{-5} sec^{-1} and 10^{-3} sec^{-1} are termed static tests. Tests in which the rates of strain range between 10^{-2} sec^{-2} and 10^2 sec^{-1} can be termed intermediate tests (these are the rates of strains usually met in various metal working processes, for instance), while the tests producing rates of strains above 10^2 sec^{-1} (and certainly of the order of magnitude of 10^3 sec^{-1} and higher) can be called dynamic in the sense that as a rule in such tests the inertia forces, the propagation of waves etc. must be considered in the analyses. Sometimes the inertia forces are considered in the analysis when rates of strain of approximately 1 sec^{-1} are reached. A review is presented by Klepaczko [1998] on recent progress in shear testing of materials at high and very high (up to 10^6 s^{-1}) strain rates. The classification of various testing machines according to the kind of test performed, i.e., tension, compression, bending, combined loading etc., is not important for the approach followed here. Kinematic and dynamic jump conditions are also given.

A theory of rate-type materials for one-dimensional case is due to Suliciu [1974]. He introduces different classes of regulated functions and introduces the concept of wave propagation for the frame of regulated functions.

The simplest diagnostic test is the uniaxial compression or tension test of a cylindrical specimen. For the description of the technical aspects of this kind of experiments see for instance Ponomarev *et al.* [1956], Ilyushin and Lenskii [1959], Dally and Riley [1965], Bell [1973], Gupta [1976], etc. Basically there are two kinds of testing machines. With the so called "hard" machines the variation in time of the length of the specimen can be prescribed, while the response of the material is known by registering the force necessary to produce this deformation. Other kind of machines called "soft" can prescribe the stress applied to the specimen (by a weight attached directly or not to the specimen), while the response of the material is observed by measuring the elongation of the specimen. Both kinds of testing machines provide some stress–strain diagrams.

The experiments made by Kolsky [1949] with copper and lead showed that the dynamic elastic modulus does not differ sensibly from the static one (Fig. 4.1.1),

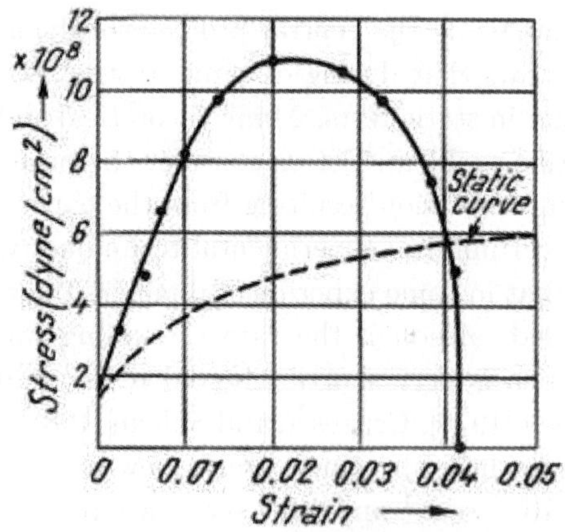

Fig. 4.1.1   Static and dynamic stress–strain curves for copper.

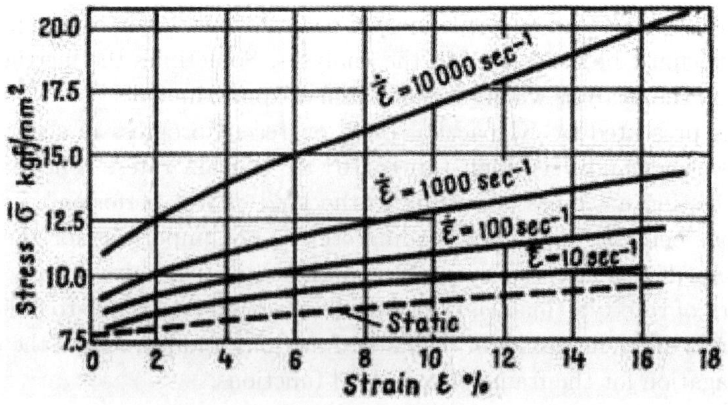

Fig. 4.1.2   Stress–strain curves for pure aluminum.

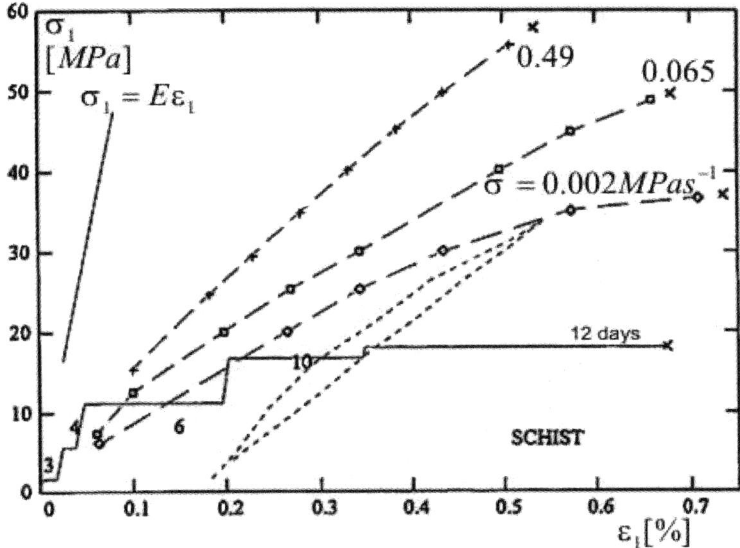

Fig. 4.1.3   Stress–strain curves for schist.

but that the whole dynamic stress–strain curve is higher that the static curve. The strain increases even when the stress is falling, and it follows that the dynamic constitutive equation for copper is time-dependent. The static stress–strain curve does not seem to be rate dependent.

In Fig. 4.1.2 are given the stress–strain curves for pure aluminum as obtained with various rates of strains by Hauser *et al.* [1960] (reproduced here from Thomsen *et al.* [1968]). If the rates of strains are high (over 10^2 sec^{-1}, say), the inertia forces cannot be neglected and the wave propagation phenomena must be considered in the analysis of the experiment.

For rocks the influence of the strain rate on both loading and unloading is shown in Fig. 4.1.3 for two loading rates $|\dot{\sigma}| = 300$ kgf cm^{-2} min^{-1} and $|\dot{\sigma}| = 1.33$ kgf cm^{-2} min^{-1} both in loading and unloading. Thus the stress–strain curve is non-linear during loading and unloading. One can see also the important hysteresis loops during unloading and the fact that the whole curve is influenced by the loading rate.

In Fig. 4.1.4 we give the variation in time of the strain and of the surface angle for copper (Bell [1968]) for an impact velocity $V_0 = 1200$ cm/sec, measured at 5.1 cm from the impacted end. For high impact velocity the strain is higher, and that can not be explained by the classical theory.

Another parameter which can be measured at the impacted end of the bar is the stress. Generally the stress can be measured with the help of piezoelectric crystal wafers placed at the impacted face. Filbey [1961] has used an X-cut quartz crystal wafer 2.12 cm in diameter and 0.25 mm thick. A typical result for two

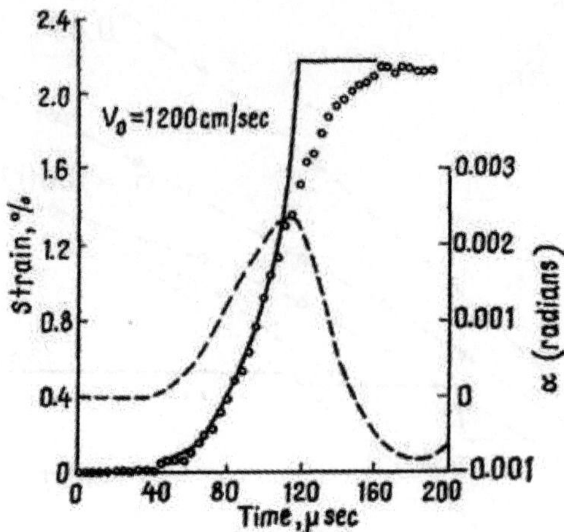

Fig. 4.1.4   Variation of strain and surface angle in time for copper.

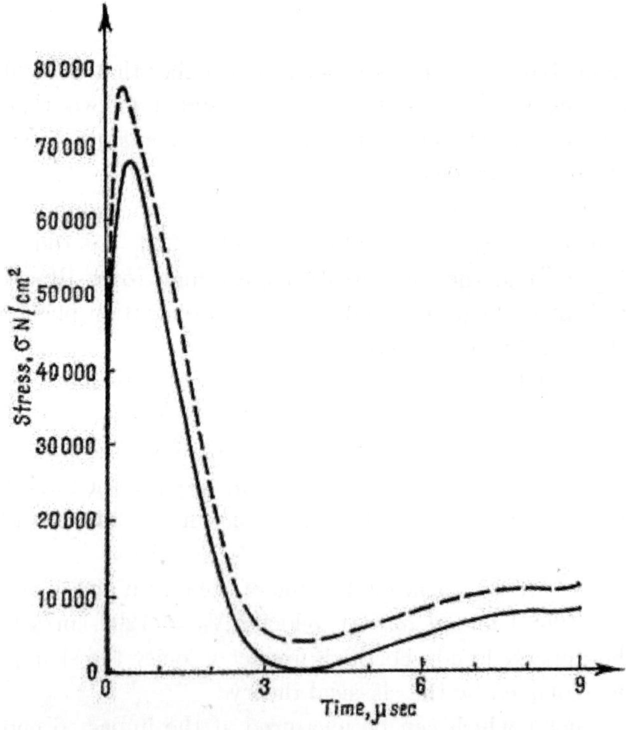

Fig. 4.1.5   Peak stress at impacted end.

experiments with symmetric impact is given in Fig. 4.1.5 (Filbey [1961]) who used the impact velocity $V_0 = 7100$ cm/sec. The decrease is in microseconds to become nearly constant for a relatively long period of time (Fig. 4.1.6) (Filbey [1961]). The medium value of the peak stress is 71043 N cm^{-2}. The stress "plateau" which is reached after several microseconds is at the level 13160 N cm^{-2}. The decrease of the stress which follows afterwards is due to the unloading.

Another experiment which played an important role in the development of the theory was the following one carried out by Bell [1951]. A medium carbon steel

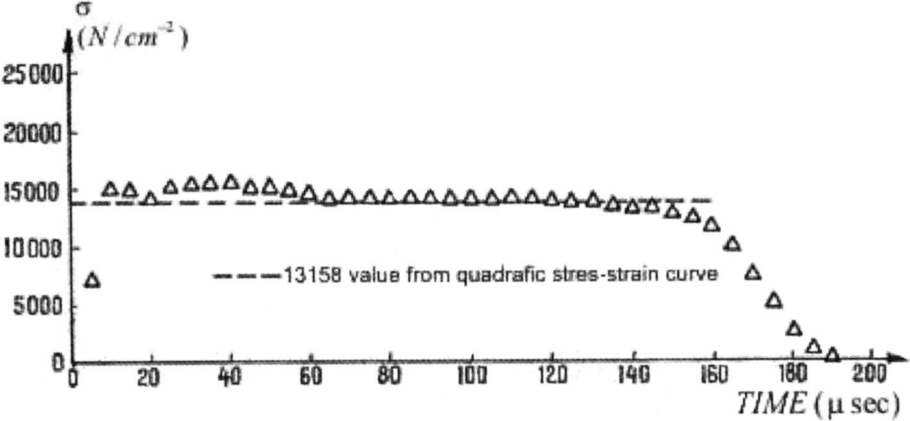

Fig. 4.1.6   Stress plateau.

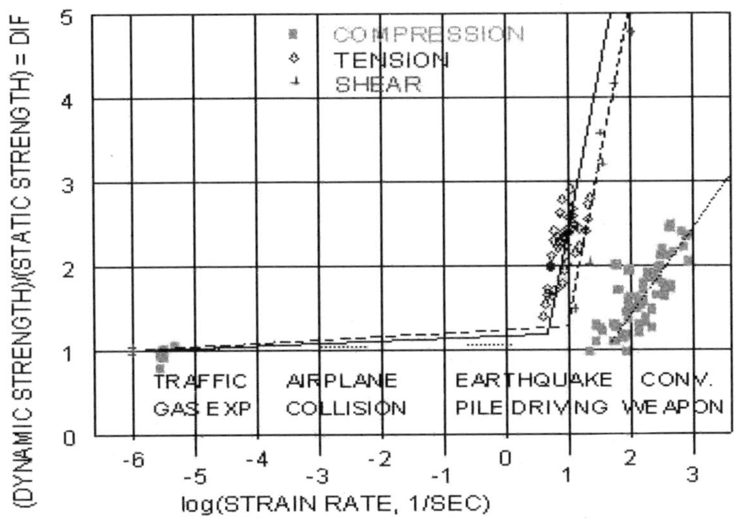

Fig. 4.1.7   General strain rate effects on mortar strength.

long bar (about 183 cm in length and consisting of portions 0.64 or 0.35 cm in diameter) was subjected to a static extension. A dynamic loading was superposed. The experimental results obtained by Bell have shown that the first waves which propagate in the bar are propagating with the bar velocity c_0. Similar tests were done afterwards by several authors.

Similar dynamic loading tests followed by a dynamic reloading have been done by Alter and Curtis [1956] on lead specimens and afterwards by Bell and Stein [1962] on aluminum. The first waves were propagating with the bar velocity. Similar tests in which the hitter was a composed bar made from two distinct materials (possessing distinct densities) have been carried out by other authors.

The general trend in the unconfined tests of cementations material data is a very slight increase in strength from a low strain rate of 1.0×10^{-7}/s up to a strain rate of approximately 1.0/s for tensile strength properties, 10/s for shear strength properties and 100/s for compressive strength properties. Above these

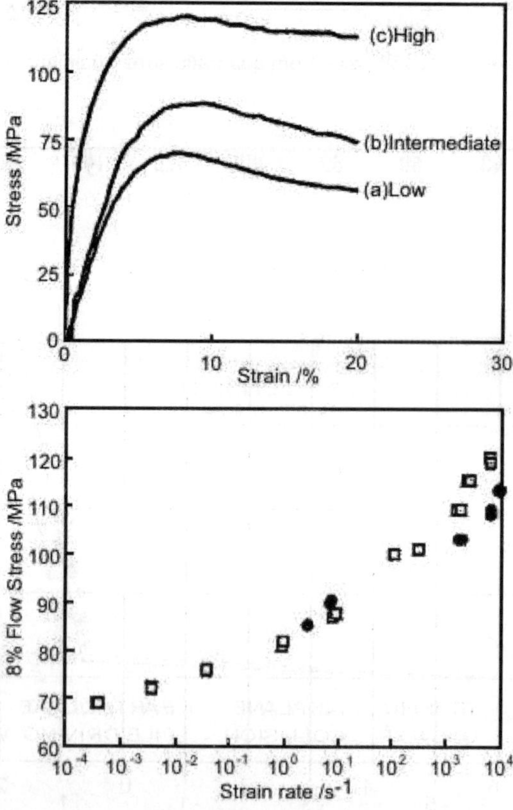

Fig. 4.1.8 Typical stress–strain curves for polycarbonate at 23°C for strain rates of (a) 0.000368 s^{-1}, (b) 8.43 s^{-1}, and 6000 s^{-1}; and measured 8% flow stress as a function of logarithmic strain rate for polycarbonate at 23°C (specimen thickness: $\simeq$ 4.4 mm, • 1.5 mm).

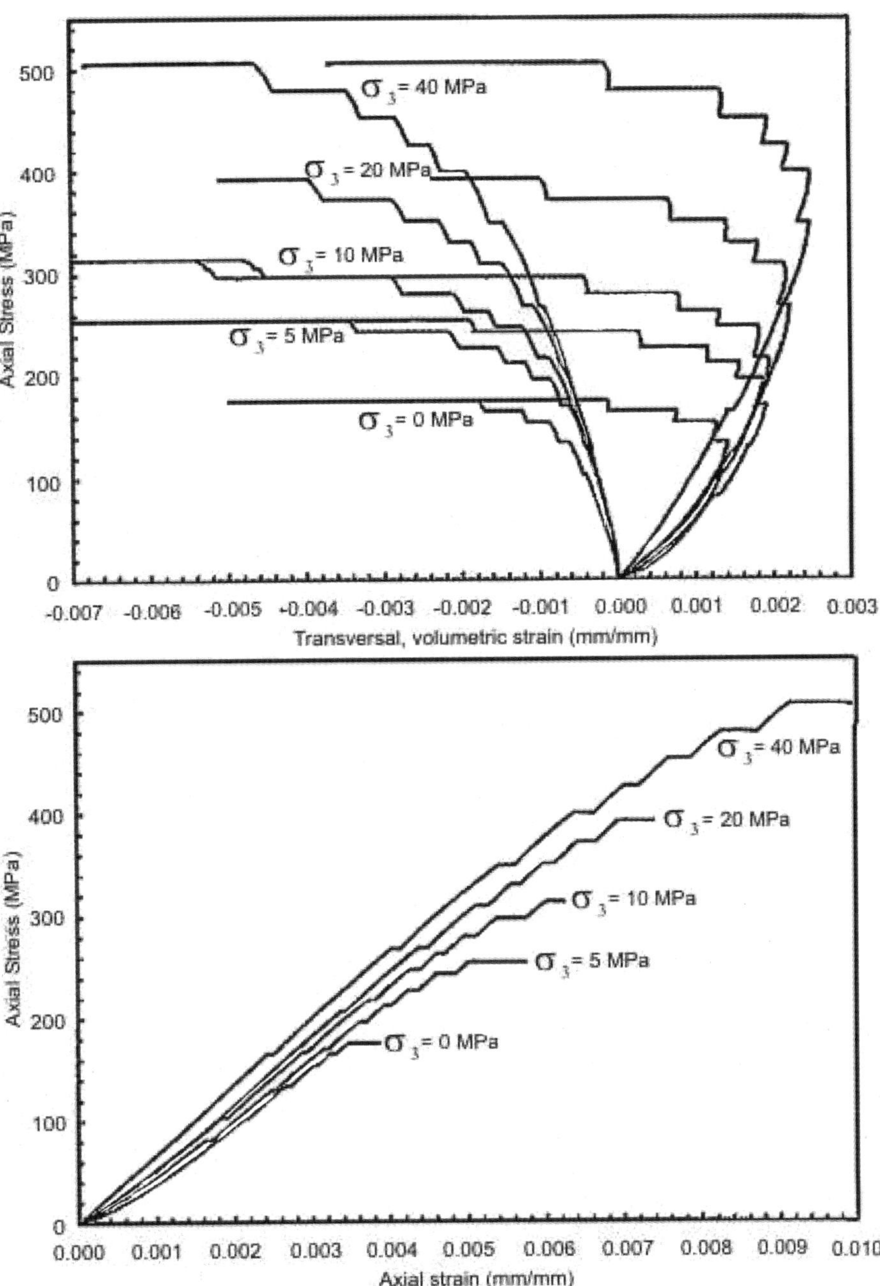

Fig. 4.1.9   Tree axial curves for granite (Maranini and Yamaguchi [2001]).

critical strain rates a rather abrupt increase in strength occurs with a slope of ($\log_{10}$ strength/$\log_{10}$ strain rate) of approximately one-third. Representative strain rate effects on unconfined strength properties reported by Schmidt and Ross [1999] are given in Fig. 4.1.7.

See also the paper of Bertholf and Karnes [1975] on the same subject. Very high strain-rate response of a NiTi shape-memory alloy vas studied by Nemat-Nasser *et al.* [2005]. They adopt a split Hopkinson bar test to get such high strain rates of about 10,000/s. The result obtained is an increase of the true stress at high strain rates. The assumption of uniform stress and strain distribution within a split Hopkinson pressure bar specimen is assumed incorrect at high impact velocities (Dioh *et al.* [1995]), and that the propagation of plastic wave fronts within the split-Hopkinson-pressure-bar are responsible for the dependence of flow stress measurements on specimen thickness at high strain rates.

The same topic is used in Dioh *et al.* [1993]; they make tests studying the thickness of the specimens in the split Hopkinson pressure bar experiments. Various thermoplastics have been studied. For instance in Fig. 4.1.8 is showing some of the results. The "high" strain rates is completely above the other. Also in the lower figure one can see that the thick specimens are above the small one, at least for higher strain rates.

Some other curves for granite measured in three axial stresses are shown in Fig. 4.1.9, from Maranini and Yamaguchi [2001]. The small original plateau is creep in a few minutes done before an unloading is performed to measure the elastic parameters (see Ch. 2). But changing the confining pressure changes the whole stress–strain curve. Also, the volumetric stress–strain curves are shoving first compression and afterwards dilatation. The compression is trying to disappear as the confining pressure is increased significantly, but not the creep by dilatancy.

Similar results are mentioned by very many authors.

4.2   The Constitutive Equation

Many authors have proposed various constitutive equations in order to describe the mechanical properties of materials that exhibit the rate effect. All have started from the assumption, suggested by experimental evidence, that, even while the stress is being continuously increased, there is no longer a one-to-one correspondence between stress and strain as prescribed by a finite stress–strain relation.

The first was Ludwik [1909] and then Prandtl [1928] who observed that for a fixed plastic strain the corresponding stress is higher, the greater the average rate of strain at which the experiment is performed. Ludwik postulated the following logarithmic equation

$$\sigma = \sigma_1 + \sigma_0 \ln \left(\frac{\dot{\varepsilon}^P}{\dot{\varepsilon}_0^P} \right).$$

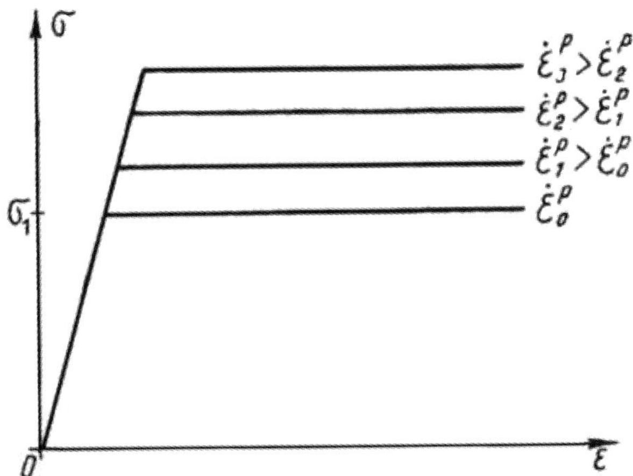

Fig. 4.2.1   Influence of the rate of strain on the stress–strain curve.

Where σ_1, σ_0 and $\dot{\varepsilon}_0^P$ are material constants. Here σ_1 is the yield stress corresponding to the strain rate $\dot{\varepsilon}_0^P$ (Fig. 4.2.1). If the strain rate is greater than $\dot{\varepsilon}_0^P$, the stress corresponding to a certain plastic strain is greater, σ_0 is the measure of this increment.

The reference configuration for such constitutive equation is the actual one, since the strain is not present.

Many simple models describing the rate effect were inspired by this formula. If the elastic part of the strain can be neglected with respect to the plastic one, then the simplest of these models can be written as

$$\dot{\varepsilon} = \begin{cases} \dfrac{\sigma - \sigma_Y}{3\eta} & \text{if } \sigma > \sigma_Y, \\ 0 & \text{if } 0 \le \sigma \le \sigma_Y, \end{cases}$$

where η is the viscosity coefficient. This model can be called Bingham model (Bingham [1922]). In it $\dot{\varepsilon}$ depends on the overstress.

This model was used for various pastes, dough, vaseline, paints, mud etc., but also for various solid bodies as metals, rocks etc. The coefficient η can be found experimentally in the following way. From here we obtain

$$3\eta = \frac{\sigma_2 - \sigma_1}{\dot{\varepsilon}_2 - \dot{\varepsilon}_1}.$$

Generally, for most solid bodies η is decreasing with increasing $\dot{\varepsilon}$, and therefore it is only within certain ranges of variation that η can be considered to be constant.

If the material is work-hardening, and if the conventional quasi-static relation is $\sigma = f(\varepsilon)$, then Malvern [1951a, 1951b] proposed the use of the expression:

$$\sigma = f(\varepsilon) + a \ln(1 + b\dot{\varepsilon}^P).$$

Here a and b are material constants. Solving this equation with respect to the rate of strain, we can write this relation as:

$$\dot{\varepsilon}^P = \frac{1}{b}\left[\exp\left(\frac{\sigma - f(\varepsilon)}{a}\right) - 1\right]. \qquad (4.2.1)$$

The reference configuration is now the initial one. This equation suggest a generalization in the sense that the plastic rate of strain must be a function of the overstress $\sigma - f(\varepsilon)$, that is, of the difference between the dynamic and the actual static plastic strain. Thus

$$E\dot{\varepsilon}^P = F(\sigma - f(\varepsilon)). \qquad (4.2.2)$$

We can write now

$$\dot{\varepsilon} = \begin{cases} \dfrac{\sigma - f(\varepsilon)}{3\eta} & \text{if } \sigma > f(\varepsilon), \\[2mm] 0 & \text{if } 0 < \sigma \le f(\varepsilon), \end{cases}$$

where $\sigma = f(\varepsilon)$ is the work-hardening condition. To give some values for η from creep curves of rock-salt (see Baroncea *et al.* [1977]) η is of the order 10^{15} Nm^{-2} sec (Poise). For schist it is 10^{12} Poise while for other rocks it may be higher (10^{17}–10^{18} Poise for argillaceous schist, see Vyalov [1978]). For ice from 10^{10} to 10^{15} Poise. For quasistatic deformations ($\dot{\varepsilon} \approx 1$ sec^{-1}) for mild steel (Cristescu [1977]) η is of the order of magnitude of 10^7 Nm^{-2} sec, but for faster deformation ($\dot{\varepsilon} \approx 100$ sec^{-1}) it is of the order 10^6 Nm^{-2} sec. For even very high rate of strains ($\dot{\varepsilon} > 10^3$ sec^{-1}) Cambell [1973] gives for various metals values of the order 10^3 Poise. The values given above are only illustrative.

Since in most cases the elastic part of the strain is connected with the stress by the Hooke's law

$$E\varepsilon^E = \sigma,$$

the full constitutive equation can be written in the form

$$E\dot{\varepsilon} = \dot{\sigma} + F(\sigma - f(\varepsilon)).$$

With $F(z)$ satisfying the properties

$$F(z) > 0 \quad \text{if } z > 0,$$

$$F(z) = 0 \quad \text{if } z \le 0.$$

This property is related to the loading/unloading criteria.

More generally, the plastic rate of strain can be considered a function of the stress and strain

$$E\dot{\varepsilon}^P = g(\sigma, \varepsilon),$$

and it yields the constitutive equation

$$E\dot{\varepsilon} = \dot{\sigma} + g(\sigma, \varepsilon).$$

This form of the constitutive equation has been used by Malvern [1951a, 1951b].

Various form of such constitutive equation was used by very many authors. Sokolovski [1948a, 1948b] used it for perfectly plastic materials with

$$g(\sigma, \varepsilon) = kF(|\sigma| - \sigma_Y),$$

where σ_Y is the static yield stress. Ting and Symonds [1964] have used the expression

$$g(\sigma) = D\left(\frac{\sigma}{\sigma_Y} - 1\right)^q$$

with D and q material constants, specific for the material considered. Gilman [1960] suggested

$$g(\sigma) = \dot{\varepsilon}_0 \exp(-A/\sigma),$$

where $\dot{\varepsilon}_0$ and A are material constants. Some generalizations are due to Perzina [1963]

$$g(\sigma) = \sum_{\alpha=1}^{N} A_\alpha \left[\exp\left(\frac{\sigma}{\sigma_Y} - 1\right)^\alpha - 1\right],$$

$$g(\sigma) = \sum_{\alpha=1}^{N} B_\alpha \left(\frac{\sigma}{\sigma_y} - 1\right)^\alpha, \quad \text{etc.}$$

All the functions and constants which occur in various expressions of the function g are, sometimes, time and temperature dependent. One can compute the rate of strain locally, the averaging is not possible. The elastic constants are independent on the rate of strain. The reference configuration is always the actual one.

The basic idea of the constitutive equations given above is that the plastic rate of strain is a function only of the dynamic overstress. This leads to the conclusion that in the plastic range the stress–strain curves for various constant rates of plastic straining are parallel curves.

As will be shown later on (see Béda [1962a]) these curves are not always parallel. Thus, instead of (4.2.3) one must look for a quasi-linear constitutive equation of the form

$$\frac{\partial \sigma}{\partial t} = \varphi(\sigma, \varepsilon)\frac{\partial \varepsilon}{\partial t} + \psi(\sigma, \varepsilon) \tag{4.2.3}$$

for both plastics (Cristescu [1964]), and for metals (Cristescu [1963, 1966]). Variants have been used by many authors. Suliciu [1966] have shown when the constitutive equation (4.2.3) can be written in the form

$$\sigma = H(\varepsilon, t).$$

In (4.2.3) $\varphi(\sigma, \varepsilon)$ which can also possibly depend on time, is the measure of the instantaneous response of the material to an increase of stress, while $\psi(\sigma, \varepsilon)$ which can also possibly depend explicitly on time, will be the measure of the non-instantaneous response of the material. It will be assumed that $\varphi > 0$.

Of course, φ and ψ may also comprise various characteristic constants of the material (yield stress, viscosity coefficient, etc.) which are to a great extent temperature dependent.

In order to solve the problem of propagation of longitudinal waves in thin bars, a solution must be obtained for the following first order quasi-linear system, containing three unknown functions σ, v and ε

$$\frac{\partial \sigma}{\partial t} = \varphi(\sigma, \varepsilon)\frac{\partial \varepsilon}{\partial t} + \psi(\sigma, \varepsilon),$$

$$\frac{\partial v}{\partial t} = \frac{1}{\rho}\frac{\partial \sigma}{\partial \varepsilon}, \tag{4.2.4}$$

$$\frac{\partial v}{\partial x} = \frac{\partial \varepsilon}{\partial t}.$$

The second equation is the equation of motion and the third one the compatibility equation; v is the particle velocity, and ρ is the constant density. The system (4.2.4) possesses three families of characteristic curves

$$dx/dt = \pm c(\sigma, \varepsilon) = \pm\sqrt{\varphi(\sigma, \varepsilon)/\rho}, \tag{4.2.5}$$

and

$$dx = 0. \tag{4.2.6}$$

Along the curves (4.2.5) the differential relations

$$d\sigma = \pm\rho c(\sigma, \varepsilon)\, dv + \psi(\sigma, \varepsilon)\, dt \tag{4.2.7}$$

are satisfied while along the lines (4.2.6) are satisfied the relations

$$d\sigma = \varphi(\sigma, \varepsilon)\, d\varepsilon + \psi(\sigma, \varepsilon)\, dt. \tag{4.2.8}$$

Thus the system (4.2.4) is totally hyperbolic. Instead of integrating the system (4.2.4) it is possibly to integrate the equivalent differential relations (4.2.7) and (4.2.8) along the characteristic curves (4.2.5) and (4.2.6) respectively. The upper and lower signs correspond to each other.

In the problem considered a perturbation produced at the end of the bar will propagate along the bar by waves whose velocity is $c(\sigma, \varepsilon)$. There will be some stationary discontinuities [Eq. (4.2.6)]. In order to analyze such phenomena in detail, the dynamic and kinematics compatibility conditions must be taken into account. If $[F]$ is used to denote the jump of a certain function F across a wave front the compatibility conditions are

$$\left[\frac{\partial \sigma}{\partial t}\right] = \varphi\left[\frac{\partial \varepsilon}{\partial t}\right], \qquad \left[\frac{\partial \sigma}{\partial x}\right] dx + \left[\frac{\partial \sigma}{\partial t}\right] dt = 0,$$

$$\left[\frac{\partial v}{\partial t}\right] = \frac{1}{\rho}\left[\frac{\partial \sigma}{\partial x}\right], \qquad \left[\frac{\partial \varepsilon}{\partial x}\right] dx + \left[\frac{\partial \varepsilon}{\partial t}\right] dt = 0, \tag{4.2.9}$$

$$\left[\frac{\partial v}{\partial x}\right] = \left[\frac{\partial \varepsilon}{\partial t}\right], \qquad \left[\frac{\partial v}{\partial x}\right] dx + \left[\frac{\partial v}{\partial t}\right] dt = 0.$$

The left three relations are obtained from Eqs. (4.2.4). The remaining three relations express the condition that the interior differentials $d\sigma$, $d\varepsilon$ and dv are continuous across a wave front. From Eq. (4.2.9) it follows that across any line (4.2.5), the derivatives of σ, ε and v are discontinuous while across a line (4.2.6) all the derivatives in the system (4.2.4) are continuous except the derivatives $\partial\varepsilon/\partial x$ which may be discontinuous.

It should be noted that the function $\varphi(\sigma,\varepsilon)$, corresponding to the instantaneous response of the material, is the only function involved in the expressions for the velocity of propagation and the dynamic compatibility conditions. The function $\psi(\sigma,\varepsilon)$ describing the non-instantaneous response or time effects is only involved in the differential relations (4.2.6) and (4.2.7). Depending on the explicit expression of the function $\varphi(\sigma,\varepsilon)$, the velocity $c(\sigma,\varepsilon)$ can increase or decrease when the stress is varying.

For the integration method, we assume the bar to be of length l and at time $t = 0$ the ends of the bar are at $x = 0$ and $x = l$. The boundary conditions must be formulated as follows:

$$\text{at } x = 0 \quad \text{and} \quad t > 0, \quad \text{we know either } v(0,t) \text{ or } \sigma(0,t). \tag{4.2.10}$$

Similarly:

$$\text{at } x = l \quad \text{and} \quad t > 0, \quad \text{we know either } v(l,t) \text{ or } \sigma(l,t). \tag{4.2.11}$$

The initial conditions are:

$$\text{at } t = 0 \quad \text{and} \quad 0 \leq x \leq l \quad \text{we know } v(x,0), \ \sigma(x,0) \text{ and } \varepsilon(x,0). \tag{4.2.12}$$

If the bar is initially at rest or moving rigidly then the functions prescribed by the conditions (4.2.12) are known along the characteristic line of positive slope passing through $x = 0$ and along the characteristic line of negative slope passing through $x = l$, i.e., along the first wave fronts which propagate in the bar. Otherwise one must first solve a Cauchy initial value problem, and then pass to the Goursat problem in the corners.

If the striking body moves with respect to the material points of the bar, then a mixed boundary value problem must be solved in a domain with a floating boundary. That was the case with the initial tests done in dynamic plasticity. In this case the impacting body must move along the bar with a velocity smaller than the velocity of the waves.

4.3 Instantaneous Plastic Response

In the previous section it was assumed that during the entire process of elastic-viscoplastic deformation a single constitutive equation written in quasi-linear differential form could be used. This procedure can generally be used for plastics, certain soils or rocks, rubbers, certain metals, etc. However, in some problems for

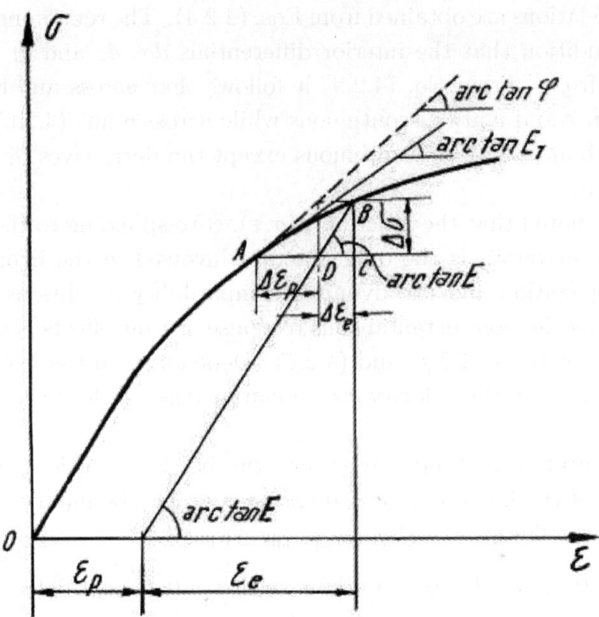

Fig. 4.3.1   Stress–strain relationship for $\psi = 0$.

some materials, especially for the majority of metals, the total strain is always composed of two components: the elastic reversible component and the plastic irreversible one (Fig. 4.3.1), with the total strain bigger than 0.2% say:

$$\varepsilon = \varepsilon^E + \varepsilon^P \tag{4.3.1}$$

where $\varepsilon^E = \sigma/E$. Both components increase during the loading of a work-hardening material, but each of them corresponds to different physical phenomenon and is therefore related to the stress by different constitutive equations, though of course these two phenomena are very tightly interconnected.

It should also be noted that the plastic part of the strain is in fact a visco-plastic component. Thus, there is no longer any question of a one-to-one correspondence between stress and strain, even in the loading process. The stress–strain relation is certainly not unique, since it depends on the loading rate. Thus, in this case the plastic part of the strain has a more general meaning than in classic inviscid plasticity theory. The exact definition, of the elastic component of the strain, can only be given when unloading is defined.

Let us now examine the constitutive equation for elastic-plastic work-hardening materials (Cristescu [1963]). If, at a point A in the plastic region (Fig. 4.3.1) the stress is increased by an amount $\Delta\sigma$, this impulse will propagate in the bar and produce an increase $\Delta\varepsilon^E$ of the elastic part of the strain and an increase $\Delta\varepsilon^P$ of the plastic part of the strain. The former will be related to the stress increment by Hooke's law, which will be written in a differential form

$$\partial\sigma/\partial t = E(\partial\varepsilon^E/\partial t)\,. \tag{4.3.2}$$

The plastic component of the strain increment is generally related to the stress increment by a quasi-linear differential relation of the form

$$\frac{\partial\sigma}{\partial t} = \varphi(\sigma,\varepsilon)\frac{\partial\varepsilon^P}{\partial t} + \psi(\sigma,\varepsilon)\,, \tag{4.3.3}$$

where the functions φ and ψ may depend on some other constants describing the mechanical properties of the material (yield limit, viscosity coefficient, work-hardening modulus, etc.). We assume that $\varphi > 0$.

If the material is in elastic state, only elastic waves will propagate with the velocity

$$c_E = \sqrt{E/\rho}\,. \tag{4.3.4}$$

If the material is plastic/rigid only constitutive equation (4.3.3) is added to the equation of motion. In this case, only pure plastic waves will propagate, the velocity of their fronts being

$$c_P = \sqrt{\varphi(\sigma,\varepsilon)/\rho}\,. \tag{4.3.5}$$

If however, the material is elastic-plastic, the following system of equations will govern the motion

$$
\begin{aligned}
\frac{\partial\sigma}{\partial t} - E\frac{\partial\varepsilon^E}{\partial t} &= 0\,, \\[2mm]
\frac{\partial\sigma}{\partial t} - \varphi(\sigma,\varepsilon)\frac{\partial\varepsilon^P}{\partial t} &= \psi(\sigma,\varepsilon)\,, \\[2mm]
\frac{\partial v}{\partial x} - \frac{\partial\varepsilon^E}{\partial t} - \frac{\partial\varepsilon^P}{\partial t} &= 0\,, \\[2mm]
\rho\frac{\partial v}{\partial t} - \frac{\partial\sigma}{\partial x} &= 0\,.
\end{aligned}
\tag{4.3.6}
$$

The characteristic curves of the system are

$$dx/dt = \pm c(\sigma,\varepsilon) = \pm\sqrt{\varphi E/\rho(\varphi + E)}\,, \tag{4.3.7}$$

and

$$dx = 0 \quad \text{(twice)}\,. \tag{4.3.8}$$

Along the characteristics (4.3.7) the differential relations

$$d\sigma = \pm\rho c\{\pm(\psi/\varphi)\}\,cdt + dv \tag{4.3.9}$$

are satisfied, while along the lines (4.3.8) are satisfied the relations

$$
\begin{aligned}
d\sigma &= Ed\varepsilon^E\,, \\[1mm]
d\sigma &= \varphi(\sigma,\varepsilon)\,d\varepsilon^P + \psi(\sigma,\varepsilon)\,dt\,.
\end{aligned}
\tag{4.3.10}
$$

The upper and lower signs in (4.3.7) and (4.3.10) correspond to each other.

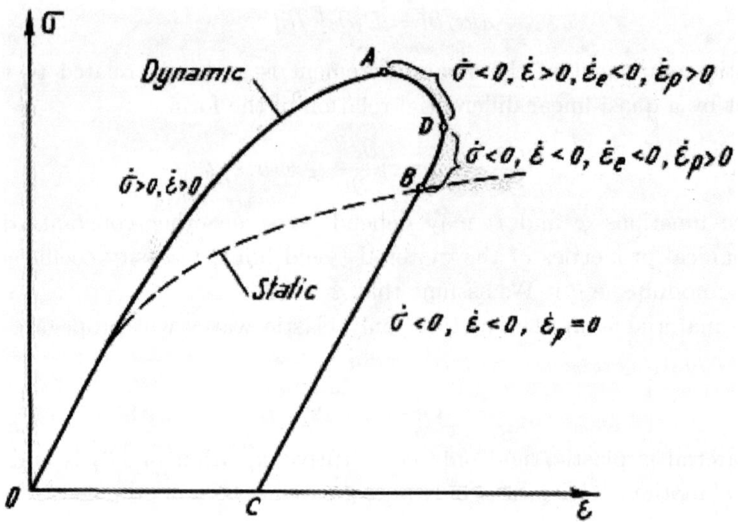

Fig. 4.3.2   A dynamic stress–strain curve showing the "overstress".

Three separate propagation velocities play a part in this problem: c_E, c_P and the velocity c of the elastic-plastic waves. These three velocities are clearly related as follows:

$$\frac{1}{c^2} = \frac{1}{c_E^2} + \frac{1}{c_P^2}. \tag{4.3.11}$$

This formula can also be written in the form

$$c_P = \frac{c}{\sqrt{1 - c^2/c_E^2}}. \tag{4.3.12}$$

Thus, the velocity of the elastic waves c_E can be considered as a "limiting velocity" for the elastic-plastic waves. It is the highest possible velocity of propagation in the material considered and an invariant of the wave propagation phenomenon. From (4.3.12) it follows immediately that $c_P \cong c$ when $c \ll c_E$. This formula can be used to find c_P, if c and c_E are obtained by measurements.

Generally we have

$$c_E > c_P \geq c.$$

The equality $c_P = c$ holds when the material is plastic/rigid and work-hardening. In this case $c^2 = E_1(\sigma, \varepsilon)/\rho$, where $E_1(\sigma, \varepsilon)$ is the variable work-hardening modulus. A dynamic stress–strain curve showing the possible variations during tests is shown in Fig. 4.3.2. The overstress is something composed of several components.

Interesting information can be obtained from the dynamic and kinematics compatibility conditions

$$\left[\frac{\partial \sigma}{\partial t}\right] - E\left[\frac{\partial \varepsilon^E}{\partial t}\right] = 0, \qquad \left[\frac{\partial \sigma}{\partial x}\right] dx + \left[\frac{\partial \sigma}{\partial t}\right] dt = 0,$$

$$\left[\frac{\partial \sigma}{\partial t}\right] - \varphi\left[\frac{\partial \varepsilon^P}{\partial t}\right] = 0, \qquad \left[\frac{\partial \varepsilon^E}{\partial x}\right] dx + \left[\frac{\partial \varepsilon^E}{\partial t}\right] dt = 0,$$

$$\left[\frac{\partial v}{\partial x}\right] - \left[\frac{\partial \varepsilon^E}{\partial t}\right] - \left[\frac{\partial \varepsilon^P}{\partial t}\right] = 0, \qquad \left[\frac{\partial \varepsilon^P}{\partial x}\right] dx + \left[\frac{\partial \varepsilon^P}{\partial t}\right] dt = 0,$$

$$\rho\left[\frac{\partial v}{\partial t}\right] - \left[\frac{\partial \sigma}{\partial x}\right] = 0, \qquad \left[\frac{\partial v}{\partial x}\right] dx + \left[\frac{\partial v}{\partial t}\right] dt = 0. \tag{4.3.13}$$

From (4.3.13) it follows that across the characteristic lines (4.3.7) all the derivatives involved in the system (4.3.13) are discontinuous. Thus (4.3.7) are wave fronts which affect all the unknown functions and certainly both parts ε^E and ε^P of the strain. Across the lines (4.3.8) all the derivatives except for the derivatives $\partial \varepsilon^E/\partial x$ and $\partial \varepsilon^P/\partial x$ are continuous.

From (4.3.13) we may easily obtain:

$$c_E^2[\partial \varepsilon^E/\partial t] = c_P^2[\partial \varepsilon^P/\partial t]. \tag{4.3.14}$$

This is the relation between the jump of the derivatives of the two components of the strain, and the velocities of propagation c_E and c_P. Since in the majority of cases, $c_E^2 \gg c_P^2$, it follows that $[\partial \varepsilon^E/\partial t] \ll [\partial \varepsilon^P/\partial t]$, and thus in many cases, if the material is in an elastic-plastic state and the work-hardening modulus is much smaller than the elastic modulus, the elastic part of the discontinuity carried by the elastic-plastic wave, can be neglected in comparison with the plastic part. Since $c_E = \text{const.}$, all the considerations concerning the relation (4.3.14) depend only on the function $\varphi(\sigma, \varepsilon)$.

We shall now discuss the special case when $\psi = 0$. In this case the differential relations (4.3.9) reduce to the well-known relations

$$d\sigma = \pm \rho \, c \, dv,$$

while from (4.3.10) it follows that

$$E d\varepsilon^E = \varphi \, d\varepsilon^P.$$

Since in general $E \gg \varphi$, it again follows that $d\varepsilon^E \ll d\varepsilon^P$ along (4.3.8). This relation or its equivalent

$$d\varepsilon = (1 + \varphi/E) \, d\varepsilon^P$$

represents the relation between the increments of the components of the strain in the case $\psi = 0$.

If $\psi \neq 0$ the relation between the components of the strain along (4.3.8) is more complicated and follows directly from (4.3.10)

$$d\varepsilon = \left(1 + \frac{\varphi}{E}\right) d\varepsilon^P + \frac{\psi}{E} \, dt.$$

This relation can also be used to examine cases when certain components or the time influence can be neglected. It may be obtained directly from the first two relations (4.3.6) by eliminating $\partial\sigma/\partial t$.

The conclusions are the following. If the constitutive (4.3.2) and (4.3.3) are used, then both the elastic and plastic components of the strain propagate by the same type of wave with the velocity (4.3.7). However both components also possess stationary discontinuities. It should be remembered that the plastic constitutive equation (4.3.3) involves both instantaneous and non-instantaneous plastic response terms.

A similar theory for elastic-plastic-viscoplastic stress waves was developed also by Tanimoto [1994]. The endochronic theory of plasticity to show the effect of the strain rate on the inelastic behavior of structural steel under cyclic loading condition is formulated by Chang *et al.* [1989]. From the endochronic theory (Valanis [1972]) one is taking

$$d\zeta^2 = P_{ijkl}\, d\varepsilon_{ij}^p\, d\varepsilon_{kl}^p - g^2\, dt^2\,.$$

By using a 4th order isotropic tensor for P the previous relation can be written

$$f^2\, dZ^2 = h^2\, d\varepsilon_{ij}^p\, d\varepsilon_{kl}^p - g^2\, dt^2\,,$$

and this equation can by written as

$$g^2\, dt^2 = F^2 f^2\, dZ^2$$

in which

$$F = \sqrt{h^2\frac{(\sigma_{ij}' - \alpha_{ij})(\sigma_{ij}' - \alpha_{ij})}{(s_y^0 f)^2} - 1}\,.$$

The viscoplastic flow condition, equivalent to the overstress of viscoplasticity is

$$h^2\frac{(\sigma_{ij}' - \alpha_{ij})(\sigma_{ij}' - \alpha_{ij})}{(s_y^0 f)^2} - 1 > 0\,.$$

All the other not explained notations are constants. The authors are also making some comparison of the analytical solutions and the experimental results.

The same problem was considered experimentally by Zhao [1997]. The classical split Hopkinson pressure bar tests are used, together with the yield stress dependency on strain rate and temperature

$$\frac{\sigma_Y}{T} = A\left[\ln\left(\frac{\dot\varepsilon}{\dot\varepsilon_0}\right) + \frac{Q}{RT}\right]$$

where A, Q, R are constant coefficients and T is the absolute temperature. Sets of tests of 230 s^{-1} to 1650 s^{-1} were considered for specimens of diameter 10 mm and length 10 mm. The results are shown in Fig. 4.3.3.

The aluminum 1100-0 was tested by Pao and Gilat [1989]. They were tested for the plastic strain rate sensitivity with temperature as

$$\dot\varepsilon_{ij}^P = 2\eta\exp\left(-\frac{H_0}{kT}\right)\sinh\left(\frac{v^*(\bar\tau - \tau^*)}{kT}\right)\frac{\sigma_{ij}'}{\bar\tau}$$

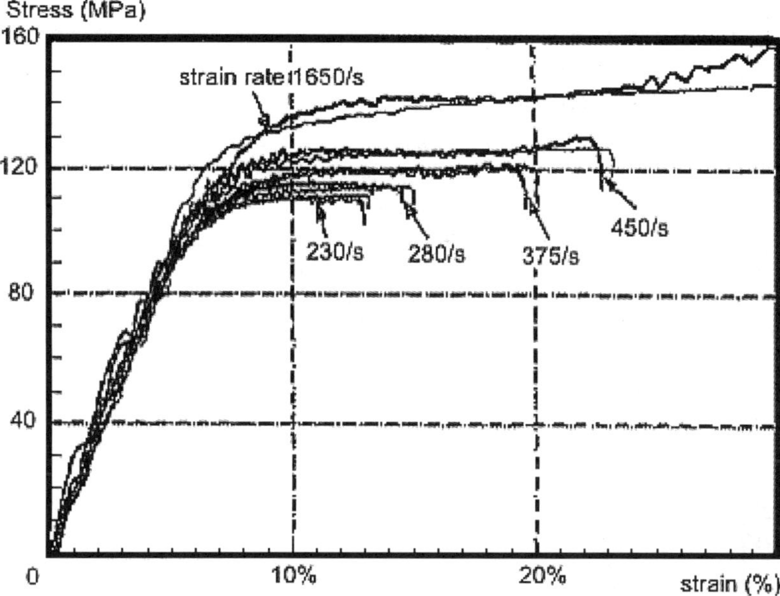

Fig. 4.3.3 Comparison between stress–strain curves derived from real tests and those from simulated tests.

were H_0, η, v^*, and τ^* are determined from results of tests in which the material is deformed at various constant strain rates and temperatures. At $T = 0$, $\bar{\tau} = \tau^* + H_0/v^*$. Other values of material constants are given. They have used the same boundary conditions as given below and the results are given in Fig. 4.3.4. Comparison with the tests of Bell is done on three distances from the impacted end. The comparison seems good.

The existence and properties of the free energy function compatible with the second law of thermodynamics in one-dimensional rate-type semilinear viscoelasticity is analyzed by Faciu and Mihailescu-Suliciu [1987]. Necessary and sufficient conditions are given such that a free energy as a function of strain and stress exists and is unique, that it is non-negative function and possesses a monotony property with respect to the equilibrium curve. A bound in energy for the smooth solutions of certain initial and boundary value problems with respect to the input data is established when the equilibrium curve is a non-monotonic curve.

Bounds in energy for the smooth solutions of initial and boundary value problems for semi-linear hyperbolic systems were given by Făciu and Simion [2000]. They prove that the integral energy identities for smooth solutions are still valid for weak solutions. Energy estimates of the weak solutions of the same problems are given. The way weak and strong discontinuities in the input data affect the regularity of the total energy of the solution is discussed. The uniqueness of the weak solutions is derived.

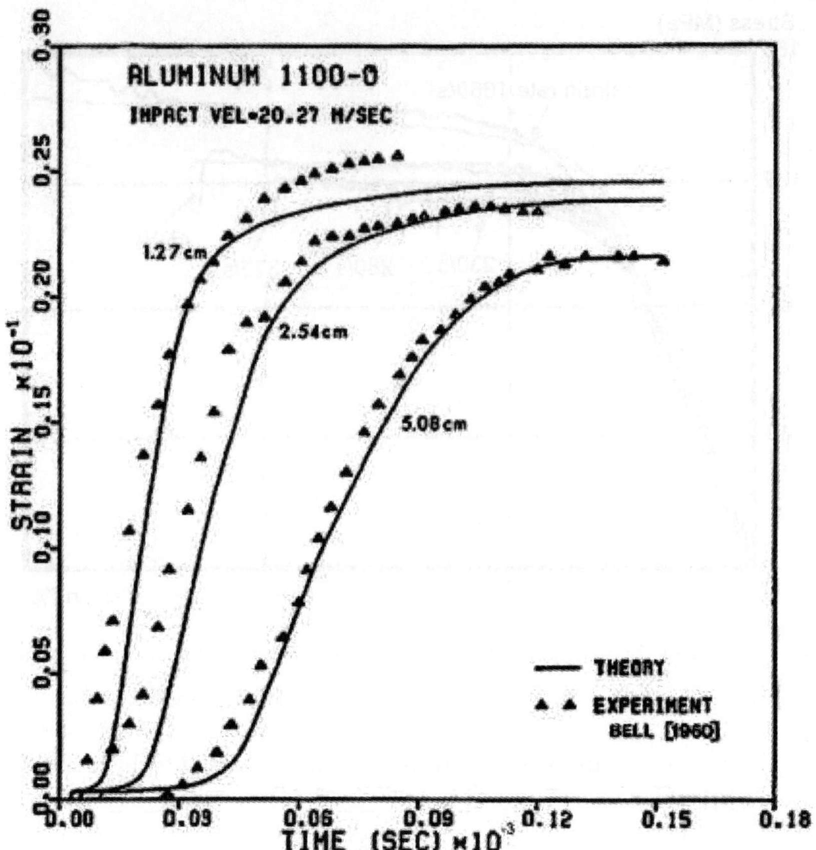

Fig. 4.3.4   Variation of strain with time at three indicated distances.

A nonhomogeneous and nonautonomous 2×2 quasilinear system of first order describing rate-type materials is considered by Manganaro and Valenti [1993] within the framework of similarity reduction procedures. The constitutive equation is of the form

$$\frac{\partial v}{\partial t} = \phi(v, \varepsilon)\frac{\partial \varepsilon}{\partial t} + \psi(v, \varepsilon)\,.$$

Classes of exact similarity solutions to the governing model as well as functional forms for the material response function involved have been determined. Reduction to linear form have been carried out for different forms of constitutive laws. In another article by Menganaro and Meleshko [2002], the method of differential constrains is applied for systems written in Riemann variables. Generalized simple waves are considered. This class of solutions can be obtained by integrating a system of ordinary differential equations. The above rate-type models is one of the systems studied.

A constitutive law of the form

$$\dot{\sigma} = \mathbf{E}\varepsilon(\dot{\mathbf{u}}) + \mathbf{G}(\sigma, \varepsilon(\mathbf{u}), \beta)$$

where $\mathbf{E}$ is a fourth order tensor, and β is the damage field was considered by Chau *et al.* [2002]. When $\beta = 1$ the material is undamaged, while the value $\beta = 0$ indicates the stage of complete damage. The mechanical damage of the material, caused by excessive stress or strain, is described by a damage function whose evolution is modeled by an inclusion of parabolic type. The authors provide a variation formulation for the mechanical problem and sketch a proof of the existence of a unique weak solution to the model. They introduce and study a fully discrete scheme for the numerical solutions of the problem. An optimal order error estimate is derived for the approximate solutions under suitable solution regularity. Numerical examples are presented to show the performance of the method.

The contact problems considered by Fernández *et al.* [2003] involve the large class of elastic-viscoplastic materials. They focus on the finite element approximations which involve nonmatching meshes on the contact part. Such a configuration in which the nodes inherited from the discretizations of the bodies may not coincide arises in computational situations when the different bodies are independently meshed and/or there is an initial distance between the bodies and/or an evolution process is considered. The first aim of the authors is to study the convergence of finite element methods and then to carry out the numerical experiments associated with the theoretical studies.

A simply non-linear anelastic model describing the acceleration of the relaxation process in the presence of vibrations was considered by Drăgănescu and Căpălnăşan [2003]. An analytic variational iteration method is used in order to study the relaxation process. Also a method of identification of the parameters of the system from the experimental data is given.

4.4 Numerical Examples

In the numerical examples of the rate-type constitutive equations under consideration were used in the form

$$\dot{\varepsilon} = \left\{ \frac{1}{E} + \chi\Phi(\sigma, \varepsilon) \right\} \dot{\sigma} + \Psi(\sigma, \varepsilon), \qquad (4.4.1)$$

with

$$\chi = \begin{cases} 0 & \text{if } \sigma \leq f(\varepsilon) \text{ or } \dot{\sigma} < 0, \\ 1 & \text{if } \sigma = f(\varepsilon) \text{ and } \dot{\sigma} > 0. \end{cases} \qquad (4.4.2)$$

The coefficient χ is used so that the constitutive equation is used during loading and unloading.

The characteristic lines are

$$\frac{dx}{dt} = \pm c(\sigma) = \sqrt{\frac{E}{\rho(1 + E\Phi)}}\,, \tag{4.4.3}$$

$$dx = 0\,,$$

the differential relations satisfied along them

$$d\sigma = \pm \rho c\, dv - \rho c^2 \Psi(\sigma, \varepsilon)\,,$$

$$d\sigma = E d\varepsilon^E\,, \tag{4.4.4}$$

$$d\varepsilon^P = \Phi(\sigma, \varepsilon)\, d\sigma + \Psi(\sigma, \varepsilon)\, dt\,.$$

The coefficient function $\Psi(\sigma, \varepsilon)$ describing the non-instantaneous response was assumed to depend linearly on the overstress (Cristescu [1972a, 1972b], Cristescu and Cristescu [1973]):

$$\Psi(\sigma, \varepsilon) = \begin{cases} \dfrac{k(\varepsilon)}{E}[\sigma - f(\varepsilon)] & \text{if } \sigma > f(\varepsilon) \text{ and } \varepsilon \geq \dfrac{\sigma}{E}\,, \\ 0 & \text{if } \sigma \leq f(\varepsilon)\,. \end{cases} \tag{4.4.5}$$

For the relaxation boundary we write

$$f(\varepsilon) = \begin{cases} \sigma_Y & \text{if } \varepsilon \leq \varepsilon_Y\,, \\ \beta(\varepsilon + \varepsilon_0)^{1/\alpha} & \text{if } \varepsilon > \varepsilon_Y \end{cases} \tag{4.4.6}$$

or some variants of this law. The examples given are for aluminum (1100°F aluminum annealed for two hours at 590°C and then furnace cooled — see Bell [1968]). For this material the relations

$$\sigma = E\varepsilon \qquad \text{if} \quad \sigma \leq \sigma_Y\,,$$

$$\sigma = \beta(\varepsilon + \varepsilon_0)^{1/\alpha} \quad \text{if} \quad \sigma > \sigma_Y$$

were used, where $E = 7038000$ N/cm^2 and $\rho_0 = 2.7$ g/cm^3, and therefore $c_0 = 5080$ ms^{-1}. For the other constants, the following numerical values were used (Cristescu and Bell [1970] and Cristescu [1972b]):
for the quasi-static stress–strain curve,

$$\sigma_{Y_0} = 703.80 \text{ N/cm}^2\,, \quad \beta = 22908.0 \text{ N/cm}^2\,, \quad \alpha = \frac{8}{3}\,;$$

for the dynamic stress–strain curve,

$$\sigma_{Y_1} = 212.18 \text{ N/cm}^2\,, \quad \beta = 38640.0 \text{ N/cm}^2\,, \quad \alpha = 2\,;$$

for the translated dynamic stress–strain curve,

$$\sigma_{Y_2} = 759.00 \text{ N/cm}^2\,, \quad \beta = 38640.0 \text{ N/cm}^2\,, \quad \alpha = 2\,, \quad \varepsilon_0 = 0.000278\,.$$

The true yield stress for the aluminum under consideration is in fact, and the stress–strain curve is obtained from the dynamic one by a translation along the strain axis (this defines ε_0). The stress–strain curve obtained by this procedure is quite close to the dynamic one.

The expression for k is an expression showing a small value for small strains but becomes greater, nearly constant, for higher values. Thus

$$k(\varepsilon) = k_0 \left[1 - \exp\left(-\frac{\varepsilon}{\bar{\varepsilon}}\right)\right] \qquad (4.4.7)$$

with k_0 and $\bar{\varepsilon}$ material constants. Various variants of this constitutive law were used.

Concerning the slope of the instantaneous curve we assume that it lies between the elastic slope and that of the relaxation boundary

$$E \geq \frac{1}{\Phi(\sigma, \varepsilon) + 1/E} > f'(\varepsilon). \qquad (4.4.8)$$

Further $\Phi(\sigma, \varepsilon)$ was chosen to match a certain set of experimental data. Two forms were used. Starting from a cubic equation we have used

$$\Phi(\varepsilon) = \chi \frac{3[\varepsilon - \varepsilon_Y - \varepsilon^* + (a/3E)^{3/2}]^{2/3}}{a} - \frac{1}{E} \, ,$$

where

$$a = m + n\sqrt{\varepsilon}$$

with m, n material constants and ε^* is a threshold strain.

Some other value is

$$\Phi(\sigma, \varepsilon) = \chi \frac{\gamma}{E} \left[3 \left(\frac{E}{p + q\sqrt{\varepsilon}}\right)^3 \left(\frac{\sigma}{E}\right)^2 - 1 \right]$$

where p, q and γ are material constants. It is quite difficult to determine the velocity of propagation experimentally, and thus Φ.

The numerical examples are for the impact of two bars, the "hitter" and the "specimen". For the specimen we have

$$t = 0, \quad 0 < x \leq l : \sigma = \varepsilon = v = 0$$

for the initial conditions.

For the boundary conditions the end of the bar $x = l$ is assumed free

$$t \geq 0, \quad x = l : \sigma = 0.$$

At $t = 0$ the bar is impacted at the end $x = 0$ by another bar identical with the specimen. Thus, for the hitter we have

$$t = 0, \quad -l \leq x < 0 : \sigma = \varepsilon = 0, \quad v = V.$$

Thus for $t = 0$ the end $x = 0$ of the specimen is subjected to a sudden jump. This jump was handled in two ways. First, it was assumed that the impact velocity is fast but smoothly transferred to the end of the specimen. A convenient very short time interval $0 \leq t < t_m$ was chosen, in which this smooth increase of velocity at the impacted end of the specimen takes place from zero up to $v_{\max}$. For simplicity, it

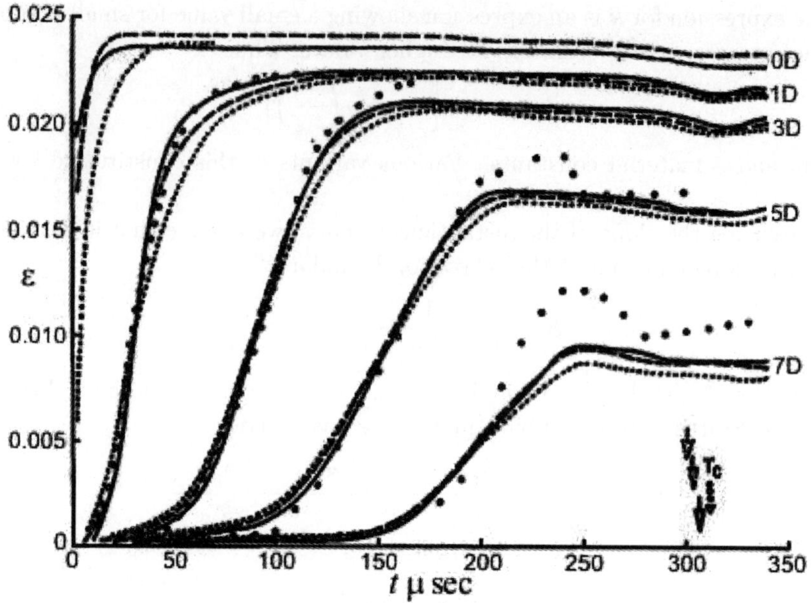

Fig. 4.4.1   Variation of the strain at various sections along the bar, and time of contact.

was assumed in the program that the velocity at the end of the specimen increases according to a linear law. Thus, the first kind of boundary conditions were

$$x = 0 \begin{cases} 0 \leq t < t_m : v = \dfrac{t}{t_m} v_{\max}\,, \\[2mm] t_m \leq t \leq T_c : v_H = v_S\,, \\[2mm] T_c < t : \sigma_S = 0\,, \end{cases}$$

where t_m is a conveniently chosen time interval (in most examples given here $t_m = 0.5\ \mu$s) and $v_{\max} = V/2$. T_c is a computed time, called time of contact. In the time interval $t_m \leq t \leq T_c$ the ends $x = 0$ of the two bars are moving together. σ_S is the specimen stress which is computed as is T_c.

Several examples have been computed. In Fig. 4.4.1 is given the variation of the strain at various sections along the bar. Also have been given also the time of contact and a classical solution. One can see that the rate type solutions are quite correct. In Fig. 4.4.2 is given the variation of the stress at the impacted end. After a peak, follows a plateau. Again the experimental data are quite well reproduced. The decrease of the stress is due to unloading. In Fig. 4.4.3 is shown the variation of the deformation at the 9D, i.e., close to the free end, while in Fig. 4.4.4 is given the variation of the displacement in time at various cross-sections.

From the above one can see that a plateau for strains can be predicted for a finite distance (Suliciu [1972]). Daimaruya and Naitoh [1983] have also considered

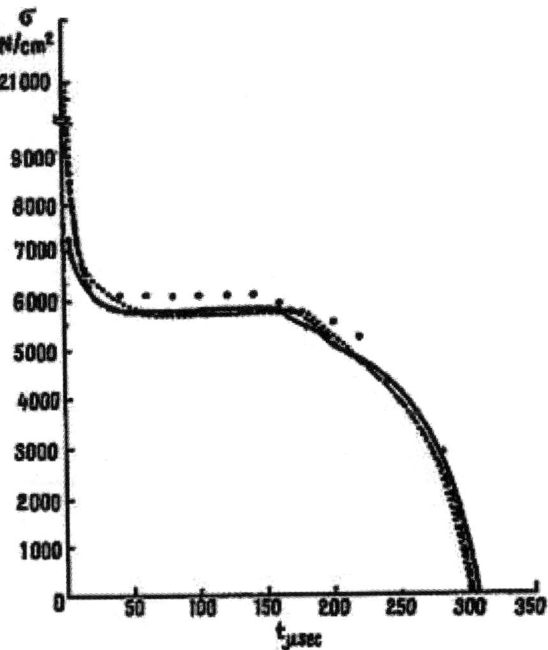

Fig. 4.4.2   Variation of the stress at the impacted end.

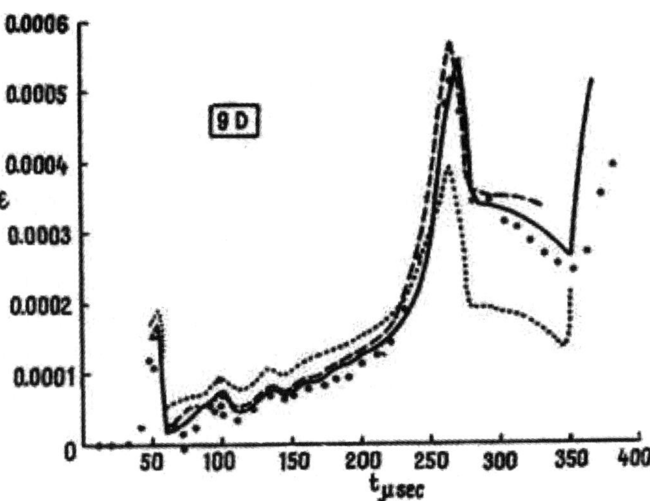

Fig. 4.4.3   Variation of the strain in time at $9D$.

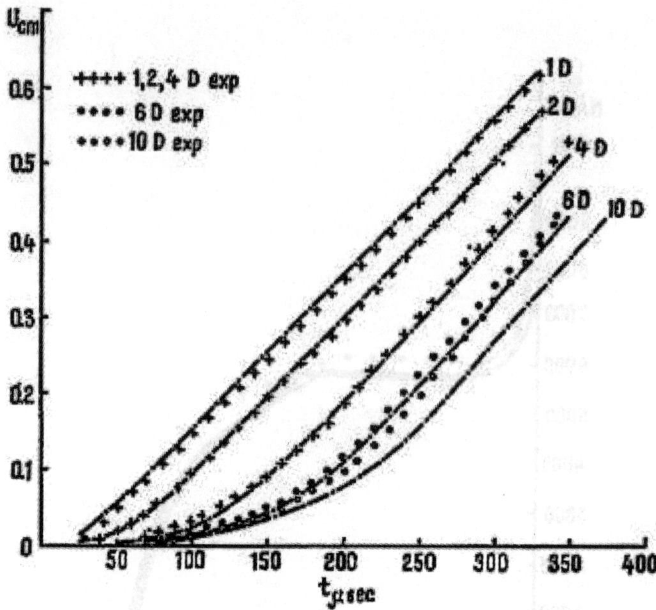

Fig. 4.4.4   Comparison of the displacement-time curves.

the plateau problem; they have shown numerically taking the model (4.2.2) for hardened aluminum $f(\varepsilon) = 6.89(20 - 0.01/\varepsilon)$ MPa, with the values of the constants $E = 68.8$ GPa, $\rho = 2.67 \times 10^3$ kg/m^3, $k = 10^6$ sec^{-1} and impact velocity $V = 15$ m/sec, that a plateau of uniform strain adjacent to the impact end can exist. For larger value of k, the plateau will be shorter. Various variants of the constitutive equation as well as examples are considered by Cristescu and Suliciu [1982]. For instance

$$\dot{\sigma} = E\varepsilon(\dot{u}) + G(\sigma, \varepsilon(u), \beta)$$

where β is a function which satisfies an ordinary differential equation as

$$\dot{\beta} = \varphi(\sigma, \varepsilon(u), \beta),$$

(see Ionescu and Sofonea [1993], Fernández [2004]). Some other authors (Luo *et al.* [2003]) have used the same model to introduce the isothermal rheological forming of high-strength alloying parts with complicated curved surface. They have used various speeds and found that for the lower forming velocity ($v = 0.022$ mm/s) the non-uniformity of temperature field is smaller than at higher speeds.

4.5   Other Papers

A first study about the effect of temperature gradients on the propagation of elasto-plastic waves is due to Francis and Lindholm [1968]. It is question of the propagation of an extensional elastoplastic wave through a long thin bar heated at the end to produce a continuously decreasing temperature profile. The temperature distribution is approximated as an exponential function.

Analytical solutions are obtained by Shaw and Cozzarelli [1971] for stress, velocity, and strain at the wave front in a suddenly loaded semi-infinite rod of a material with a linear instantaneous response and nonlinear inelastic response. The material properties are assumed to depend on position directly or through a dependence on a prescribed non-uniform temperature field. Detailed solutions are obtained for two examples, a nonlinear viscoelastic material with temperature dependent parameters and a rate-sensitive plastic material which may have temperature dependent parameters and yield point.

The dynamic mechanical behavior of titanium in shear was studied by Lawson and Nicholas [1972]. They have used a constitutive equation of the form

$$ G\frac{\partial \gamma}{\partial t} = \frac{\partial \tau}{\partial t} + g(\tau, \gamma) \, . $$

They found that the torsional plastic waves in pure titanium can be predicted from the rate-dependent theory of plastic wave propagation. The excellent agreement between theoretical and experimental results for stresses at the end of a long tube subjected to a torsional pulse demonstrates the consistency of the entire testing procedure.

The problem of impact on a nonlinear viscoelastic rod of finite length is considered by Mortell and Seimour [1972]. They have considered a constitutive law of the form

$$ \dot{\sigma} = \phi(\varepsilon)\dot{\varepsilon} + \psi(\sigma, \varepsilon) \, . $$

It is shown that, in the high frequency or geometrical acoustics limit, the disturbance in the rod may be represented as the superposition of two modulated simple waves traveling in opposite directions which do not interact in the body of the material.

A comparison with the experiments is done by Kuriyama and Kawata [1973]. They consider three types of constitutive equation; that is, the Karman-type theory, the Malvern-type theory, and the Cristescu-type theory. The constitutive equation is written

$$ E\dot{\varepsilon} - \dot{\sigma} = h(\sigma, \varepsilon) \, . $$

They find that the front of the stress wave travels always with the elastic-wave velocity and the plateau of the uniform plastic strain remains in the neighborhood of the impact end. They find that the Johnston–Gilman-type of constitutive equation is matching better.

Suliciu *et al.* [1973] have developed a class of general solutions for wave propagation in a rate-type material. These solutions, obtained by invoking equation splitting, demonstrate elastic-like behavior for the kinematics variables u and ε although the material is inelastic. This observation serves as a warning to those carrying out dynamic experiments involving only axial loadings. Measurements of strain and velocity may not reveal the true characteristics of materials. An extended solution for transverse motion is presented and experiments involving transverse deflection are highly recommended because these results do provide information about σ and therefore the rate dependencies.

A one-dimensional problem has been considered by Nicholson and Phillips [1978]. They have considered elastic/viscoplastic/plastic materials of the form

$$\dot{\varepsilon} = \frac{\dot{\sigma}}{E_0} + \frac{\dot{\sigma}}{E_p} + \eta \left\langle \sigma - \sigma_0 - k \left(\varepsilon - \frac{\sigma}{E_0} \right) \right\rangle$$

where σ_0 are the quasi-static yield stress and all the other are constants. Various variants of the constitutive equation are considered for stress impact and velocity impact. Then the Laplace transform is used, and near the impact end the solution is obtained in terms of some special functions. For an elastic/viscoplastic material, a region of uniform strain arises at the impact end more rapidly for velocity impact than for stress impact. Also for an elastic/viscoplastic/plastic material, the plastic strain makes no contribution at the impact end for velocity impact, but it does for stress impact.

For aluminum and copper rods, for a wide range of striking velocities and rods lengths, the general applicability of one-dimensional non-viscous dynamic plasticity theory is confirmed (Dawson [1972]).

A model of the kind used in this chapter vas proposed by Suliciu [1981] to describe the Savart–Masson effect by a rate-type constitutive equation. The viscosity coefficient has strong variations in some regions of the stress–strain plane that lie above the equilibrium curve $\sigma = f(\varepsilon)$.

Cernocky [1982] has considered four constitutive equations of viscoplasticity and made a comparison between them. Each theory is fitted to the same stress–strain data, and both analytical and numerical methods are employed to highlight similarities and differences between their predictions. Different manifestations of strain-rate and stress-rate history effects predicted by the theories are compared, and the theories are shown to share a significant qualitative bias between responses to stress and strain-controlled loading.

Some dynamic tests on aluminum titanium and steel were done by Nicholas [1981]. He has found a rate sensitivity of 10 to 10^2 s^{-1}. In another papers Nicholas *et al.* [1987a, b] a finite difference computer code is used to numerically simulate uniaxial strain waves generated from plate impact experiments. The incremental flow law of Bodner and Partom and the Perzyna law are compared with experiments. Numerical computations using a finite difference computer code illustrate that the decay of the amplitude of the propagating precursor elastic wave is a consequence of material strain-rate dependence.

The possibility of describing rubber-like materials by means of a rate-type viscoelastic constitutive equation is due to Mihailescu-Suliciu and Suliciu [1987]. The problem of generation of one-dimensional tensile shock waves in rubber-like materials is studied numerically and compared to the exact elastic nonlinear solution and the steady wave solution. It is shown that a rate-type semilinear visco-elastic model can describe the steepening of the wave during its propagation and a "thickness" of the wave is naturally incorporated. An energetic criterion for the numerical stability is also introduced.

An experimental technique is described by Chhabildas and Swegle [1989] which uses anisotropic crystals to generate dynamic pressure-shear loading in materials. The coupled longitudinal and shear motion generated upon planar impact of anisotropic crystal can be transmitted into a specimen bonded to the rear surface of crystal, and monitored using velocity interferometer techniques.

In a paper by Li and Clifton [1981] one is presenting stress–strain curves for aluminum sandwiched between hard steel plates. The shear strain rates go up to and sustain these rates for a duration of 10^{-6} s. These stress–strain curves are much higher than the classical ones.

Damage is considered by Ziegler [1992] to be a time-variant imperfection of the yielding material, analogously to the growth of porosity of a voided material with given initial porosity. Also plastic strain is a distortion of the perfect elastic background structure. The internal variable approach is transformed into a numerical routine of computational dynamic viscoplasticity.

Starting from the phenomena not well described as the increase of stress with strain rate, the propagation speed of the stress wave corresponding to a large strain is slow, and others, Tanimoto *et al.* [1993] try to explain it by introducing an elastic-plastic-viscoplastic constitutive equation of the form

$$\rho\frac{\partial v}{\partial t} = \frac{\partial \sigma}{\partial x},$$

$$\frac{\partial v}{\partial x} = \frac{\partial \varepsilon}{\partial t},$$

$$\frac{\partial \varepsilon}{\partial t} = \begin{cases} \left\{ \dfrac{1}{E} + \dfrac{2}{3\zeta}\left(\dfrac{\sigma_E}{\sigma} - 1\right)^{k_1} \right\} \dfrac{\partial \sigma}{\partial t} + \dfrac{2}{3\eta}\left(\dfrac{\sigma}{\sigma_S} - 1\right)^{k_2}\sigma, & \sigma > \sigma_S \\ \dfrac{1}{E}\dfrac{\partial \sigma}{\partial t}, & \sigma \leq \sigma_S, \end{cases}$$

with the various constants given as: $E = 103.5$ GPa, $\sigma_S = 102.4$ MPa, $\zeta = 11.77$ MPa, $\eta = 1.177 \times 10^3$ Pa s, $k_1 = 0.85$, $k_2 = 5.4$, $\rho = 8.391 \times 10^{-9}$ kgr/mm^3. The approach is also numerical, as given above. Several figures are given to show the effect. We give only two. In Fig. 4.5.1 is given the variation of stress in the distance from the impact end. In Fig. 4.5.2 is given the variation of strain with distance from the impacted end. Close to the impacted end the strain is constant (plateau) and this value is increasing in time. That is not in agreement with the tests, but the authors are trying to find various arguments.

An approach is developed to generate exact solutions to a hyperbolic model of the form

$$u_t + v_x = 0, \quad v_t + h(v)u_x = f(v)$$

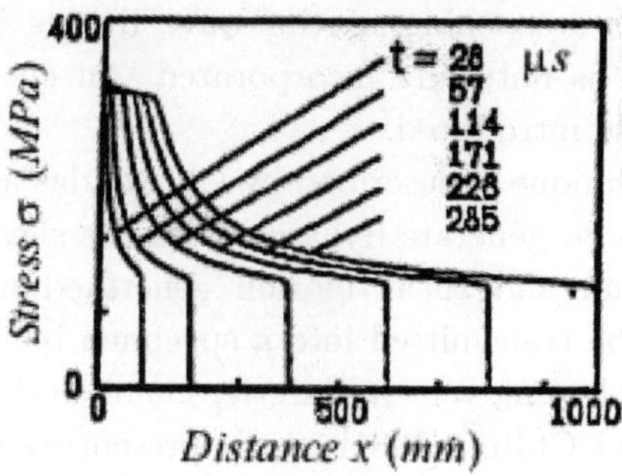

Fig. 4.5.1   Stress-distance relationship.

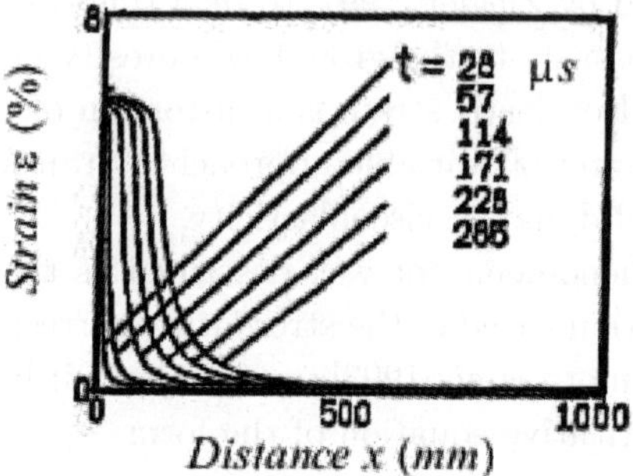

Fig. 4.5.2   Strain-distance relationship.

by Fusco and Manganaro [1994a]. The leading idea of the analysis which is carried on herein is to require the model in point to be considered with a pair of additional equations defining Riemann invariants-like quantities along the characteristic curves. Hence, classes of model constitutive laws allowing a reduction procedure to hold are characterized. Later on, a nonhomogeneous system of partial differential equations of first order involving two dependent and two independent variables is considered also by Fusco and Manganaro [1994b]. The same authors (Fusco and Manganaro [1996]) consider the system

$$u_t - v_x = 0, \quad \varepsilon_t - u_x = 0, \quad v_t - \phi(t, \varepsilon, v)u_x = \psi(t, \varepsilon, v)$$

for which solutions are given. The same kind of equations vas also considered by Tabov [1996]. He obtains easy to verify criterion for the existence of such solutions and clearly stated conditions on the initial data providing the existence and uniqueness of the solutions.

The system

$$v_t - \sigma_x = 0, \quad \sigma_t - \phi(t, \sigma)v_x = \psi(t, \sigma)$$

is proposing to be reduced to linear form 2×2 non-homogeneous and non-autonomous first order quasi-linear systems by Currò and Valenti [1996]. The system through the use of a variable transformation is reduced to a homogeneous and an autonomous form which can be linearized by means of hodograph transformation.

A general algorithm of implicit stress integration in viscoplasticity is presented by Kojic [1996]. The algorithm is first applied to isotropic metals obeying the von Mises yield condition with mixed hardening and then to orthotropic metals.

A theory of viscoplasticity bounded between two distinct rate-independent generalized plasticity models was considered by Auricchio [1997]. Main features of the model are the following: (1) it approaches the two rate-independent generalized plasticity models for the case of fast and slow loading conditions, respectively; (2) it properly describes loading-unloading-reloading conditions for any loading rate; (3) it reproduces the experimentally observed difference between the dynamic and the static yielding conditions. One has in mind metals, polymers, geomaterials, and shape-memory alloys. Both the time-continuous and the time-discrete version of the model are discussed. Several examples are given.

The main objective of the paper by Lodygowski and Perzina [1997] is the investigation of adiabatic shear band localized fracture phenomenon in inelastic solids during dynamic loading processes. This kind of fracture can occur as a result of an adiabatic shear band localization generally attributed to a plastic instability implied by micro-damage and thermal softening during dynamic plastic flow processes.

A review is presented by Klepaczko [1998] on recent progress in shear testing of materials at high and very high strain rates. Some experimental technique are discussed which allow for materials testing in shear up to $10^6 1/s$. More information is provided on experimental techniques based on the Modified Double Shear specimen loaded by direct impact. This technique has been applied so far to test a variety of materials, including construction, armor and inoxidable steels, and also aluminum alloys. The double shear configuration has also been applied to test sheet metals, mostly used in the automotive industry, in a wide range of strain rates. A new experimental configuration which can be applied for experimental studies of adiabatic shear propagation and high speed machining is discussed. One is also showing that for steel at lower impact velocities, the energy to the final localization increases up to 681 MJ/m^3; however, at impact velocities higher than 100 m/s, the energy drops considerably, to the value of ~ 8.0 MJ/m^3. Thus, the energy drop is slightly less than a hundred times.

Results of pressure-shear plate impact experiments on an alumina ceramic (AD85) containing a significant amount of an oxide glass phase were done by Sundaram and Clifton [1998]. Under high strain-rate loading, the ceramic (porosity of approximately 9%) shows low shear strength and a strong strain softening behavior. The material is modeled as an elastic-visco-plastic solid. The collapse of the voids causes a strong hydrostatic relaxation which brings down the pressure.

A finite element technique is presented by Yokoyama [2001] for the analysis of one-dimensional torsional plastic waves in thin-walled tube. Three different non-linear constitutive relations deduced from elementary mechanical models are used to describe the shear stress–strain characteristics of the tube material at high rates of strain (10^3 s^{-1}). It is question of elasto-plastic model of Karman-Duwez, the elasto-viscoplastic model of Sokolovsky–Malvern and the elasto-viscoplastic-plastic model of Cristescu–Lubliner. The resulting incremental equations of torsional motion for the tube are solved by applying a direct numerical integration technique in conjunction with the constitutive relations. A comparison with experimental results shows that the strain-rate dependent solutions show a better agreement with the experimental results than the strain-rate independent solutions.

The superplastic deformation and cavitations damage characteristics of a modified aluminum alloy are investigated at a temperature from 500 to 550°C by Khaleel *et al.* [2001]. The experimental program consists of uniaxial tension tests and digital image analysis for measuring cavitations. The experiments reveal that evolution of damage is due to both nucleation and growth of voids. A viscoplastic model for describing deformation and damage in this alloy is developed based on a continuum mechanics framework. The model includes the effect of strain hardening, strain rate sensitivity, dynamic and static recovery, and nucleation and growth of voids.

Altenhof and Ames [2002] study the strain rate effects for aluminum and magnesium steering wheel armatures when they are subjected to dynamic impact tests. Two geometrically different steering wheel armatures, a three spoke proprietary aluminum alloy armature and a four spoke magnesium alloy armature, underwent experimental impact testing. The strain rates are not very high: 17 s^{-1} for aluminum and 46 s^{-1} for magnesium armature.

Stress–strain response under constant and variable strain-rate is studied for selected models of inelastic behavior by Lubarda *et al.* [2003]. The derived closed-form solutions for uniaxial loading enable simple evaluation of the strain-rate effects on the material response. Various one-dimensional models have been written. In the case of parabolic hardening, the flow stress is related to the plastic strain

$$\sigma^p = Y + (\hat{\sigma} - Y)\frac{\varepsilon^p}{\hat{\varepsilon}^p}\left(2 - \frac{\varepsilon^p}{\hat{\varepsilon}^p}\right)$$

where Y is the initial yield stress, and $\hat{\varepsilon}^p$ is the plastic strain at the apex of the plastic stress–strain curve with the stress σ^p. For the hyperbolic type of hardening we have

$$\sigma^p = \hat{Y} - \frac{\hat{Y} - Y}{1 + \varepsilon^p/\hat{\varepsilon}^p}.$$

For viscoelastic-plastic model, the stress is

$$\sigma = \sigma^*\left[1 - \exp\left(-\frac{E\varepsilon}{\sigma^*}\right)\right], \quad \varepsilon = -\frac{\sigma^*}{E}\ln\left(1 - \frac{\sigma}{\sigma^*}\right).$$

In Fig. 4.5.3 are given three curves corresponding to this case. Here $\sigma^* = \eta\dot{\varepsilon}$ with η the viscosity coefficient. In a similar way one has considered several curves including the viscoelastic-elastoplastic model.

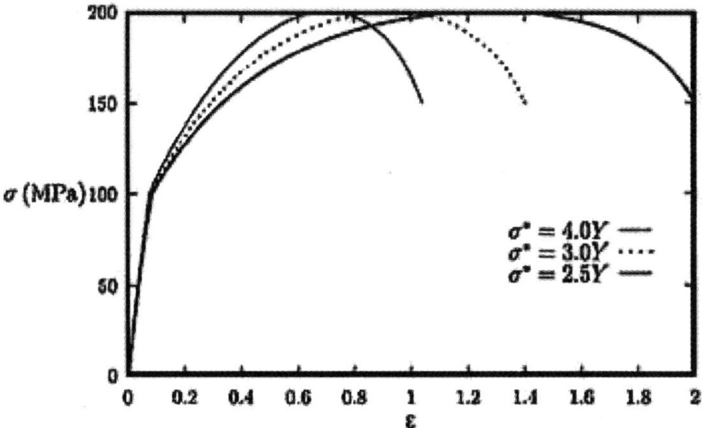

Fig. 4.5.3   The stress–strain curves for viscoelastic-plastic model in the case of parabolic hardening.

The split Hopkinson pressure bar, or Kolsky's apparatus was re-analyzed by Zhao and Gary [1996] and by Zhao [2003]. The same system of equation

$$\frac{\partial \varepsilon}{\partial t} = \frac{1}{E}\frac{\partial \sigma}{\partial t} \qquad\qquad \text{if } \sigma \le \sigma_s \,,$$

$$\frac{\partial \varepsilon}{\partial t} = \frac{1}{E}\frac{\partial \sigma}{\partial t} + g(\sigma,\varepsilon) \quad \text{if } \sigma > \sigma_s$$

is used with

$$g(\sigma,\varepsilon) = \frac{(1 + E_t/E)\,\sigma - \sigma_s - E_t\varepsilon}{\eta}\,.$$

He is proposing to calculate directly the dispersive relation for the wave dispersion corrections and to use the elastic simulation to determine more accurately correspondence between the wave beginnings. A method to increase measuring duration is proposed. An identification technique, especially for non-metallic materials, based on an inverse calculation method is also presented.

The Kolsky torsion bar was used by Hu and Feng [2004] to study the behavior of polymer melts at high shear rates from 10^2 s^{-1} to 10^4 s^{-1} and temperature up to 300°C. To give an example on Fig. 4.5.4 is given one case considered. The important influence of the strain rate is evident.

Constitutive modeling of the high strain rate behavior of interstitial-free steel was considered by Uenishi and Teodosiu [2004]. The strain rates are quite high of 1000 s^{-1}. To have an idea, we reproduce in Fig. 4.5.5 several curves from that paper. In order to simulate the high strain rate tensile test, and that of temperature, the Johnson–Cook model is used

$$\bar\sigma(\bar\varepsilon, \dot{\bar\varepsilon}, T) = (A + B\bar\varepsilon^n)\left[1 + C\ln\left(\frac{\dot{\bar\varepsilon}}{\dot\varepsilon_0}\right)\right]\left[1 - \left(\frac{T - T_{\text{room}}}{T_{\text{ref}} - T_{\text{room}}}\right)^m\right]$$

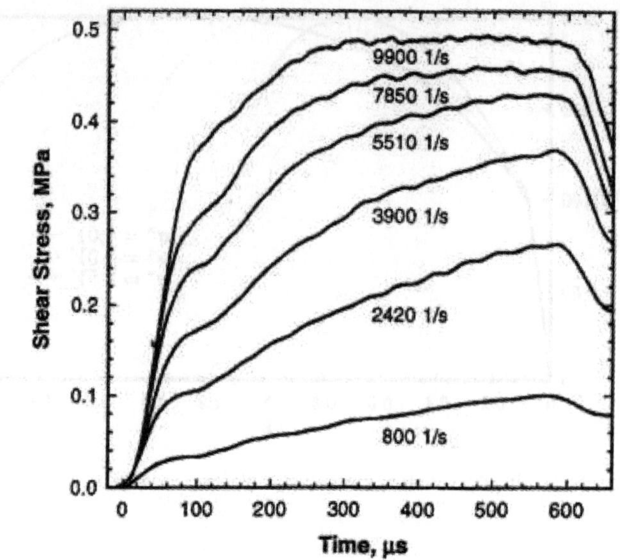

Fig. 4.5.4   Shear stress profiles for various shear rates and 190°C initial temperature.

where $\bar{\sigma}$, $\bar{\varepsilon}$ and $\dot{\bar{\varepsilon}}$ are the equivalent tensile stress, the equivalent tensile strain and the equivalent tensile strain rate, respectively. The first term represents the work hardening behavior at the reference strain rate $\dot{\varepsilon}_0$ and at the room temperature. The second and third terms express the effects of strain rate and temperature, respectively. The parameters A, B, n, C and m are material parameters and T denotes the absolute temperature. T_{room} and T_{ref} are the absolute room temperature and melting temperature of the material, in the case 293 and 1809 K, respectively. A FE analysis has been made under the assumption that the process is adiabatic.

Stoffel [2004] [2005] has considered an evolution of plastic zones in dynamically loaded plates using different elastic-viscoplastic laws. He has used the Chaboche [1989] low developed for non-linear hardening

$$\dot{\varepsilon}_{ij}^p = \frac{3}{2}\dot{p}\frac{\sigma_{ij}' - X_{ij}'}{II_{\sigma'}(\sigma_{rs}' - X_{rs}')},$$

$$\dot{p} = \left\langle \frac{II_{\sigma'}(\sigma_{ij}' - X_{ij}') - R - k}{K} \right\rangle^n,$$

$$\dot{X}_{ij} = \frac{2}{3}a\dot{\varepsilon}_{ij}^p - sX_{ij}\dot{p}; \quad \dot{R} = b_1(b_2 - R)\dot{p}$$

with the abbreviations ε_{ij}^p, p, σ_{ij}, X_{ij}, R denoting the plastic strain tensor, equivalent plastic strain, second Piola–Kirchhoff stress tensor, backstress tensor and isotropic hardening. The yield limit k and the material parameters a, s, b_1, b_2,

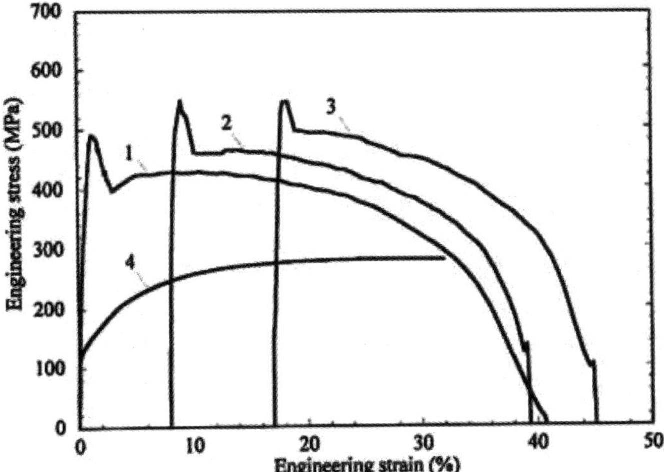

Fig. 4.5.5 Stress–strain curves of an IF steel at different strain rates, including strain rate jump tests: (1) monotonic tensile test at 1000 s^{-1}; (2) tensile test at 1000 s^{-1} after a tensile pre-strain of 8% at 0.001 s^{-1}; (3) after a tensile pre-strain of 16% at 0.001 s^{-1}; (4) monotonic tensile test at 0.001 s^{-1}.

n, K must be obtained from tension tests. A separation of isotropic and kinematic hardening was not possible so that in the Chaboche model kinematic hardening is assumed.

Another model considered is the Tanimura [1979] one, which can be expressed as

$$\dot{\varepsilon}^p_{ij} = \frac{3}{2} r \dot{p} \frac{\sigma'_{ij}}{II_{\sigma'}(\sigma'_{ij})} \, ,$$

$$\dot{p} = e^{\frac{-S_0}{(1/\sqrt{3})II_{\sigma'}(\sigma'_{ij}) - \tau^*}}$$

with the material parameters S_0, r, τ^* to be determined from uni-axial tension tests. This is an overstress model without hardening.

The third model is the Bodner–Partom [1975] model. This is a model disregarding the yield stress. The model is

$$\dot{\varepsilon}^p_{ij} = \frac{3}{2} \dot{p} \frac{\sigma'_{ij}}{II_{\sigma'}(\sigma'_{rs})} \, ,$$

$$\dot{p} = \frac{2}{\sqrt{3}} D_0 e^{-\frac{1}{2}\left(\frac{R+D}{II_{\sigma'}(\sigma'_{rs})}\right)^{2n} \frac{n+1}{n}} \, ,$$

$$D = X_{ij} \frac{\sigma_{ij}}{II_{\sigma'}(\sigma_{rs})} \, , \quad \dot{R} = m_1 (R_1 - R) \sigma'_{ij} \dot{\varepsilon}^p_{ij} \, ,$$

$$\dot{X}_{ij} = m_2 \left(\frac{3}{2} D_1 \frac{\sigma'_{ij}}{II_{\sigma'}(\sigma'_{rs})} - X_{ij} \right) \sigma'_{kl} \dot{\varepsilon}^p_{kl} \, ,$$

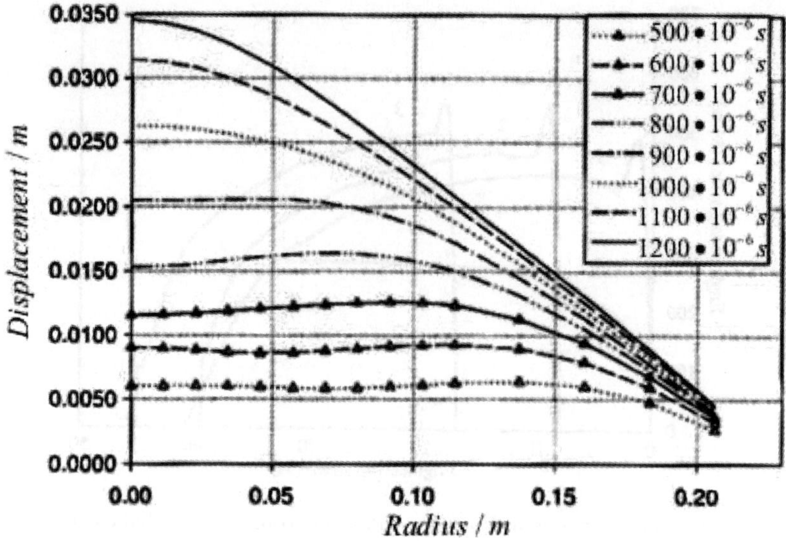

Fig. 4.5.6   Conical shape of the plate.

where the material parameters n, D_0, D_1, R_1, m_1, and m_2 have to be identified from tension tests.

These three models are applied by Stoffel for the motion of a thin (thickness 2 mm) circular membrane of 138 mm and 553 mm. The motion of such a plate at various times is shown in Fig. 4.5.6. Trying to describe the motion of such plates the author was obliged to consider the elastic material properties in the simulations. Using the Chaboche model elastic, hardening and viscous material properties can be separated from each other. This allows in the material parameter identification procedure a precise adaptation of the viscoplastic law to the uni-axial tension tests. In the Bodner–Partom model elastic properties are not assumed and hardening as well as viscous behaviors are combined with each other. In the Tanimura model the hardening is neglected. With this described advantage in the case of the elastic-viscoplastic Chaboche model simulations using this law lead to the most precise results.

A novel technique for time-resolved detection and tracking of interfacial and matrix fracture in layered materials is due to Minnaar and Zhou [2004]. The measurements at multiple locations allow the speeds at which interfacial crack front or matrix cracking/delamination front propagates to be determined. The results show that the speed of delamination or the speed of matrix cracking/delamination increases linearly with impact velocity.

In another paper Colak [2004] consider small strain, isotropic, viscoplasticity theory based on overstress, modified to model the complex cyclic hardening behavior under proportional and non-proportional loading. The theory is without yield surface and no loading/unloading conditions. The theory consists of two tensors valued state variables, which are in equilibrium stress and the kinematic stress, and a scalar isotropic stress with a flow law.

The propagation of stress pulses is studied by Ostoja-Starzewski [1995] in one-dimensional piecewise constant microstructure of grains with power-law elastic response, having randomness present in constitutive moduli and grain lengths. A study is conducted of the sensitivity of loading and unloading shock waves.

The flow law is (see Krempl [1998], development of viscoplasticity theory based on overstress Krempl [1988], Sütçü and Krempel [1989]):

$$\dot{\boldsymbol{\varepsilon}}' = \dot{\boldsymbol{\varepsilon}}^E + \dot{\boldsymbol{\varepsilon}}^P = \frac{1+v}{E}\dot{\boldsymbol{\sigma}}' + \frac{3}{2}\frac{\boldsymbol{\sigma}' - \mathbf{g}}{Ek[\Gamma]}$$

where g is the deviatoric part of the equilibrium stress G,

$$\Gamma^2 = \frac{3}{2}(\boldsymbol{\sigma}' - \mathbf{g}) : (\boldsymbol{\sigma}' - \mathbf{g})$$

and the viscosity function

$$k = k_1\left[1 + \frac{\Gamma}{k_2}\right]^{-k_3}$$

with k_1, k_2 and k_3 material constants. The equilibrium stress is introduced to represent the defect structure of the material. The growth law for the equilibrium stress consists of elastic, inelastic, dynamic recovery term and kinematic hardening contributions:

$$\dot{\mathbf{g}} = \frac{\Psi[\Gamma]}{E}\left(\dot{\boldsymbol{\sigma}}' + \frac{\boldsymbol{\sigma}' - \mathbf{g}}{k} - \frac{\Gamma}{k}\frac{\mathbf{g} - \mathbf{f}}{A}\right) + \left(1 - \frac{\Psi[\Gamma]}{E}\right)\dot{\mathbf{f}}$$

where Ψ is the shape function which affects the transition between initial quasi elastic behavior and inelastic flow. It is related as

$$\Psi[\Gamma] = C_1 + \frac{C_2 - C_1}{\exp[C_3\Gamma]}$$

where C_1, C_2 and C_3 are material constants. The positive, decreasing shape function Ψ is bounded by $1 > \Psi[\Gamma]/E > E_t/E$. The growth laws for the kinematic stress f is given by

$$\dot{\mathbf{f}} = \frac{E_t}{E}\frac{(\boldsymbol{\sigma}' - \mathbf{g})}{k[\Gamma]}$$

where E_t is the tangent modulus at the maximum inelastic strain. The state variable A, with the dimension of stress, grows according to

$$\dot{A} = A_c[A_f - A]\sqrt{\frac{2}{3}\dot{p}}$$

where A_c and A_f are material constants. The equivalent inelastic strain rate $\dot{p}$ is

$$\dot{p} = \sqrt{\dot{\boldsymbol{\varepsilon}}'^P : \dot{\boldsymbol{\varepsilon}}'^P}.$$

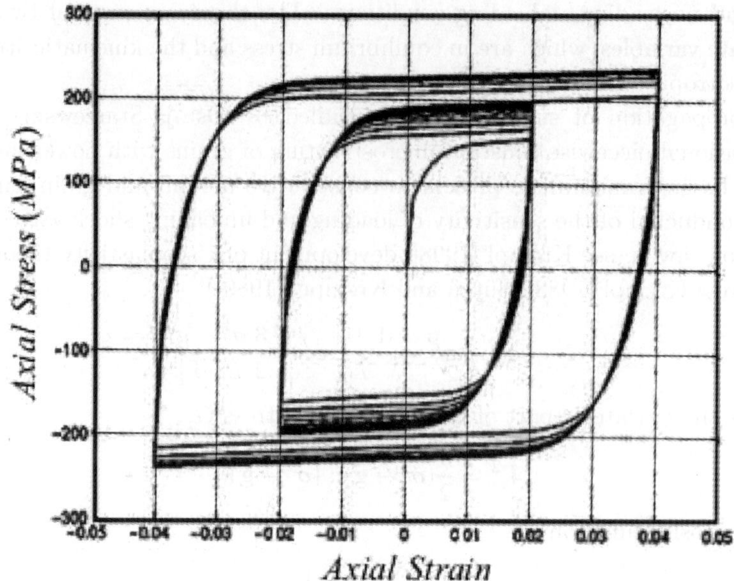

Fig. 4.5.7   Uniaxial dual-amplitude step-up test using modified viscoplasticity. Further hardening
is observed after the increase in strain amplitude. The strain rate is $\dot{\varepsilon} = 1 \times 10^{-5}$ s^{-1}.

Since an extra hardening is observed and that is rate independent, one is using
the Krempel [1998] constitutive equation. Also to include the strain amplitude
effect as well as the loading path effect, the non-proportionality measure defined by
Tanaka [1994] is used. The forth order tensor

$$\dot{C}_{ijkl} = c_c(u_{ij}u_{kl} - C_{ijkl})\,\dot{p}$$

with c_c a material constant, is introduced. $\mathbf{u} = \dot{\varepsilon}^P/\dot{p}$, and the non-proportionality
measure is

$$\Phi = \sqrt{\frac{C_{ijkl}C_{ijkl} - u_{ij}C_{klij}C_{klmn}u_{mn}}{C_{prst}C_{prst}}}.$$

The center of inelastic strain range, Y is modeled by Tanaka [1994],

$$\dot{\mathbf{Y}} = r_y(\varepsilon'^P - \mathbf{Y})\,\dot{p}$$

where r_y is a material constant. The size of the memory surface q is $q = |\varepsilon'^P - \mathbf{Y}|$
where $|\ |$ indicates magnitude.

The theory was applied to various loadings. For instance, for uniaxial dual
amplitude step-up test is shown in Fig. 4.5.7.

The mathematical study of the problem of unilateral contact between an elastic-
viscoplastic body and a rigid frictionless foundation is due to Sofonea [1997]. The
constitutive law is

$$\dot{\sigma} = E\dot{\varepsilon} + G(\sigma, \varepsilon)$$

where E and G are constitutive functions. For G one is using

$$G(\sigma, \varepsilon) = \begin{cases} -k_1 F_1(\sigma - f(\varepsilon)) & \text{if } \sigma > f(\varepsilon) \\ 0 & \text{if } g(\varepsilon) \le \sigma \le f(\varepsilon) \\ k_2 F_2(g(\varepsilon) - \sigma) & \text{if } \sigma < g(\varepsilon) \end{cases}$$

where $k_1, k_2 > 0$ are viscosity constants and F_1, F_2 are increasing functions with $F_1(0) = F_2(0) = 0$. The variation formulation of the problem is given and the existence and uniqueness result for the displacement and stress field is obtained. A fixed point method was used in order to obtain the existence and uniqueness of the solution. The continuous dependence of the solution with respect to the input data as well as the stability result is then analyzed.

Matos and Dodds [2002] have considered the initiation of brittle fracture triggered by a transgranular cleavage mechanism typical of that exhibited by ferritic steels operating in the ductile-to-brittle transition region. The plastic strain is of the form

$$\dot{\varepsilon}^{vp} = \begin{cases} D\left[\left(\dfrac{q}{\sigma_e}\right)^{\gamma} - 1\right] & \dfrac{q}{\sigma_e} > 1 \\ 0 & \dfrac{q}{\sigma_e} \le 1 \end{cases}$$

where D and γ denote material parameters, q denotes the rate-dependent uniaxial tensile stress and σ_e denotes the static uniaxial tensile stress.

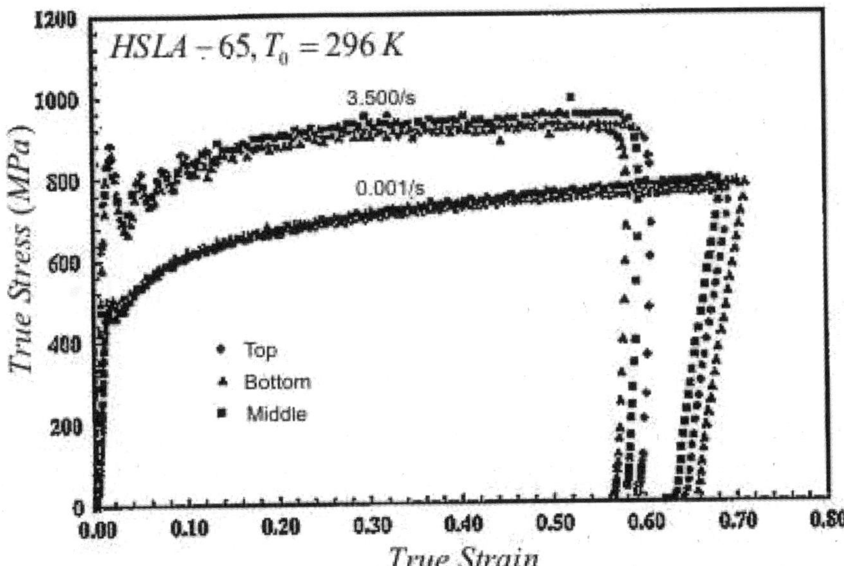

Fig. 4.5.8   Stress–strain curves of samples cut from an HSLA-65 plate at indicated directions and locations.

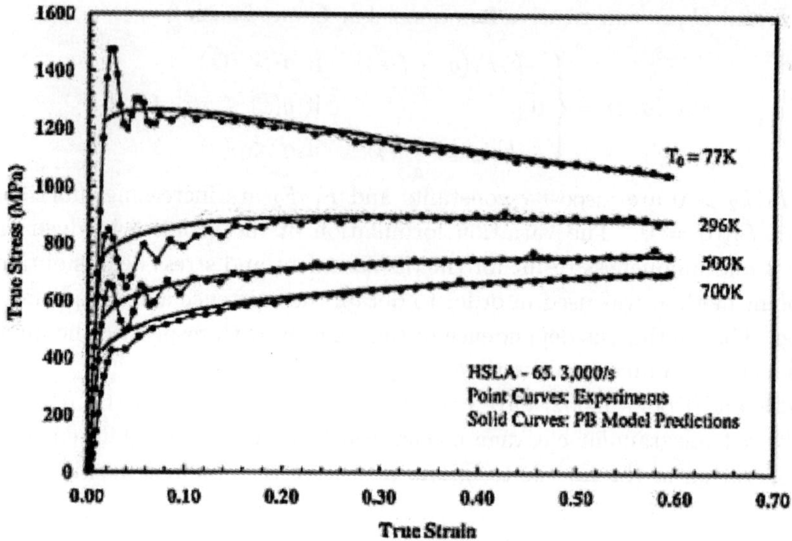

Fig. 4.5.9   Comparison of PB-model predictions with experimental results at a strain rate of 3000/s and indicated initial temperatures.

A paper in which one is discussing the Hopkinson technique for the dynamic recovery experiments is due to Nemat-Nasser *et al.* [1991]. In a recent paper Nemat-Naser and Guo [2005] study the thermomechanical response of high-strength low-alloy steel (HSLA-65) in uniaxial compression tests. True strains exceeding 60% are achieved in these tests, over the range of strain rates from 10^{-3}/s to about 8500/s, and at initial temperatures from 77 to 1000 K. To show the influence of the strain rate is given the Fig. 4.5.8. Even the elastic parameters seem to be influenced. They apply the Johnson–Cook model and their own (PB-model). For their own model, it is for $T \leq T_c$:

$$\tau = 760\gamma^{0.15} + 1450\left\{1 - \left[-10.6 \times 10^{-5}T\ln\frac{\dot{\gamma}}{4 \times 10^8}\right]^{1/2}\right\}^{3/2}$$

where $T = T_0 + 0.267 \int_0^\gamma \tau d\gamma$ and for $T > T_c$ we have $\tau = 760\gamma^{0.15}$ where

$$T_c = \left(-10.16 \times 10^{-5}\ln\frac{\dot{\gamma}}{4 \times 10^8}\right)^{-1}.$$

Figure 4.5.9 is showing the theory adapted to the experimental data.

The compressive stress–strain behavior of a NiTi shape memory alloy has been determined by Chen *et al.* [2001], over strain rates of 10^{-3}–7.5×10^2 s^{-1}. A split Hopkinson pressure bar with a pulse shaping technique was used to perform valid dynamic tests. Experimental results show that, the plateau stress of the shape memory alloy is strain rate dependent. The material is returning to its original length upon unloading.

Forrestal *et al.* [2003] present a Hopkinson bar technique to evaluate the performance of accelerometers that measure large amplitude pulses, such as those experienced during projectile penetration tests. An aluminum striker bar impacts a thin Plexiglas or copper disk placed on the impact surface of an aluminum incident bar. The Plexiglas or copper disk pulse shaper produces a nondispersive stress wave that propagates in the aluminum incident bar and eventually interacts with a tungsten disk at the end of the bar. A quartz stress gage is placed between the aluminum bar and tungsten disk, and an accelerometer is mounted to the free end of the tungsten disk. An analytical model shows that the rise time of the incident stress pulse in the aluminum bar is long enough and the tungsten disk length is short enough that the response of the tungsten disk can be accurately approximated as rigid-body motion. They measure stress at the aluminum bar-tungsten disk interface with the quartz gage and they calculate rigid-body acceleration of the tungsten disk from Newtonian's Second Law and the stress gage data. In addition they measure strain-time at two locations on the aluminum incident bar to show that the incident strain pulse is nondispersive and they calculate rigid-body acceleration of the tungsten disk from a model that uses this strain-time data. Thus, they can compare accelerations measured with the accelerometer and accelerations calculated with models that use stress gage and strain gage measurements. They show that all three acceleration-time pulses are in very close agreement for acceleration amplitudes to about 20,000 G.

In a recent paper Gómez-del Río *et al.* [2005] have tested dynamic tensile composites at law temperature using split Hopkinson pressure bar. The temperature is 20 and $-60°$C. The results of the dynamic tests showed little influence of temperature and strain rate on the tensile strength of a unidirectional laminate loaded in the fiber direction; in contrast, the strength increases appreciably in the transverse direction at low temperature and high strain rate. Regarding strain rate, a slight increase in tensile strength was observed under dynamic loading.

In another paper Müller *et al.*[2002] have a repetitive thermodynamics on viscoplastic body, since moulding plastic deformation is accompanied by heating. They have three such loadings. They show that the deformation in time, and temperature, is tending towards a single loading.

In the paper by Tsai and Prakash [2005] is studied the normal plate impact experiments on 2-D layered material targets to understand the role of material architecture and material inelasticity in governing the elastic precursor decay and late-time wave dispersion. In order to understand the effects of layer thickness and the distance of wave propagation on elastic precursor decay and late-time dispersion several targets with various layer and target thicknesses are employed. Moreover, in order to understand the effects of material inelasticity both elastic-elastic and elastic-viscoelastic bilaminates are utilized. The results of the study indicate that the structure of acceleration waves is strongly influenced by impedance mismatch of the layers constituting the laminates, density of interfaces, distance of wave propagation, and material inelasticity. The speed of elastic precursor is independent of the impedance mismatch of the individual laminae constituting the bilaminates and is equal to the average wave speed within the bilaminates. The speed of the late-time dispersion wave is observed to decrease with an increase in impedance mismatch; however, it is found to be independent of the density of interfaces, i.e., the number of layers in a given thickness laminate. The decay of the elastic precursor is observed to increase with an increase in impedance mismatch, the density of interfaces, and the distance of wave propagation. The rise-time of the late-time dispersion wave increases with an increase in impedance mismatch; however, it is observed to decrease with an increase in the density of interfaces. The frequency of oscillations of the late-time dispersive wave is observed to decrease with an increase in impedance mismatch; however, it is observed to increase with an increase in the density of interfaces.

A model is presented by Ubachs *et al.* [2005] to describe the mechanical behavior of microstructure dependent materials. The model describes the microstructure evolution using a phase field approach and employs a constitutive model dependent on the phase field to account for the underlying microstructure. It has been demonstrated by applying it to eutectic tin-led solder, a material whose mechanical characteristics are strongly influenced by its continuously evolving microstructure. For the constitutive behavior the elasto-viscoplastic Perzyna model has been adopted. The Perzyna model allows the use of different hardening functions, which makes it suitable for a large variety of materials. It is coupled to the phase field model through the model parameters which are taken dependent on the mass fraction field resulting from the solution of the phase field equations. In this way the microstructure is accounted for using a continuum mechanics approach.

Mixed-mode open-notch flexure, anti-symmetric loaded end-notched flexure and center-notched flexure specimens were used to investigate dynamic mixed I/II mode delamination fracture using a fracturing split Hopkinson pressure bar by Wosu *et al.* [2005]. An expression for dynamic energy release rate is formulated and evaluated. The experimental results show that dynamic delamination increases linearly with mode mixing.

In several papers by Krempl and Ho [2001], Ho and Krempl [2001], [2002] is developed a complicated theory based on overstress, to describe the rate sensitivities inside and outside the dynamic strain aging regime. The compressible and incompressible, isotropic, small strain viscoplasticity theory based on overstress is given.

An experimental paper by Bourne [2006], is devoted to the plate impact to load in uniaxial strain and to show that a lower threshold exists for the propagation of fracture from the impact face. This is 4GPa in soda-lime glass. The velocity of impact is 1.5–2.5 km/s.

Bibliography

Altenhof W. and Ames W., 2002, Strain rate effects for aluminum and magnesium alloys in finite element simulations of steering wheel armature impact tests, *Fatigue Fract. Engng. Mater. Struct.* **25**, 1149–1156.

Alter B. E. K. and Curtis C. W., 1956, Effect of strain rate on the propagation of a plastic strain pulse along a lead bar, *J. Appl. Phys.* **27**, 1079–1085.

Auricchio F., 1997, A viscoplastic constitutive equation bounded between two generalized plasticity models, *Int. J. Plasticity* **13**, 8–9, 697–721.

Baroncea A., Cristescu and Glodeanu E., 1977, A model for the description of the deformation of rocksalt, *St. Cerc. Mec. Appl.* **36**, 1, 31–40 (in Roumanian).

Béda G., 1962a, *Acta. Tech. Acad. Sci. Hung.* **38**, 161–187.

Bell J. F., 1951, Propagation of plastic waves in prestressed bars, *U.S. Navy Tech. Rep. No. 5*, Baltimore, The Johns Hopkins University.

Bell J. F., 1968, The phisics of large deformation of crystalline solids, *Springer Tracts in Natural Philosophy*, Vol. 14, Springer-Verlag, Berlin.

Bell J. F., 1973, The experimental foundations of solid mechanics, in *Hadbuch der Physik* (ed. S. Flugge), Vol. Via/1, Springer-Verlag, Berlin.

Bell J. F. and Stein A., 1962, The incremental loading wave in the pre-stressed plastic field, *J. Mécan.* **1**, 4, 395–412.

Bertholf L. D. and Karnes C. H., 1975, Two-dimensional analysis of the split Hopkinson pressure bar system, *J. Mech. Phys. Solids* **23**, 1–19.

Bingham E. C., 1922, *Fluidity and Plasticity*, McGraw-Hill, New York.

Bodner S. R. and Partom Y., 1975, Constitutive equations for elasto-viscoplastic strain-hardening materials, *ASME J. Appl. Mech.* **42**, 385–389.

Bourne N. K., 2006, On the impact and penetration of soda-lime glass, *Int. J. Imp. Engineering* (in press).

Campbell J. D., 1970, *Dynamic Plasticity of Metals*, Springer Verlag, 92 pp.

Campbell J. D., 1973, Dynamic plasticity-macroscopic and microscopic aspects, *Materials Science and Engineering* **12**, 1, 3–21.

Cernocky E. P., 1982, An examination of four viscoplastic constitutive theories in uniaxial monotonic loading, *Int. J. Solids Structures* **18**, 11, 989–1005.

Chaboche J. L., 1989, Constitutive equations for cyclic plasticity and cyclic viscoplasticity, *Int. J. Plasticity* **5**, 247–302.

Chang K. C., Sugiura K. and Lee G. C., 1989, Rate-dependent material model for structural steel, *J. of Engeenering Mach.* **115**, 3, 465–474.

Chau O., Fernández-García J. R., Han W. and Sofonea M., 2002, A frictionless contact problem for elastic-viscoplastic materials with normal compliance and damage, *Com. Met. Appl. Mech. Engi.* **191**, 44, 5007–5026.

Chen W. W., Wu Q. P., Kang J. H. and Winfree N. A., 2001, Compressive superelastic behaviour of a NiTi shape memory alloy at strain rates of 0.001–750 s^{-1}, *Int. J. Solids Struct.* **38**, 8989–8998.

Chhabildas L. C. and Swegle J. W., 1989, Dynamic pressure-shear loading of using anisotropic crystals, *J. Appl. Phys.* **51**, 9, 4799–4807.

Colak O. U., 2004, A viscoplasticity theory appied to proportional and non proportional cyclic loading at small strains, *Int. J. Plasticity* **20**, 1387–1401.

Cristescu C. and Cristescu N. D., 1973, A numerical method to describe unloading in dynamic plasticity, *2nd Int. Conf. on "Structural Mechanics in Reactor Technology"*, Berlin. L **4/8**, 1–14.

Cristescu N. D., 1963, On the propagation of elastic-plastic waves in metallic rods, *Bull. Acad. Polon. Sci.* **11**, 129–133.

Cristescu N. D., 1964, Some problems in the mechanics of extensible strings, in: Kolsky H. and Prager W. (eds.), *Symp. IUTAM Stress Waves in Anelastic Solids*, Providence, 1963 (Springer Verlag, Berlin–Göttingen–Heidelberg) 118–132.

Cristescu N. D., 1966, About the propagation of elastic/plastic waves in thin rods, in: *Dynamics of Mashines*, Proc. Conf. Liblice-Praga, 1963, 87–96.

Cristescu N. D., 1967, *Dynamic Plasticity*, North Holland Pub., Amsterdam.

Cristescu N. D., 1972a, *Rate-Type Constitutive Equations in Dynamic Plasticity*, Noordhof International Publishing, Leyden, 287–310.

Cristescu N. D., 1972b, A procedure for determining the constitutive equations for materials exhibiting both time-dependent and time-independent plasticity, *Int. J. Solids Structures* **8**, 511–531.

Cristescu N. D., 1977, Speed influence in wire drawing, *Rev. Roumaine de Mécanique Appliquée* **22**, 3, 391–399.

Cristescu N. D. and Bell J. F., 1970, On unloading in the symmetrical impact of two aluminum bars, in: *In Elastic Behaviour of Solids*, Kannienen M. F., Adler W. F., Rosenfield A. R. and Jaffee R. I. (eds.), McGraw Hill, New York, 397–421.

Cristescu N. D. and Suliciu I., 1982, *Viscoplasticity*, Martinus Nijhoff Pub.,307 pp.

Currò C. and Valenti G., 1996, A linearization procedure for quasi-linear non homogeneous and non-autonomous first-order systems, *Int. J. Non-Linear Mechanics* **31**, 3, 377–386.

Daimaruya M. and Naitoh M., 1983, On the existence of a strain plateau in the strain-rate dependent theory of Malvern for plastic wave propagation, *J. Appl. Mech., Trans. ASME* **50**, 3, 678–679.

Dally J. W. and Riley W. F., 1965, *Experimental Stress Analysis*, McGraw Hill, New York.

Dawson T. H., 1972, On the applicability of one-dimensional non-viscous dynamic plasticity theory, *Int. J. Mech. Sci.* **14**, 43–48.

Dioh N. N., Leevers P. S. and Williams J. G., 1993, Thickness effects in split Hopkinson pressure bar tests, *Polymer* **34**, 20, 4230–4234.

Dioh N. N., Ivankovic A., Leevers P. S. and Williams J. G., 1995, Stress wave propagation effects in split Hopkinson pressure bar tests, *Proc. Roy. Soc. London* **449**, 187–204.

Drăgănescu Gh. E. and Căpălnăaşan V., 2003, Nonlinear relaxation phenomena in polycristalline solids, *Int. J. Nonlinear Sci. Num. Sci.* **4**, 219–225.

Făciu C. and Mihailescu-Suliciu M., 1987, The energy in one-dimensional rate-type semilinear viscoelastcity, *Int. J. Solids Sructures* **23**, 11, 1505–1520.

Faciu C. and Simion N., 2000, Energy estimates and uniqueness of the weak solutions of initial-boundary value problems for semilinear hyperbolic systems, *Z. Angev. Math. Phys.* **51**, 792–805.

Fernández J. R., Hild P. and Viaňo J. M., 2003, Numerical approximation of the elastic-viscoplastic contact problem with non-matching meshes, *Numer. Math.* **94**, 501–522.

Fernández L. R., 2004, Numerical analysis of a contact problem between two elastic viscoplastic bodies with hardening and nonmatching meshes, *Finite Elements in Analysis and Design* **40**, 771–791.

Filbey G. L., 1961, *Intensive Plastic Waves* Ph.D. dissertation, The Johns Hopkins University, Baltimore.

Forrestal M. J., Togami T. C., Baker W. E. and Frew D. J., 2003, Performance evaluation of accelerometers used for penetration experiments, *Experimental Mechanics* **43**, 1, 90–96.

Francis P. H. and Lindholm U. S., 1968, Effect of temperature gradients on the propagation of elastoplastic waves, *J. Appl. Mech. Trans. ASME 35*, 3, 441–448.

Fusco D. and Manganaro N., 1994a, Riemann invariants-like solutions for a class of rate-type materials, *Acta Mechanica* **105**, 23–32.

Fusco D. and Manganaro N., 1994b, A method for determining exact solutions to a class of nonlinear models based on introduction of differential constrains, *J. Math. Phys.* **35**, 7, 3659–3669.

Fusco D. and Manganaro N., 1996, A method for finding exact solutions to hyperbolic systems of first-order PDEs, *IMA J. Appl. Math.* **57**, 3, 223–242.

Gilman J. J., 1960, Physical nature of plastic flow and fracture, in: Lee E. H. and Symonds P. S. (eds.), *Plasticity, Proc. Second. Symp. on Naval Structural Mechanics*, Providence, 1960 (Pergamon Press, New York–Oxford–London–Paris) 44–99.

Goldsmith W., 2001, *Impact, the Theory and Physical Behavior of Colliding Solids*, Dover, N.Y.

Gómez-del Río T., Barbero E., Zaera R. and Navarro C., 2005, Dynamic tensile behaviour at low temperature of CFRP using a split Hopkinson pressure bar, *Composites Science and Technology* **65**, 61–71.

Gupta Y. M., 1976, Shear measurements in shock-loaded solids, *Appl. Phys. Lett.* **29**, 11, 694–697.

Hauser F. E., Simmons J. A. and Dorn J. E., 1961, Strain rate effects in plastic wave propagation, in: Shewmon P. G. and Zackay V. F. (eds.), *Response of Metals to High Velocity Deformation, Proc. Conf. Colorado* (Interscience Publ., New York, London) 93–110.

Ho K. and Krempl E., 2001, The modeling of unusual rate sensitivities inside and outside the dynamic strain aging regime, *Trans. ASME, J. Engng. Mat. Tech.* **123**, 28–35.

Ho K. and Krempl E., 2002, Extension of the viscoplasticity theory based on overstress (VBO) to capture non-standard rate dependence in soils, *Int. J. Plasticity* **18**, 851–872.

Hu Y. and Feng R., 2004, On the use of a Kolsky torsion bar to study the transient large-strain response of polymer melts at high shear rates *J. Appl. Mech.* **71**, 441–449.

Ilyushin A. A. and Lenskii V. S., 1959, *Strength of Materials*, Fizmatghiz, Moscow (in Russian).

Ionescu I. and Sofonea M., 1993, *Functional and Numerical Methods in Viscoplasticity*, Oxford University Press, Oxford, 265 pp.

Khaleel M. A., Zbib H. M. and Nyberg E. A., 2001, Constitutive modeling of deformation and damage in superplastic materials, *Int. J. Impact Engn.* **17**, 277–296.

Klepaczko J. R., 1998, Remarks on impact shearing, *J. Mech. Phis. Solids* **46**, 10, 2139–2153.

Kojic M., 1996, The Governing Parameter Method for implicit integration of viscoplastic constitutive relations for isotropic and orthotropic metals, *Computational Mechanics* **19**, 49–57.

Kolsky H., 1949, An investigation of the mechanical properties of materials at very high rates of loading, *Proc. Phys. Soc. B* **62**, 676–700.

Krempl E., 1988, Phenomenological modeling of viscoplasticity, *Revue Phys. Appl. Phisics Abstracts* **23**, 331–338.

Krempl E., 1998, Creep-Plasticity Interaction. Lecture Notes. *CISM* No. 187, Udine, Italy, p. 74.

Krempl E. and Ho K., 2001, Inelastic Compressible and Incompressible, Isotropic, Small Strain Viscoplasticity Theory Based on Overstress (VBO), *Lemaitre Handbook of Materials Behaviour Models*, Sec. 5.6, 336–348.

Kuriyama S. and Kawata K., 1973, Propagation of stress wave with plastic deformation in metal obeying the constitutive equation of the Johnston–Gilman type, *J. Appl. Phys.* **44**, 8, 3445–3454.

Lawson J. E. and Nicholas T., 1972, The dynamic mechanical behavior of titanium in shear, *J. Mech. Phys. Solids* **20**, 65–76.

Li C. H. and Clifton R. J., 1982, Dynamic stress–strain curves at plastic shear strain rate, in *Shock Waves in Condensed Matter — 1981*, Nellis W. J., Seaman L. and Graham R. A. (eds.), American Institute of Physics, New York, 360–366.

Lodygowski T. and Perzyna P., 1997, Numerical modeling of localized fracture of inelastic solids in dynamic loading processes, *Int. J. Numer. Meth. Engng.* **40**, 4137–4158.

Lubarda V. A., Benson D. J. and Meyers M. A., 2003, Strain-rate effects in rheological models of inelastic response, *Int. J. Plasticity* **19**, 8, 1097–1118.

Ludwik P., 1909, Über den Einfluss der Deformationsgeschwindigkeit bei bleibenden Deformationen mit besonderer Berücksichtigung der Nachwirkungserscheinungen, *Physik. Zeitschrift* **10**, 411–417.

Luo Y., Mori T., Li S. and Luo S., 2003, Isothermal rheological forming of high strength alloying part with complicated curved surface, *Materials Sci. Forum* **437–438**, 337–340.

Malvern L. E., 1951a, Plastic wave propagation in a bar of metal exhibiting a strain rate effect, *Quart. Appl. Math.* **8**, 4, 405–411.

Malvern L. E., 1951b, The propagation of longitudinal waves of plastic deformation in a bar of material exhibiting a strain rate effect, *J. Appl. Mech.* **18**, 2, 203–208.

Manganaro N. and Valenti G., 1993, Group analysis and linearization procedure for a nonautonomous model describing rate-type materials, **J. Math. Phys. 34**, 4, 1360–1369.

Manganaro N. and Meleshko S., 2002, Reduction procedure and generalized simple waves for systhems written in Riemann variables, *Nonlinear Dynamics*, **30**, 87–102.

Maranini E. and Yamaguchi T., 2001, A non-associated viscoplastic model for the behavior of granite in triaxial compression, *Mech. of Mater.* **33**, 5, 283–293.

Matos C. G. and Dodds R. H. Jr., 2002, Probabilistic modeling of weld fracture in steel frame connections Part II: seismic loading, *Engineering Structures* **24**, 687–705.

Mihailescu-Suliciu M. and Suliciu I., 1987, On tensile shock waves in rubber-like materials, *J. Appl. Mech.* **54**, 498–502.

Minnaar K. and Zhou M., 2004, A novel technique for time-resolved detection and tracking of interfacial and matrix fracture in layered materials, *J. Mech. Phys. Solids* **52**, 2771–2799.

Mortell M. and Seymour B. R., 1973, Pulse propagation in nonlinear viscoelastic rod of finite length, *SIAM J. Appl. Math.* **22**, 2, 209–224.

Müller I., Sahota H.-S. and Villaggio P., 2002, On the thermodynamics of repetitive visco-plastic moulding, *Z. Angew. Math. Phys.* **53**, 1139–1149.

Nemat-Nasser S., Isaacs J. B. and Starrett J. E., 1991, Hopkinson techniques for dynamic recovery experiments, *Proc. R. Soc. Lond.* A **435**, 371–391.

Nemat-Nasser S., Choi J. Y., Guo W. G. and Isaacs J. B., 2005, Very high strain-rate response of a NiTi shape-memory alloy, *Mechanics of Materials* **37**, 287–289.

Nemat-Nasser S. and Guo W. G., 2005, Thermomechanical response of HSLA-65 steel plates: experiments and modeling, *Mechanics of Materials* **37**, 379–405.

Nicholas T., 1981, Tensile testing of materials at high rates of strain, *Experimental Mechanics* **21**, 5, 177–185.

Nicholas T., Rajendran A. M. and Grove D. J., 1987a, Analytical modeling of precursor decay in strain-rate dependent materials, *Int. J. Solids Struct.* **12**, 1601–1614.

Nicholas T., Rajendran A. M. and Grove D. J., 1987b, An offset yield criterion from precursor decay analysis, *Acta Mechanica* **69**, 1–4, 205–218.

Nicholson D. W. and Phillips A., 1978, On the impact end in logitudinal dynamic plastic wave propagation, *Acta Mechanica* **29**, 75–92.

Ostoja-Starzewski M., 1995, Wavefront propagation in a class of random microstructures-II, Non-linear elastic grains, *Int. J. Non-Linear Mechanics* **30**, 6, 771–781.

Pao Y. H. and Gilat A., 1989, Modeling 1100-0 aluminum over a wide range of temperatures and strain rates, *Int. J. Plasticity* **5**, 183–196.

Perzina P., 1963, The study of the dynamical behaviour of rate sensitive plastic materials, *Arch. Mech. Stos.* **15**, 113–130.

Ponomarev S. D., Biderman V. L., Likharev K. K., Makushin V. M., Malinin N. N. and Feodos'ev V. I., 1956, Resistance calculus in construction of machines, *Mashigiz Moscow*, *Tom* III, 553–580 (in Russian).

Prandtl L., 1928, Ein Gedankenmodell zur kinematischen Theorie der festen Körper, *Z. Angew. Math. Mech.* **8**, 85–106.

Schmidt M. L. and Ross C. A., 1999, Shear strength of concrete under dynamic loads, *ASME Pressure Vessel and Piping Conf.* Boston, MS, August 1999.

Shaw R. P. and Cozzarelli F. A., 1971, Wave-front stress relaxation in a one dimensional nonlinear inelastic material with temperature and position dependent properties, *J. Appl. Mech.* **38**, 1, 47–50.

Sofonea M., 1997, On a contact problem for elastic-viscoplastic bodies, *Nonlinear Analysis, Theory, Methods & Applications* **29**, 9, 1037–1050.

Sokolovskii V. V., 1948a, Propagation of elastic-visco-plastic waves in bars, *Docl. Akad. Nauk SSSR* **60**, 5, 775–778 (in Russian).

Sokolovskii V. V., 1948b, Propagation of elastic-visco-plastic waves in bars, *Pricl. Mat. Meh.* **12**, 3, 261–280 (in Russian).

Stoffel M., 2004, Evolution of plastic zones in dynamically loaded plates using different elastic-viscoplastic laws, *Int. J. Solids and Structures* **41**, 6813–6830.

Stoffel M., 2005, Sensitivity of simultaneous depending on material parameter variations, *Mech. Res. Com.* **32**, 332–336.

Suliciu I., 1966, private comunication.

Suliciu I., 1972, Additional coments on Suliciu, Malvern and Cristescu's paper. "Remarks concerning the "plateau" in dynamic plasticity", *Archives of Mechanics* **27**, 4, 665–667.

Suliciu I., 1974, Classes of discontinuous motions in elastic and rate-type materials, *Arch. Mech. Stos.* **26**, 4, 675–699.

Suliciu I., 1981, On the Savart–Masson effect, *J. Appl. Mech. Trans. ASME* **48**, 426–428.

Suliciu I., Lee S. Y. and Ames W. F., 1973, Nonlinear traveling waves for a class of rate-type materials, *J. Math. Anal. Appl.* **42**, 2, 313–322.

Sundaram S. and Clifton R. J., 1998, The influence of a glassy phase on the high strain rate response of a ceramic, *Mechanics of Materials* **29**, 233–251.

Sütçü M. and Krempel E., 1989, A stability analysis of the uniaxial viscoplasticity theory based on overstress, *Computational Mechanics* **4**, 401–408.

Tabov I., 1996, Riemann invariants for a class of partial differential systems, *Differential Equations* **32**, 11, 1550–1553.

Tanaka E., 1994, A nonproportionality parameter and a cyclic viscoplastic constitutive model taking into account amplitude dependencies and memory effects of isotropic harden, *European J. of Mechanics* **13**, 2, 155–173.

Tanimoto N., 1994, One-dimensional propagation speed of an elastic-plastic viscoplastic stress wave, *Nuclear Eng. and Design* **150**, 2–3, 275–280.

Tanimoto N., Fukuoka H. and Fujita K., 1993, One-dimensional numerical analysis of a bar subjected to longitudinal impulsive loading, *ISME International Journal*, Series A, **36**, 2, 137–145.

Tanimura S., 1979, A practical constitutive equation covering a wide range of strain rates, *Int. J. Eng. Sci.* **17**, 997–1004.

Thomsen E. G., Yang C. T. and Kobayashi S., 1968, *Mechanics of Plastic Deformation in Metal Processing*, Macmillan Co., London.

Ting T. C. T. and Symonds P. S., 1964, Longitudinal impact on visco-plastic rods linear stress–strain rate law, *J. Appl. Mech.* **31**, Ser. E, 199–207.

Tsai L. and Prakash V., 2005, Structure of weak shock waves in 2-D layered material systems, *Int. J. Solids Struct.* **42**, 724–750.

Ubachs R. L. J. M., Schreurs P. J. G. and Geers M. G. D., 2005, Phase field dependent viscoplastic behaviour of solder alloys, *Int. J. Solids Sructures* **42**, 2533–2558.

Uenishi A. and Teodosiu C., 2004, Constitutive modeling of the high strain rate behaviour of interstitial-free steel, *Int. J. Plasticity* **20**, 915–936.

Valanis K. C., 1972, Irreversible thermodinamics of continuous media, Internal Variable Theory, *CISM Courses and Lectures No. 77, International Centre for Mechanical Sciences*, Springer, Wien.

Vyalov S. S., 1978, Rheological foundations for soil mechanics, *Vysh. Shkola, Moscow* (in Russian).

Wosu S. N., Hui D. and Dutta P. K., 2005, Dynamic mixed-mode I/II delamination fracture and energy release rate of unidirectional graphite/epoxy composites, *Eng. Fracture Mech.* **72**, 1531–1558.

Yokoyama T., 2001, Finite element computation of torsional plastic waves in a thin walled tube, *Archive of Applied Mechanics* **71**, 6–7, 359–370.

Zhao H., 1997, A study of specimen thickness effects in the impact tests on polymers by numeric simulations, *Polymer* **39**, 5, 1103–1106.

Zhao H., 2003, Material behaviour characterization using SHPB techniques, tests and simulations, *Computers & Structures* **81**, 12, 1301–1310.

Zhao H. and Gary G., 1996, On the use of SHPB techniques to determine the dynamic behavior of materials in the range of small strains, *Int. J. Solids Structures* **33**, 23, 3363–3375.

Ziegler F., 1992, Developments in structural dynamic viscoplasticity including ductile damage, *ZAMM* **72**, 4, T5–T15.

Chapter 5

Mechanics of Extensible Strings

5.1 Introduction

The problem was considered for a long time. But generally only static problems or problems involving nearly quasi static conditions were studied (various cables existing in applications). Rubber cables or synthetic strings are elongating very much when impacted, even with quite small forces. But in some modern textile machines the speed of operation is quite high (knitting, weaving, sewing, carding machines). Also in elastic cables used to break airplanes on ships, the speed is quite high. In dynamic problems the cables are influenced what concerns the yield limit, the work-hardening modulus, the breaking stress and the elastic modulus.

Impact velocities of the order of some hundred meters per second or even a few kilometers per second, when the component of the impact velocity ranges between the velocities of propagation of transverse and longitudinal waves, occur less often in practice and lead to difficult mathematical problems, as will be pointed out below. One must also take into account that textile materials cannot resist transverse impacts if the latter exceed a certain limit; for each textile yarn there is a certain limiting transverse impact velocity which at once breaks the string at the point of impact. These limiting velocities, generally, of the order of many hundred meters per second.

From the mechanical point of view, that is the simplest problem in which two kind of waves propagate and reflect together, influencing one another at every moment. This motion can be studied both experimentally and theoretically, because both kinds of wave can be observed experimentally during their propagation. The problem can be studied experimentally very well, for the two waves propagating and influencing each other. On the other hand, the mechanics of strings closely resembles many other problems of dynamic plasticity from the mechanical and mathematical points of view.

We are not giving here a complete literature. The beginning of the literature is given in Cristescu [1967]. Some other papers are mentioned in the exposure. The equations of motion of a string in plane motion are written by Rakhmatulin

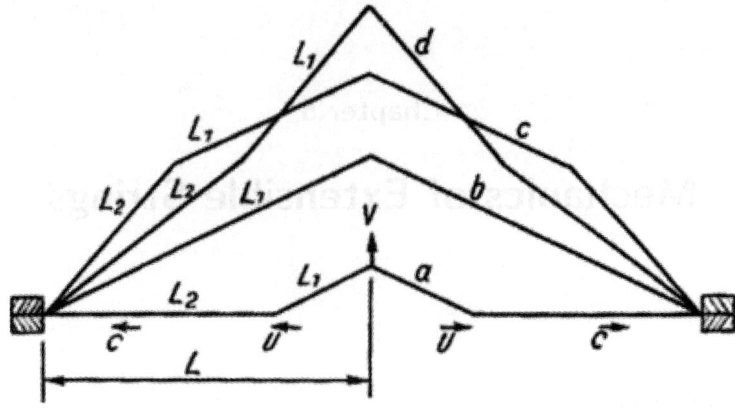

Fig. 5.1.1　Typical configuration of a yarn specimen after transverse impact.

and Shapiro [1955] in order to determine the dynamic relation between stress and strain for strings. The impact of strings by punctual bodies of finite mass, when the velocity of impact is variable, was studied by Rakhmatulin [1951] and Riabova [1953], and by some other authors. In the paper by Smith *et al.* [1956] a textile yarn segment about half a meter long is clamped at each end and impacted transversely at midpoint with a velocity of about 70 m/sec. The stress–strain curve for this yarn is obtained by measuring a high speed photographic record of the motion of the yarn. A typical configuration of the yarn specimen after transverse impact is indicated in Fig. 5.1.1.

It is assumed that the transverse waves always propagate more slowly and in fact are reduced to a single shock wave which modifies the shape of the yarn. By measuring the lengths defined in Fig. 5.1.1 on the frames of the photograph record, the average strain as defined by

$$\bar{\varepsilon} = \frac{L_1 + L_2 - L}{L}$$

is used. This manner of defining strain as an average strain, in the presence of wave propagation, is certainly a restrictive one. By measuring also the velocity of propagation of the transverse wave, dynamic stress–strain curves are obtained for high-tenacity nylon, fortisan, and fiberglass. The stress–strain curves for nylon obtained by Smith *et al.* [1956] are reproduced in Fig. 5.1.2 for various strain rate. The important conclusion is that for such materials the increasing rate of strain tends to steepen the slope of the stress–strain curve at the origin. A theory of these phenomena, assuming the yarn to be infinitely long, has been developed by Smith *et al.* [1958], and Smith *et al.* [1960]. The velocities of impact range between 1400 m/sec for undrawn nylon to 5000 m/sec for high tenacity rayon and glass-fiber. In Fig. 5.1.3 is given the stress–strain curves for acetate yarns as obtained by Smith *et al.* [1961]. All these yarns show an appreciable rate influence. The most strongly

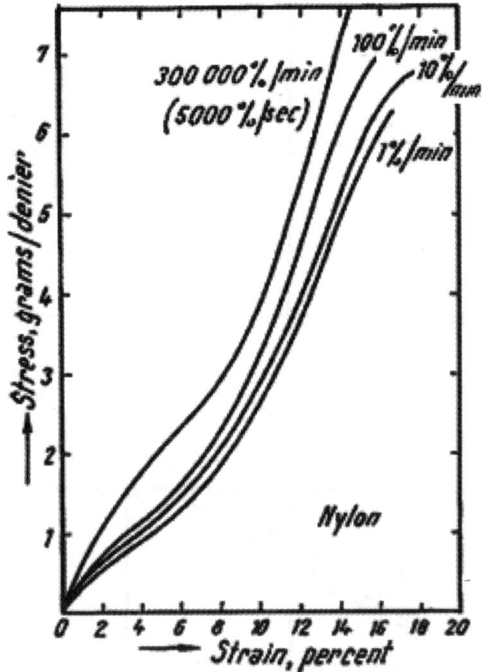

Fig. 5.1.2   Stress–strain curves for Nylon.

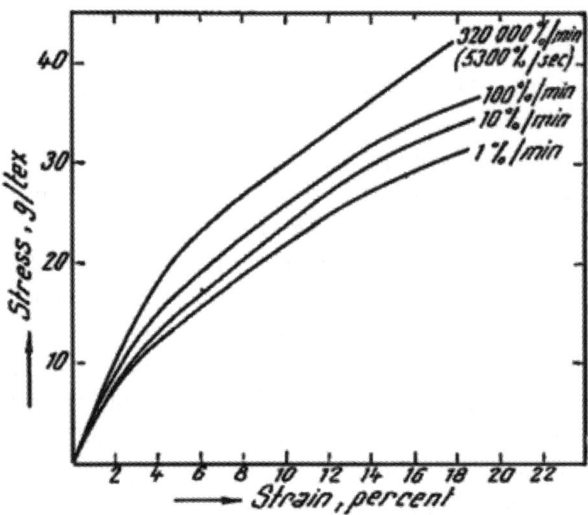

Fig. 5.1.3   Stress–strain curves for braided silk.

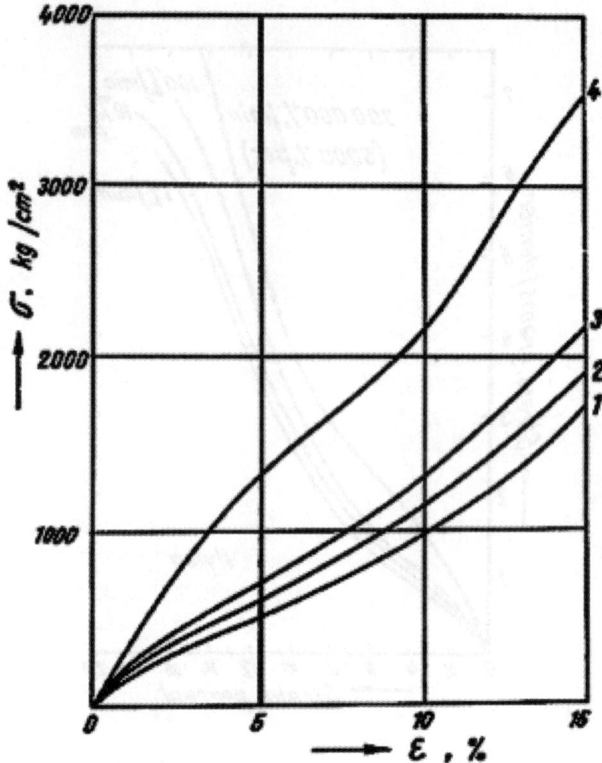

Fig. 5.1.4   Stress–strain curves for kapron for the following strain rates: (1) 1.2×10^{-3} sec^{-1}, (2) $1.2 \times (10^{-2}$ sec^{-1}, 3) 1.2×10^{-1} sec^{-1}, (4) (280 sec^{-1}.

influenced factor is the yield stress, but generally the whole stress–strain curve is rate dependent.

Another experimental technique that of impacting yarns with rifle bullets, was developed by Smith *et al.* [1963]. The transverse velocity of impact is now bigger: up to 700 m/sec. The effect of air drag on the motion of yarn is also analyzed by Smith *et al.* [1964]. In another paper Fenstermaker and Smith [1965] have described experiments in which transversely impacted filaments are observed by flash photography. Polyester yarns are used. The strain distribution is obtained at various times after impact and at various impact velocities. These experiments revealed important time effects (creep and relaxation effects) within 50 sec after impact. See also the papers of Smith and Fenstermaker [1967], Smith *et al.* [1965].

Longitudinal impact with a bullet were performed by Lewis [1957]; the velocity of the bullet is 170 m/sec. Similar experiments are reported by Victorov *et al.* [1966] for kapron and other yarns, tested at room temperature for rate of strain ranging between 10^{-3} and 10^2 sec^{-1}. An increase of the rate of strain raises the stress–strain curve appreciably.

Some other authors have also reported results on transverse impact of yarns by bullets. But impact grater than 460 m/sec produce a sudden rupture of the yarn and generally all yarns are strongly rate dependent.

5.2 Equations of Motion

In order to obtain the equations of motion (Cristescu [1967]) of an extensible string moving in three-dimensional space, we use a curvilinear coordinate s taken along the string and starting from an arbitrary coordinate origin. This Eulerian coordinate is sometimes called the "laboratory" coordinate or the "actual" coordinate, corresponding to the deformed string at the actual moment t. The tension or total force in the string is denoted by $\mathbf{T}$. It is a variable factor, both along the string and in time. In extensible, perfectly flexible strings, the tension in every point is always directed along the tangent to the string, which will be represented below as a line.

Consider a portion of the string comprised between the curvilinear coordinates s_1 and s_2 (Fig. 5.2.1). This portion of the string will be subjected now only to the tension, but possibly also, to various external forces (X, Y, Z), which are proportional to the length of the portion of the string considered. For this portion we can write

$$\left(T\frac{\partial x}{\partial s}\right)\bigg|_{s_2} - \left(T\frac{\partial x}{\partial s}\right)\bigg|_{s_1} + \int_{s_1}^{s_2} X\,ds = \int_{s_1}^{s_2} \rho\frac{\partial^2 x}{\partial t^2}\,ds$$

and two similar equations for y and z. Here ρ is the density of the string, which is function of the two variables s and t. The projection of $\mathbf{T}$ on the three considered axes are $T(\partial x/\partial s)$, $T(\partial y/\partial s)$ and $T(\partial z/\partial s)$, T being the modulus of $\mathbf{T}$. The previous equation can be written in the form

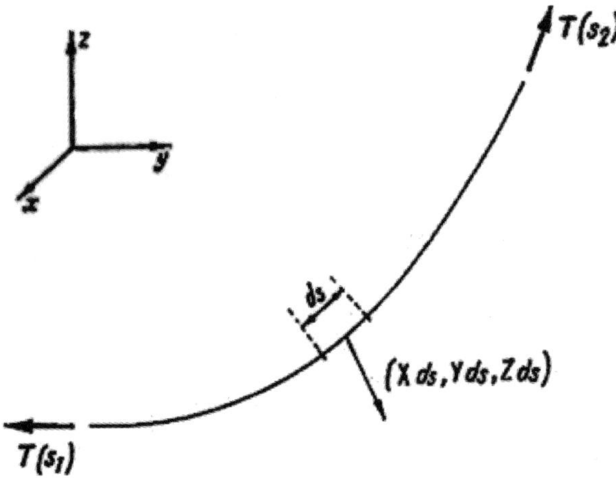

Fig. 5.2.1   Element of a string.

$$\int_{s_1}^{s_2} \left[\frac{\partial}{\partial s}\left(T\frac{\partial x}{\partial s}\right) + X - \rho\frac{\partial^2 x}{\partial t^2}\right] ds = 0$$

or taking into account that s_1 and s_2 are arbitrary

$$\frac{\partial}{\partial s}\left(T\frac{\partial x}{\partial s}\right) + X = \rho\frac{\partial^2 x}{\partial t^2}\,, \tag{5.2.1}$$

and two similar equations for y and z.

It is easy to replace the curvilinear coordinate s by a Lagrangean coordinate s_0.

The unknown function x, y, z will depend on the variable s_0 through the intermediary of s, and also on the time t: $x(s(s_0), t)$ etc. Thus for a fixed value of t, the formulae

$$\frac{\partial x}{\partial s_0} = \frac{\partial x}{\partial s}\frac{ds}{ds_0}\,, \quad \text{etc.} \tag{5.2.2}$$

will apply.

The strain of the string is defined by

$$\varepsilon = \frac{ds - ds_0}{ds_0} = \sqrt{\left(\frac{\partial x}{\partial s_0}\right)^2 + \left(\frac{\partial y}{\partial s_0}\right)^2 + \left(\frac{\partial z}{\partial s_0}\right)^2} - 1\,. \tag{5.2.3}$$

The last expression is obtained using the formulae (5.2.2). Thus these formulae can be written

$$\frac{\partial x}{\partial s_0} = (1+\varepsilon)\frac{\partial x}{\partial s}\,, \quad \frac{\partial y}{\partial s_0} = (1+\varepsilon)\frac{\partial y}{\partial s}\,, \quad \frac{\partial z}{\partial s_0} = (1+\varepsilon)\frac{\partial z}{\partial s}\,. \tag{5.2.4}$$

From the law of conservation of mass, it follows that

$$\rho\,ds = \rho_0\,ds_0$$

where ρ_0 is the density of the string in the initial non-deformed state. Thus, the relation between the actual density ρ and the initial one ρ_0 is

$$\rho_0 = \rho(1+\varepsilon)\,. \tag{5.2.5}$$

Taking into account the relation

$$ds = (1+\varepsilon)\,ds_0 \tag{5.2.6}$$

as well as (5.2.4) and (5.2.5), the equations of motion (5.2.1) can be written as

$$\frac{\partial}{\partial s_0}\left(\frac{T}{1+\varepsilon}\frac{\partial x}{\partial s_0}\right) - \rho_0\frac{\partial^2 x}{\partial t^2} + (1+\varepsilon)X = 0 \tag{5.2.7}$$

with two similar equations for y and z.

To these equations must be added the constitutive equation.

5.3 The Finite Constitutive Equation: Basic Formulae

The constitutive equation used most in dynamic problems is the finite constitutive equation, of the form

$$T = T(\varepsilon)\,. \tag{5.3.1}$$

This function is generally a monotonic (sometimes strictly monotonic) increasing function.

Taking into account (5.3.1) together with the system (5.2.7) and the relation obtained from (5.2.3) by differentiating with respect to s_0, we obtain

$$x_{s_0}\frac{\partial x_{s_0}}{\partial s_0} + y_{s_0}\frac{\partial y_{s_0}}{\partial s_0} + z_{s_0}\frac{\partial z_{s_0}}{\partial s_0} - (1+\varepsilon)\frac{\partial \varepsilon}{\partial s_0} = 0\,,$$

$$\frac{T}{1+\varepsilon}\frac{\partial x_{s_0}}{\partial s_0} - \rho_0\frac{\partial x_t}{\partial t} + x_{s_0}\frac{(1+\varepsilon)dT/d\varepsilon - T}{(1+\varepsilon)^2}\frac{\partial \varepsilon}{\partial s_0} + (1+\varepsilon)X = 0\,, \tag{5.3.2}$$

with two similar relations for y and z. For the sake of simplicity the notations $x_{s_0} = \partial x/\partial s_0$, $x_t = \partial x/\partial t$, etc. has been used.

In order to find the type of the system, let us find the type of the characteristic lines. For this purpose, the relations

$$dx_{s_0} = \frac{\partial x_{s_0}}{\partial s_0}\,ds_0 + \frac{\partial x_{s_0}}{\partial t}\,dt\,, \quad dx_t = \frac{\partial x_t}{\partial s_0}\,ds_0 + \frac{\partial x_t}{\partial t}\,dt\,, \tag{5.3.3}$$

and similar ones for y and z, will be combined with (5.3.2).

The system (5.3.2) contains four equations with four unknown functions x, y, z, and ε, and two independent variables s_0 and t. It is clear that the strain ε can be eliminated using (5.2.3). However, to facilitate subsequent mechanical interpretation it is more advantageous to retain ε as an unknown function. The ten relations (5.3.2), (5.3.3) contain ten derivatives of the form $\partial x_{s_0}/\partial s$, $\partial x_{s_0}/\partial t$, etc. By solving with respect to these derivatives we obtain

$$\frac{\partial \varepsilon}{\partial s_0} = \frac{\xi x_{s_0} + \eta y_{s_0} + \zeta z_{s_0}}{(1+\varepsilon)[\rho_0(ds_0/dt)^2 - dT/d\varepsilon]}\,,$$

$$\frac{\partial x_{s_0}}{\partial s_0} = -\frac{(1+\varepsilon)^3[dT/d\varepsilon - \rho_0(ds_0/dt)^2]\xi}{(1+\varepsilon)^3[\rho_0(ds_0/dt)^2 - T/(1+\varepsilon)][\rho_0(ds_0/dt)^2 - dT/d\varepsilon]} \tag{5.3.4}$$

$$+ \frac{x_{s_0}[(1+\varepsilon)dT/d\varepsilon - T](\xi x_{s_0} + \eta y_{s_0} + \zeta z_{s_0})}{(1+\varepsilon)^3[\rho_0(ds/dt)^2 - T/(1+\varepsilon)][\rho_0(ds_0/dt)^2 - dT/d\varepsilon]}\,,$$

and similar expressions for the other second order derivatives which occur in (5.3.2). In (5.3.4) the notations

$$\xi = \rho_0 \left(dx_{s_0} \frac{ds_0}{dt} - dx_t \right) \frac{1}{dt} + (1 + \varepsilon)X \,,$$

$$\eta = \rho_0 \left(dy_{s_0} \frac{ds_0}{dt} - dy_t \right) \frac{1}{dt} + (1 + \varepsilon)Y \,, \tag{5.3.5}$$

$$\zeta = \rho_0 \left(dz_{s_0} \frac{ds_0}{dt} - dz_t \right) \frac{1}{dt} + (1 + \varepsilon)Z$$

have been used.

From (5.3.4) it follows (Cristescu [1954]) that the system (5.3.2) is totally hyperbolic since it has four distinct families of characteristic curves. Two of them are defined by the differential relations

$$\frac{ds_0}{dt} = \pm \left[\frac{T}{\rho_0(1 + \varepsilon)} \right]^{1/2} \,, \tag{5.3.6}$$

while the other two by the relations

$$\frac{ds_0}{dt} = \pm \left[\frac{1}{\rho_0} \frac{dT}{d\varepsilon} \right]^{1/2} \,. \tag{5.3.7}$$

Thus all the characteristic lines are generally curved lines whose slope depends on T and ε. From the mechanical point of view, the characteristics lines represent wave fronts. There will be two kinds of waves: the characteristic lines (5.3.6) represent wave fronts which propagate with the velocity

$$c_I(\varepsilon) = \left[\frac{T}{\rho_0(1 + \varepsilon)} \right]^{1/2} \,, \tag{5.3.8}$$

while the characteristic lines (5.3.7) represent wave fronts which propagate with a velocity

$$c_{II}(\varepsilon) = \left[\frac{1}{\rho_0} \frac{dT}{d\varepsilon} \right]^{1/2} \,. \tag{5.3.9}$$

If (5.3.6) is taken into account, it follows from (5.3.4) (Cristescu [1960, 1961]) that the following differential relations

$$dx_t = \pm c_I(\varepsilon)\, dx_{s_0} + \frac{1}{\rho_0}[(1 + \varepsilon)X - x_{s_0}F]\, dt - \frac{x_{s_0}}{1 + \varepsilon}(\pm c_I(\varepsilon)\, d\varepsilon - dv) \,,$$

$$\tag{5.3.10}$$

and four similar relations for y and z, are satisfied on these lines. Along the characteristics (5.3.7) are satisfied the differential relations

$$\pm c_{II}(\varepsilon)\, d\varepsilon - dv + \frac{1 + \varepsilon}{\rho_0}\, F dt = 0 \,. \tag{5.3.11}$$

Here the notations

$$dv = \frac{1}{1 + \varepsilon}(x_{s_0}\, dx_t + y_{s_0}\, dy_t + z_{s_0}\, dz_t) \,,$$

$$\tag{5.3.12}$$

$$F = \frac{1}{1 + \varepsilon}(Xx_{s_0} + Yy_{s_0} + Zz_{s_0})$$

have been used, where v is the projection on the tangent to the string of the velocity of the corresponding particle of the string. The differential dv is computed along a non-specified characteristic line. F is the projection on the tangent to the string of the external forces which act on the portion of the string considered, of unit length.

The upper and lower signs in (5.3.6) and (5.3.10) and in (5.3.7) and (5.3.11) correspond to each other.

Let us write the "jump conditions", by denoting by $[A]$ the jump of the function A across a characteristic line. Writing the relations (5.3.3) for either side of a characteristic line, and taking into account that the differentials dx_{s_0}, dx_t, etc. are computed along a characteristic line and are in fact "interior derivatives" which are continuous across such a line, we obtain for the kinematics compatibility conditions:

$$\left[\frac{\partial x_{s_0}}{\partial s_0}\right]\frac{ds_0}{dt} + \left[\frac{\partial x_s}{\partial t}\right] = 0, \quad \left[\frac{\partial x_t}{\partial s_0}\right]\frac{ds_0}{dt} + \left[\frac{\partial x_t}{\partial t}\right] = 0 \tag{5.3.13}$$

with four similar relations for y and z. In (5.3.13) the rate ds_0/dt must be replaced by one of the two expressions given by (5.3.6) or (5.3.7), depending of the characteristic line considered. If the equations of motion (5.3.2) are written for either side of a characteristic line and the difference is taken, and if (5.3.13) is also taken into account, we obtain

$$x_{s_0}\left[\frac{\partial x_{s_0}}{\partial s_0}\right] + y_{s_0}\left[\frac{\partial y_{s_0}}{\partial s_0}\right] + z_{s_0}\left[\frac{\partial z_{s_0}}{\partial s_0}\right] - (1+\varepsilon)\left[\frac{\partial \varepsilon}{\partial s_0}\right] = 0,$$

$$\left\{\frac{T}{1+\varepsilon} - \rho_0\left(\frac{ds_0}{dt}\right)^2\right\}\left[\frac{\partial x_{s_0}}{\partial s_0}\right] + \frac{(1+\varepsilon)dT/d\varepsilon - T}{(1+\varepsilon)^2}x_{s_0}\left[\frac{\partial \varepsilon}{\partial s_0}\right] = 0, \tag{5.3.14}$$

with two other relations for y and z. These are the dynamic compatibility equations (Cristescu [1954]).

In order to see what happens on each characteristic line, we combine (5.3.6) and (5.3.14), to get

$$x_{s_0}\left[\frac{\partial x_{s_0}}{\partial s_0}\right] + y_{s_0}\left[\frac{\partial y_{s_0}}{\partial s_0}\right] + z_{s_0}\left[\frac{\partial z_{s_0}}{\partial s_0}\right] = 0 \quad \text{and} \quad \left[\frac{\partial \varepsilon}{\partial s_0}\right] = 0. \tag{5.3.15}$$

Similarly, from (5.3.7) and (5.3.14) [taking into account also (5.2.2) and (5.2.3)] we obtain:

$$\frac{\lfloor \partial x_{s_0}/\partial s_0 \rfloor}{x_s} = \frac{\lfloor \partial y_{s_0}/\partial s_0 \rfloor}{y_s} = \frac{\lfloor \partial z_{s_0}/\partial s_0 \rfloor}{z_s} = \left[\frac{\partial \varepsilon}{\partial s_0}\right]. \tag{5.3.16}$$

From (5.3.15) follows that the waves which propagate with the velocity (5.3.8) are *transverse* waves which affect only the shape of the string but do not affect the longitudinal strain (because $[\partial \varepsilon/\partial s_0] = 0$). The waves which propagate with the velocity (5.3.9) are *longitudinal* waves which produce only propagable elongations of the string, but do not change its shape. The jumps $[\partial x_{s_0}/\partial s_0]$, etc. of the second derivatives of the coordinates are simply the projections on the coordinate axes of the discontinuity of $\partial \varepsilon/\partial s_0$; the relation

$$\left[\frac{\partial x_{s_0}}{\partial s_0}\right]^2 + \left[\frac{\partial y_{s_0}}{\partial s_0}\right]^2 + \left[\frac{\partial z_{s_0}}{\partial s_0}\right]^2 = \left[\frac{\partial \varepsilon}{\partial s_0}\right]^2$$

is also satisfied.

Thus the projection of the discontinuity on the tangent to the string will propagate through longitudinal waves while the orthogonal component of this discontinuity will propagate through transverse waves. Both types of waves influence one another.

The equation of motion of strings considered to be made of a phase-transforming material like a shape-memory alloy, was done by Purohit and Bhattacharya [2003]. They study the phase boundaries, the driving force acting on them and the kinetic relation governing their propagation. The paper presents also a numerical method for studying general initial and boundary value problems in strings.

5.4   The Order of Propagation of Waves

The order of propagation depends on the initial and boundary conditions, as well on the constitutive equation.

Suppose that, at the end of the string, the strain increases monotonically and that at each moment of time both types of waves are generated. The order of propagation of waves in the string will be examined for a material whose constitutive equation is represented in Fig. 5.4.1 (Cristescu [1954, 1958]). Such constitutive equation does not correspond to any specific material and has been selected only by way of example.

From (5.3.8) and (5.3.9) it follows that $c_I < c_{II}$ for those tension and strain states which are represented by points on the segments OA or BC. In this case the transverse waves propagate more slowly than the longitudinal ones. On the segments AB and CD the inequality $c_I > c_{II}$ holds and in this case the transverse waves propagate faster than the longitudinal waves. Finally there are certain points A, B, C on the diagram, and sometimes even certain segments on this diagram, where the differential relation

$$(1 + \varepsilon)\frac{dT}{d\varepsilon} = T \tag{5.4.1}$$

is satisfied. In these points, or portions of the diagram, $c_I = c_{II}$ and hence both waves generated at the end of the string propagate with the same velocity. The relation (5.4.1) is satisfied over an entire portion of the tension–strain diagram for synthetic strings (Miklowitz [1947], Marshall and Thomson [1954]). Generally $c_I \neq c_{II}$ and we will consider this case.

If the slope of the diagram is zero in some portions (for example $A'B'$) then for the corresponding states of tension and strain the propagation of longitudinal waves is no longer possible, and $c_{II} = 0$. If at a certain moment, the tension at the end of the string begins to increase once more (for instance starting at the point

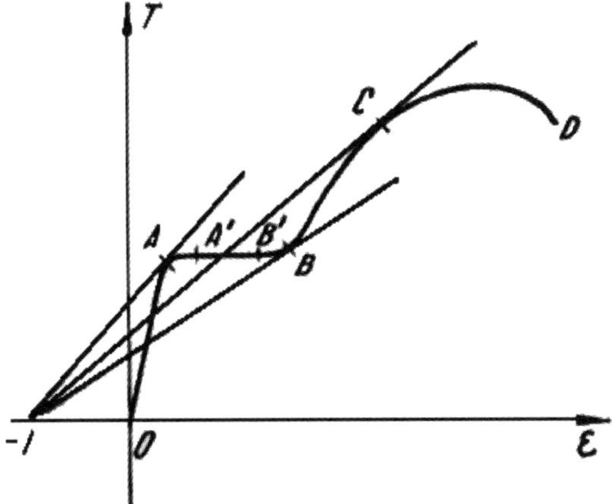

Fig. 5.4.1 Typical tension–strain curve. At points A, B, C the velocities of propagation of transverse and longitudinal waves are equal.

B') the longitudinal waves can propagate again. The case when the longitudinal waves propagate first and the transverse ones (which do not produce elongations) later, is the simplest case and has bee studied by many authors (see Rahmatulin and Demianov [1961]). All these authors considered strings which were initially rectilinear and in which the two types of waves propagate successively in distinct portions of the string.

5.5 Boundary and Initial Conditions

We shell now formulate the boundary and initial conditions for a general three-dimensional motion of the extensible string (Cristescu [1964]). One can formulate the boundary conditions for fixed or moving points of the string. For instance in problems concerning the impact of strings by moving bodies, the body which strikes the string can contact the same material point of the string permanently or can move along it (floating boundary). Thus both stationary and moving boundaries have to be considered.

The initial conditions can also be formulated in various manners. In the case of a Cauchy problem, for $t = 0$ and $0 \leq s_0 \leq l$ the following functions must be prescribed

$$\varepsilon(s_0, 0), v(s_0, 0), x(s_0, 0), \ldots, x_{s_0}(s_0, 0), \ldots, x_t(s_0, 0), \ldots, \qquad (5.5.1)$$

where l is the initial length of the string. Thus, at the initial moment, the position of the string, the velocity of each of its points, and the strain distribution along the string must be known. It is assumed that the functions (5.5.1) possess continuous first order derivatives. In this case (Courant [1962] Ch. V), within a certain band

$0 \leq t \leq h$, the system (5.3.2) possesses a unique solution with continuous derivatives, provided that h is sufficiently small and the coefficients in (5.3.2) as well as the data are Lipschitz-continuous. The solution can be further extended for $t > h$ as long as the continuity conditions, which must be satisfied by the coefficients, hold.

Not all the functions in (5.5.1) can be prescribed arbitrarily since these functions must satisfy certain compatibility conditions, the equations given in the previous paragraphs.

If the boundary conditions always affect the same material points of the string (stationary boundaries) the boundary conditions have to be prescribed in the $s_0 Ot$ plane along the straight lines $s_0 = 0$ and $s_0 = l$. If the bodies which strike the string, or move its ends, move along the string, then the boundary conditions must be prescribed along floating boundaries. Such a problems arises for instance, if a string is wrapped around a bobbin. Another example is the safety cables used in the Saturn V rockets which have sent people to the Moon. It is known that in this case, in order to avoid the danger of fire, a cable was attached to the rocket, close to the top (about 100 m), and the other end was attached to a pole on the ground. On this cable was attached a cable, isolated against fire. When testing the system, one has observed that when the cage was going down, that the cable is almost able to wrap up around the pulley. In other words, when going down, the pulley was approaching the velocity of the transverse waves traveling in the cable.

In all those cases the motion along the string of the impacting body is smaller than the velocity of propagation of the transverse c_I or longitudinal waves. Thus, in most cases, a mixed boundary problem must be solved. If the striking body moves along the string with a velocity higher than c_{II} or c_I then certain discontinuities of the shape or of the strain of the string propagate, and these would soon produce an entanglement or rupture of the string.

The elastic string. If the string is elastic, the constitutive equation (5.3.1) becomes

$$T = E\varepsilon \,, \tag{5.5.2}$$

where E is an elastic constant. The velocities of propagation of the two waves are (Cristescu [1951])

$$c_I(\varepsilon) = \left[\frac{E\varepsilon}{\rho_0(1+\varepsilon)} \right]^{1/2} \quad \text{and} \quad c_{II} = \left[\frac{E}{\rho_0} \right]^{1/2}. \tag{5.5.3}$$

Thus, the longitudinal waves propagate with a constant velocity. The equality (5.4.1) is impossible and hence the two waves always propagate with distinct velocities. For each ε, $c_{II} > c_I(\varepsilon)$.

The linear work-hardening string. In this case the constitutive equation (5.3.1) consists of two relations

$$\begin{aligned} T &= E\varepsilon & &\text{if } T \leq T_Y \,, \\ T &= T_Y + E_1(\varepsilon - \varepsilon_Y) & &\text{if } T \geq T_Y \,, \end{aligned} \tag{5.5.4}$$

where T_Y is the yield tension of the material considered and E_1 is the constant work-hardening modulus. If the tension does not exceeds the yield tension the situation coincides with that described above. If $T > T_Y$ the velocities of propagation become

$$c_I(\varepsilon) = \left[\frac{T_Y + E_1(\varepsilon - \varepsilon_Y)}{\rho_0(1+\varepsilon)}\right]^{1/2} \quad \text{and} \quad c_{II} = \left[\frac{E_1}{\rho_0}\right]^{1/2}. \qquad (5.5.5)$$

It should be noted that in the neighborhood of the yield point, the velocity c_I varies continuously with ε, while c_{II} suffers a jump from $c_{II} = (E/\rho_0)^{1/2}$ for $T < T_Y$, to $c_{II} = (E_1/\rho_0)^{1/2}$ for $T > T_Y$. Generally c_{II} is smaller in the plastic domain than in the elastic one and is a constant throughout this first domain. Depending on the sizes of E_1 and T_Y, three possibilities arise

$$E_1 < \frac{T_Y}{1+\varepsilon_Y} \quad \text{and} \quad c_{II} < c_I\,,$$

$$E_1 = \frac{T_Y}{1+\varepsilon_Y} \quad \text{and} \quad c_{II} = c_I\,, \qquad (5.5.6)$$

$$E_1 > \frac{T_Y}{1+\varepsilon_Y} \quad \text{and} \quad c_{II} > c_I\,.$$

All of these possibilities are encountered in practice. The last of them leads to certain difficulties since, in this case, longitudinal shock waves can appear. The second case has already been discussed above; it corresponds to the case when both types of waves propagate with the same velocity, when the characteristic curves (5.3.6) and (5.3.7) coincide and the last three equations of the system (5.3.2) reduce to

$$E_1 \frac{\partial x_{s_0}}{\partial s_0} - \rho_0 \frac{\partial x_t}{\partial t} + (1+\varepsilon)X = 0\,,$$

with similar equations for y and z. Through the intermediary of ε, this system of equations becomes a system of three partially coupled differential equations. If external forces are absent the three equations are no longer partially coupled and each of them can be integrated separately.

Inextensible strings. Another possible case is that of a string which has suffered a former deformation. The strain is then generally distributed non-homogeneously along the string but remains constant in time in each section of the string. In this case the system (5.3.2) reduces to three semi-linear non-coupled equations

$$\Phi(s_0)\frac{\partial^2 x}{\partial s_0^2} - \rho_0 \frac{\partial^2 x}{\partial t^2} + \frac{\partial x}{\partial s_0}\frac{d\Phi(s_0)}{ds_0} + (1+\varepsilon(s_0))X(s_0,t) = 0\,,$$

with two similar ones for y and z. Here $\Phi(s_0) = T(\varepsilon)/(1+\varepsilon)$ is a known function of s_0.

The integration of such equations do not entail special difficulties. If the strain is uniformly distributed along the string $\varepsilon = \text{const.}$, and the equation reduces to an equation with constant coefficients

$$c_I^2 \frac{\partial^2 x}{\partial s_0^2} - \frac{\partial^2 x}{\partial t^2} + \frac{1+\varepsilon}{\rho_0} X(s_0,t) = 0 \,,$$

with two others for y and z. Here c_I and ε are constants. The string behaves as an inextensible string which may possibly undergo appreciable transverse motions.

5.6   Numerical Examples

The numerical examples are obtained with integration along the four characteristic lines. Details are given in Cristescu [1967] and will not be given here. Examples are given for semi-infinite string, for finite string, for finite elastic string. For the extensible cable used to break high speed moving bodies when at the impacted end we have

$$M d\mathbf{V} = A\mathbf{T}dt$$

where $\mathbf{V}$ is the velocity of the body, A is the area of the transverse section of the string and $\mathbf{T}$ is the stress at the end of the string (see Fig. 5.6.1):

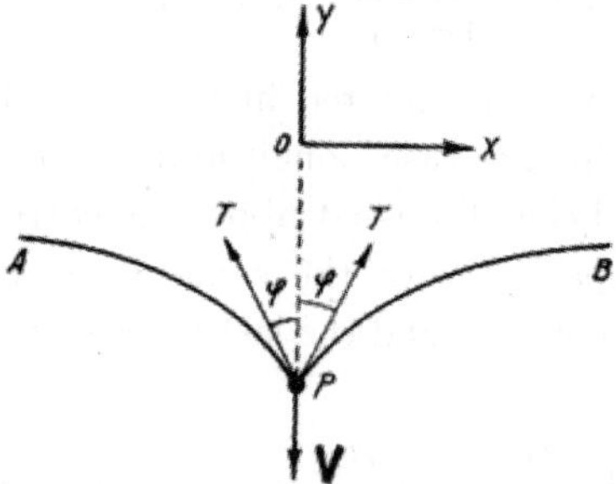

Fig. 5.6.1   Impact of a rectilinear cable with a fast moving body.

The shape of the cable is shown in Fig. 5.6.2. Were given also vertical velocity, horizontal velocity, strain distribution, longitudinal velocity, and variation of strain along the first transverse wave front and the length of the cable at various times.

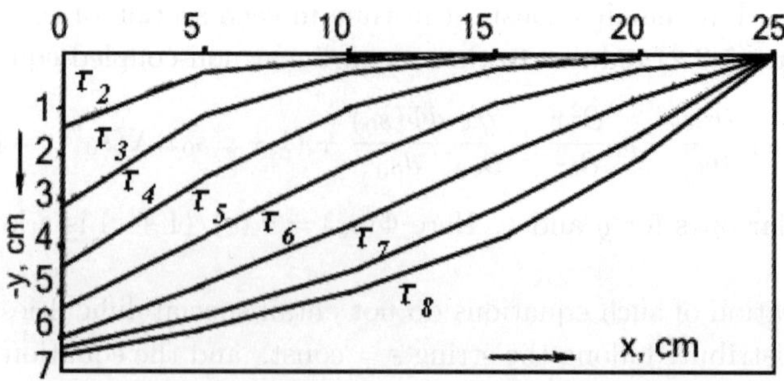

Fig. 5.6.2   Shape of the cable at various times.

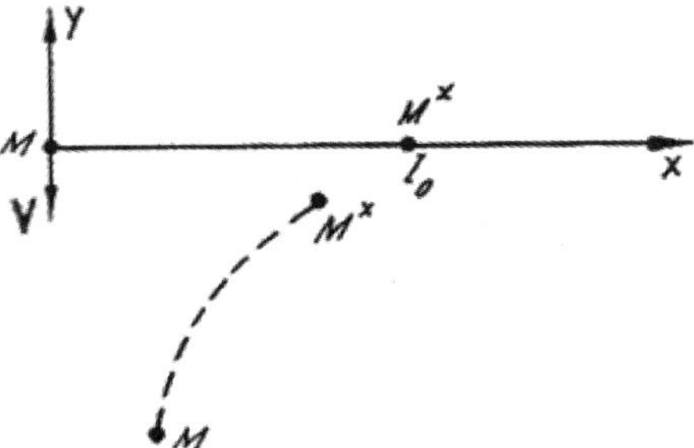

Fig. 5.6.3   Motion of two bodies connected by an extensible cable; Schematic representation of two positions of the system.

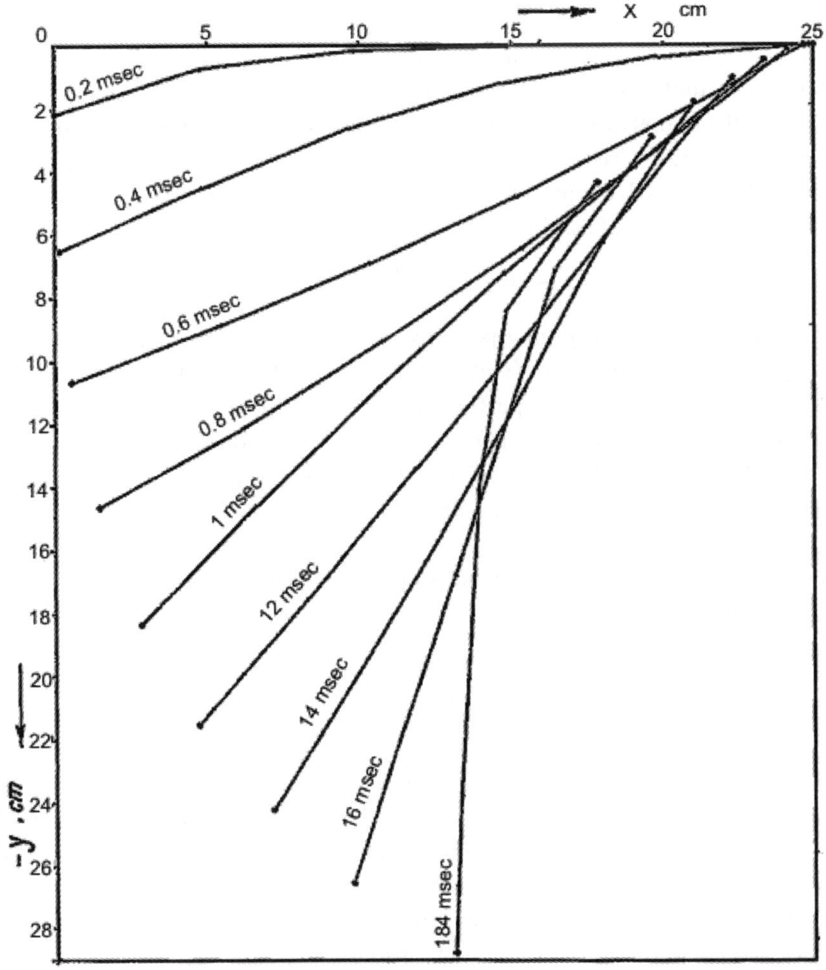

Fig. 5.6.4   Successive positions of the system for the case $M^* = 2M$.

The motion of two bodies connected by an extensible cable was also considered. The mass of the two bodies were $M^* = 2M$ or $2M^* = M$ (Fig. 5.6.3). The successive positions of the system for the case $M^* = 2M$ are given in Fig. 5.6.4. Were also given the strain profile at various time, the characteristic field showing the propagation of the first longitudinal and the first transverse wave fronts in the cable and several reflections from the two ends of the cable, the equalization of the vertical velocities of the two bodies, the strain profiles at various times and the variation of the length of the cable.

5.7 Rate Influence

Finite constitutive equations cannot be used to describe the mechanical properties of synthetic strings for which the rate influence is very large. Thus, in many cases, various classes of differential constitutive equations, which can be applied to different kinds of materials, must be considered. Some such differential constitutive equations will be analyzed below in connection with the problem of rapid motion of extensible strings (Cristescu [1965]).

The first constitutive equation which will be considered is that which is quasi-linear both in the rate of strain and in the rate of tension:

$$\frac{\partial \varepsilon}{\partial t} = g(T, \varepsilon) \frac{\partial T}{\partial t} + f(T, \varepsilon), \tag{5.7.1}$$

where the functions f and g may possibly depend explicitly on s_0, t, as well as on various characteristic constants (yield tension, etc.). The function $g(T, \varepsilon)$ describes the instantaneous response properties of the material, at a certain increment of the tension, while the function $f(T, \varepsilon)$ describes the non-instantaneous response properties. Relations of the type (5.7.1) can be used both for loading processes and for non-linear unloading processes. Various special cases of the relation (5.7.1) will be considered below.

The constitutive equation must be supplemented by the equations

$$x_{s_0} \frac{\partial x_{s_0}}{\partial s_0} + y_{s_0} \frac{\partial y_{s_0}}{\partial s_0} + z_{s_0} \frac{\partial z_{s_0}}{\partial s_0} - (1 + \varepsilon) \frac{\partial \varepsilon}{\partial s_0} = 0,$$

$$\frac{T}{1 + \varepsilon} \frac{\partial x_{s_0}}{\partial s_0} - \rho_0 \frac{\partial x_t}{\partial t} + \frac{x_{s_0}}{1 + \varepsilon} \frac{\partial T}{\partial s_0} - \frac{x_{s_0} T}{(1 + \varepsilon)^2} \frac{\partial \varepsilon}{\partial s_0} + (1 + \varepsilon) X = 0, \tag{5.7.2}$$

with two other ones for y and z. (5.7.1) and (5.7.2) represent the system of equations that describes the motion of a string which satisfies a constitutive equation of the type (5.7.1). This time, there is one more unknown function T, so that there is now five unknown quantities i.e. x, y, z, ε and T.

In order to find the type of this system (5.7.1) and (5.7.2) will be combined with the six equations (5.3.3) as well as the relations

$$d\varepsilon = \frac{\partial \varepsilon}{\partial s_0} ds_0 + \frac{\partial \varepsilon}{\partial t} dt, \quad dT = \frac{\partial T}{\partial s_0} ds_0 + \frac{\partial T}{\partial t} dt. \tag{5.7.3}$$

Taken together, these relations form a system of ten equations with ten unknown derivatives $\partial x_{s_0}/\partial s_0$, $\partial x_t/\partial t$, $\partial y_{s_0}/\partial s_0$, $\partial y_t/\partial t$, $\partial z_{s_0}/\partial s_0$, $\partial z_t/\partial t$, $\partial\varepsilon/\partial s_0$, $\partial\varepsilon/\partial t$, $\partial T/\partial s_0$ and $\partial T/\partial t$. By solving this system with respect to these derivatives, we obtain:

$$\frac{\partial\varepsilon}{\partial t} = \frac{f\,dt + g\,dT + (1+\varepsilon)Fg\,ds_0 - \rho_0 g(ds_0/dt)\,dv}{[1-\rho_0 g(ds_0/dt)^2]\,dv}$$

$$\frac{\partial\varepsilon}{\partial s_0} = \frac{-(f\,dt + g\,dT - d\varepsilon) + \rho_0 g[dv - (ds_0/dt)d\varepsilon] - ((1+\varepsilon)/\rho_0)gF\,ds_0}{[1-\rho_0 g(ds_0/dt)^2]\,ds_0}$$

$$\frac{\partial T}{\partial t} = \frac{-[\rho_0(ds_0/dt)^2 f + 1]\,dT - (1+\varepsilon)F\,ds_0 + \rho_0(ds_0/dt)\,dv}{[1-\rho_0 g(ds_0/dt)^2]\,dt}\,,$$

$$\frac{\partial x_{s_0}}{\partial s_0} = \frac{[T/(1+\varepsilon) - \rho_0(ds_0/dt)^2](f\,dt + g\,dT - d\varepsilon)x_{s_0}}{-(1+\varepsilon)[T/(1+\varepsilon) - \rho_0(ds_0/dt)^2][1-\rho_0 g(ds_0/dt)^2]\,ds_0}$$

$$+ \frac{(1+\varepsilon)[1-\rho_0 g(ds_0/dt)^2][(1+\varepsilon)X\,dt - \rho_0 dx_t + \rho_0(ds_0/dt)\,dx_{s_0}]\,ds_0/dt}{-(1+\varepsilon)[T/(1+\varepsilon) - \rho_0(ds_0/dt)^2][1-\rho_0 g(ds_0/dt)^2]\,ds_0}$$

$$- \frac{x_{s_0}[1 - Tg/(1+\varepsilon)][(1+\varepsilon)F\,dt - \rho_0(dv - (ds_0/dt)\,d\varepsilon)]\,ds_0/dt}{-(1+\varepsilon)[T/(1+\varepsilon) - \rho_0(ds_0/dt)^2][1-\rho_0 g(ds_0/dt)^2]\,ds_0}\,,$$

$$(5.7.4)$$

and similarly for the other six derivatives.

Thus, in the case of the constitutive equation (5.7.1) the equations of motion of the problem possess five families of characteristic lines

$$\left(\frac{ds}{dt}\right)^2 = \frac{1}{\rho_0}\frac{T}{1+\varepsilon} = c_I^2(T,\varepsilon)\,, \tag{5.7.5}$$

$$\left(\frac{ds_0}{dt}\right)^2 = \frac{1}{\rho_0 g} = c_{II}^2(T,\varepsilon)\,, \tag{5.7.6}$$

$$s_0 = \text{const.} \tag{5.7.7}$$

Along these characteristic lines the following differential relations

$$\rho_0 dx_t = \pm\rho_0 c_I dx_{s_0} + (1+\varepsilon)X\,dt - \frac{x_{s_0}}{1+\varepsilon}\{(1-\varepsilon)F\,dt - \rho_0(dv \mp c_I d\varepsilon)\}\,, \tag{5.7.8}$$

$$\pm c_{II}(f\,dt + g\,dT) + \frac{1+\varepsilon}{\rho_0}F\,dt - dv = 0\,, \tag{5.7.9}$$

$$d\varepsilon = g\,dT + f\,dt\,, \tag{5.7.10}$$

are satisfied respectively. Thus there are two types of propagable waves, whose velocities of propagation are functions of both tension and strain. There are also stationary (non propagable) discontinuities (5.7.7). Certain information concerning the nature of the waves which appear in this case can be obtained from the dynamic and kinematics compatibility conditions which are

$$x_{s_0}\left[\frac{\partial x_{s_0}}{\partial s_0}\right] + y_{s_0}\left[\frac{\partial y_{s_0}}{\partial s_0}\right] + z_{s_0}\left[\frac{\partial z_{s_0}}{\partial s_0}\right] - (1+\varepsilon)\left[\frac{\partial \varepsilon}{\partial s_0}\right] = 0,$$

$$\left\{\frac{T}{1+\varepsilon} - \rho_0\left(\frac{ds_0}{dt}\right)^2\right\}\left[\frac{\partial x_{s_0}}{\partial s_0}\right] + \frac{x_{s_0}}{1+\varepsilon}\left[\frac{\partial T}{\partial s_0}\right] - \frac{x_{s_0}T}{(1+\varepsilon)^2}\left[\frac{\partial \varepsilon}{\partial s_0}\right] = 0,$$

$$\dots\dots\dots\dots\dots$$

$$\left[\frac{\partial \varepsilon}{\partial t}\right] - g\left[\frac{\partial T}{\partial t}\right] = 0,$$

$$-\left[\frac{\partial x_{s_0}}{\partial s_0}\right]\left(\frac{ds_0}{dt}\right)^2 + \left[\frac{\partial x_t}{\partial t}\right] = 0,$$

$$\dots\dots\dots\dots\dots$$

$$\left[\frac{\partial \varepsilon}{\partial t}\right]dt + \left[\frac{\partial \varepsilon}{\partial s_0}\right]ds_0 = 0,$$

$$\left[\frac{\partial T}{\partial t}\right]dt + \left[\frac{\partial T}{\partial s_0}\right]ds_0 = 0,$$

$$(5.7.11)$$

and four equations for y and z. The first five relations (5.7.11) are the dynamic compatibility equations while last five are the kinematics compatibility conditions. In (5.7.11) and in the following formulae, the square brackets are used to denote the jump of a certain function across a wave front.

Across the characteristic lines (5.7.5), it follows from the second, third and fourth relation (5.7.11) that

$$\left[\frac{\partial T}{\partial s_0}\right] = \frac{T}{1+\varepsilon}\left[\frac{\partial \varepsilon}{\partial s_0}\right].$$

From this relation, and taking into account the other relations (5.7.11) as well, it follows that across a line (5.7.5)

$$\left[\frac{\partial T}{\partial s_0}\right] = \left[\frac{\partial \varepsilon}{\partial s_0}\right] = \left[\frac{\partial T}{\partial t}\right] = \left[\frac{\partial \varepsilon}{\partial t}\right] = 0, \tag{5.7.12}$$

if the relation $c_I \neq c_{II}$ is satisfied. Moreover, from the first relation (5.7.11), we obtain

$$x_{s_0}\left[\frac{\partial x_{s_0}}{\partial s_0}\right] + y_{s_0}\left[\frac{\partial y_{s_0}}{\partial s_0}\right] + z_{s_0}\left[\frac{\partial z_{s_0}}{\partial s_0}\right] = 0. \tag{5.7.13}$$

Thus from (5.7.12) and (5.7.13) it follows that (5.7.5) are transverse wave fronts which affect only the shape of the string but not the tension nor the longitudinal strain. Both the equation (5.7.5) of the characteristic curves and the differential relations (5.7.8) satisfied on these characteristics formally coincide with those corresponding to the finite constitutive equation.

If one takes into account (5.7.6), it follows from (5.7.11) that across a characteristic of this family the relations

$$\frac{1}{x_{s_0}}\left[\frac{\partial x_{s_0}}{\partial s_0}\right] = \frac{1}{y_{s_0}}\left[\frac{\partial y_{s_0}}{\partial s_0}\right] = \frac{1}{z_{s_0}}\left[\frac{\partial z_{s_0}}{\partial s_0}\right] = \left[\frac{\partial \varepsilon}{\partial s_0}\right] \tag{5.7.14}$$

are satisfied if $c_I \neq c_{II}$. Thus the discontinuities of the derivatives of the coordinates are directed along the tangent to the string. From (5.7.14) one obtains conclusions similar to those of (5.3.16) in the case of the finite constitutive equation. Thus, (5.7.6) represent longitudinal wave fronts which affect the strain and tension but not the shape of the string. The differential relations (5.7.9) are different from those corresponding to the finite constitutive equation.

The characteristics (5.7.12) represent stationary discontinuities. Across such a line, all the derivatives with respect to the variable t (the interior derivatives) are continuous. From the second relation (5.7.11), it follows that:

$$\left[\frac{\partial x_{s_0}}{\partial s_0}\right] = -\frac{x_{s_0}}{T}\left[\frac{\partial T}{\partial s_0}\right] + \frac{x_{s_0}}{1+\varepsilon}\left[\frac{\partial \varepsilon}{\partial s_0}\right], \tag{5.7.15}$$

with two similar relations for y and z. Introducing this relation into the first relation (5.7.11), we obtain $[\partial T/\partial s_0] = 0$, so that the derivatives of the tension are continuous across a line $ds_0 = 0$. It follows that (5.7.15) reduces to (5.7.14) and therefore that the stationary discontinuities affect only the strain.

We can consider some special cases. The first one is if the form

$$\frac{\partial T}{\partial t} = E\frac{\partial \varepsilon}{\partial t} - G(T, \varepsilon), \tag{5.7.16}$$

where the function $G(T, \varepsilon)$ is often taken to depend linearly on the overstress

$$G(T, \varepsilon) = k[\sigma - F(\varepsilon)]. \tag{5.7.17}$$

In (5.7.17) k is a specific constant of the material while $\sigma = F(\varepsilon)$ is the static constitutive equation of the material considered. Constitutive equation of the form (5.7.16) corresponds to materials whose instantaneous response to a tension increment is purely elastic.

A second special case of the constitutive equation (5.7.1) is the constitutive equation

$$\frac{\partial \varepsilon}{\partial t} = f(T, \varepsilon), \tag{5.7.18}$$

which corresponds to materials which do not possess an instantaneous response to an increment of the tension. This case leads to very interesting mathematical problems. The particular case of the Voigt model has been mentioned by Cristescu [1952] while for a general discussion of a constitutive equation of the form

$$T = T(\varepsilon, \dot{\varepsilon}) \tag{5.7.19}$$

the reader is referred to Cristescu [1957]. Combining the equations of motion with (5.7.18), one obtains the complete system describing the motion. The characteristics of the system are

$$\left(\frac{ds_0}{dt}\right)^2 = \frac{1}{\rho_0}\frac{T}{1+\varepsilon} = c^2(T,\varepsilon)\,, \tag{5.7.20}$$

$$dt = 0\,, \tag{5.7.21}$$

$$ds_0 = 0\,. \tag{5.7.22}$$

It should be noted that the two distinct characteristic lines (5.7.6) in the present special case are superposed and coincide with the family of straight linear $dt = 0$. The characteristics (6.7.5), which are transverse wave fronts, exist in this particular case too, as for the constitutive equation (5.7.1). Thus the system is of an intermediary type.

The influence of the temperature influence has been considered by Manacorda [1958]. Let us denote by $\theta(s_0, t)$ the actual temperature of the points of the string, by θ_e the actual external temperature and by θ_* the initial uniform temperature of both string and external media. We shell also use the notation $\pi = \theta - \theta_*$ and $\pi_e = \theta_e - \theta_*$. The equations of motion, also take into account the force due to the resistance of the medium in which the string moves. This is of the form $-\Re(x_t^2 + y_t^2 + z_t^2)\mathbf{v}$ with $\Re \geq 0$ (this force is directed along the velocity of the point, but in the opposite sense). Thus

$$\rho_0\frac{\partial x_t}{\partial t} = \frac{\partial}{\partial s_0}\left(\frac{T}{1+\varepsilon}\frac{\partial x}{\partial s_0}\right) + (1+\varepsilon)(X - \Re x_t) \tag{5.7.23}$$

and two similar expressions for y and z. Attaching the Fourier law and the constitutive law of the form

$$T = T(s_0, \varepsilon, \dot{\varepsilon}, \pi)\,. \tag{5.7.24}$$

Manacorda showed that across the transverse wave fronts (5.3.6) the second order derivatives of the strain as well as those of the temperature do not suffer a jump. Thus, the transverse waves do not directly influence the strain and temperature variation.

The influence of temperature on the dynamics of extensible strings which move in a resistive medium has also been considered by Dinca [1967], who first considered a constitutive equation of the form

$$T = T(\varepsilon, \pi)\,. \tag{5.7.25}$$

The system is of hyperbolic-parabolic type. Dinca [1967] has used also instead of the Fourier law the law proposed by Kaliski [1965] and the system again becomes hyperbolic.

5.8 Shock Waves

In the previous sections the integration schemes have been given for cases when only ordinary waves propagate in the string. Thus at the end of the string, for $t = 0$, the values of the stress and of the particle velocity, prescribed according to boundary and initial conditions, must coincide.

In some particular motions of the string, one should consider the possibility that shock waves might appear. Two types of shock waves can occur, i.e. longitudinal shock waves which are propagable discontinuities of ε, T and v, and transverse shock waves which are propagable discontinuities of the slope of the string.

We shall examine the propagation of shock waves in strings which can move in space (Cristescu *et al.* [1966]). As before, it will be assumed that the strings are perfectly flexible, although to neglect the rigidity of the string when considering transverse shock waves is a fairly rash assumption. The propagation of shock waves in various particular cases, when the string moves in a single plane, has been studied by Rakhmatulin (see Rakhmatulin and Dmianov [1961]), Pérès [1953] Ch. XII, Craggs [1954], and Pavlenko [1959].

The conditions satisfied across a shock wave front can be obtained from the conditions of continuity of coordinates, from the conditions of conservations of mass and conservation of momentum. If c is used to denote the velocity of propagation of the shock wave, it follows from the conditions of continuity of coordinates that

$$[x_t] + c[(1 + \varepsilon)x_s] = 0, \tag{5.8.1}$$

with two similar conditions for y and z. These are the kinematics compatibility conditions. The vector form of this relation is

$$[\mathbf{v}] + \mathbf{c}[(1 + \varepsilon)\boldsymbol{\tau}] = 0. \tag{5.8.1a}$$

We shall denote by "+" the value of a certain function after the shock wave front has passed, and by "−" the value of the same function before this event.

The law of conservation of mass can be written as follows

$$\rho_0 ds_0 = \rho^- ds^- = \rho^+ ds^+. \tag{5.8.2}$$

Sometimes it is useful to write this relation in the form

$$\rho_0 = \rho^-(1 + \varepsilon^-) = \rho^+(1 + \varepsilon^+). \tag{5.8.3}$$

Considering the element of the string which is traversed by the shock wave front in time dt, and writing for this element, of mass $\rho_0 ds_0$, the conditions of conservation of momentum yields:

$$\rho^-(1 + \varepsilon^-)c(x_t^+ - x_t^-) = T^- x_s^- - T^+ x_s^+. \tag{5.8.4}$$

The relation (5.8.1) can be written

$$x_t^+ - x_t^- = c\{(1 + \varepsilon^-)x_s^- - (1 + \varepsilon^+)x_s^+\}. \tag{5.8.5}$$

Introducing (5.8.5) into (5.8.4) we obtain

$$\rho_0 c^2 \{(1+\varepsilon^+)x_s^+ - (1+\varepsilon^-)x_s^-\} = T^+ x_s^+ - T^- x_s^-\,,$$

or

$$\{\rho_0 c^2\{(1+\varepsilon^+) - T^+\}x_s^+ = \{\rho_0 c^2(1+\varepsilon^-) - T^-\}x_s^-\,, \tag{5.8.6}$$

and similar relations for y and z. The relations (5.8.6) link the values of various functions before and after the shock wave front has passed. They involve both the functions which describe the longitudinal motion of the string, and those which refer to the transverse motion.

By squaring the relations (5.8.6), summing, and taking into account that x_s, y_s, z_s are direction cosines which cannot be simultaneously zero, we obtain

$$\rho_0 c^2\{(1+\varepsilon^+) - T^+ = \rho_0 c^2 (1+\varepsilon^-) - T^-\,, \tag{5.8.7}$$

which yields the velocity of propagation:

$$c_{II}^2 = \frac{[T]}{\rho_0 [\varepsilon]}\,. \tag{5.8.8}$$

The formula (5.8.7) was derived by Pérès [1953] Ch. XI, and by Pavlenko [1959] for the case of plane motion; it can be used to calculate the velocity of propagation of longitudinal shock waves. This velocity is generally variable and does not coincide with the velocity of propagation of smooth waves. However, this coincidence occurs if the slope is constant on the portion of the tension–strain curve considered (elastic, linear work-hardening, etc.). If the curvature of the tension–strain curve and the jumps $[T]$ and $[\varepsilon]$ are not too big, the velocity furnished by (5.8.8) does not differ greatly from the velocity of propagation of longitudinal smooth waves $c_{II}^2 = (1/\rho_0)(dT/d\varepsilon)$.

If one introduces (5.8.7) into (5.8.6) it follows that

$$[x_s] = [y_s] = [z_s] = 0\,. \tag{5.8.9}$$

In this case, and taking (5.8.3) into account, the two kinds of relation (5.8.1) and (5.8.4) reduce to a single type of relation

$$[x_t] = -c_{II}[\varepsilon]x_s\,. \tag{5.8.10}$$

Thus the shock waves which propagate with the velocity (5.8.8) are longitudinal waves which do not change the shape of the string, while the jump of the material velocity is directed along the tangent to the string. The component of this velocity which is normal to the string remains continuous.

The conclusions developed from (5.8.7) hold only if both terms in (5.8.7) are different from zero. Let us consider the case when both these terms are zero. This yields a velocity of propagation c furnished by

$$c^2 = \frac{T^+}{\rho_0(1+\varepsilon^+)} = \frac{T^-}{\rho_0(1+\varepsilon^-)} = \frac{T^+ - T^-}{\rho_0(\varepsilon^+ - \varepsilon^-)}\,. \tag{5.8.11}$$

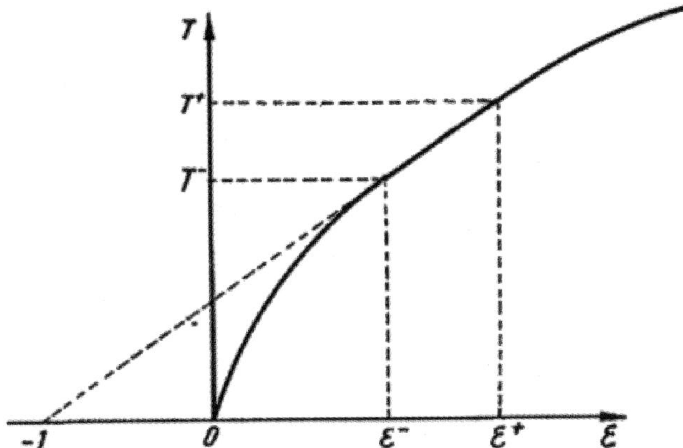

Fig. 5.8.1   The first special case when longitudinal and transverse shock waves propagate with the same velocity.

These are shock waves which produce discontinuities for all the unknown functions involved in the problem, i.e. $[\varepsilon] \neq 0$, $[T] \neq 0$, $[x_s] \neq 0$, $[y_s] \neq 0$, $[z_s] \neq 0$, $[x_t] \neq 0$, $[y_t] \neq 0$, $[z_t] \neq 0$. This case has been discussed by Pavlenco [1959]. Let us examine in what circumstances such waves are possible. A first case when such waves may arise is that shown in Fig. 5.8.1, when a whole portion of the stress–strain curve is a segment of the straight line passing trough the point $(-1, 0)$, and the jump of the tension is comprised in this segment.

A second possible case has been represented in Fig. 5.8.2, which is self-explanatory; this case is certainly not a stable one. From these special cases, one can conclude that a shock wave which simultaneously carries discontinuities for both strain and slope, occurs seldom (for very special constitutive equations and loading laws).

A third case when (5.8.11) holds is the case when

$$[\varepsilon] = 0 \quad \text{so that} \quad [T] = 0. \tag{5.8.12}$$

In this case the formula (5.8.11) can be written

$$c_I^2 = \frac{T}{\rho_0(1 + \varepsilon)}. \tag{5.8.13}$$

Introducing (5.8.12) into (5.8.1) [or into (5.8.4)] it follows that

$$[x_t] + c_I(1 + \varepsilon)[x_s] = 0. \tag{5.8.14}$$

These are transverse shock waves. Their velocity of propagation (5.8.13) always coincides with the velocity of propagation of smooth transverse waves (5.3.8). For the elastic strings the formula (5.8.13) was given by Cristescu [1951] and for the general extensible strings by Pérès [1953] Ch. XI #19.

Taking into account that the case of simultaneous propagation of discontinuities of strain and slope occurs only very seldom, in the special cases mentioned in

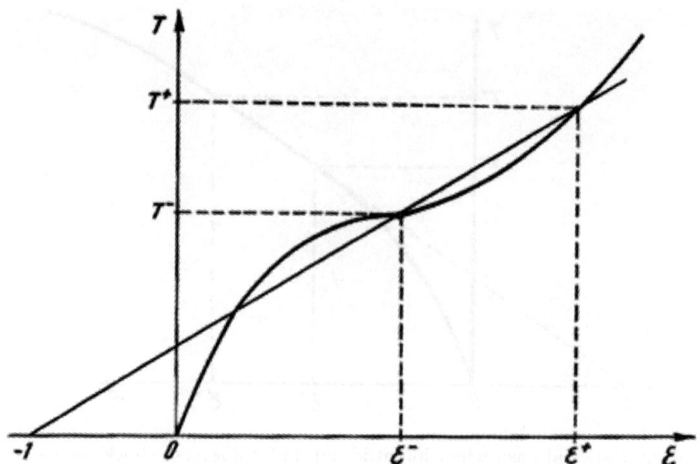

Fig. 5.8.2   The second special case (unstable) when longitudinal and transverse shock waves propagate with the same velocity.

Figs. 5.8.1 and 5.8.2, one can conclude that a shock produced at a point of the string will generally propagate by means of two kinds of shock waves, one transverse and the other longitudinal. The discontinuities of the velocity prescribed by the boundary conditions produce a jump of the strain [formula (5.8.19)] which propagates by means of a longitudinal shock wave, and a jump of slope [formula (5.8.14)] which propagates by means of a transverse shock wave. These waves propagate with the velocities (5.8.8) and (5.8.13) respectively.

5.9   Other Papers

The equations of the elastic strings were considered by many other authors.

Disturbances produced by the motion of a driver which is rigidly bonded to the edge of a plate are used by Parker and Varley [1968], to motivate parameter expansion technique which, when applied to the equations of finite elasticity, generate approximating equations which describe law-frequency deflection and stretching waves traveling along stretched elastic plates and rods in the limit when bending forces are negligible compared with membrane forces. The structure of the boundary layer at the driven edge and shock layers, where the low-frequency or filament approximations are locally invalid, are also discussed. The low-frequency equations are used to discuss the interaction between progressing finite amplitude deflection and stretch waves in the limit when the stretch rate is small compared with the angular speed of the plate. The disturbance is locally that of a pure deflection simple wave whose amplitude and frequency are modulated by slow variations in the stretch. As the stretch increases the frequency increases while the amplitude decreases. The stretch wave is also modified by deflection of the plate: the speeds of wavelets carrying constant values of stretch are always less than their values in the pure stretch simple wave.

Also, Parker [1970] considers the motion of a plane non-linearly elastic string. It is shown that extensional signals passing through a non-uniform disturbance are not carried along exact characteristics, but are carried at a considerable slower speed, even though the governing equations are non-dispersive.

The impact of composites impacted with a body will not be discussed here. We mention only the damage observed in composites, as one is impacting them, the successive impact of various laminas, and the successive damage produced (Cristescu *et al.* [1975]). That was continued by several other papers as for instance Papanicolaou *et al.* [1998], Navarro [1998] and Kook *et al.* [2000]. In these papers the figures are very close to those from impacted strings.

The Riemann and Goursat step data problems for extensible strings have been considered by Mihailescu and Suliciu [1975a, b]. They consider longitudinal discontinuities, for which $[\lambda] \neq 0$, $[\tau] = 0$, $\mathbf{U} = \lambda\boldsymbol{\tau}$ $(|\tau|) = 1$, and therefore

$$[\mathbf{U}] = [\lambda]\boldsymbol{\tau}, \quad [\mathbf{V}] + c[\lambda]\boldsymbol{\tau} = 0, \quad c^2 = \frac{[T]}{\rho[\lambda]},$$

and transversal discontinuities, for which $[\lambda] = 0$, $[\tau] \neq 0$, therefore

$$[\mathbf{U}] = \lambda[\boldsymbol{\tau}], \quad [\mathbf{V}] + c\lambda[\boldsymbol{\tau}] = 0, \quad c^2 = \frac{T}{\rho\lambda}.$$

They write the system to be solved as

$$\frac{\partial}{\partial r}\left(\frac{T}{\lambda}\mathbf{U}\right) - \rho\frac{\partial \mathbf{V}}{\partial t} = 0, \quad \frac{\partial \mathbf{U}}{\partial t} - \frac{\partial \mathbf{V}}{\partial r} = 0$$

$$T = f(\varepsilon) \geq 0, \quad \lambda = |\mathbf{U}| \geq 1, \quad \varepsilon = \lambda - 1$$

with

$$f(0) = 0$$

$$f'(\varepsilon) > 0 \quad \text{for } \varepsilon \geq 0,$$

$$f''(\varepsilon) < 0$$

where $\mathbf{U} = \partial\mathbf{x}/\partial r$, $\mathbf{V} = \partial\mathbf{x}/\partial t$, $\rho = $ const. > 0 and x is a function.

In order to ensure unique solutions for the problems considered in the paper, one is adopting the following principle. A shock discontinuity is possible only when a rarefaction fan is not possible (Suliciu [1973]). The Goursat problem in strains is solved as well as the problem in velocities and the Riemann problem. The transient response of an impulsively loaded plastic string on a plastic foundation was considered by Mihailescu-Suliciu *et al.* [1996]. The last authors consider the problem of an impulsive loading of a long rigid-plastic string resting on a rigid-plastic foundation. A closed form solution is obtained by disregarding the longitudinal motion and

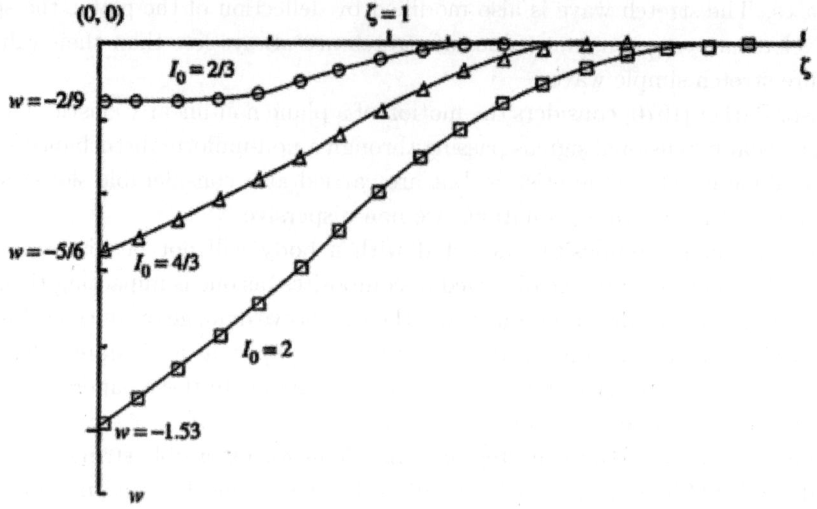

Fig. 5.9.1   Final string shape for different I_0.

considering an arbitrary large transversal motion. Expressions for the final shape
of the string are derived in terms of magnitude of the applied impulse. It is found
that the stress and the foundation reaction force are not uniquely determined, while
the shape of the string is. The final shape of the string is given in Fig. 5.9.1. Here
w is a dimensionless displacement, ζ a dimensionless variable and I_0 an impulse.

A continuous initial or initial-boundary value problem for an elastic visco-plastic
string on an elastic viscoplastic foundation is formulated by Suliciu [1996]. An
energy identity/inequality is deduced. Bounds in energy of the solutions are derived
and the stress deviation of the viscoplastic model from the plastic one is estimated.
As a farther limiting case similar results are derived for the rigid perfect plastic
case.

In the paper by Jang *et al.* [1990] the response of advanced composites to
low-velocity projectile loading was investigated. The impact failure mechanisms
of composites containing various fibers with different strength and ductility were
studied by a combination of static indentation testing, instrumented falling dart
impact testing, acoustic emission monitoring, and scanning electron microscopy.
The composites containing fibers with both high strength and high ductility demonstrate a superior impact resistance as compared to those containing fibers with high
strength (graphite fibers) or high ductility (nylon) but not both. Upon impact
loading, the composites containing polyethylene fibers usually exhibited a great
degree of plastic deformation and some level of delamination.

The paper by Wang and Vu-Khanh [1991] reports both experimental and
numerical investigations on delamination mechanisms in $[0_5, 90_5, 0_5]$ carbon
fiber/poly laminate subjected to low-velocity impact. It was found that these

composites exhibits the same damage mechanisms as epoxy-based composites, but superior delamination resistance.

Thick glass-fibre-reinforced laminates of various dimensions have been subjected to low-velocity impact with a flat-ended impactor by Zhou [1995]. Post-impact examination of damage is carried out by visual inspection, diametric cross-sectioning, and ultrasonic C-scanning. A number of damage mechanisms have been observed, and their influence on impact behavior is found to be generally dependent on the impact force or incident kinetic energy level.

By replacing the first-order law of motion (velocity given) with a second-order law (acceleration given), dynamics of e.g. flexible and rigid elastic strings is described by Uby [1996] in a similar manner. A system of nine coupled partial differential equations results, but on eliminating longitudinal motion, a smaller system is obtained. Classic vibration equations are recovered as special cases.

The free surface spalling phenomena is 2D carbon/carbon and carbon-polyimide laminates was studied by Gupta *et al.* [1996] by subjecting them to laser-produced nano-second rise-time compressive stress pulses. The compression pulse travels through the laminate normal to the plies and, upon reflection into a tension pulse from the sample's front surface, leads to fiber/matrix debondings either in the bundles within a ply or at the interplay interface. Additional information on the attenuation and the dispersion characteristics associated with the propagation of such stress pulses was also obtained for both types of composite microstructures.

High speed cine techniques have been used by Gellert *et al.* [1998] to examine the perforation of thin targets constructed of glass fiber reinforced plastic, Spectra and Kevlar composites as well as nylon and Kevlar fabrics. From the film record the kinetic, strain and (for the composites only) delamination/surface energy terms were evaluated for the rear layer of material. Simple models for the deformation of the panels were used to compare these energies, summed for all layers, with the projectile energy loss. All the energy terms are shown to be significant. The work highlights many of the features which need to be accounted for in modeling ballistic perforation of fabric and fiber reinforced composite materials.

Karbhari and Rydin [1999] have shown that resin transfer molded composites exhibit impact induced damage mechanisms and sequences different from those shown by laminated composites due to differences in layering and compaction of reinforcement. In addition to the classical modes of damage, mechanisms such as inter-bundle, inter-bundle and void pocket cracking are also seen.

In a paper denoted to the Riemann problem, Hanche-Olsen *et al.* [2002] presents the problem of Riemann for an elastic string with a linear Hooke's law. They are showing that the Riemann problem has a unique solution. A weak solution of the nonstrictly hyperbolic conservation law when the total variation of the initial data is sufficiently small is due to Reiff [2002], which examined an elastic string of infinite length in three-dimensional space, restricted to possess non-simple eigenvalues of constant multiplicity.

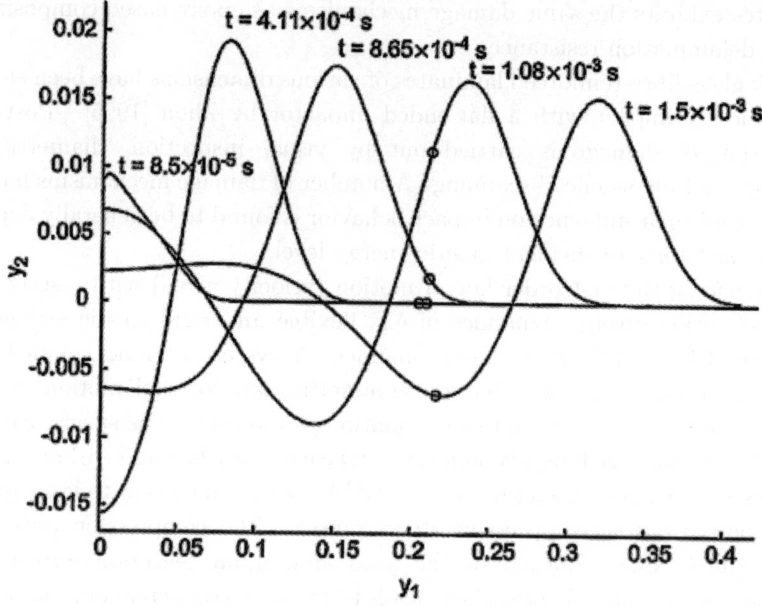

Fig. 5.9.2   Traveling wave in a whipped string with a phase boundary.

The paper of Purohit and Bhattacharya [2003] presents a theory to describe the dynamical behavior of a string made of a phase-transforming material like a shape-memory alloy. The study of phase boundaries, the driving force acting on them and the kinetic relation governing their propagation is of central concern. The paper proposes a qualitative experimental test of the notion of a kinetic relation, as well as a simple experimental method for measuring it quantitatively. It presents a numerical method for studying initial and boundary value problems in strings, and concludes by exploring the use of phase transforming strings to generate motion at very small scales. Figure 5.9.2 is showing one example of traveling waves in a whipped string; the grid size is 1.25×10^{-3} in this calculation. The string is fixed at the right end, initially stretched horizontally and then whipped or impacted at the left end $x = 0$ with the following velocity: $\dot{y}_1 = 0$ and

$$\dot{y}_2 = 100.0 \text{ m/s}, \quad 0 \leq t < 1.6 \times 10^{-4},$$

$$\dot{y}_2 = -100.0 \text{ m/s}, \quad 1.6 \times 10^{-4} \leq t < 3.2 \times 10^{-4},$$

$$\dot{y}_2 = 0.0, \quad 3.2 \times 10^{-4} \leq t < 1.2 \times 10^{-3}.$$

The traveling wave propagating through the length of the string is discernible. The breakdown of classical solutions of the planar motion of an elastic string was studied by Manfrin [2004]. The string is infinite in length and having a planar motion, in the absence of external forces. One assumes strict hyperbolicity in the sense

$$T'(r) > \frac{T(r)}{r} > 0$$

for any $r > 1$.

A novel experimental technique is developed by Minnaar and Zhou [2004] for time-resolved detection and tracking of damage in the forms of delamination and matrix cracking in layered materials such as composite laminates. The technique is non-contact in nature and uses dual or quadruple laser interferometers for high temporal resolution. Simultaneous measurements of differential displacement and velocity at individual locations are obtained to analyze the initiation and progression of interfacial fracture and/or matrix cracking/delamination in a polymer matrix composite system reinforced by graphite fibers. The measurements at multiple locations allow the speeds or matrix cracking/delamination front propagates to be determined. The results show that the speed of delamination or the speed of matrix cracking/delamination increases linearly with impact velocity.

An analytical model for studying the material failure in shear hinges formed during the dynamic plastic response of a circular plate under projectile impact is due to Chen *et al.* [2005]. Analytic solutions for the ballistic perforation performance of a fully clamped plate struck by a blunt projectile are obtained. It takes into account both the global response (plate bending) and the localized shear. Based on the estimations of the shear strain and the size of shear hinge, condition for the initiation of an adiabatic shear band is formulated. The effectiveness of the present model is demonstrated by reasonable agreement between the theoretical predictions and the available experimental data for the perforation of steel plates.

In another paper (Stoffel [2006]) an explanation for the development of conical shapes in the case of shock wave loaded plated is proposed based on the wave propagation phenomena.

Impact damage processes in composite sheet and sandwich honeycomb materials, impacted with an impactor of 5 kg from a drop hight of 1 m, with a velocity of 4,4 m s^{-1}, was done by Dear *et al.* [2006].

Experiments were conducted on aluminum plates of 1 mm thickness by using a gas gun and projectiles with blunt and hemispherical noses (Gupta *et al.* [2006]). Ballistic limit velocity was found to be higher for hemispherical projectiles than for blunt projectiles.

The ballistic impact of woven fabric was done by Naik *et al.* [2006]. Using the analytical formulation, ballistc limit, contact duration at ballistic limit, surface radious of the cone formed and the radious of the damaged zone have been predicted for typical woven fabric composites.

The numerical results of ballistic impact and perforation of woven aramid fabric are presented by Tan *et al.* [2006]. The crimp is introduced in several ways.

Influence of boundary conditions on the ballistc performance of high-strength fabric target were considered by Zeng *et al.* [2006]. Targets that are unclamped on two edges can absorb more impact energy than those with all four sides clamped. It also observed that slippage at clamped edges contributes to higher energy absorption. The impact velocity is less than 500 m/s.

In another paper (Tan and Ching [2006]) a FE model of woven fabric that reflects the orthotropic properties of the fabric, the viscoelastic nature of the yarns, the crimping of the yarns, the sliding contact betveen yarns and yarn breakage using an assembly of viscoelastic bar elements.

Bibliography

Chen X. W., Li Q. M. and Fan S. C., 2005, Initiation of adiabatic shear failure in a clamped circular plate, *Int. J. Impact Engn.* **31**, 877–893.

Courant R., 1962, *Methods of Mathematical Physics*, Vol. II, *Partial Differential Equations*, Interscience Publi., New York.

Craggs J., 1954, Wave motion in plastic-elastic strings, *J. Mech. Phis. Solids* **2**, 4, 115–124.

Cristescu N., 1951, Discontinuities in the motion of perfectly flexible elastic strings, *Commun. Acad. Rep. Populare Romane* **1**, 439–445.

Cristescu N., 1952, Discontinuities in the motion of visco-elastic, perfectly flexible strings, *Rev. Univ. C.I. Parhon Politeh. Bucuresti, Ser. Stiint. Nat.* **1**, 68–81.

Cristescu N., 1954, About the loading and unloading waves which can appear in the motion of elastic or plastic flexible strings, *Prikl. Mat. Meh.* **18**, 3, 257–264 (in Russian).

Cristescu N., 1957, On the propagation of waves in rubber, *Prikl. Mat. Meh.* **21**, 6, 797–800 (in Russian).

Cristescu N., 1958, Dynamic problems in the theory of plasticity, *Editura Tehnica* (in Rumanian).

Cristescu N., 1960, Elastic-plastic waves in strings, *Arch. Mech. Stos.* **12**, 597–615.

Cristescu N., 1961, Spatial motions of elastic-plastic strings, *J. Mech. Phys. Solids* **9**, 165–178.

Cristescu N., 1964, Some problems of the mechanics of extensible strings, *Symp. IUTAM Stress Waves in Anelastic Solids*, Providence, 1963 (Springer-Verlag, Berlin–Götthingen–Heidelberg) 118–132.

Cristescu N., 1965, Loading/unloading criteria for rate sensitive materials, *Arch. Mech. Stos.* **17**, 291–305.

Cristescu N., 1967, *Dynamic Plasticity*, North-Holland Pub. Comp., 614 pp.

Cristescu N., Dinca G. and Suliciu I., 1966, On the propagation of shock waves in extensible strings, *Analele Univ. Bucuresti, Ser. Stiint. Nat., Mat.-Mec.* **15**, 65–73.

Cristescu N., Malvern L. E. and Sierakowski R. L., 1975, in *Foreign Object Impact Damage to Composites*, *ASTM STP 568*, American Society for Testing and Materials, pp. 159–172.

Dear J. P., Lee H. and Brown S. A., 2006, Impact damage processes in composite sheet and sandwich honeycomb materials, *Int. J. Impact Engineerig* (in print).

Dinca G., 1967, Some remarks about the thermodynamics of extensible strings, *Studii Cercetari Matematice* **19**, 659–680.

Fenstermaker C. A. and Smith J. C., 1965, *Appl. Polymer Symposia* **1**, 125–146.

Gellert E. P., Pattie S. D. and Woodward R. L., 1998, Energy transfer in ballistic perforation of fiber reinforced composites, *J. Materials Sci.* **33**, 1845–1850.

Gupta N. K., Iqbal M. A. and Sekhon G. S., 2006, Experimental and numerical studies on the behavior of thin aluminum plates subjected to impact by blunt- and hemisphericsal-nosed projectiles, *Int. J. Imp. Engineering* (in press).

Gupta V., Pronin A. and Anand K., 1996, Mechanisms and quantification of spalling failure in laminated composites under shock loading, *J. Composite Mat.* **30**, 6, 722–747.

Hanche-Olsen H., Holden H. and Risebro N. H., 2002, The Riemann problem for an elastic string with a linear Hooke's law, *Quart. Appl. Math.* **60**, 4, 695–705.

Jang B. Z., Chen L. C., Hwang L. R., Hawkes J. E. and Zee R. H., 1990, The response of fibrous composites to impact loading, *Polymer Composites* **11**, 3, 144–157.

Karbhari V. M. and Rydin R. W., 1999, Impact characterization of RTM composites II: Damage mechanisms and damage, *J. Materials Sci.* **34**, 5641–5648.

Kaliski S., 1965, Wave equation of thermoelasticity, *Bull. Acad. Polon. Sci., Ser. Sci. Tech.* **13**, 211–219.

Kook J. S., Yang I. Y. and Adachi T., 2000, Characteristics of delamination in graphite/epoxy laminates under static and impact loads, *Key Engng. Materials* **183–187**, 731–736.

Lewis G. M., 1957, *Proc. Conf. Prop. Mater. At High Rates of Strain* (Inst. Mech. Engrs., London) 190–194.

Manacorda T., 1958, *Riv. Mat. Univ. Parma* **9**, 13–19.

Manfrin R., 2004, On the breakdown of classical solutions of the planar motion of an elastic string, *Bull. Sci. Math.* **128**, 253–263.

Marshall I. and Thomson A. B., 1954, The cold drawing of high polimers, *Proc. Roy. Soc.* A **221**, 541–557.

Mihailescu M. and Suliciu I., 1975a, Riemann and Goursat step data problems for extensible strings, *J. Math. Anal. Appl.* **52**, 10–24.

Mihailescu M. and Suliciu I., 1975b, Riemann and Goursat step data problems for extensible strings with non-convex stress–strain relation, *Rev. Roumaine Math. Pures. Appl.* **20**, 551–559.

Mihailescu-Suliciu M., Suliciu I., Wierzbicki T. and Fatt Hoo M. S., 1996, Transient response of an impulsively loaded plastic string on a plastic foundation, *Quart. Appl. Math.* **54**, 2, 327–343.

Miklowitz J., 1947, The initiation and propagation of the plastic zone along a tension specimen of nylon, *J. Colloid Sci.* **2**, 193–215.

Minnaar K. and Zhou M., 2004, A novel technique for time-resolved detection and tracking of interfacial and matrix fracture in layered materials, *J. Mech. Phys. Solids* **52**, 2771–2799.

Naik N. K., Shrirao P. and Reddy B. C. K., 2006, Ballistic impact behaviour of woven fabric composites: Formulation, *Int. J. Impact Engineering* (in press).

Navarro C., 1998, Simplified modelling of of the ballistic behaviour of fabrics and fibre reinforced polymeric matrix composites, *Key Engin. Mat.* **141–143**, 383–400.

Papanicolaou G. C., Blanas A. M., Pournaras A. V. and Stavropoulos C. D., 1998, Impact damage and residual strength of FRP composites, *Key Engin. Mat.* **141–143**, 127–148.

Parker D. F., 1970, Interacting waves in long stretched elastic strings, *J. Mech. Phis. Solids* **18**, 331–342.

Parker D. F. and Varley E., 1968, Interaction of finite amplitude deflection and stretching waves in elastic membranes and strings, *Quart. Journ. Mech. and Applied Math.* **21**, Pt.3, 329–352.

Pavlenko A. L., 1959, On the propagation of perturbations in a flexible string, *Izv. Akad. Nauk. SSSR, Otd. Tekhn. Nauk, Mekhn. Mashinostr.* **4**, 112–122.

Pérès J., 1953, *Mécanique Générale*, Masson, Paris.

Purohit P. K. and Bhattacharya K., 2003, Dynamics of strings made of phase transforming materials, *J. Mech. Phis. Solids* **51**, 3, 393–424.

Rakhmatulin H. A., 1951, Transverse shock with variable velocity on a flexible string, *Uch. Zap. Mosk. Univ.* **4**, 5, 154.

Rakhmatulin H. A. and Shapiro G. S., 1955, Propagation of perturbations in non linear-elastic and nonelastic media, *Izv. Akad. Nauk SSSR, Otd. Fekhn. Nauk* **2**, 68–89.

Rahmatulin H. A. and Demianov Yu. A., 1961, Strength under intensive momentary loads, *Moscow* (in Russian).

Reiff A. M., 2002, Existence of weak solutions to the elastic string equations in three dimensions, *Quart. Appl. Math.* **60**, 3, 401–424.

Riabova E. H., 1953, Transverse shock with variable velocity on a flexible string, *Vestn. Mosk. Univ.* **10**, 85–91.

Smith J. C., Blandford J. M. and Schiefer H. F., 1960, Stress–strain relationship of yarns subjected to rapid impact loading: 6. Velocities of strain waves resulting from impact, *Textile Res. J.* **30**, 10, 752–769.

Smith J. C., Crackin F. L. Mc and Schiefer H. F., 1958, Stress–strain relationship in yarns subjected to rapid impact loading: 5. Wave propagation in long textile yarns impacted transversely, *J. Res. Nat. Std.* **60**, 5, 517–534.

Smith J. C., Crackin F. L. Mc, Schiefer H. F., Sione W. K. and Towne K. M., 1956, Stress–strain relationship in yarns subjected to rapid impact loading: 4. Transverse impact tests, *J. Res. Nat. Bur. Std.* **57**, 2, 83–89.

Smith J. C. and Fenstermaker C. A., 1967, Strain-wave propagation in strips of natural rubber subjected to high-velocity transverse impact, *J. Appl. Phys.* **38**, 11, 4218–4224.

Smith J. C., Fenstermaker C. A. and Shouse P. J., 1964, Experimental determination of air drag on textile yarn struck transversely by high-velocity projectile, *J. Res. Nat. Bur. Std.* C **68**, 177–181.

Smith J. C., Fenstermaker C. A. and Shouse P. J., 1965, Stress–strain relationships in yarns subjected to rapid impact loading. 11. Strain distribution resulting from rifle bullet impact, *Textile Res. J.* **35**, 743.

Smith J. C., Fenstermaker C. A. and Shouse P. J., 1963, Experimental determination of air drag on textile yarn struck transversly by high-velocity projectile, *Textile Res. J.* **33**, 919–934.

Smith J. C., Shouse P. J., Blandford J. M. and Towne K. M., 1961, *Textile Res. J.* **31**, 721–734.

Stoffel M., 2006, Shape forming of shock wave loaded viscoplastic plates, *Mec. Res. Comm.* **33**, 35–41.

Suliciu I., 1973, Motions with discontinuités in solides — Nonlinear models, *Stud. Cerc. Mat.* **25**, 53–170 (in Rumanian).

Suliciu I., 1996, An energetic characterization of the constitutive equations for plastic strings on plastic foundations, *Constitutive Relation in High/Very High Strain Rates*, IUTAM Symposium, Noda, Japan, Oct. 16–19, 1995. Spriger Vlg.

Tan V. B. C. and Ching T. W., 2006, Computational simulation of fabric armour subjected to ballistic impacts, *Int. J. Impact Engineering* (in press).

Tan V. B. C., Shim V. P. W. and Zeng X., 2006, Modelling crimp in woven fabrics subjected to ballistic impact, *Int. J. Impact Engineering* (in print).

Uby L., 1996, Strings, rods, vortices and wave equations, *Proc. R. Soc. Lond.* **452**, 1531–1543.

Victorov V. V., Dobrovolsckii I. P., Cristrescu N., Kovalenco T. S., Stepanov P. D. and Shapiro G. S., 1966, Study of some dynamic properties of some polymers, *Proc. Symp. on the Prop. of Elastic-plastic Waves in Cont. Med.*, Baku, 1964, 103–108.

Wang H. and Vu-Khanh T., 1991, Impact-induced delamination in $[0_5, 90_5, 0_5]$ carbon fiber/polyetheretherketone composite laminates, *Polymer Eng. Sci.* **31**, 18, 1301–1309.

Zeng X. S., Shim V. P. W. and Tan V. B. C., 2006, Influence of boundary conditions on the ballistic performance of high-strength fabric targets, *Int. J. Impact Engineering* (in press).

Zhou G., 1995, Damage mechanisms in composite laminates impacted by a flat-ended impactor, *Composites Sci. Tech.* **54**, 267–273.

Chapter 6

Flow of a Bigham Fluid

This chapter is devoted to the motion of a Bingham body through various devices; through a tube with or without friction at the wall, flow in a viscosimeter, flow in the field of natural slopes, etc. Plastic flow through conical converging dies will not be described (Cristescu [1975] [1976], Fu and Loo [1995]). For general concept on strain rate intensity factor see Aleksandrov *et al.* [2003].

6.1 Flow of a Bingham Fluid Through a Tube

We consider a tube of length l and of radius R (Buckingham [1921]). We work in cylindrical coordinates.

At $z = 0$ we have (Fig. 6.1.1)

$$\sigma = -p \tag{6.1.1}$$

that is a pressure is applied.

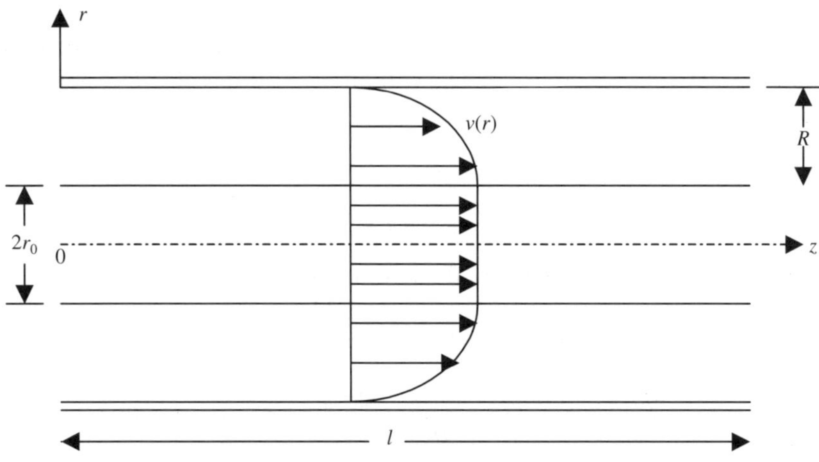

Fig. 6.1.1   The flow in a tube.

251

We assume "telescopic flow" i.e.,

$$v_r = v_\theta = 0\,, \quad v_z = v = f(r)\,. \tag{6.1.2}$$

The rate of deformation tensor in cylindrical coordinates are

$$D_{rr} = \frac{\partial v_r}{\partial r}\,, \quad D_{\theta\theta} = \frac{v_r}{r} + \frac{1}{r}\frac{\partial v_\theta}{\partial\theta}\,, \quad D_{zz} = \frac{\partial v_z}{\partial z}\,, \quad D_{\theta z} = \frac{1}{2}\left(\frac{1}{r}\frac{\partial v_z}{\partial\theta} + \frac{\partial v_\theta}{\partial z}\right),$$

$$D_{rz} = \frac{1}{2}\left(\frac{\partial v_r}{\partial z} + \frac{\partial v_z}{\partial r}\right), \quad D_{r\theta} = \frac{1}{2}\left(\frac{\partial v_\theta}{\partial r} - \frac{v_\theta}{r} + \frac{1}{r}\frac{\partial v_r}{\partial\theta}\right).$$

From here for our problem we have

$$D_{rr} = D_{\theta\theta} = D_{zz} = D_{\theta z} = D_{r\theta} = 0\,, \quad D_{rz} = \frac{1}{2}f'(r)\,. \tag{6.1.3}$$

The constitutive equation for a Bingham (viscoplastic/rigid) fluid is:

$$\sigma_{ij} = \sigma\delta_{ij} + k_{ij} + 2\eta D_{ij} \tag{6.1.4}$$

with the plastic part

$$D_{ij} = 2\lambda k_{ij} \quad \text{if} \quad k_{ij}k_{ij} = 2k^2\,. \tag{6.1.5}$$

If only

$$k_{ij}k_{ij} < 2k^2 \quad \text{we have} \quad D_{ij} = 0\,.$$

From (6.1.3) and (6.1.5) we obtain

$$D_{rz} = 2\lambda k_{rz} \quad \text{and} \quad k_{rz}^2 = k^2 = \tau_Y^2\,. \tag{6.1.6}$$

Thus for the only non zero component:

$$k_{rz} = k_{zr} = \tau_Y\,(\text{sign}\,D_{rz}), \quad \text{all other } k_{ij} = 0\,. \tag{6.1.7}$$

We observe now that $f' < 0$ or $(\partial v_z/\partial r) < 0$ since v_z decreases with increasing r. Therefore $\text{sign}\,D_{rz} = -1$. Since all $D_{ij} = 0$ all $\sigma'_{ij} = 0$, besides D_{rz} and σ_{rz}. We have thus

$$\sigma_{rr} = \sigma_{\theta\theta} = \sigma_{zz} = \sigma\,, \quad \sigma_{\theta z} = \sigma_{r\theta} = 0\,,$$

$$\sigma_{rz} = \sigma_{zr} = -\tau_Y + \eta f'(r)\,. \tag{6.1.8}$$

The equations of motion in cylindrical coordinates are:

$$\frac{\partial\sigma_{rr}}{\partial r} + \frac{1}{r}\frac{\partial\sigma_{r\theta}}{\partial\theta} + \frac{\partial\sigma_{zr}}{\partial z} + \frac{\sigma_{rr} - \sigma_{\theta\theta}}{r} + \rho(b_r - a_r) = 0\,,$$

$$\frac{\partial\sigma_{r\theta}}{\partial r} + \frac{1}{r}\frac{\partial\sigma_{\theta\theta}}{\partial\theta} + \frac{\partial\sigma_{z\theta}}{\partial z} + \frac{2\sigma_{r\theta}}{r} + \rho(b_\theta - a_\theta) = 0\,,$$

$$\frac{\partial\sigma_{rz}}{\partial r} + \frac{1}{r}\frac{\partial\sigma_{z\theta}}{\partial\theta} + \frac{\partial\sigma_{zz}}{\partial z} + \frac{\sigma_{rz}}{r} + \rho(b_z - a_z) = 0\,. \tag{6.1.9}$$

Introducing (6.1.8) in (6.1.9)$_1$ and (6.1.9)$_2$, and by disregarding the body forces and the accelerations, we have:

$$\frac{\partial \sigma_{rr}}{\partial r} = 0 \Rightarrow \frac{\partial \sigma}{\partial r} = 0,$$

$$\Rightarrow \sigma(z) \tag{6.1.10}$$

$$\frac{\partial \sigma_{\theta\theta}}{\partial \theta} = 0 \Rightarrow \frac{\partial \sigma}{\partial \theta} = 0,$$

thus σ depends on z alone.

Introducing (6.1.8) in (6.1.9)$_3$ we obtain:

$$\frac{d\sigma}{dz} = -\eta \left(f'' + \frac{f'}{r} \right) + \frac{\tau_Y}{r}, \tag{6.1.11}$$

where σ depends on z alone and f depends on r alone. This by integrating with respect to z we get

$$\sigma = \left[-\eta \left(f'' + \frac{f'}{r} \right) + \frac{\tau_Y}{r} \right] z + C \tag{6.1.12}$$

with the boundary conditions at the end of the tube (independent on r, thus constant in the hole cross section):

$$z = 0 : \sigma = \sigma_0 \Rightarrow C = \sigma_0$$

$$z = l : \sigma_l = 0. \tag{6.1.13}$$

Thus for $z = l$, from (6.1.12), by integration with respect to r, we obtain:

$$-\eta f' r + \tau_Y r + \frac{\sigma_0}{l} \frac{r^2}{2} + C_1 = 0 \tag{6.1.14}$$

or

$$\eta f' = \tau_Y + \frac{\sigma_0}{l} \frac{r}{2} + \frac{C_1}{r}. \tag{6.1.15}$$

Comparing now this equation with the last (6.1.8) we get:

$$\sigma_{rz} = \frac{\sigma_0}{l} \frac{r}{2} + \frac{C_1}{r}. \tag{6.1.16}$$

From the shearing flow symmetry it follows that for

$$r = 0, \quad \frac{\partial v}{\partial r} = 0 \Rightarrow f' = 0 \Rightarrow D_{rz}|_{r=0} = 0 \Rightarrow \sigma_{rz}|_{r=0}.$$

Thus, around $r = 0$ exists a cylinder of radius r_0, where $k_{ij} k_{ij} < 2k^2$, i.e., is rigid and for

$$r = r_0 : \sigma_{rz} = -\tau_Y \tag{6.1.17}$$

the condition $k_{ij} k_{ij} = 2k^2$ is just satisfied. Introducing (6.1.17) in (6.1.16) for $r = r_0$ we have

$$-\tau_Y = \frac{\sigma_0}{l} \frac{r_0}{2} + \frac{C_1}{r_0}. \tag{6.1.18}$$

From the global equilibrium condition of rigid cylinder of radius r_0 (the inner core): $\pi r_0^2 \sigma_0 = -\tau_Y (2\pi r_0) l$ or

$$\tau_Y = -\frac{r_0 \sigma_0}{2l}\,. \tag{6.1.19}$$

Introducing (6.1.19) in (6.1.18) we get $C_1 = 0$.

Returning to (6.1.15) by integration:

$$-\eta v + \tau_Y r + \frac{\sigma_0}{4l} r^2 + C_2 = 0\,. \tag{6.1.20}$$

Flow with wall adherence.

We assume now that the material <u>adheres</u> on the wall, i.e.,

$$r = R : v = 0\,. \tag{6.1.21}$$

By introducing (6.1.21) in (6.1.20) we obtain $\tau_Y R + (\sigma_0/4l)R^2 = -C_2$ and from (6.1.20):

$$\eta v = \tau_Y (r - R) + \frac{\sigma_0}{4l}(r^2 - R^2)\,. \tag{6.1.22}$$

If we put here $v = 0$ we get for r two solutions

$$r_1 = R\,, \quad r_2 = -R + 2r_0\,.$$

The second solution is impossible, since we must have $r_0 > R$.

The velocity of the rigid internal nucleus is obtained from (6.1.22) for $r = r_0$:

$$v_0 = -\frac{\sigma_0}{4l\eta}(r_0 - R)^2\,. \tag{6.1.23}$$

The radius of the nucleus is [from (6.1.19)]:

$$r_0 = -\frac{2l\tau_Y}{\sigma_0} = \frac{2l\tau_Y}{\Delta p} \quad \text{with} \quad \sigma_0 < 0\,. \tag{6.1.24}$$

If $\tau_Y \to 0$ from (6.1.20) follows $r_0 \to 0$ and (6.1.22) becomes a parabolic distribution of velocities of a viscous fluid, i.e., for fixed l and Δp, r_0 is a measure of "plasticity".

Some conclusions:

- r_0 increases with l; when r_0 increases towards R, l increases towards $l_{MAX} = (R\Delta p/2\tau_Y)$, and from (6.1.23) $v_0 \to 0$.
- r_0 increases with τ_Y (decrease of temperature, decrease of humidity, irradiation, etc.) when τ_Y increases its maximal value is $\tau_{YMAX} = (R\Delta p/2l)$, from (6.1.24) $r_0 \to R$, and from (6.1.23) $v_0 \to 0$.
- r_0 decreases with the increasing σ_0, i.e., the inner core radius decreases with increasing pressure. If $\sigma_0 \searrow \Rightarrow r_0 \nearrow$, i.e., $\sigma_0 \searrow$ towards $\sigma_{0min} = -2l\tau_Y/R$ and from (6.1.24) $r_0 \to R$ and from (6.1.23) $v_0 \to 0$.
- The maximum value of r_0 is $R = (2l\tau_Y/\Delta p)$, when the whole body becomes rigid and from (6.1.23) $v = 0$ since the fluid adhere on the wall and $v_0 \to 0$.
- Flow is possible only if $r_0 < R$ or $(2l\tau_Y/\Delta p) < R$ or $\Delta p > (2l\tau_Y/R)$ is the limit pressure difference for the flow to take place.

- Δp must be bigger if:

$$\begin{cases} \text{the tube is longer (bigger } l) \\ \tau_Y \text{ is bigger (lower temperature or dryer paste)} \\ \text{smaller } R \end{cases}$$

If the fluid does not adhere to the wall, i.e., the slip is possible, see below.

Volume flux.

The volume flux of the fluid comprised between r_0 and R is:

$$\frac{dQ_1}{dt} = \int_{r_0}^{R} v 2\pi r\, dr = \frac{2\pi}{\eta} \left\{ \begin{array}{c} -\dfrac{R^2 \sigma_0}{4l}\dfrac{R^2 - r_0^2}{2} + \dfrac{\sigma_0}{4l}\dfrac{R^4 - r_0^4}{4} \\[2ex] -\tau_Y \left(R\dfrac{R^2 - r_0^2}{2} - \dfrac{R^3 - r_0^3}{3} \right) \end{array} \right\} \tag{6.1.25}$$

with v from (6.1.22), or

$$\frac{dQ_1}{dt} = -\frac{\pi \sigma_0 R^4}{8\eta l}\left[1 - \frac{4}{3}\frac{r_0}{R} - \frac{5}{3}\left(\frac{r_0}{R}\right)^4 + 4\left(\frac{r_0}{R}\right)^3 - 2\left(\frac{r_0}{R}\right)^2 \right]. \tag{6.1.26}$$

The volume flux Q_2 of the portion which remain rigid is [with v from (6.1.23)]:

$$\frac{dQ_2}{dt} = \int_0^{r_0} v 2\pi r\, dr = -\frac{2\pi\sigma_0}{4l\eta}(r_0 - R)^2 \int_0^{r_0} r\, dr = -\frac{\pi\sigma_0}{4l\eta}(r_0 - R)^2 r_0^2 . \tag{6.1.27}$$

If at $t = 0$, $Q = 0$, the total volume flux is $Q = Q_1 + Q_2$, and

$$\frac{Q}{t} = -\frac{\pi\sigma_0 R^4}{8l\eta}\left[1 - \frac{4}{3}\frac{r_0}{R} + \frac{1}{3}\left(\frac{r_0}{R}\right)^4 \right] \tag{6.1.28}$$

and with r_0 from (6.1.24):

$$\frac{Q}{t} = -\frac{\pi\sigma_0 R^4}{8l\eta}\left[1 + \frac{4}{3}\frac{2}{R}\frac{l\tau_Y}{\sigma_0} + \frac{1}{3}\left(\frac{2l\tau_Y}{R\sigma_0}\right)^4 \right] \tag{6.1.29}$$

or

$$\frac{Q}{t} = -\frac{\pi\sigma_0 R^4}{8l\eta}\left[1 - \frac{2}{3}\frac{2l\tau_Y}{R\sigma_0} + \frac{1}{3}\left(\frac{2l\tau_Y}{R\sigma_0}\right)^2 \right]\left(1 + \frac{2l\tau_Y}{R\sigma_0} \right)^2 \tag{6.1.30}$$

which is the formula of Buckingham [1921]–Reiner [1926]. For $\tau_Y = 0$ this formula reduces to the one for viscous fluids.

If we introduce variables which do not depend on the tube dimensions

$$V = \frac{4Q/t}{R^3\pi}, \qquad P = -\frac{R\sigma_0}{2l} \tag{6.1.31}$$

(6.1.30) is written

$$V = \frac{P}{\eta}\left[1 - \frac{2}{3}\frac{\tau_Y}{P} + \frac{1}{3}\left(\frac{\tau_Y}{P}\right)^2 \right]\left(1 - \frac{\tau_Y}{P} \right)^2 \tag{6.1.32}$$

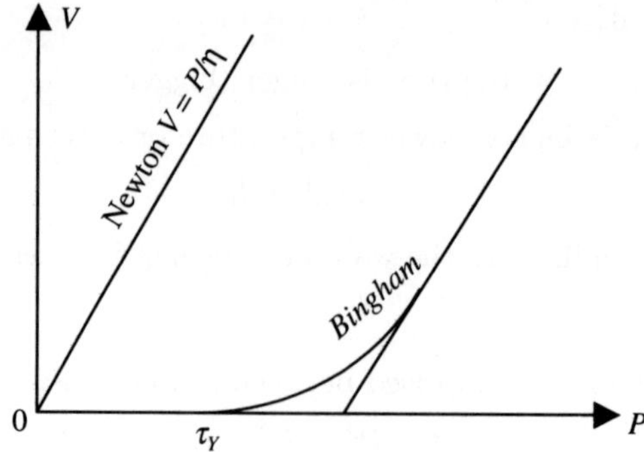

Fig. 6.1.2   The difference between the two formulas.

for $\tau_Y \to 0$ we obtain the solution for Newtonian viscosity

$$V = \frac{P}{\eta} \quad \text{or} \quad \frac{Q}{t} = \frac{\pi R^4 \sigma_0}{8\eta l}$$

i.e., the Hagen–Poiseuille formula (Fig. 6.1.2).

Flow with friction at the wall.
Viscous friction.

For a general Bingham material, the fact that for $r_0 \to R$ the flow is no more possible is a restriction which we have to remove. Generally, when r_0 increases, v_0 decreases. This could be accepted up to a certain limit.

Let us assume that for

$$r_0 \to R : \; v \to v_f \tag{6.1.33}$$

where v_f would be a limit velocity due to friction between the Bingham body and the wall. In this case, by replacing (6.1.22) by (6.1.33) we get

$$\eta(v - v_f) = \tau_Y(r - R) + \frac{\sigma_0}{4l}\left(r^2 - R^2\right). \tag{6.1.34}$$

An estimation of the friction shearing stress τ_f can be obtained by assuming $r_0 = R$ and considering the motion of the material as a rigid body: $\pi R^2 \sigma_0 = 2\pi R l \tau_f$ or

$$\tau_f = \frac{R\sigma_0}{2l}. \tag{6.1.35}$$

Let us further assume that the friction takes place in a "boundary layer" of thickness h, and that the property of this boundary layer can be approximated by a Newtonian linear viscous flow $\boldsymbol{\sigma}' = 2\mu\boldsymbol{D}$, where μ is a viscosity coefficient for the boundary layer. For the $\sigma_{rz} = \tau_f$ component we get

$$\tau_f = \mu \frac{v_f}{h} \quad \left(\frac{\partial v_f}{\partial r} \approx \frac{v_f}{h}\right) \tag{6.1.36}$$

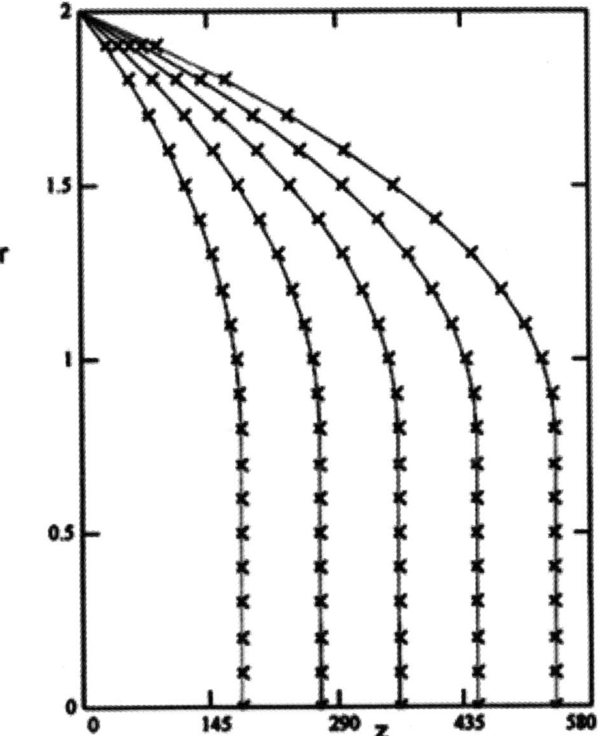

Fig. 6.1.3   The flow of a Bingham body with adherence at the wall.

where h is the thickness of the boundary layer (assumed to be 10^{-n} cm, with $n = 3$ or 4, say). Combining (6.1.36) and (6.1.35):

$$v_f = \frac{R\sigma_0 h}{2l\mu} \tag{6.1.37}$$

which is to be introduced in (6.1.34).

Examples are given in Fig. 6.1.3 for $v = 0$ at $r = R$ and in Fig. 6.1.4 for $v = v_f$ at $r = R$. The examples are computed for $R = 2$ cm, $l = 20$ cm, $\tau_Y = 20$ kPa, $h = 10^{-2}$ cm, $\sigma_0 = 1$ Mpa, $\mu = 0.01$ kPa s and $\eta = 1$ s^{-1}.

Viscoplastic friction.

Any other friction law could be considered, for instance the friction law proposed for metal plasticity (Cristescu [1975])

$$\tau = m\sqrt{II_{\sigma'}} \tag{6.1.38}$$

where m is a "friction factor" $0 \leq m \leq 1$, with $m = 1$ corresponding to adherence and $m = 0$ for no friction. For Bingham model this law becomes:

$$\tau = m(\tau_Y + 2\eta\sqrt{II_D}) . \tag{6.1.39}$$

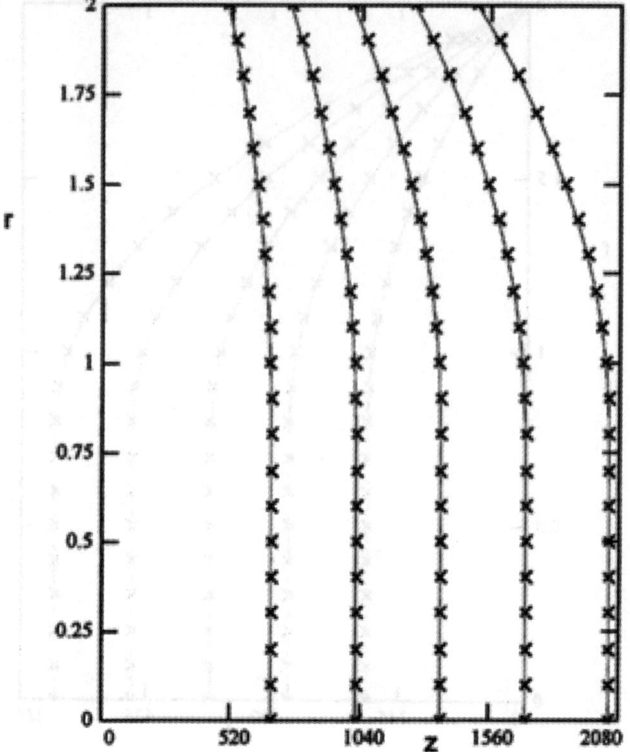

Fig. 6.1.4   The flow of a Bingham body with friction at the wall.

For our case we have

$$\tau = m\left(-\tau_Y + 2\eta D_{rz}\right).\tag{6.1.40}$$

Thus the slide takes place when D_{rz} at the wall reaches the value

$$D_{rz} = \frac{1}{2}f'(r) = \frac{1}{2\eta}\left(\tau_Y + \frac{\sigma_0}{2l}r\right),\quad \sigma_{rz} = \frac{\sigma_0}{2l}r,\tag{6.1.41}$$

and at the wall the law (6.1.40) becomes:

$$\tau = m\frac{\sigma_0}{2l}R = m\sigma_{rz}|_R.\tag{6.1.42}$$

Thus the sliding starts when the applied pressure reaches the

$$\sigma_0 = \frac{\tau}{m}\frac{2l}{R} = \frac{2l}{R}\sigma_{rz}\bigg|_R\tag{6.1.43}$$

and, again:

$$\tau = m\sigma_{rz}|_R.\tag{6.1.44}$$

Thus, more pressure is needed if the contact surface fluid/tube is longer, if the tube is thinner, and if m is bigger.

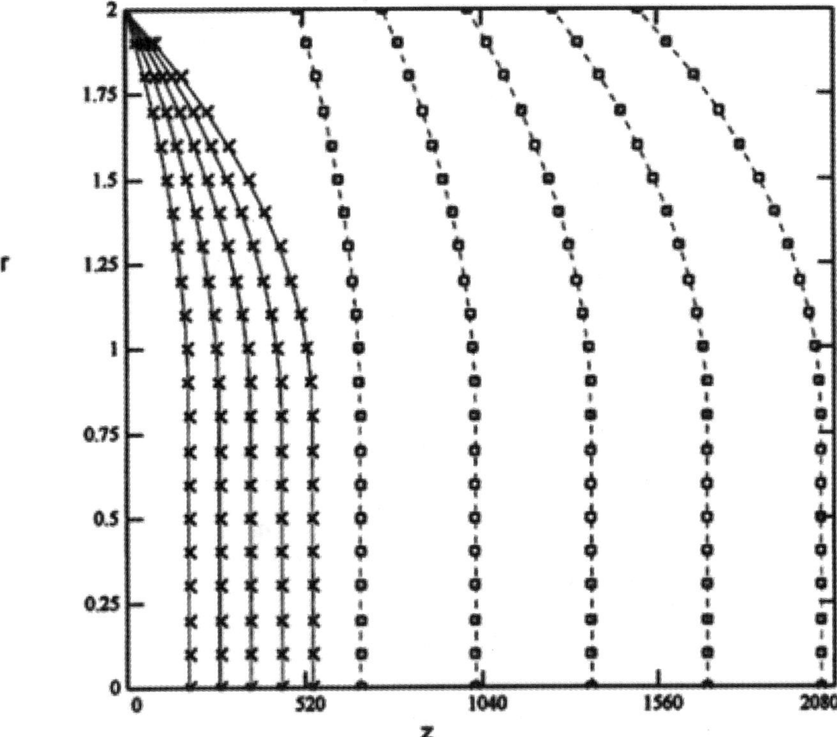

Fig. 6.1.5   Motion in the case of a viscoplastic friction.

In the case when a viscous layer does not exist any more, and $r_0 = R$, $\sigma_{rz}|_R = (\sigma_0/2l)R$, and $\sigma_{rz}|_R$ increases with σ_0.

When σ_0 reaches a certain limit, satisfying (6.1.44) the plug moves. Figure 6.1.5 is an example.

6.2   Flow of a Bingham Fluid Between Two Circular Concentric Cylinders (Reiner [1960])

Let us consider two circular concentric cylinders of height h. The inner one has the radius R_i and is either kept fixed and a torque M is measured, or is rotating with angular velocity $\dot{\omega}_i$. The external cylinder, of radius $R_e > R_i$ is either rotating with the angular velocity $\dot{\omega}_e$ or is fixed. Let us consider the stationary motion of a fluid between these two cylinders (Fig. 6.2.1). This is a typical motion taking place in viscometers. If $\dot{\omega}_i = 0$ the viscometer is of Couette–Hatschek type. If $\dot{\omega}_e = 0$ we get the principle of the Searle viscometer. The fixed cylinder (or mobile) can be used also for the measurement of the torque induced by the rotating fluid.

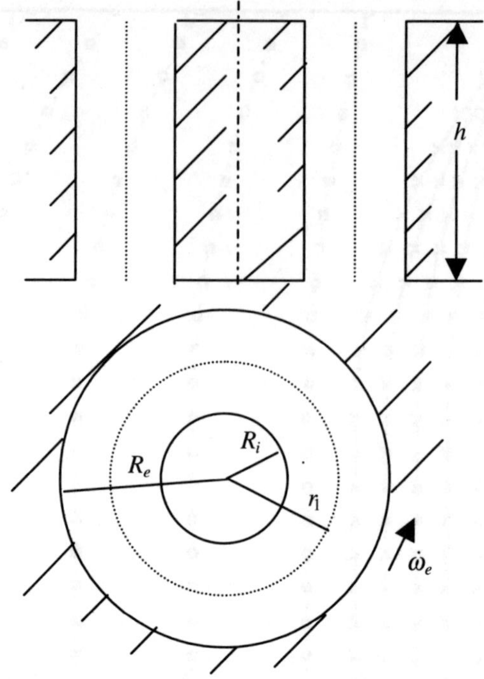

Fig. 6.2.1   Flow of Bingham body between two cylinders.

We use cylindrical spatial coordinates r, θ, z. We assume the flow field of the form

$$v_r = v_z = 0, \quad v_\theta = v = f(r). \tag{6.2.1}$$

The components of the rate of deformation tensor in cylindrical coordinates become in this case

$$D_{r\theta} = \frac{1}{2}\left[\frac{\partial v_\theta}{\partial r} - \frac{v_\theta}{r}\right] = \frac{1}{2}\left[f' - \frac{f}{r}\right],$$

$$D_{rr} = D_{\theta\theta} = D_{zz} = D_{rz} = D_{\theta z} = 0. \tag{6.2.2}$$

Since every single particle is in motion

$$v = f = r\dot\omega(r) \tag{6.2.3}$$

where the angular velocity $\dot\omega$ depends on r. The rate of deformation becomes

$$D_{r\theta} = \frac{r}{2}\frac{d\dot\omega}{dr}. \tag{6.2.4}$$

The *Newtonian viscous fluid* is the first case to consider.

$$\sigma_{r\theta} = \eta r \frac{d\dot\omega}{dr} \tag{6.2.5}$$

and

$$\sigma_{rr} = \sigma_{\theta\theta} = \sigma_{zz} = -p \quad \text{and} \quad \sigma_{z\theta} = \sigma_{zr} = 0. \tag{6.2.6}$$

The equilibrium equations becomes for this case (disregarding the body forces and the accelerations):

$$-\frac{\partial p}{\partial r} + \frac{1}{r}\frac{\partial \sigma_{r\theta}}{\partial \theta} = 0,$$

$$\frac{\partial \sigma_{r\theta}}{\partial r} - \frac{1}{r}\frac{\partial p}{\partial \theta} + \frac{2\sigma_{r\theta}}{r} = 0,$$

$$-\frac{\partial p}{\partial z} = 0. \tag{6.2.7}$$

Taking into account that the problem is with axial symmetry ($\partial/\partial\theta = 0$) it follows that $p = \text{const.}$ (the atmospheric pressure). From $(6.2.7)_2$ follows

$$\sigma_{r\theta} = \frac{C_1}{r^2} \tag{6.2.8}$$

and with (6.2.5):

$$\eta\frac{d\dot\omega}{dr} = \frac{C_1}{r^2} \tag{6.2.9}$$

i.e.,

$$\eta\dot\omega = \frac{C_1}{2r^2} + C_2. \tag{6.2.10}$$

The two constants are determined from the boundary conditions:

$$t > 0 \begin{cases} r = R_i : \dot\omega = \dot\omega_i, \\ r = R_e : \dot\omega = \dot\omega_e. \end{cases} \tag{6.2.11}$$

Thus

$$C_1 = 2\eta(\dot\omega_e - \dot\omega_i)\left(\frac{1}{R_i^2} - \frac{1}{R_e^2}\right)^{-1}$$

$$C_2 = 2\eta\frac{\dot\omega_e R_e^2 - \dot\omega_i R_i^2}{R_e^2 - R_i^2}. \tag{6.2.12}$$

Introducing in (6.2.10) we get:

$$\dot\omega = \frac{\dot\omega_i R_i^2(R_e^2/r^2 - 1) - \dot\omega_e R_e^2(R_i^2/r^2 - 1)}{R_e^2 - R_i^2} \tag{6.2.13}$$

while the tangential stress follows from (6.2.8):

$$\sigma_{r\theta} = \frac{2\eta}{r^2}\frac{\dot\omega_e - \dot\omega_i}{1/R_i^2 - 1/R_e^2}. \tag{6.2.14}$$

The torque exerted on a cylindrical surface of radius r and of height h can be estimated as

$$M = \sigma_{r\theta}(2\pi rh)\,r \tag{6.2.15}$$

i.e., with (6.2.14):

$$M = 4\eta\pi h \frac{\dot\omega_e - \dot\omega_i}{1/R_i^2 - 1/R_e^2} \, .$$

Thus the torque is independent on r.

In the particular case when the <u>internal cylinder is fixed</u>:

$$\dot\omega_i = 0 \, ,$$

$$\dot\omega = \dot\omega_e \frac{R_e^2(1 - R_i^2/r^2)}{R_e^2 - R_i^2} \, ,$$

$$M = 4\eta\pi h \frac{\dot\omega_e}{1/R_i^2 - 1/R_e^2} \, . \tag{6.2.16}$$

The torque is measured either on the internal (fixed) cylinder, or on the external one.

In the particular case when the <u>external cylinder is fixed</u>;

$$\dot\omega_e = 0 \, ,$$

$$\dot\omega = \dot\omega_i \frac{R_i^2(R_e^2/r^2 - 1)}{R_e^2 - R_i^2} \, ,$$

$$M = 4\eta\pi h \frac{-\dot\omega_i}{1/R_i^2 - 1/R_e^2} \, . \tag{6.2.17}$$

The *Bingham fluid.*
The constitutive equation is now:

$$\sigma_{r\theta} = \tau_Y + \eta r \frac{d\dot\omega}{dr} \quad \text{for } \sigma_{r\theta} > \tau_Y \, ,$$

$$\dot\omega = \text{const.} = 0 \quad \text{for } \sigma_{r\theta} \leq \tau_Y \, . \tag{6.2.18}$$

The other stress components satisfy (6.2.6) with $p = $ const. From $(6.2.7)_2$ follows again (6.2.8) and therefore with (6.2.18):

$$\frac{\tau_Y}{r} + \eta \frac{d\dot\omega}{dr} = \frac{C_1}{r^3} \, , \tag{6.2.19}$$

or after integration

$$\eta\dot\omega = -\frac{C_1}{2r^2} - \tau_Y \ln r + C_2 \, . \tag{6.2.20}$$

Let us assume now that the <u>internal cylinder is fixed</u> and the <u>other one rotates</u> with the angular velocity $\dot\omega_e$ prescribed. Thus for

$$t > 0 \begin{cases} r = R_i : \dot\omega_i = 0 \, , \\ r = R_e : \dot\omega = \dot\omega_e \, . \end{cases} \tag{6.2.21}$$

From $(6.2.21)_1$ and (6.2.20) follows

$$C_2 = \frac{C_1}{2R_i^2} + \tau_Y \ln R_i$$

and (6.2.20) becomes

$$\eta\dot{\omega} = \frac{C_1}{2}\left(\frac{1}{R_i^2} - \frac{1}{r^2}\right) - \tau_Y \ln\frac{r}{R_i}\,. \tag{6.2.22}$$

The constant C_1 can be determined if the torque M is measured on one of the cylinders. We have

$$C_1 = \frac{M}{2\pi h}\,, \tag{6.2.23}$$

where h is the height of the portion of the cylinder in contact with the fluid. Thus (6.2.22) becomes, if M is known,

$$\eta\dot{\omega} = \frac{M}{4\pi h}\left(\frac{1}{R_i^2} - \frac{1}{r^2}\right) - \tau_Y \ln\frac{r}{R_i} \tag{6.2.24}$$

and for the outer cylinder:

$$\eta\dot{\omega}_e = \frac{M}{4\pi h}\left(\frac{1}{R_i^2} - \frac{1}{R_e^2}\right) - \tau_Y \ln\frac{R_e}{R_i} \tag{6.2.25}$$

which is the formula due to Reiner and Riwlin [1927]. In the particular case $\tau_Y \to 0$ we obtain again $(6.2.17)_3$ for the viscous fluid.

From (6.2.15) follows that for a constant value of M, the tangential stress $\sigma_{r\theta}$ is bigger when r is smaller. On the other hand the flow takes place only if $\sigma_{r\theta} > \tau_Y$, i.e., [from (6.2.15)]:

$$M > M_0 = \tau_Y\, 2\pi R_i^2 h\,. \tag{6.2.26}$$

If

$$M < M_0 \tag{6.2.27}$$

the Bingham fluid is stressed, but no flow is possible. If however (6.2.26) is satisfied, then a portion of the material in the neighborhood of the inner cylinder is flowing. If the value of the measured moment is M_1, the radius r_1 of the portion of the material which is in motion can be obtained from (6.2.15) for $\sigma_{r\theta} = \tau_Y$:

$$r_1 = \sqrt{\frac{M_1}{2\pi h \tau_Y}}\,. \tag{6.2.28}$$

The condition that the whole material in $R_i \leq r \leq R_e$ be in motion (i.e., $r_1 = R_e$) is

$$M \geq M_e = 2\pi R_e^2 h \tau_Y\,. \tag{6.2.29}$$

If M_1 and r_1 are measured (6.2.28) can be used to determine τ_Y. Afterwards from (6.2.25) follows η.

If the material comprised between $r_1 \leq r < R_e$ rotates as a rigid, the formula (6.2.24) becomes

$$\eta\dot{\omega}_1 = \frac{M_1}{4\pi h}\left(\frac{1}{R_i^2} - \frac{1}{r_1^2}\right) - \tau_Y \ln\frac{r_1}{R_i} \tag{6.2.30}$$

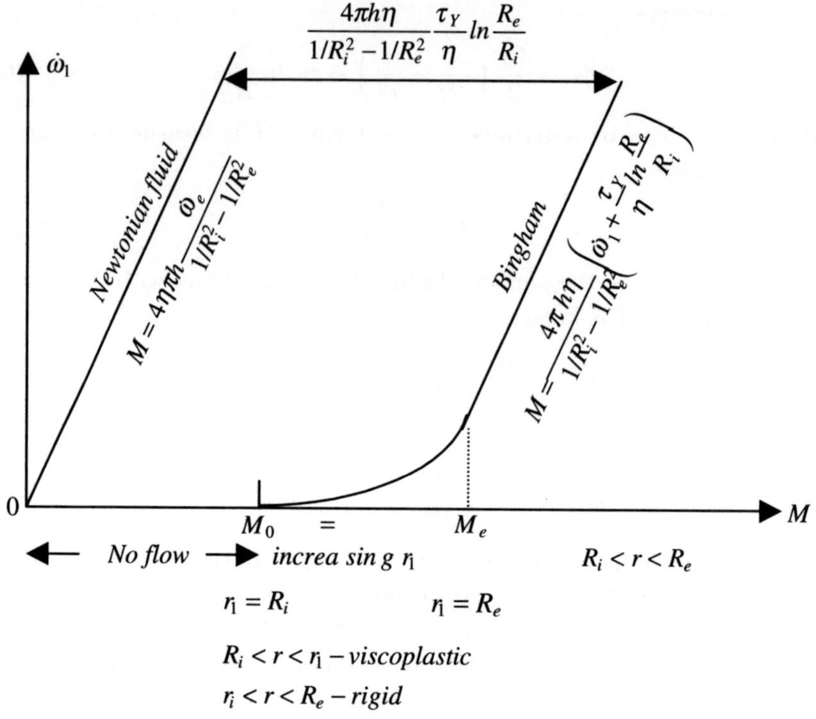

Fig. 6.2.2   The relationship between M and $\dot{\omega}$.

and for the torque

$$M_1 = \frac{4\pi h\eta}{1/R_i^2 - 1/r_1^2}\left(\dot{\omega}_1 + \frac{\tau_Y}{\eta}\ln\frac{r_1}{R_i}\right).\tag{6.2.31}$$

In Fig. 6.2.2 is showing the different relations between M and $\dot{\omega}$ which results from the formulae.

$$\frac{4\pi h\eta}{1/R_i^2 - 1/R_e^2}\frac{\tau_Y}{\eta}\ln\frac{R_e}{R_i}$$

is a measure of the Bingham effect.

It is sometimes convenient to introduce dimensionless variables, independent on the geometry of the apparatus

$$V = 2\frac{\dot{\omega}_i}{1 - R_i^2/R_e^2},\qquad P = \frac{M}{2R_i^2\pi h}.\tag{6.2.32}$$

With these variables one can rewrite (6.2.24) to become

$$V = \frac{P}{\eta}\left[1 - \frac{\tau_Y}{P}\frac{\ln(R_i/R_e)^2}{1 - (R_i/R_e)^2}\right].\tag{6.2.33}$$

The relationship $P \sim V$ defines the properties of the material called consistency.

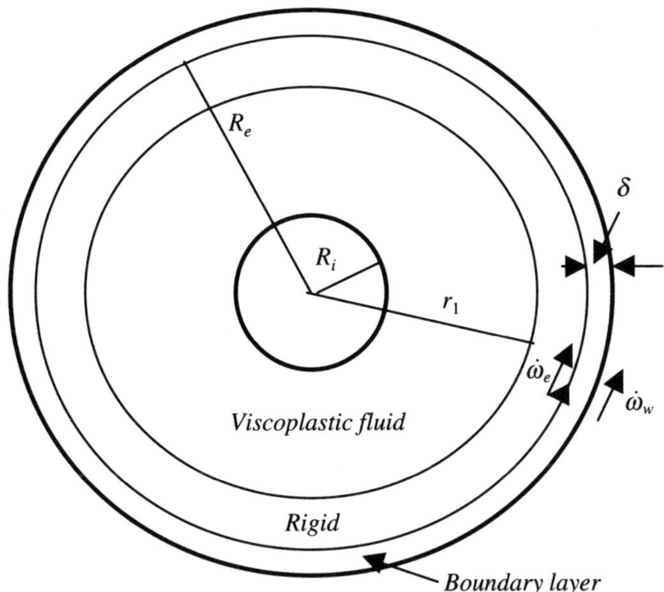

Fig. 6.2.3   The motion with a boundary layer.

For $\tau_Y \to 0$ the model becomes a viscous fluid. For $\eta \to 0$ the model becomes a perfectly plastic one.

Boundary layer. Static/dynamic dry friction.

As the torque or $\dot{\omega}_e$ is increased, there is a limit value of r_1 when the powder (the Bingham body) in contact with the wall, will slip (Fig. 6.2.3). This will happen when the shear transmitted to the cylindrical surface of radius r_1 equals or surpasses the frictional shearing force exerted on the cylinder of radius R_e : $2\pi r_1 h\tau_Y = 2\pi R_e h\tau_f$, where τ_f is the frictional shearing stress. Thus (see Fig. 6.2.4)

$$r_1 \tau_Y = R_e \tau_f \, . \tag{6.2.34}$$

So long as

$$\tau_f > \frac{r_1 \tau_Y}{R_e} \tag{6.2.35}$$

the fluid will adhere to the wall. If however, when r_1 is increasing towards r_{1f}

$$\tau_f \leq \frac{r_{1f} \tau_Y}{R_e} \tag{6.2.36}$$

the rigid portion of the fluid of thickness $R_e - r_{1f}$ will slip along the wall.

Let us assume that in a very thin layer of thickness δ (see Fig. 6.2.3) the friction law can be described by a relationship of the form

$$\tau_f = m(\tau_Y - 2n\sqrt{II_D}) \tag{6.2.37}$$

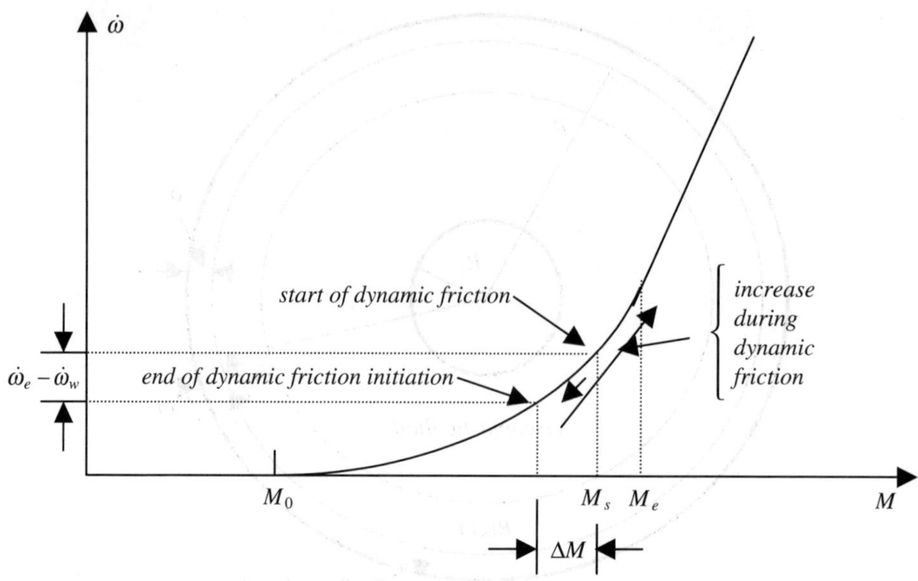

Fig. 6.2.4   Explanation of the text.

where m is a "friction factor" with $0 \leq m \leq 1$, II_D is the second invariant of the rate of deformation tensor of the material in the boundary layer, and n is a "dynamic friction factor" expressing a measure between the "static" and "dynamic" frictions. If no slip takes place then $II_D = 0$ and the "static" friction law is

$$\tau_f = m\tau_Y . \tag{6.2.38}$$

The value $m = 0$ corresponds to perfect lubrication, when $\tau_f = 0$. From (6.2.34) follows $r_1 = 0$ and therefore the whole volume of fluid comprised between $R_i \leq r \leq R_e$ is not moving at all and remain rigid. For $m = 1$ we get the other extreme case; again from (6.2.34) follows $r_1 = R_e$, i.e., when $\dot{\omega}$ (or M) increases it is ultimately possible that the thickness of the rigid layer be reduced to zero thickness, when $M \to M_e$ (Fig. 6.2.5).

In the general case the law (6.2.37) will be used to describe the friction when the slip takes place. Since in II_D a single nonzero component is involved

$$D_{r\theta} = \frac{r}{2} \frac{d\dot{\omega}}{dr} \tag{6.2.39}$$

we will write this component for the thin layer of thickness δ as nonzero component is involved

$$D_{r\theta} \cong \frac{R_e}{2} \frac{\dot{\omega}_w - \dot{\omega}_e}{\delta} \tag{6.2.40}$$

where $\dot{\omega}_w$ is the angular velocity of the external viscometer cylinder, $\dot{\omega}_e$ is the "external" angular velocity of the layer $r = R_e$ after sliding (Fig. 6.2.6). The

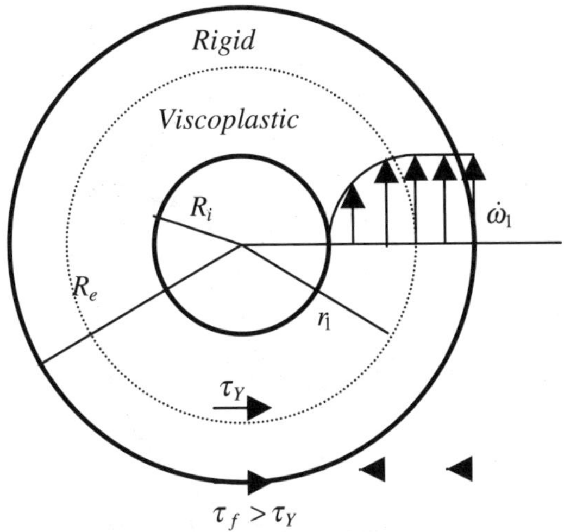

Fig. 6.2.5 Velocity profile for $r_1 < R_e$.

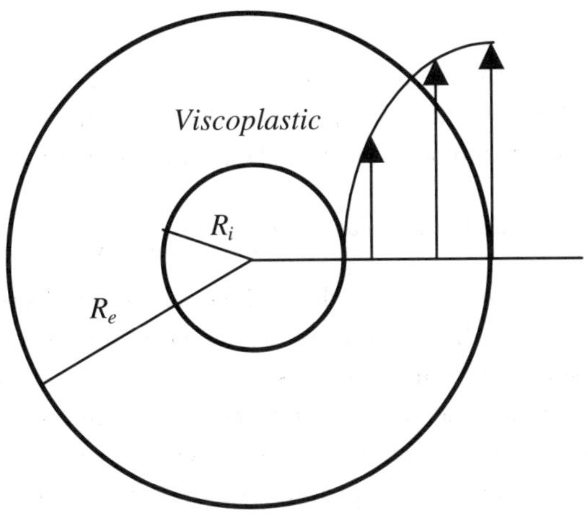

Fig. 6.2.6 Velocity profile for $r_1 = R_e$.

friction law (6.2.37) becomes

$$\tau_f = m\left(\tau_Y - nR_e\frac{\dot{\omega}_w - \dot{\omega}_e}{\delta}\right) \tag{6.2.41}$$

from where follows

$$\dot{\omega}_e = \dot{\omega}_w + \left(\frac{\tau_f}{m} - \tau_Y\right)\frac{\delta}{nR_e}. \tag{6.2.42}$$

Here since during dynamic friction $\tau_f < m\tau_Y$ follows that $\dot{\omega}_e < \dot{\omega}_w$. δ depends on the material and fluid/viscometer interface, and n is the measure of the difference

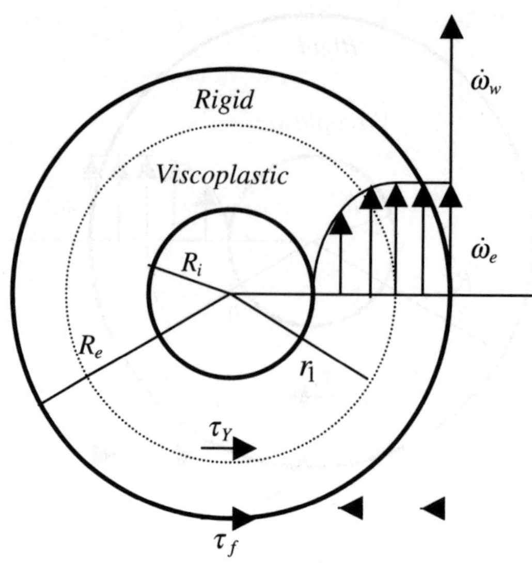

Fig. 6.2.7   Velocity profile for the case of sliding with dry dynamic friction.

between static and dynamic frictions. In the case of "static" friction, when (6.2.38) holds, from (6.2.42) follows $\dot{\omega}_e = \dot{\omega}_w$. (6.2.42) is giving the value of $\dot{\omega}_e$ to be introduced in the boundary condition (6.2.21). The quantity

$$\Delta\dot{\omega} = \dot{\omega}_e - \dot{\omega}_w = \left(\frac{\tau_f}{m} - \tau_Y\right)\frac{\delta}{nR_e} \tag{6.2.43}$$

is the sudden decrease of $\dot{\omega}_e$ and as such produces a sudden decrease ΔM of M [see (6.2.31)]. From the value M_s reached just before slip have started, M decreases to $M_s - \Delta M$ during slip. The jump decrease of M produces a jump decrease of r_1 [see (6.2.28)], i.e., a sudden increase of the thickness of the rigid portion of the fluid (Fig. 6.2.7). If $\dot{\omega}$ or M are further increased, r_1 will continue to increase again.

If ΔM is measured, one can determine $\Delta\dot{\omega}$ and thus δ/n, assuming that m is determined by other experimental procedures. If δ is obtained by physical considerations, n follows. Generally we need the ratio δ/n, only.

6.3   A Model for Slow Motion of Natural Slopes
(Cristescu *et al.* [2002])

Introduction.

Gravitational creep flow, in the sense of deformation in time along a slope under gravitational forces, is often encountered in nature. Examples include: submarine landslides, debris flow, coal slurries, drilling mud, etc. Bedrock landslides have long been recognized as significant geomorphic processes in mountainous topography.

Some landslides are deep-seated and involve long-term displacement at low strain rates. Landslide velocities extend from millimeters per year, to hundreds of kilometers per hour. Conventional stability analysis, which treats the geologic material in the slope as a rigid perfectly plastic body, may provide information on the safety factor of stable mass of soil. To describe this a variety of nonlinear viscous models have been considered. To overcome some of limitations of the viscous flow models, Desai *et al.* [1995] developed an elasto-viscoplastic model. Based on field observations, it was assumed that a shear zone exists up to a certain depth, while the bottom layer under the shear zone remains rigid and at rest. It was supposed that in the shear zone the material obeys an associated viscoplastic constitutive equation while the influence between the rigid bottom and the shearing zone was described by a non associated viscoplastic constitutive equation. The postulated thickness of each of these zones was based on field observations. The "interface" model developed by Desai *et al.* [1995] was further refined by Samtani *et al.* [1996]; finite element implementation of the interface element was described along with verification of the model with respect to the field behavior of a creeping natural slope. The model parameters are determined from laboratory tests.

Below is presented a new model describing gravitational motion along a slope. First, an analysis of gravitational consolidation is performed using an elastic-viscoplastic constitutive equation for dry sand (Cristescu [1991]). Based on this analysis, it can be concluded that the assumption of a linear variation of density with depth applies. This hypothesis is supported by most experimental data (see for example Desai *et al.* [1995]). Next, a non homogeneous Bingham model, with yield stress depending on the current density, is considered for the geologic material in the slope. Although the density could be considered to be linearly dependent on depth, it is shown that a linear variation of the yield stress with depth is physically unacceptable.

Here we give a procedure for the determination of all of the material parameters involved in the proposed model from in situ measurements. More specifically, we use only density measurements and measured displacement of the upper boundary of the shear zone corresponding to any time interval Δt, with all of the other available inclinometer records being further used for validation purposes.

Formulation of the problem.

A granular material deposited on an inclined slope compacts slowly in time due to its own weight. The geometry of the natural slope and the related coordinate system is shown in Fig. 6.3.1. θ is the slope angle, the x-axis along the slope direction, y-axis normal to the x-axis and situated in the slope plane. The compaction characteristic can be described with a non associated viscoplastic constitutive equation of the form (see Ch. 2):

$$\boldsymbol{D} = \frac{1}{2\eta} \left\langle 1 - \frac{W(t)}{H(\boldsymbol{\sigma})} \right\rangle \frac{\partial F}{\partial \boldsymbol{\sigma}} \tag{6.3.1}$$

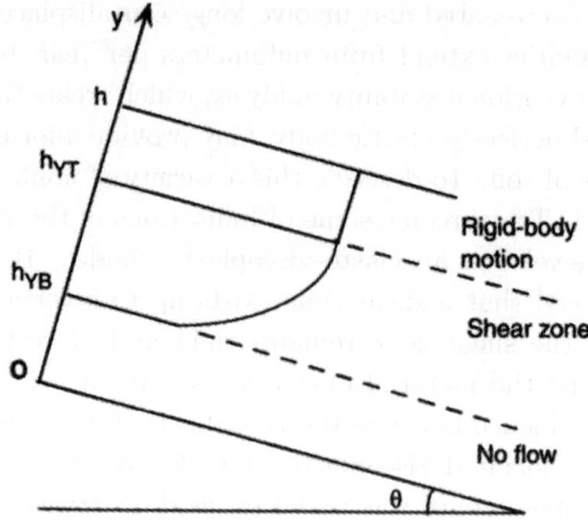

Fig. 6.3.1 Schematic flow along a slope.

where $\mathbf{D}$ is the irreversible strain rate tensor, $H(\boldsymbol{\sigma})$ is the yield function, $F(\boldsymbol{\sigma})$ is the viscoplastic potential, $\boldsymbol{\sigma}$ is the Cauchy stress tensor, $W(t) = \int_{t_0}^{t} \boldsymbol{\sigma} \cdot \mathbf{D}dt$ is the irreversible stress work per unit volume, which is used here as an internal state variable, η is a viscosity coefficient and the bracket $\langle A \rangle$ denotes the positive part of function A, i.e., $\langle A \rangle = A$ if $A > 0$ and $\langle A \rangle = 0$ if $A \leq 0$.

If we assume that the gravitational stress state to which the granular material is subjected on the slope is essentially constant in time, Eq. (6.3.1) can be easily integrated to give the variation in time of the volumetric strain ε_v as

$$\varepsilon_V(t) = \frac{\langle 1 - W(t_0)/H(\boldsymbol{\sigma})\rangle \partial F/\partial \sigma}{(1/H)(\partial F/\partial \sigma) \cdot \sigma} \left\{ 1 - \exp\left[\frac{1}{H}\frac{\partial F}{\partial \sigma} \cdot \sigma \left(\frac{t_0 - t}{2\eta} \right) \right] \right\} \quad (6.3.2)$$

where t_0 is an initial reference time, t is the current time, and σ is the mean stress. Since the density is related to the volumetric strain through

$$\rho(t) = \frac{\rho(t_0)}{1 - \varepsilon_V(t)} \quad (6.3.3)$$

we can determine at each depth the variation with time of the density. However, to obtain the density variation from Eq. (6.3.3), we need to estimate the stress distribution within the layer of bulk material.

Consider a layer of material of constant depth h deposited horizontally ($\theta = 0$ in Fig. 6.3.1). At an arbitrary depth y, the vertical stress is

$$\sigma_{yy}(t) = \int_{h}^{y} \rho(t)gdy = -\rho(t_0)g(h - y) - p_h \quad (6.3.4)$$

where p_h stands for the pressure exerted at the top surface $y = h$. For bodies of large lateral extent, assuming a plane strain state, Hooke's law and isotropy, give the other stress components

$$\sigma_{xx} = \sigma_{zz} = \frac{v}{1 - v}\sigma_{yy} \quad (6.3.5)$$

where v is the Poisson's ratio. The stress given by Eq. (6.3.5) correspond to $t = t_0$ in Eq. (6.3.2) [i.e., the initial data for the viscoplastic Eq. (6.3.1)]. Next we give an example of application of Eqs. (6.3.2)–(6.3.5) for dry sand. The explicit expressions of the yield function, H, and viscoplastic potential, F, used in Eq. (6.3.2) are for dry sand. The material parameters involved in the expressions of these constitutive functions were determined from triaxial compression data reported in Hettler *et al.* [1984]. From Eq. (6.3.2) it follows that the density at any depth is an increasing function of time and tends asymptotically towards a limit value, say $\rho(\infty)$. Figure 6.3.2 shows the variation of the bottom density with time for $h = 6$ m, $v = 0.36$, and $\eta = 20$ Poise. Figure 6.3.3 shows the plot of the ultimate limit values of the density versus the depth y corresponding to $v = 0.3$.

A linear law of variation of the limit density $\rho(\infty)$ with depth approximates the data well

$$\rho = \frac{\rho_h - \rho_0}{h} y + \rho_0 \tag{6.3.6}$$

where $\rho_0 = \rho(0)$ is the density at the bottom, $y = 0$ after compaction and $\rho_h = \rho(h)$ is the density at the top, $y = h$. To capture the influence of the compaction history on the creep flow of the material deposited on the slope, a non homogeneous Bingham model is proposed.

$$D = \begin{cases} 0 & \text{if } II_{\sigma'} \leq k^2(\rho) \\ \dfrac{1}{2\eta(\rho)} \left\langle 1 - \dfrac{k(\rho)}{\sqrt{II_{\sigma'}}} \right\rangle \sigma' & \text{if } II_{\sigma'} > k^2(\rho) \end{cases} \tag{6.3.7}$$

where D is the irreversible strain rate tensor, $D_{ij} = 1/2(v_{i,j} + v_{j,i})$, $\mathbf{v}$ the the velocity field, $\sigma' = \sigma - 1/3(\text{tr}\sigma)\mathbf{1}$ is the stress deviator, and $II_{\sigma}' = (1/2)\sigma'_{ij}\sigma'_{ij}$ is the second invariant of the stress deviator.

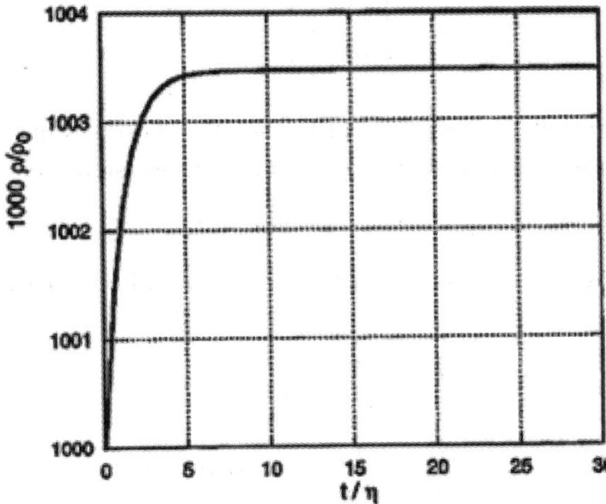

Fig. 6.3.2 Evolution in time of the bottom. Density as predicted by (6.3.3).

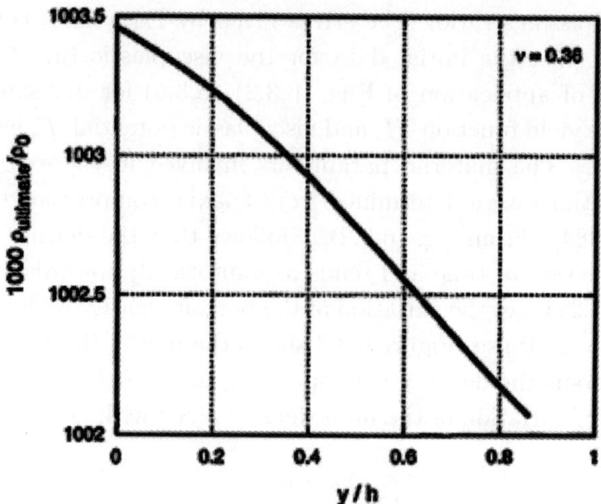

Fig. 6.3.3   Variation of the ultimate density (i.e., density at $t \to \infty$) with depth as predicted by Eq. (6.3.3) ($\eta = 20$ Poise, $h = 6$ m).

In this model both the yield stress $k(\rho)$ and the viscosity coefficient $\eta(\rho)$ are considered to be functions of the actual density; k is given in stress units and η in Poise.

We make the usual kinematics assumption that the particles move in the slope direction (see Fig. 6.3.1) with a velocity depending on depth only, i.e.,

$$v_x = V(y), \quad v_y = 0, \quad v_z = 0 \tag{6.3.8}$$

hence, the strain rate tensor is

$$\boldsymbol{D} = \frac{1}{2}\frac{\partial V}{\partial y}\begin{bmatrix} 0 & 1 & 0 \\ 1 & 0 & 0 \\ 0 & 0 & 0 \end{bmatrix}. \tag{6.3.9}$$

We assume also that the depth, h, of the stratum is constant. Using Eqs. (6.3.9) and (6.3.7), we obtain (stresses are negative in compression) the following:

$$\sigma_{xx} = \sigma_{yy} = \sigma_{zz} - p \quad \text{and} \quad \sigma_{xz} = \sigma_{yz} = 0 \tag{6.3.10}$$

where $-p$ is the mean pressure. Substituting Eq. (6.3.10) in Cauchy equilibrium equations we find

$$-\frac{\partial p}{\partial x} + \frac{\partial \sigma_{xy}}{\partial y} + \rho g \sin\theta = 0,$$

$$-\frac{\partial p}{\partial y} + \frac{\partial \sigma_{xy}}{\partial x} - \rho g \sin\theta = 0,$$

$$\frac{\partial p}{\partial z} = 0. \tag{6.3.11}$$

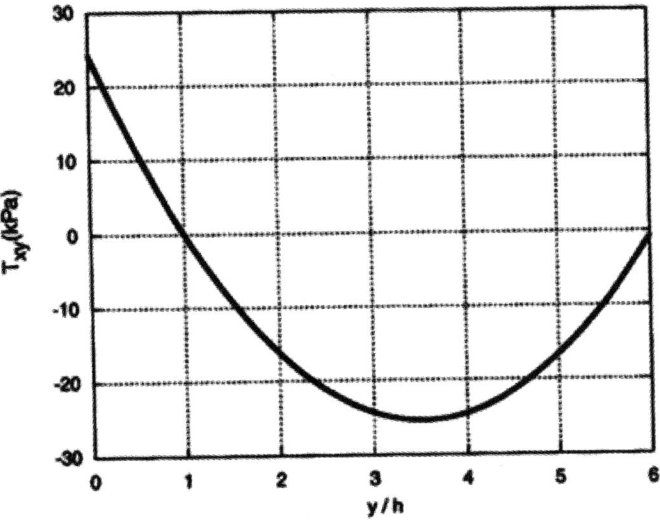

Fig. 6.3.4   Variation of shearing stress with depth.

In the proposed constitutive Eq. (6.3.7), the yield stress is a function of the actual density. Since the density varies with depth, we further assume that the yield stress is a function of depth only. Thus the model exhibits transverse isotropy, i.e., the material is isotropic in any plane, $y = $ constant (the group of symmetry of the material consists of all of the rotations about the y direction). Because the yield limit is assumed to depend solely on depth, it follows that the shear stress σ_{xy} is a function of y only. Hence,

$$p(y) = -g\cos\theta \int_h^y \rho(y)\,dy \tag{6.3.12}$$

and the shear stress is

$$\sigma_{xy} = -g\sin\theta \int_h^y \rho(y)\,dy. \tag{6.3.13}$$

For the specific law of linear variation of ρ with depth, Eq. (6.3.6), and the boundary condition $\sigma_{xy}|_{y=h} = 0$, we further obtain

$$\sigma_{xy} = g\sin\theta \left\{ (h-y)\left[\rho_0 + \frac{(\rho_h - \rho_0)(h+y)}{2h}\right]\right\}. \tag{6.3.14}$$

Figure 6.3.4 shows the variation of σ_{xy} with depth corresponding to $\theta = 20°$, $h = 6$ m, $\rho_0 = 1.4$ g/cm^3, and $\rho_h = 1.002$ g/cm^3. The minimum value of σ_{xy} is reached at a point that is outside the interval $0 \le y \le h$, while the maximum value is reached at the bottom

$$\sigma_{xy}(0) = \frac{\rho_0 - \rho_h}{2} hg\sin\theta. \tag{6.3.15}$$

This value is an increasing function of the stratum thickness h, and of the stratum slope θ. Similarly, we obtain

$$p(y) = g \cos \theta \left[\frac{\rho_h - \rho_0}{2h} (h^2 - y^2) + \rho_0 (h - y) \right]. \tag{6.3.16}$$

Its maximum value is reached at the bottom:

$$p(0) = \frac{\rho_0 + \rho_h}{2} gh \cos \theta.$$

We note that p is a decreasing function of θ.

Incipient shear flow.

According to the constitutive law Eq. (6.3.7), flow takes place if and only if $\sigma_{xy}(y) > k(y)$ i.e., when the shear stress surpasses the yield limit. The threshold condition for the flow initiation is therefore $\sigma_{xy}(y) = k(y)$. While most data support a linear variation of density with depth [see Eq. (6.3.6)], a linear variation of the yield stress with depth is physically unacceptable. A rigorous proof of this statement is presented in the next section. For the time being, we will postulate a power law variation of the yield stress with depth

$$k(y) = (k_0 - k_h) \left(1 - \frac{y}{h} \right)^n + k_h \tag{6.3.17}$$

where $k_0 = k(0)$ is the yield stress at the bottom, $k_h = k(h)$ the yield stress at the top, and $n > 1$ is a material constant. As an illustration of the capabilities of the model, in the following we describe the theoretical pattern of movements corresponding to $n = 2$ in the power law Eq. (6.3.17). The condition for flow initiation $\sigma_{xy}(y) = k(y)$ is

$$\left[\frac{G}{2h} (\rho_0 - \rho_h) + \frac{k_h - k_0}{h^2} \right] y^2 + \left[\frac{2}{h} (k_0 - k_h) - G\rho_0 \right] y$$

$$+ \frac{1}{2} Gh(\rho_0 + \rho_h) - k_0 = 0 \tag{6.3.18}$$

where $G = g \sin \theta$.

The case $(G/2h)(\rho_0 - \rho_h) - (k_h - k_0)/h^2 = 0$ corresponding to an overall depth of the moving slope given by $h = 2(k_h - k_0)/G(\rho_0 - \rho_h)$ is physically unacceptable and will not be considered. For $h < 2(k_h - k_0)/G(\rho_0 - \rho_h)$, the discriminant of Eq. (6.3.18), considered as a second order polynomial in y, is

$$\Delta = \frac{1}{h^2} [h^2 G^2 \rho_h^2 + 2hG(\rho_0 - \rho_h)k_h + 4k_h(k_h - k_0)]. \tag{6.3.19}$$

For $h = $ const., the discriminant of $\Delta(\theta) = 0$, as a second order polynomial in h is

$$\Delta_1 = 4k_h h^2 [k_h(\rho_0 - \rho_h)^2 + 4(k_0 - k_h)\rho_h^2]. \tag{6.3.20}$$

Since $\rho_0 \geq \rho_h$ and $k_0 \geq k_h$, it follows that $\Delta_1 \geq 0$ and that the equation $\Delta(\theta) = 0$ has only one positive root. Furthermore, the equation $\Delta(\theta) = 0$ has a solution in $(0, \pi/2)$ if and only if: $h > H$, where

$$H = \frac{-(\rho_0 - \rho_h)k_h + \sqrt{k_h[k_h(\rho_0 - \rho_h)^2 + 4\rho_h^2(k_0 - k_h)]}}{g\rho_h^2}. \tag{6.3.21}$$

In conclusion, if $h < H$, then $\sigma_{xy}(y) < k(y)$ for any $y \in [0, h]$, i.e., no shear flow is possible. If $h > H$ then:

- Incipient shearing motion initiates at the depth

$$h_{YI} = \frac{g\rho_0 \sin\theta_i - (2/h)(k_0 - k_h)}{2[((\rho_0 - \rho_h)/2h)\,g\sin\theta_i - (k_0 - k_h)/h^2]}$$

where θ_i is the unique solution in $(0, \pi/2)$ of the equation $\Delta(\theta) = 0$; $\sigma_{xy}(y) < k(y)$ for any $y \neq h_{YI}$; the top stratum $0 < y < h_{YI}$ is a rigid plug in motion, while the bottom stratum is rigid and at rest.

- For slope inclination, $\theta > \theta_i$, $\Delta(\theta > 0)$, Eq. (6.3.18) has two distinct solutions h_{YB} and h_{YT}. In this case, $\sigma_{xy}(y) > k(y)$ for $h_{YB} < y < h_{YT}$ and $\sigma_{xy}(y) < k(y)$ otherwise. The pattern of movement of the slope is as depicted in Fig. 6.3.1. There is a shear zone $h_{YB} < y < h_{YT}$, its lower and upper boundaries being the solutions of Eq. (6.3.18), i.e.,

$$h_{YB} = \frac{-\rho_0 g \sin\theta + (2/h)(k_0 - k_h) - \sqrt{\Delta}}{2[((\rho_0 - \rho_h)/2h)g\sin\theta - (k_0 - k_h)/h^2]},$$

$$h_{YT} = \frac{-\rho_0 g \sin\theta + (2/h)(k_0 - k_h) + \sqrt{\Delta}}{2[((\rho_0 - \rho_h)/2h)g\sin\theta - (k_0 - k_h)/h^2]}. \tag{6.3.22}$$

As an example, Figs. 6.3.5(a)–6.3.5(c) show the variation of the shear stress $\sigma_{xy}(y)$ and of the yield limit $k(y)$ with depth corresponding to a slope inclination $\theta = \theta_i/2$, $\theta = \theta_i$ and $\theta = 1.2\theta_i$, respectively; $h = 20$ m, $k_0 = 48$ kPa, $k_h = 8$ kPa, $\rho_0 = 1.3$ g/cm^3, and $\rho_h = 1.2$ g/cm^3. For this set of values of the constitutive parameters: $\theta_i = 8.4°$. The thickness of the top rigid stratum depends strongly on the magnitude of the yield stress k_h. If k_h is very small, the thickness of this stratum becomes very small (i.e., if $k_h \to 0$ then $h_{YT} \to 0$). The top rigid stratum $h_{YT} \leq y \leq h$ moves with the velocity $v(h_{YT})$ of the layer $y = h_{YT}$ ("plug flow"), while the bottom rigid stratum $0 \leq y \leq h_{YB}$ is at rest. The case when this bottom layer may slide along the bottom bedrock will be discussed in the section "Slip with friction at the bottom". Note that from Eq. (6.3.18) it follows that there is an inclination, $\theta = \theta_C$, such that the lower boundary of the shearing zone coincides with the bottom layer $y = 0$,

$$\sin\theta_C = \frac{k_0}{((\rho_h + \rho_0)/2)gh} \tag{6.3.23}$$

the upper boundary $y = h_{YT}$ of the shearing zone being

$$h_{YT} = 2h\left[1 - \frac{k_0 \rho_h}{2k_0 \rho_h - k_h(\rho_0 + \rho_h)}\right].$$

Let us briefly analyze some limit cases. If the yield stress at the bottom of the stratum is excessively high, then the incipient shear flow takes place at the free surface. Indeed, if in Eq. (6.3.22) $k_0 \to \infty$ then $h_{Yi} \to h$. Note that if the flow condition $\sigma_{xy}(y) = k(y)$ is satisfied at $y = h$, i.e., motion starts at the free surface,

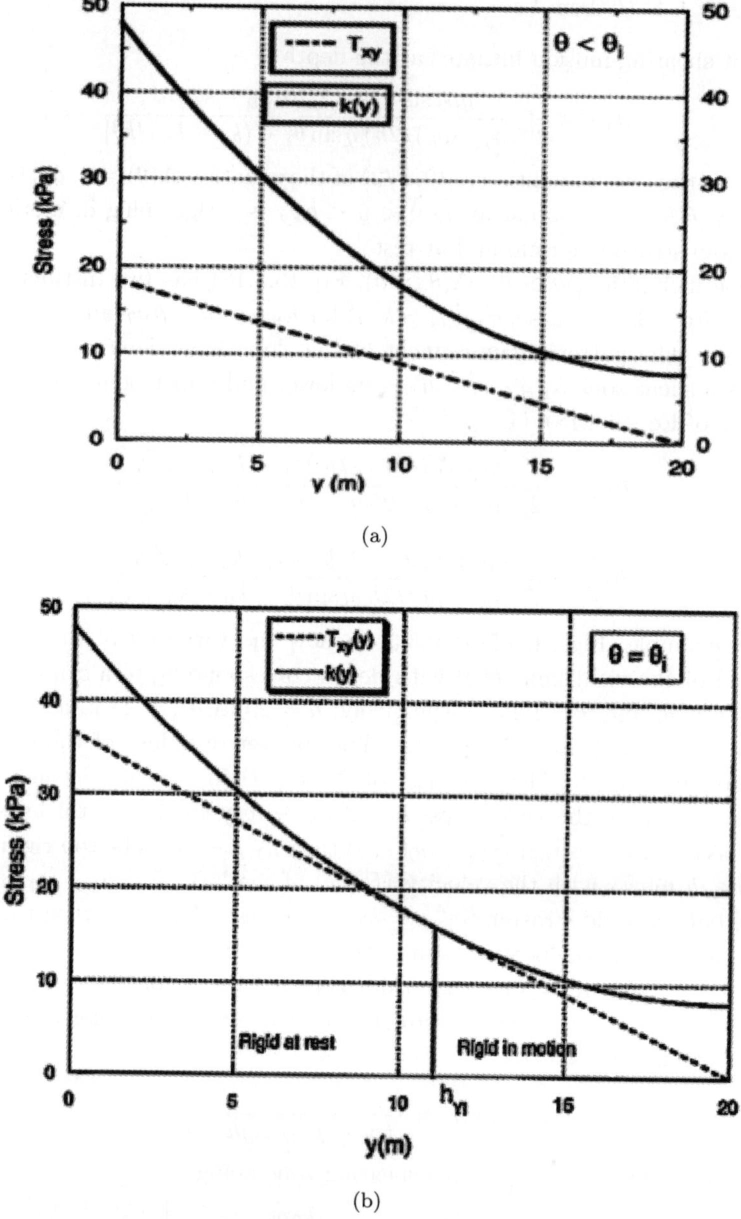

Fig. 6.3.5 Variation of the shear stress σ_{xy} and yield stress k ($n = 2$) with depth for different slope inclination θ: (a) $\theta < \theta_i$, the yield condition is not satisfied; no shear or rigid motion is possible. (b) $\theta = \theta_i$, shear is possible at depth $y = h_{YI}$; the layer $h_{YI} < y < h$ is in a rigid body motion. (c) $\theta > \theta_i$, in the layer $h_{YB} < y < h_{YI}$, $|\sigma_{xy}(y)| > k(y)$ so the material is shearing; for $y > h_{YB}$ the material is at rest.

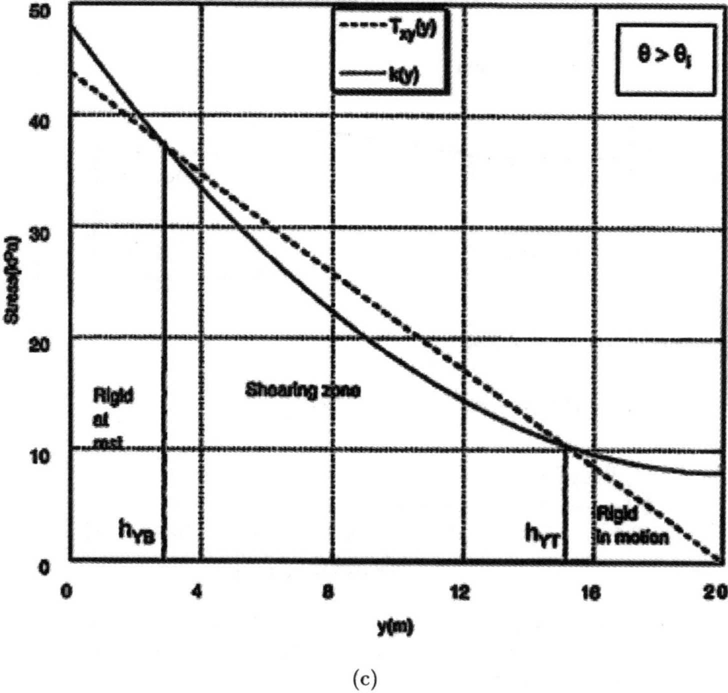

(c)

Fig. 6.3.5 (*Continued*).

then the yield stress at the free surface approaches zero and the material is free flowing. Generally, even after flow has initiated, the material will further compact if the stress to which it is subjected is below the compressibility/dilatancy boundary of the material. Compaction results in an increase of the yield limit, which causes a decrease in the displacement rate to extremely low values. Conversely, a decrease in yield stress and an increase in the unit weight of the soil may be caused, for instance, by heavy rainfalls or sudden snow melting, water-level change, volcanic eruption, or earthquake shaking. This in turn, results in a higher displacement rate. Another possible trigger of shear flow may be an additional pressure added in a short time interval to the top boundary of the stratum (a thick layer of snow, for instance), which will raise the curve $\sigma_{xy} = \sigma_{xy}(y)$ is such that it will intersect $k = k(y)$.

In our analysis, we have assumed that the depth h of the stratum is constant and θ is steadily increased. However, a similar solution can be obtained if the inclination θ of the slope is constant, and the thickness h is steadily increased.

Determination of $k(y)$.

Several methods can be used to determine k. A procedure based on in situ measurements will be given in the section "Example of slow motion of a natural slope".

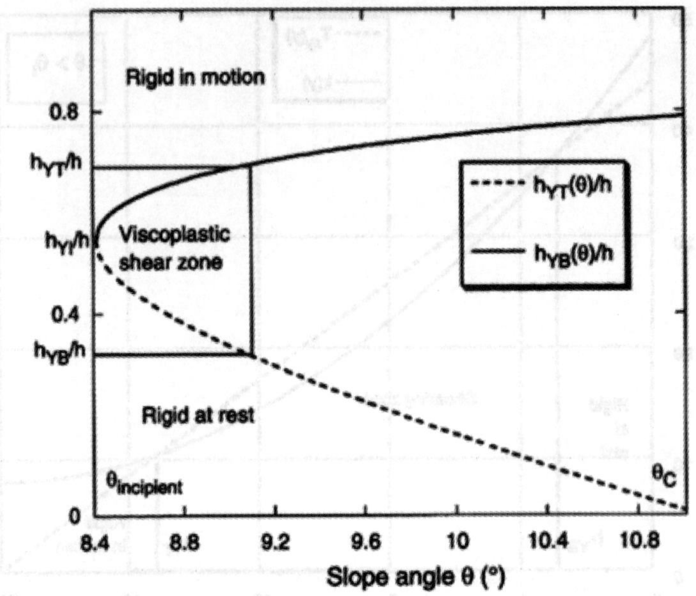

Fig. 6.3.6   Floating boundaries between the rigid and viscoplastic domains when the slope θ ranges from θ_i (slope angle corresponding to incipient shear flow) to θ_C (corresponding to viscoplastic motion up to the bottom of stratum).

In the following, we present an inverse method based on laboratory measurements of the motion of the stratum for different chute inclinations θ.

Specifically, the necessary data are:

- the measured value of the depth h_{Yi}, where the shear motion initiates and the corresponding inclination θ_i;
- h_{YB} and h_{YT} for successive increasing values of θ ($\theta > \theta_i$), [see Eq. (6.3.22)]. Obviously, the slope of σ_{xy} as well as the locations of the vertical lines $y = h_{YB}(\theta)$ and $y = h_{YT}(\theta)$ (see Fig. 6.3.6) is changing with increasing θ; and
- the critical value θ_C (or of h_{YT}) corresponding to the case when the whole stratum is in motion [see Eq. (6.3.23)].

Using the measured values of $h_{YB}(\theta)$ and $h_{YT}(\theta)$, respectively, we then compute $\sigma_{xy}(h_{YB}(\theta))$ and $\sigma_{xy}(h_{YT}(\theta))$ to get $k(h_{YB})$ and $k(h_{YT})$, respectively. These data points are then used to approximate $k(y)$. Once $k(y)$ is determined, the evolution in time of the boundaries between the domains of rigid motion and of viscoplastic deformation corresponding to a given law of variation of the chute angle with time can be obtained using Eqs. (6.3.22) and (6.3.23). As an example, Fig. 6.3.6 shows the variation with θ of the upper rigid-viscoplastic boundary $h_{YT}(\theta)$ and lower rigid-viscoplastic boundary $h_{YB}(\theta)$ for $\theta \in (\theta_i, \theta_C)$ and the following set of values of the constitutive parameters: $k_0 = 48$ kPa, $k_h = 8$ kPa, $n = 2$, $\rho_0 = 1.3$ g/cm^3,

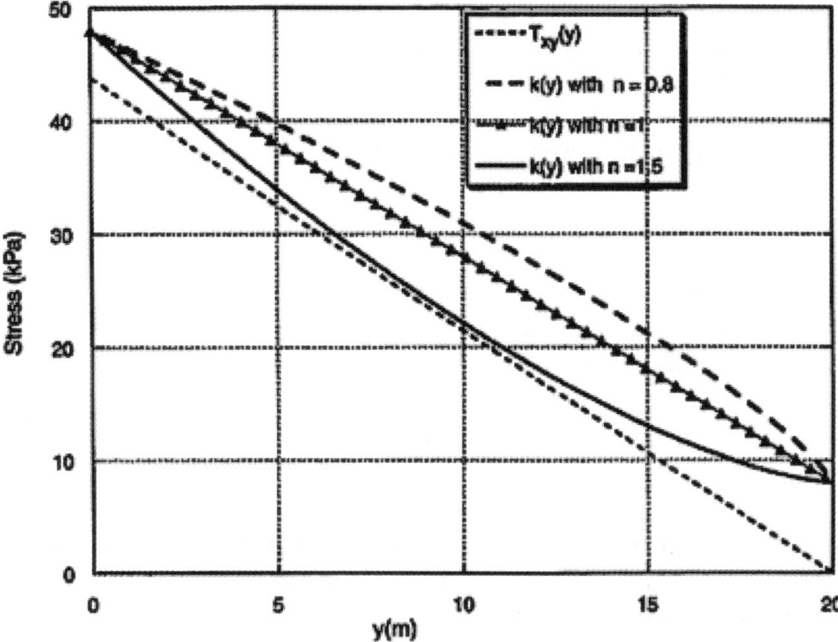

Fig. 6.3.7 Variation of the shear stress σ_{xy} and yield stress $k(y)$ for three values of the exponent n [see Eq. (6.3.17)]; only $n > 1$ is physically possible.

$\rho_h = 1.2$ g/cm^3. The initial common point of these boundaries is at $h = h_{YI}$ and $\theta = \theta_i$, $h_{YT}(\theta)$ is an increasing function (i.e., the thickness of the rigid stratum is decreasing) while $h_{YB}(\theta)$ is a decreasing function (i.e., the thickness of the lower rigid stratum decreases). Note that the stress distribution $\sigma_{xy}(y)$ given by Eq. (6.3.14) is general and does not involve any constitutive assumptions. We have postulated a power law variation ($n > 1$) of the yield stress $k(y)$ with depth [see Eq. (6.3.17)]. In the following we show that a linear variation ($n = 1$) or $n < 1$ is physically unacceptable. Indeed, for the non homogeneous Bingham material D_{xy} is proportional to the overstress $\sigma_{xy} - k$ and thus reaches a maximum value at a location that corresponds to a maximum in overstress. If a linear dependence is assumed (see Fig. 6.3.7, case $n = 1$) then for $h = $ const., and θ gradually increased, $\sigma_{xy}(y)$ intersects $k(y)$ always at $y = 0$. Thus, shear flow would always initiate at the bottom. Though this is a possible particular case, generally it is expected that incipient shear flow would start at some other depth, mainly if h were very large. The same arguments hold against the choice of a power law variation with $n < 1$ (see Fig. 6.3.7). Only for $n > 1$ (see Fig. 6.3.7), incipient shear flow can start anywhere between 0 and h. Generally, the yield stress may depend on other parameters such as humidity, temperature, and particle size. Here we assume that all of these parameters influence ρ, which in turn affects k.

Velocity field.

According to the proposed model Eq. (6.3.7), in the shear zone ($II_{\sigma'} > k^2$),

$$2\eta(y)D_{xy} = [\sigma_{xy} - k(y)](\operatorname{sign}\sigma_{xy}) \tag{6.3.24}$$

and since from Eq. (6.3.14) we have $\operatorname{sign}\sigma_{xy} = +1$, we obtain

$$D_{xy} = \frac{1}{2\eta(y)}[\sigma_{xy}(y) - k(y)] \quad \text{for } h_{YB} < y < h_{YT},$$

$$D_{xy} = 0 \quad \text{for } 0 \le y \le h_{YB} \quad \text{and} \quad h_{YT} \le y \le h, \tag{6.3.25}$$

$\sigma_{xy}(y) - k(y)$ being the "overstress".

Substituting Eqs. (6.3.9), (6.3.14), (6.3.15) in (6.3.17) we obtain

$$\frac{dV}{dy} = \frac{1}{\eta}\left\langle g\sin\theta\left[\frac{\rho_h - \rho_0}{2h}(h^2 - y^2) + \rho_0(h - y)\right] - \frac{k_0 - k_h}{h^n}(h - y)^n - k_h\right\rangle. \tag{6.3.26}$$

For simplicity, we assume that the viscosity η is constant. If there is no slip at the bottom, i.e., $V(h_{YB}) = 0$, then after integration we obtain the velocity field in the viscoplastic zone $h_{YB} < y < h_{YT}$

$$V(y) = \frac{1}{\eta}\left\{g\sin\theta\left[\frac{\rho_h - \rho_0}{2h}\left(h^2 y - \frac{1}{3}y^3\right) + \rho_0\left(hy - \frac{1}{2}y^2\right)\right]\right.$$

$$\left. + \frac{k_0 - k_h}{(n+1)h^n}(h - y)^{n+1} - k_h y\right\} + C \tag{6.3.27}$$

where

$$C = -\frac{1}{\eta}\left\{g\sin\theta\left[\frac{\rho_h - \rho_0}{2h}\left(h^2 h_{YB} - \frac{1}{3}h_{YB}^3\right) + \rho_0\left(hh_{YB} - \frac{1}{2}h_{YB}^2\right)\right]\right.$$

$$\left. + \frac{k_0 - k_h}{(n+1)h^n}(h - h_{YB})^{n+1} - k_h h_{YB}\right\}.$$

For the particular case when $h_{YB} = 0$, $C = -(1/\eta)((k_0 - k_h)/(n+1))h$. As an example Fig. 6.3.8 shows the velocity profile Eq. (6.3.27) after $\Delta t = 100$ days for: $\theta = 10.81°$, $h = 20$ m, $k_0 = 48$ kPa, $k_h = 8$ kPa, $\rho_0 = 1.3$ g/cm^3, $\rho_h = 1.2$ g/cm^3, and $\eta = 1.48 \times 10^6$ Poise. The bottom layer $0 \le y \le h_{YB}$ of the granular material remains rigid and at rest ($V = 0$). The top layer $h_{YT} \le y \le h_{YB}$ moves as a rigid with the velocity $V(h_{YT})$ of the layer $y = h_{YT}$.

Note that if velocity data are available and if no slip takes place at the bottom, η can be estimated from Eq. (6.3.27). Both the exponent n in the expression of $k(y)$ and the viscosity coefficient η influence the velocity profiles. Setting

$$\theta = 10.81°, \ h = 20 \text{ m}, \ k_0 = 48 \text{ kPa}, \ k_h = 8 \text{ kPa},$$

$$\rho_0 = 1.3 \text{ g/cm}^3, \ \rho_h = 1.2 \text{ g/cm}^3, \ n = 2, \ \eta = 1.48 \times 10^6 \text{ Poise},$$

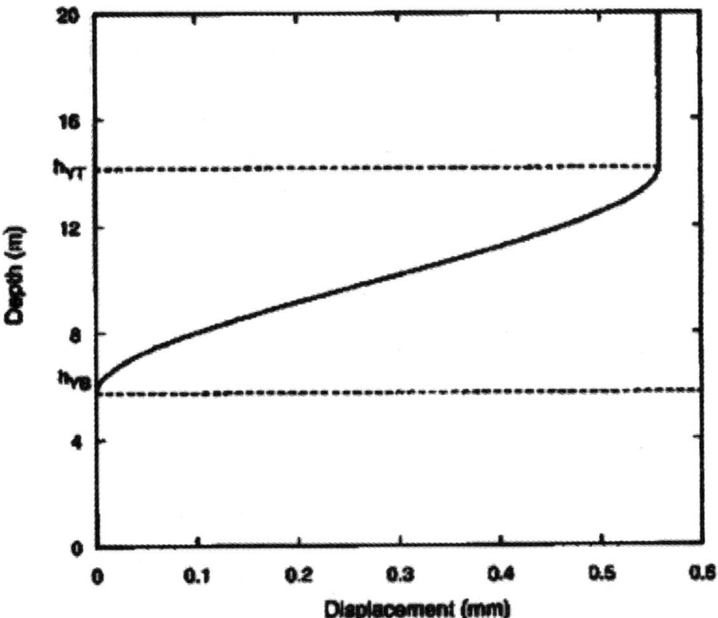

Fig. 6.3.8 Computed velocity profile corresponding to $\theta = 10.81°$; viscoplastic flow develops in the layer $h_{YB} < y < h_{YT}$ ($h_{YB} = 6.32$ m, $h_{YT} = 13.9$ m).

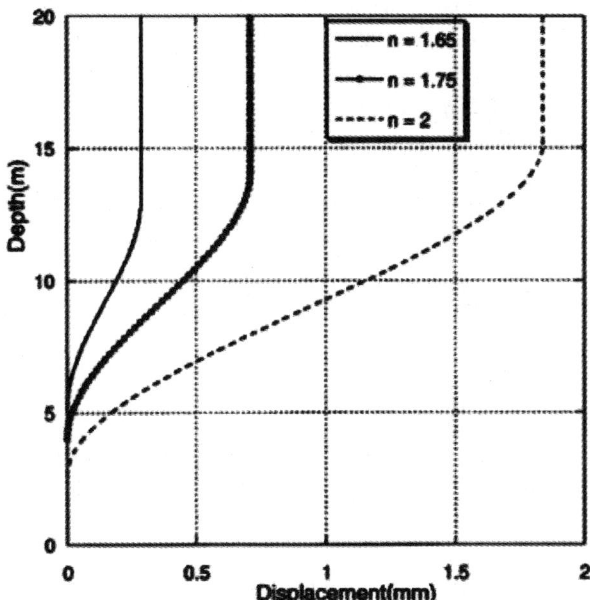

Fig. 6.3.9 Influence of the parameter n on the velocity profile.

for the lower η we obtain a larger shear deformation zone, a thinner rigid layer, and more severe displacements (see Fig. 6.3.8). Figure 6.3.9 shows the influence of n when all of the other constitutive parameters are fixed (same numerical values as in the previous example). Using Eq. (6.3.27) we can also compute the total volume of granular material moving down the slope. The total volume of granular material moving down between $h_{YB} \leq y \leq h_{YT}$ in a width interval Δz in the z-direction and in a unit time interval (the flux) is

$$Q_S = \Delta z \int_{h_{YB}}^{h_{YT}} u(y)\, dy \tag{6.3.28}$$

to which we have to add the material that is in rigid body motion in the zone $h_{YT} \leq y \leq h$, i.e.,

$$Q_T = \Delta z \int_{h_{YT}}^{h} u(h_{YT})\, dy = (\Delta z)u(h_{YT})(h - h_T)\,. \tag{6.3.29}$$

Thus, for a unit width interval in the z-direction ($\Delta z = 1$), and $n = 2$, we obtain

$$Q_S = \frac{1}{\eta} \left\{ g\sin\theta \left[\frac{\rho_h - \rho_0}{h}\left(h^2\frac{y^2}{2} - \frac{y^4}{12}\right) + \rho_0\left(h\frac{y^2}{2} - \frac{y^3}{6}\right)\right] \right. \\ \left. + \frac{k_0 - k_h}{12h^2}(y-h)^4 - k_h\frac{y^2}{2} \right\}\Bigg|_{h_Y}^{h_T} + C\,y\big|_{h_Y}^{h_T}$$

$$= \frac{1}{\eta} \left\{ g\sin\theta \left[\frac{\rho_h - \rho_0}{h}\left(h^2\frac{h_T^2}{2} - \frac{h_T^4}{12}\right) + \rho_0\left(h\frac{h_T^2}{2} - \frac{h_T^3}{6}\right)\right] \right. \\ \left. + \frac{k_0 - k_h}{12h^2}(h_T-h)^4 - k_h\frac{h_T^2}{2} \right\} + Ch_T$$

$$- \frac{1}{\eta} \left\{ g\sin\theta \left[\frac{\rho_h - \rho_0}{h}\left(h^2\frac{h_Y^2}{2} - \frac{h_Y^4}{12}\right) + \rho_0\left(h\frac{h_Y^2}{2} - \frac{h_Y^3}{6}\right)\right] \right. \\ \left. + \frac{k_0 - k_h}{12h^2}(h_Y-h)^4 - k_h\frac{h_Y^2}{2} \right\} - Ch_Y\,. \tag{6.3.30}$$

Slip with friction at the bottom.

In the following, we further examine how frictional effects at the interface influence the pattern of movement. Most authors consider Coulomb friction at the interface,

$$\tau = \mu\sigma_n\,. \tag{6.3.31}$$

According to Eq. (6.3.31), the tangential stress τ, at any point, is proportional to the normal pressure σ_n, and is directed opposite to the relative motion between the soil and the bedrock. The coefficient of friction μ is taken as a constant for a given soil and is considered to be independent of the velocity. It is worthwhile to note that Coulomb's law describes correctly the mechanics of friction between two rigid bodies. To accurately reproduce friction between a rigid body and a deformable body, different friction law should be considered. To describe friction between a rigid and a perfectly plastic body, the following law was proposed on a great number of tests (Levanov *et al.* [1976]):

$$\frac{\tau}{\tau_Y} = k_s \left[1 - \exp\left(a \frac{\sigma}{\sigma_Y} \right) \right] \tag{6.3.32}$$

where $a < 0$, $0 \leq k_s \leq 1$, and τ_Y and σ_Y are the yield stress in simple shear and uniaxial compression, respectively. Generally, k_s characterize the rugosity of the hard surface and ranges from 0 to 1. The limit $k_s = 1$ corresponds to very high rugosity when no slip is possible, and a viscoplastic deformation takes place up to the rigid surface. The limit value $k_s = 0$ corresponds to ideal smooth contact and "perfect" lubrication, when sliding is taking place at the smallest applied shearing stress. Note that for normal pressures much lower than σ_Y, the law Eq. (6.3.32) reduces to Coulomb's law Eq. (6.3.31), while for high pressures (generally for $\sigma > \sigma_Y$) when the material is fully plastic Eq. (6.3.32) reduces to

$$\tau = m\tau_Y \tag{6.3.33}$$

with m constant, i.e., a constant shear stress is assumed at the interface, irrespective of the normal pressure. Since the maximum shear a material can withstand according to the von Mises yield criterion is τ_Y, it follows that $0 \leq m \leq 1$: (i) for no friction, $m = 0$; (ii) for the case of complete adherence to the wall (no slip), $m = 1$, the material in contact with the wall is deforming plastic. The friction law Eq. (6.3.33) is widely used in plastic forming of metals to describe friction between the die (rigid body) and deforming material (Avitzur [1968]). A generalization of the friction law Eq. (6.3.33) to the case of contact between a rigid body and a viscoplastic material was proposed by Cristescu [1975]

$$\tau = m\sqrt{II_{\sigma'}} \tag{6.3.34}$$

where $II_{\sigma'}$ is the second invariant of the stress deviator. In the case where the viscoplastic material is of Bingham type, the friction law Eq. (6.3.34) becomes

$$\tau = m(k + 2\eta\sqrt{II_D}) \tag{6.3.35}$$

where k is the yield stress of the material, η is the viscosity, II_D is the second invariant of the stretching tensor D and $0 \leq m \leq 1$. In the following, we consider friction law Eq. (6.3.35) between the material in the slope and the bedrock (assumed to be a rigid body). This law reproduces the fact that the frictional shear stress is identical with the shearing yield stress.

Let as assume that the maximum shear stress that the bottom surface, $y = 0$ can hold without relative motion is mk_0, $m < 1$, then $\sigma_{xy}|_{y=0} = mk_0$ is first reached at $\theta = \theta_s$

$$\sin\theta_s = \frac{2mk_0}{(\rho_0 + \rho_h)hg} \tag{6.3.36}$$

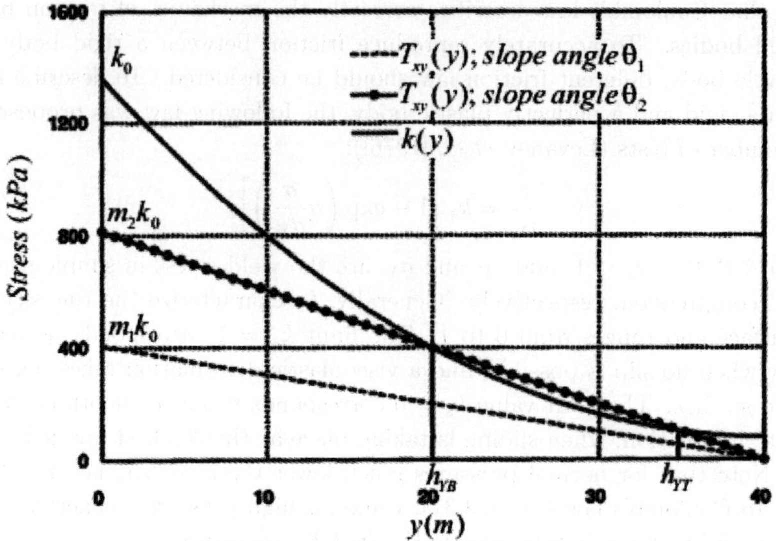

Fig. 6.3.10   The angle at which sliding initiates and the extent of the shear zone for different friction conditions at the interface $y = 0$.

[see Eq. (6.3.15)] and the thickness of the rigid plug sliding along the bottom surface, $y = 0$ is $h_{YB}(\theta_s)$ [see Eq. (6.3.22)]. In conclusion, if the friction law Eq. (6.3.35) is assumed at the interface and $\theta \geq \theta_s$, then there will be a rigid plug at the bottom (see Fig. 6.3.10), i.e., the viscoplastic deformation zone cannot extend to the bottom. If $m \to 1$ (very rough bottom surface) it follows that the sliding velocity $V_s \to 0$ and the zone of viscoplastic flow extends to the bottom. In the case where θ is fixed and h is steadily increased, the depth at which sliding starts is

$$h_s = \frac{2mk_0}{(\rho_0 + \rho_h)g \sin \theta}.$$

Similar arguments hold, if somewhere in the stratum $0 \leq y \leq h$ there exists a weaker plane $y = y_s$ along which the maximum shear stress is much smaller than in the neighborhood layers. In other words, there exists a plane $y = y_s$ of yield stress $k(y_s)$ smaller than that of the stress above it. As an example, see the case study reported by Alonso *et al.* [1993] where a thin marl layer has weaker shear properties than all of the other materials in the moving stratum.

Example of slow motion of a natural slope.

Laboratory experiments are far from being representative of long-term in situ behaviors. Hence, we propose a procedure for the identification of the model parameters based on in situ measurements: density measurements and inclinometer recordings.

The coefficients of the proposed model are: the viscosity η, the densities ρ_0 and ρ_h, as well as the parameters involved in the law of variation of the yield stress with depth: k_0, k_h and n. From measurements of the density of the slope material at the surface and at two other depths we can estimate ρ_h and the slope of the density line. One inclinometer record gives the values of the upper limit, h_{YT}, the lower limit of the shear zone h_{YB} and h (the extent of the domain of interest). Given the slope angle θ, ρ_0, and ρ_h from Eq. (6.3.14), we calculate $\sigma_{xy}(h_{YT})$ and $\sigma_{xy}(h_{YB})$. Since the yield condition is satisfied at h_{YT} and h_{YB}, it follows that $k(h_{YT}) = \sigma_{xy}(h_{YT})$ and $k(h_{YB}) = \sigma_{xy}(h_{YB})$. Next, using Eq. (6.3.17) we calculate

$$n = \frac{\ln[(k_h - k(h_{YB}))/(k_h - k_{YT})]}{\ln[(h - h_{YB})/(h - h_{YT})]},$$

$$k_0 = k_h + \frac{k(h_{YB}) - k(h_{YT})}{(1 - (k(h_{YB})/h)^n - (1 - (k(h_{YT})/h)^n},$$

$$k_h = k_{YB} - (k_0 - k_h)\left(1 - \frac{k(h_{YB})}{h}\right)^n. \tag{6.3.37}$$

Further, the measured displacement of the upper boundary of the stress zone, $u(h_{YT})$, corresponding to any time interval Δt, permits the determination of the viscosity coefficient η. Indeed, assuming $V(h_{YB}) = 0$ (i.e., the strata below the shear zone are at rest) from Eq. (6.3.27) we obtain

$$\eta = \frac{\Delta t}{u(h_{YT})} \left[g\sin\theta \left\{ \frac{\rho_h - \rho_0}{2h}\left[h^2(h_{YT} - h_{YB}) - \frac{1}{3}(h_{YT}^3 - h_{YB}^3)\right] + \rho_0 \left[h(h_{YT} - h_{YB}) - \frac{1}{2}(h_{YT}^2 - h_{YB}^2) \right] \right\} + \frac{k_0 - k_h}{(n+1)h^n}[(h - h_{YT})^{n+1} - (h - h_{YB})^{n+1}] - k_h(h_{YT} - h_{YB}) \right] \tag{6.3.38}$$

with $u(h_{YT})$ the measured displacement of the upper boundary of the shear zone. Hence, all of the material parameters are obtained from exact formulae.

In the following, we apply the proposed model to the description of the slow movement of a natural slope, the Villarbeney landslide in Switzerland (data after Samtani *et al.* [1996]). Creep flow data over one year were recorded at two locations E_1 and E_2 about 250 m apart. At both locations it was observed that there is a top layer, which is moving essentially as a rigid body. Under this layer, another layer exists where shearing flow takes place. At even deeper layers no motion was recorded. Taking the origin of the y axis at the bottom.

Table 6.3.1 Model parameter n at location E_1.

k_h (kPa)	n
10	1.14
20	1.31
30	1.55
42.5	2

At location E_1

$$h_{YT}(= 6m) \leq y \leq h(= 18m) \qquad \text{top layer in rigid motion}$$

$$h_{YB}(= 1m) \leq y \leq h_{YT}(= 6m) \qquad \text{slow creep shearing takes place}$$

$$0 \leq y \leq h_{YB}(= 1m) \qquad \text{rigid at rest.} \qquad (6.3.39)$$

At location E_2

$$h_{YT}(= 2m) \leq y \leq h(= 8m) \qquad \text{top layer in rigid motion}$$

$$h_{YB}(= 0.5m) \leq y \leq h_{YT}(= 2m) \qquad \text{slow creep shearing takes place}$$

$$0 \leq y \leq h_{YB}(= 0.5m) \qquad \text{rigid at rest.} \qquad (6.3.40)$$

At the location of borehole E_1, the base slope angle is $\theta = 14°$, while at location E_2, the base slope angle is $\theta = 17°$. Samtani *et al.* [1996] also give an empirical linear law for the variation with depth of the unit weight of the form

$$\gamma(h - y) = \gamma_h + \zeta(h - y) \qquad (6.3.41)$$

where $\gamma(h - y)$ is the unit weight at the depth $h - y$ below the surface, γ_h is the unit weight at the surface, and ζ is a material parameter. At borehole E_1, $\gamma_h = 21.85$ kN/m^3 and $\zeta = 0.0024$ kN/m^4, while at borehole E_2: $\gamma = 21.6$ kN/m^3 and $\zeta = 0.024$ kN/m^4. The data thus support the linear law of variation of density with depth Eq. (6.3.6) that we have obtained using the constitutive Eq. (6.3.2). Using Eqs. (6.3.39) and (6.3.41), we obtain $k(h_{YB}) = 90.637$ kPa and $k(h_{YT}) = 63.824$ kPa at E_1, and $k(h_{YB}) = 47.989$ kPa and $k(h_{YT}) = 38.281$ kPa at E_2. The exponent n and k_0 in the power law variation of yield stress with depth can be estimated using Eq. (6.3.37). However, the ground yield stress k_h at both locations was not available. As an example in Table 6.3.1, we give the values of this experiment corresponding to several values of the ground yield stress k_h. We assume $k_h = 30$ kPa at location E_1 and using Eq. (6.3.37) we obtain $n = 1.55$. Next, we calculate the viscosity coefficient η using Eq. (6.3.38) and the measured displacement of the upper boundary of the shear zone $u(h_{YT}) = 17.5$ mm corresponding to $\Delta t = 100$ days. We obtain $\eta = 1.523 \times 10^9$ Poise. All of the other available inclinometer records at location E_1 are further used for validation purposes. Figures 6.3.11(a)–6.3.11(d) show a comparison between the data (symbols) and the predicted displacement profiles (lines) at location E_1 corresponding to time intervals 148, 196, 260, and

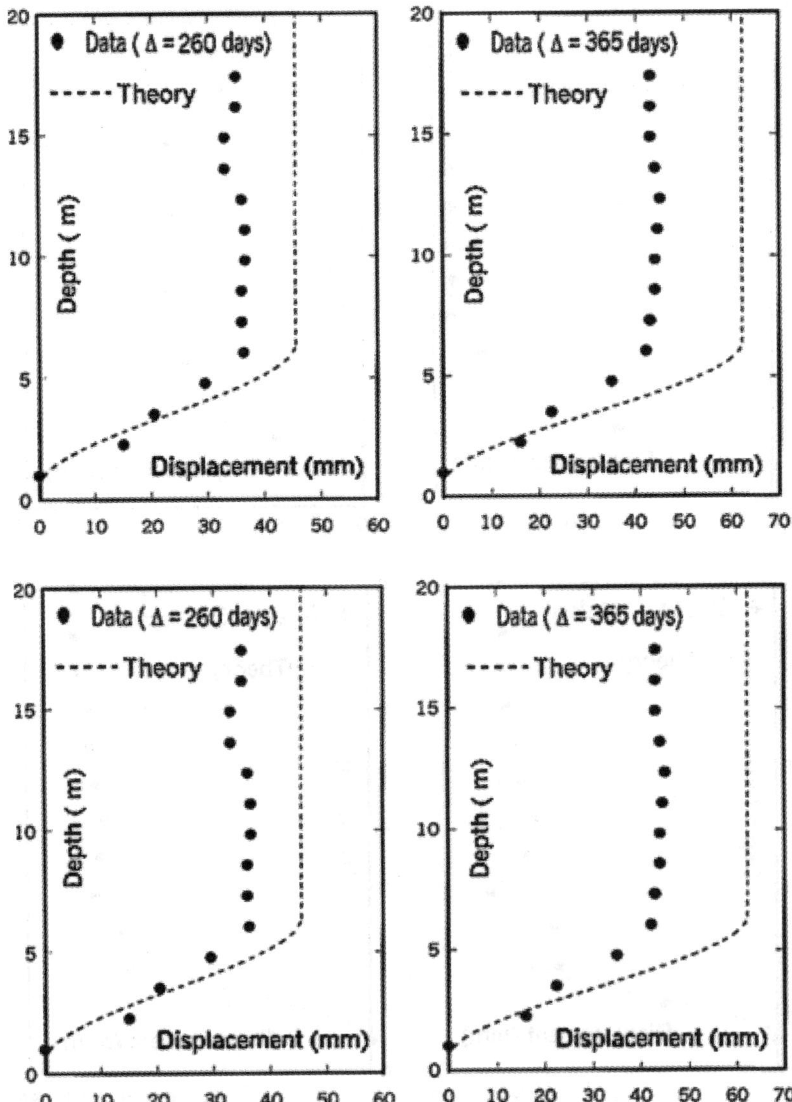

Fig. 6.3.11 Villaberney landslide. Experimental displacement profiles (symbols) and model predictions (dashed lines) at location E_1 after a period of: $\Delta t = 148$, 196, 260 and 356 days. The displacement after $\Delta t = 100$ days, which was used for the determination of the model parameters, is not shown.

356 days. At location E_2 we assume that $k_h = 22$ kPa, and to calculate η we use the measured displacement of the upper boundary of the shear zone, $u(h_{YT}) = 6.3$ mm, corresponding to $\Delta t = 100$ days. The obtained values are: $\eta = 3.852 \times 10^8$ Poise and $n = 2$. The model predictions versus the data, which was not used for identification

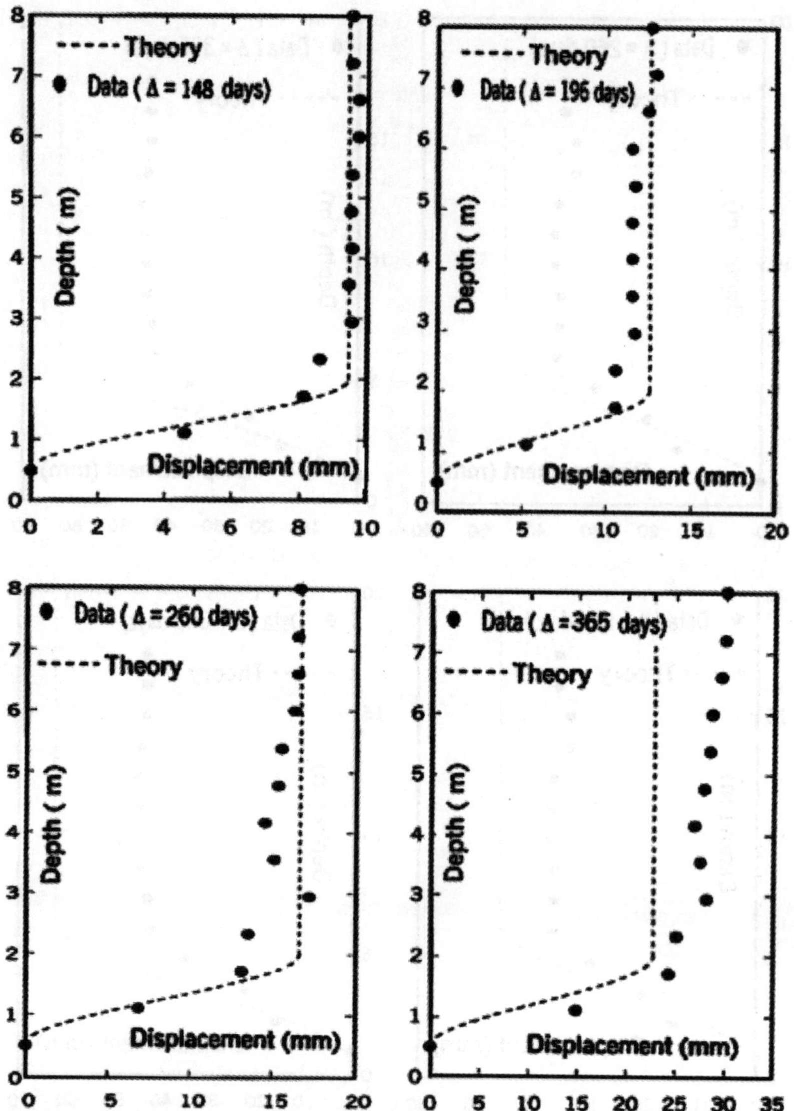

Fig. 6.3.12   Villaberney landslide. Experimental displacement profiles (symbols) and model predictions (dashed lines) at location E_2 after a period of: $\Delta t = 148$, 196, 260 and 356 days. The displacement profile after $\Delta t = 100$ days, which was used for the determination of the model, is not given.

of the model parameters (readings at $\Delta t = 148$, 196, 260 and 365 days), are shown in Figs. 6.3.12(a)–6.3.12(d). The model predictions seem reasonable; however for an accurate determination of the exponent n, we would have needed the value of the yield stress k_h at the surface.

Conclusion.

The non homogeneous Bingham model proposed accounts for the effect of gravitational consolidation on the material response and allows for a variation of density and yield stress with depth. A criterion for shear flow initiation is formulated and the ensuing motion is described. Generally, at a certain depth there is a shear region; the top layer above it is in rigid motion (unless the cohesion is zero). The layer below the shear region is either rigid and at rest or in rigid body motion (plug sliding along the bedrock). Velocity profiles for several slip conditions at the interface with the bedrock are given. A procedure for determining the constitutive parameters using only the measured displacement of the upper boundary of the shear zone, $u(h_{YT})$, corresponding to any time interval Δt, and density measurements was given. Finally, the model was used to predict the observed field behavior of the natural slope from the Villarbeney landslide in Switzerland, and Fosso San Martino (inclinometers readings after Bertini *et al.* [1984]) (Cazacu and Cristescu [2000]). Let as note that records of a single inclinometer were used for the determination of the model, whereas all the other available records were utilized to test the predictive capabilities of the model.

Laboratory tests have been undertaken by Petley and Allison [1997] to elucidate the behavior of deep-seated landslides. In deep-seated failures deformation has been reported at depths of up to 250 m. In the movement zone, owing to the weight of the overburden and the surrounding stress environment, conventional soil mechanics cannot be used to explain effectively associations between the landslip activity and the deformation mechanisms operating within the mowing mass. A number of tests were undertaken, the most significant focusing on the transition between ductile and brittle behavior. The results presented in this paper identify a transitional phase of behavior in which creep-like movement will manifest itself at the base of a deep-seated landslide as growth of microcracks.

In the Mount Stuart batholith, Washington, Paterson and Miller [1998] have examined magmatic fabrics patterns around stopped blocks in tonalite located near the pluton roof and completed 1:1 scale, three-dimensional mapping of fabric patterns around stopped blocks in diorite located ≈ 1000 m below the roof.

There are many places in the world where we have a sliding of several millimeters in 100 days, say. They are producing a recurrent instability phenomenon. Slow motion develops along a deep-seated sliding surface, involving a volume of rock between 22 and 35 million cubic meters. For instance, at Rosone in Italy, such a slide was studied by Forlati *et al.* [2001]. This rock slope is affected by recurrent instability phenomena. The slow movement develops along a deep-seated sliding surface, involving a volume of rock between 22 and 35 million cubic meters. A series of non-linear, time dependent analyses has been carried out through the finite element method. A visco-plastic constitutive law, allowing for strain softening effects.

The flow of a Bingham fluid taking into account inhomogeneous yield limit of the fluid was considered by Hild *et al.* [2002]. After setting the general three dimensional problem, the blocking property is introduced. Then they focus on necessary and sufficient conditions such that blocking of fluid occurs. The anti-plane flow in two dimensional and one dimensional cases is considered.

More fine properties dealing with local stagnant regions as well as local regions where the fluid behaves like a rigid body are obtained in dimension one. Some other variants of Bingham bodies including non-linear Bingham as $\tau = \tau_y + K(\dot{\gamma})^n$ were considered by Talmon and Huisman [2005] to describe the flow of drilling fluids.

6.4   How to Measure the Viscosity and Yield Stress

Introduction.

There are many apparatuses used today to measure the viscosity of fluids and/or yield stress. Their principles and theories are presented in many books, as for instance in Bird *et al.* [1987]. Generally all these devices need a significant volume of liquid in order to determine its viscosity. However, for some applications like biomedical, fluid supply can be extremely limited. For that reason, we have recently developed a falling cylinder viscometer that requires only a tiny amount of fluid (about 20 μl) in order to measure viscosity. This viscometer is based on a falling ball viscometer of Tran-Son-Tay *et al.* [1988]. The advantage of falling cylinder versus the falling ball is that the generated shear rate is better defined. Falling cylinder viscometers have been used for over 75 years, but remarkably the problem of a cylinder of finite length falling inside a cylindrical tube has not been solved yet in simple closed form solution, involving a magnetic field as well.

It appears that Pochettino [1914] was the first to study experimentally the passage from "solid state" to the "fluid state" using a method of falling cylinder in order to study the mechanical properties of tar. Bridgman [1926] used a falling cylinder to determine the "relative viscosities" of fluids subjected to high pressures. He devised the apparatus and analyzed the various possible experimental errors, mainly related to the "inertia effect", i.e., the time needed for the cylinder to reach a steady-state falling velocity. His apparatus did not give the absolute viscosity, but only the relative viscosity. The relative viscosities for a variety of fluids, for several temperatures and pressures were determined. It was found that viscosity increases with pressure. Using a falling cylinder type viscometer, viscosities of methane and propane at low temperatures and high pressures were determined by Huang *et al.* [1966], and viscosities of methane, ethane, propane and n-butane by Swift *et al.* [1960]. A theoretical analysis of the laminar fluid in the annulus of a falling cylinder viscometer was made by Lohrenz *et al.* [1960].

The falling cylinder viscometer was analyzed by Ashare *et al.* [1965] for both Newtonian and non-Newtonian fluids; approximate expression for axial non-Newtonian flow in the annuli was developed. They assumed that, since annular slit is small, it can be regarded as a plane slit. A theoretical analysis for the falling cylinder viscometer for Bingham fluid and power law model, is due to Eichstadt and Swift [1966], while the influence of the eccentricity on the terminal velocity of the cylinder was considered by Chen *et al.* [1968].

A falling magnetic stainless steel slug was developed by Mc Duffie and Barr [1969] to measure viscosities between 1 and 10^4 P at pressure up to 3500 Kg/cm^2 and temperature between $-60°$ and $100°$C. The motion of the slug was determined with a differential transformer that moved along the tube. An automatic falling cylinder viscometer for high pressures was also developed by Irving and Barlow [1971]; the sinker was either a solid cylinder or one with a central hole, and the fall time is detected inductively by a series of coils along the viscometer tube. Viscosity in the range 0.01–3000 P was determined. Another kind of falling coaxial cylinder viscometer for lower viscosity fluids as paints, was constructed by Chee *et al.* [1976], where a weighted rod is falling into a closed-end concentric cylinder.

A laser Doppler technique was used to measure the velocity of a falling-slug in a high pressure viscometer by Dandridge and Jackson [1981]; the viscosities of two polyisobutenes have been determined as function of pressure. Viscosities in excess of 10^7 Pa s have been determined by this method. An improved version of the viscometer was presented by Chan and Jackson [1985]. Measurements of the relative viscosity of aqueous solutions for various temperatures and pressures up to 120 MPa are due to Tanaka *et al.* [1988]; they found that viscosity increases almost linearly with pressure. Tanaka *et al.* [1994] later used a laser beam that passed through a pair of sapphire windows to a phototransistor to determine the falling time of the cylinder. A special designed falling cylinder viscometer was used by Kiran and Sen [1992] to measure high-pressure viscosities in the 10 to 70 MPa range and temperatures from 310 to 450 K, of n-butane, n-pentane, n-hexane and n-octane (see also Kiran and Sen [1995]). Kiran and Gokmenoglu [1995] also published viscosity values for Polyethylene solutions subjected to high pressures.

Chen and Swift [1972] analyzed the "entrance" and "exit" effects in a falling cylinder viscometer for creeping and non-creeping flow, giving a numerical correction for incompressible Newtonian fluid. End effects occurring in the falling cylinder viscometer were also analyzed by Wehbeh and Hussey [1993]; experimental data for closed and open tubes were also presented. A theoretical and experimental study allowing for the prediction of end effects is due to Gui and Irvine [1994]; the flow field is obtained numerically (see also Gui and Irvine [1996]). A superposition technique was utilized by Park and Irvine [1995] to account for the end effects of a flat ends falling cylinder viscometer; the technique is applicable for Newtonian fluids. Park and Irvine [1997] also gave a method to simultaneously determine the density and viscosity of the liquid by using needles of three distinct densities. The derivation of the "exact" solution for the motion of a liquid flowing past a falling cylinder with a frontal spherical end was analyzed by Borisov [1998] with some assumptions concerning the front shape of the cylinder and liquid velocity. The finite element study of a uniform flow past a needle in a cylindrical tube for materials having different constitutive equations is due to Phan-Thien *et al.* [1993].

In the chapter we present a simple theory of the flow of a viscous fluid in a falling cylinder viscometer according to Cristescu *et al.* [2002], and Cristescu [2005]. The velocity profile is obtained in finite form for both open tube and closed tube. Also in finite form is obtained a formula for the determination of the viscosity coefficient. These formulae contain also a term describing the influence of a magnetic field on the motion of the falling cylinder and of the fluid. How this term is used in viscosity measurements will be described in future papers. Since all the obtained formulae are in finite form, a parametric study (determination of the influence of various factors involved) of the fluid flow in the falling cylinder viscometer can be done quite easily. The comparison of the determination of the viscosity parameter according to the present theory with viscosity determined with a cone-plate viscometer is quite good.

Microrheometer. The most common viscometers and rheometers presently in use are those based on the measurement of stress on a fixed surface while a parallel or opposing surface is moved with a known applied strain rate. However, as already mentioned, these devices require fairly large sample volumes. Often it is difficult, if not impossible, to collect large volumes of samples for testing. In response to this chalange, Tran-Son-Tay *et al.* [1988] developed an acoustically tracked falling ball

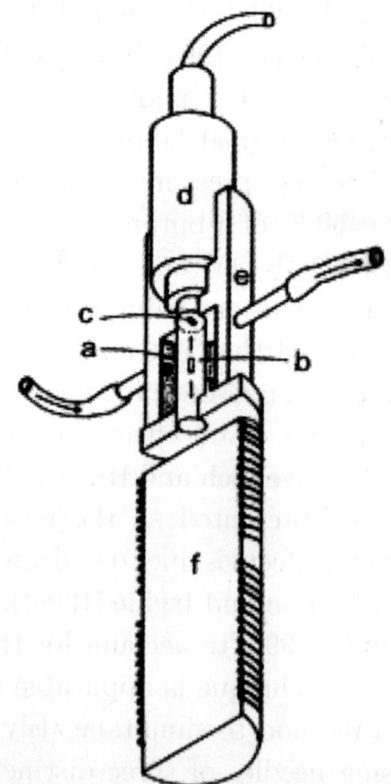

Fig. 6.4.1 Schematic of Microrheometer (a-sample tube, b-cylinder, c-piezoelectric crystal, d-ultrasound transducer, e-water jacket, f-electromagnet).

rheometer (Microrheometer). The Microrheometer (Fig. 6.4.1) is a unique device that was designed to measure the viscosity and viscoelasticity of small samples of biological fluids by using a spherical steel ball that is concentrically located in a small cylindrical tube with a volume of about 20 μl. The ball falls either under the force of gravity or is levitated and oscillated with a force produced by a magnetic field. The position of the ball is tracked by ultrasonic pulse-echo method. Furthermore, the small volume permits accurate temperature control and rapid temperature changes to be effected in the sample under study. Modifications have been made to the original setup so that a constant shear rate will be exhibited across the fluid, that is, a falling cylinder is used in place of a ball.

A 20 μl sample is loaded, by retrograde injection, into a disposable glass tube with an inner diameter of 1.6 mm and a height of 10 mm. Once loaded, the tube is centered inside a cylindrical, Plexiglas, water-jacket chamber. The ultrasonic transducer is housed at the bottom center of the Plexiglas chamber. An O-ring forms a watertight seal with the base of the sample tube and with the transducer. A plastic cap is screwed into place at the top of the chamber and provides a seal that completely protects the sample. A schematic of the Microrheometer is given in Fig. 6.4.1. The exact size of the tube and falling cylinder are given in Fig. 6.4.2. One can see that the falling cylinder is very large, as compared with the tube. That is very important for the approach.

A small electromagnet coupled with a micromanipulator is used to position and drop a 1.215 mm diameter cylinder in the center of the tube, where the strongest echo occurs. The pulse-echo mode is used to locate and track the falling cylinder. A single sound pulse is transmitted into the fluid medium by pulsing an ultrasound

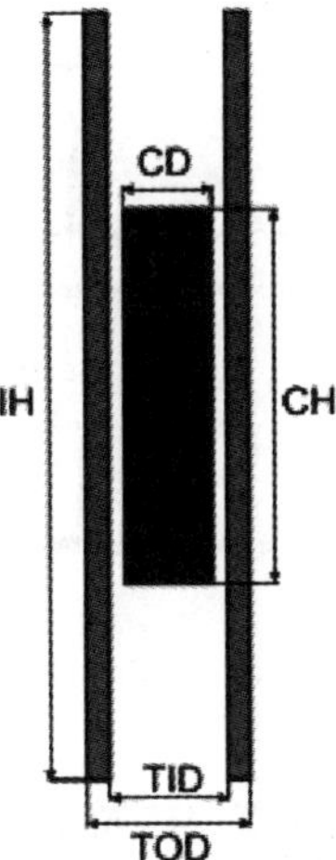

Fig. 6.4.2 The dimensions of the falling cylinder and tube; CD-cylinder diameter 1.215 mm, CH-cylinder height 4.90 mm, TID-tube inner diameter 1.61 mm, TOD-tube outer diameter 2.21 mm, IH-tube height 10.0 mm.

transducer that also acts as a receiver. Any returning echoes from the cylinder cause a voltage rise across the transducer that is amplified by the ultrasonic pulse/receiver unit. The time for the sound waves to travel to and back from the cylinder, and the speed of sound in the fluid are measured. From that information, the distance between the transducer and the cylinder, i.e., the location of the falling cylinder is determined.

Two types of measurements can be performed with our Microrheometer: (1) a speed-of-sound evaluation, and (2) a steady-state viscosity. The parameters of interest are c, the local speed of sound in the medium and η, the bulk suspension viscosity.

In order to measure the speed of sound in the fluid, the instrument needs to be calibrated. To accomplish this, the sample chamber is first filled to the top with distilled water and then capped with a glass cover slip to assure a fixed sample height h.

Once the chamber height is determined, the water is removed and replaced by the fluid sample. The local speed of sound in the medium c, is then determined. The fluid viscosity is determined from the speed of sedimentation of the falling cylinder. Details of the theory, including the use of a magnetic force to pull on the

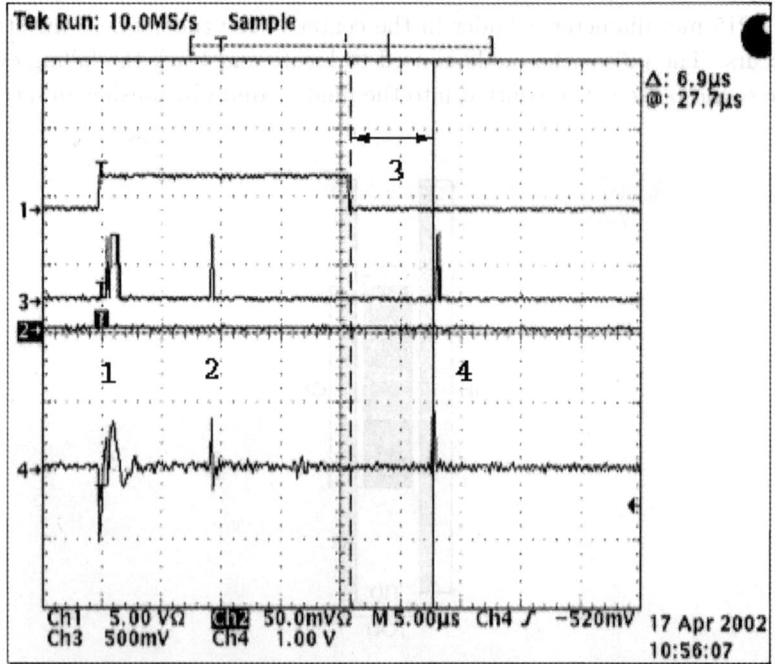

Fig. 6.4.3   Main bang in the large signal on the left, the echo of the cylinder is the smaller signal on the right (at 3).

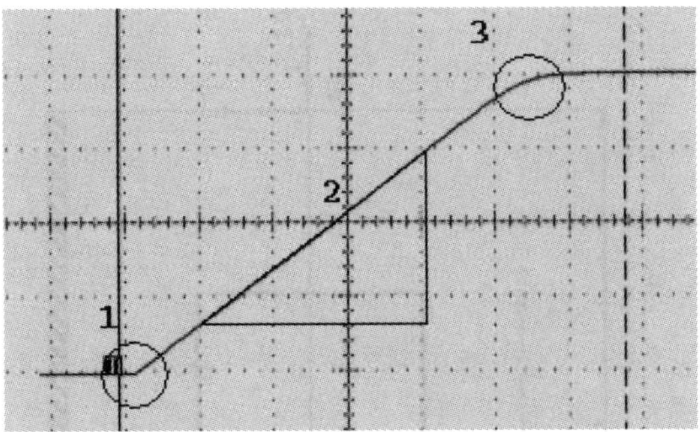

Fig. 6.4.4 Falling of the cylinder with constant velocities (on 3). The measurements starts at point 1 and end at 3, where the cylinder reaches the bottom.

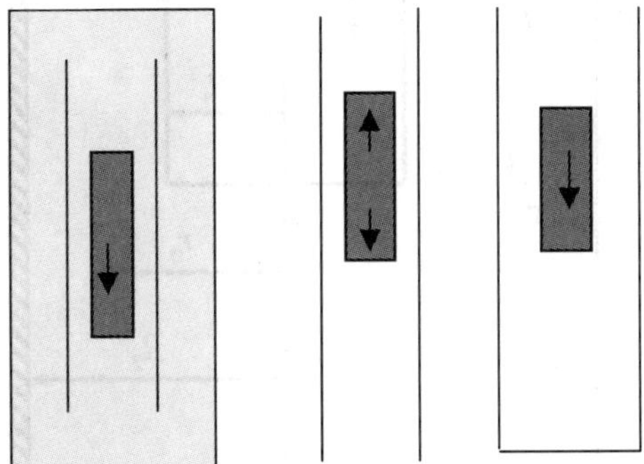

Fig. 6.4.5 Three approaches of the theory: falling in an infinite tube, moving up or down under a pressure gradient, and falling in a closed tube.

cylinder, are provided in the following section. The main bang followed by the pulse reflected from the top of the cylinder is shown in Fig. 6.4.3. The time for the sound waves to travel to and back from the cylinder, and the speed of sound in the fluid is measured. That is shown also in Fig. 6.4.4 where is shown that on the portion considered, the reading is quite linear.

That is shown in Fig. 6.4.5. That is why the theory is based on three successive approaches. First we study the flow in an infinite open tube (see the first Fig. 6.4.5). Then is studied the flow in an infinite tube under a pressure gradient (second Fig. 6.4.5). Finally at the end we study the flow in a finite tube, having in mind that a part of the fluid is flowing up while some is flowing down with the cylinder (last Fig. 6.4.5).

Theory. In what follows we use cylindrical coordinates (see Fig. 6.4.6) and the notation shown in this figure. We use: ρ_c-cylinder density, ρ-fluid density, v_1-velocity of the cylinder.

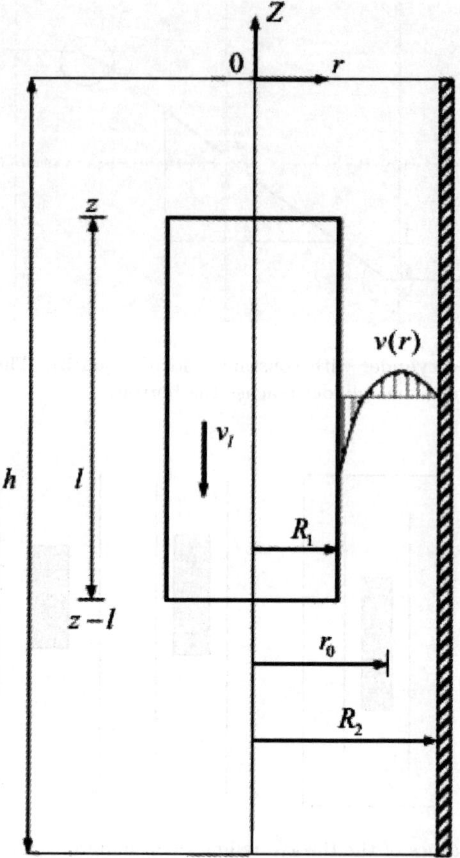

Fig. 6.4.6   Schematic of the falling cylinder viscometer and notations used.

Assumptions made are:

A1. The flow is laminar and telescopic; the velocity components are

$$v_r = v_\theta = 0, \quad v_z = v = f(r) \tag{6.4.1}$$

with $r = \sqrt{x^2 + y^2}$.

A2. The boundary conditions at the cylinder wall:

$$t \geq 0, \quad r = R_1 : v(R_1) = -v_1 \tag{6.4.2}$$

i.e. the fluid adders to the cylinder wall.

A3. The boundary condition at the tube wall

$$t \geq 0, \quad r = R_2 : v(R_2) = 0 \tag{6.4.3}$$

i.e., the fluid adheres to the tube wall.

A4. The fluid is a Newtonian viscous fluid

$$T'_{ij} = 2\eta D_{ij} \tag{6.4.4}$$

where T'_{ij} is the Cauchy stress deviator, and η is the viscosity coefficient, which may depend on z, if the fluid is non-homogeneous.

Taking into account (6.4.1) the only non-zero strain rate is

$$D_{rz} = \frac{1}{2}f'(r) \tag{6.4.5}$$

and from (6.4.4) we get for the stresses

$$T_{rr} = T_{\theta\theta} = T_{zz} = \sigma\,, \quad T_{\theta z} = T_{\theta r} = 0\,, \quad T_{rz} = \eta(z)f'(r)\,. \tag{6.4.6}$$

Since v is varying with r and assuming that this variation is of the kind shown in Fig. 6.4.6, we expect for T_{rz}:

$$T_{rz} = \eta(z)\frac{\partial v}{\partial r} \begin{cases} \text{positive if} \ \ R_1 < r < r_0 \\ \text{negative if} \ \ r_0 < r < R_2 \\ \quad 0 \ \ \text{for} \ \ r = r_0 \end{cases} \tag{6.4.7}$$

if at $r = r_0$ the velocity reaches a maximum and the stress T_{rz} is zero.

From the equilibrium equations written in cylindrical coordinates, since the problem is with cylindrical symmetry, follows for our case

$$\frac{\partial \sigma}{\partial r} = \frac{\partial \sigma}{\partial \theta} = 0\,, \quad \frac{\partial T_{rz}}{\partial r} + \frac{\partial \sigma}{\partial z} + \frac{T_{rz}}{r} + \rho b_z = 0 \tag{6.4.8}$$

where b_z is the body force component. Thus $\sigma(z)$ depends on z alone, and from the last equation (6.4.8) and (6.4.6) we get

$$\eta\frac{d(rf')}{dr} + \left(\frac{d\sigma}{dz} + \rho b_z\right)r = 0\,. \tag{6.4.9}$$

Integrating once with respect to r, we get

$$T_{rz} = -\left(\frac{d\sigma}{dz} + \rho b_z\right)\frac{r}{2} + \frac{C_1}{r}\,. \tag{6.4.10}$$

Integrating a second time with respect to r we obtain

$$\eta f + \left(\frac{d\sigma}{dz} + \rho b_z\right)\frac{r^2}{4} = C_1 \ln r + C_2\,. \tag{6.4.11}$$

The integration constants can be determined from the boundary conditions A2 and A3

$$C_1 = \frac{\eta v_1 + (d\sigma/dz + \rho b_z)(R_2^2 - R_1^2)/4}{\ln(R_2/R_1)}\,,$$

$$C_2 = \left(\frac{d\sigma}{dz} + \rho b_z\right)\frac{R_2^2}{4} - \frac{\eta v_1 + (d\sigma/dz + \rho b_z)(R_2^2 - R_1^2)/4}{\ln(R_2/R_1)}\ln R_2\,. \tag{6.4.12}$$

Thus the velocity distribution is obtained as

$$\eta v = \left(\frac{d\sigma}{dz} + \rho b_z\right)\left(\frac{R_2^2 - r^2}{4}\right) + \frac{1}{\ln(R_2/R_1)}\left[\eta v_1 + \left(\frac{d\sigma}{dz} + \rho b_z\right)\frac{R_2^2 - R_1^2}{4}\right]\ln\frac{r}{R_2}$$

$$(6.4.13)$$

if the viscosity coefficient is known and $d\sigma/dz$ is the pressure gradient along the tube.

Let us write now a global equilibrium condition for the cylinder: the projections on the z-axis of all forces acting on the cylinder are zero; i.e., *gravitational force + buoyancy force + shearing force on the cylinder wall* $= 0$. Thus

$$T_{rz}|_{R_1} = \frac{R_1}{2}g(\rho_c - \rho + m) \qquad (6.4.14)$$

with m the "local density" of the magnetic force. With (6.4.9) this relation becomes

$$\eta v_1 = \left(\frac{d\sigma}{dz} + \rho b_z\right)\left[\frac{R_1^2}{2}\ln\frac{R_2}{R_1} - \frac{R_2^2 - R_1^2}{4}\right] + \frac{R_1^2}{2}\ln\frac{R_2}{R_1}[g(\rho_c - \rho + m)]$$

$$(6.4.15)$$

which is a relation between ηv_1, $(d\sigma/dz) + \rho b_z$ and m.

We can examine now the velocity profile for open tube. First let us consider an open vertical tube (or an "infinite" tube) in which can move a heavy cylinder

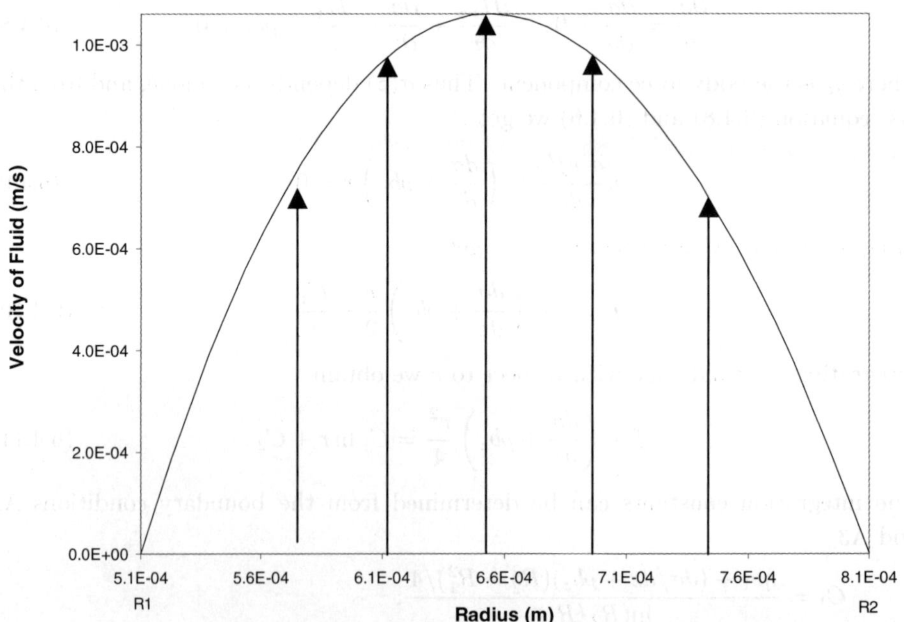

Fig. 6.4.7 Velocity of the fluid moving upward under the pressure gradient (6.4.16) which keeps the cylinder stationary.

and a pressure gradient exists. Let us assign various values to the pressure gradient $d\sigma/dz$ in order to find out the various possible velocity profiles.

Example 1. From (3.14) follows that if the pressure gradient satisfies the relation

$$\frac{d\sigma}{dz} + \rho b_z = \frac{-(R_1^2/2)\ln(R_2/R_1)[g(\rho_c - \rho + m)]}{(R_1^2/2)\ln(R_2/R_1) - (R_2^2 - R_1^2)/4}$$

the cylinder is stationary ($v_1 = 0$). In other words under this pressure gradient the shearing forces of the fluid flowing upwards will keep stationary the cylinder. Figure 6.4.7 shows an example computed for $\eta = 0.1014$ N s/m^2 (101.4 P) and

$$\frac{d\sigma}{dz} + \rho b_z = 9.36 \times 10^4 \text{ Pa m}^{-1}. \tag{6.4.16}$$

The velocity distribution shown was obtained with (6.4.12).

Example 2. If the pressure gradient is greater than that obtained from (3.16), the shearing force of the fluid flowing up is able to move upwards the cylinder. In order to give an example, Fig. 6.4.8 is showing the velocity profile of the fluid for $(d\sigma/dz) + \rho b_z = 1.2 \times 10^5$ Pa $\times$ m^{-1}. If this term reaches the value 1×10^6 Pa $\times$ m^{-1} the velocity of the fluid in contact with the cylinder is equal with that of the cylinder itself. This case can be obtained from (3.12) by putting the condition that the maximum velocity of the fluid be reached for $r = R_1$.

Example 3. This example considers the most common case when term $(d\sigma/dz) + \rho b_z$ has a smaller value than that obtained from (3.15). Figure 6.4.9 shows an example obtained with the value $(d\sigma/dz) + \rho b_z = 8.0 \times 10^4$ Pa m^{-1} and same η as before. In this case most of the fluid is moving upwards, but a thin layer of fluids neighboring the cylinder, the fluid is moving downwards. In the present example $|v_1| = 0.000528$ m/s (down) is obtained from (6.4.14).

Example 4. The last example corresponds to very small values for $(d\sigma/dz) + \rho b_z$. Let us consider the case when the pressure gradient is zero $d\sigma/dz = 0$. Introducing this value in (6.4.14) together with $\rho g = 9.807 \times 10^3$ (kg/m^2 s^2) and same value for η as above, we get $v_1 = -0.00363$ m/s (down). The velocity profile follows from (6.4.12). This time the falling cylinder induces into the fluid a shearing stress which will put it to flow downwards (see Fig. 6.4.10). The velocity of the fluid in contact with the tube is certainly zero.

Velocity profile for closed tube. Let us consider now the case when the bottom of the tube is closed and the pressure gradient is generated by the falling cylinder. We can write down the condition that the volume of the fluid displaced by the falling cylinder is equal to the volume of the fluid flowing (up and down) between the cylinder and the tube:

$$\int_{R_1}^{R_2} v 2\pi r \, dr = v_1 \pi R_1^2. \tag{6.4.17}$$

 Dynamic Plasticity

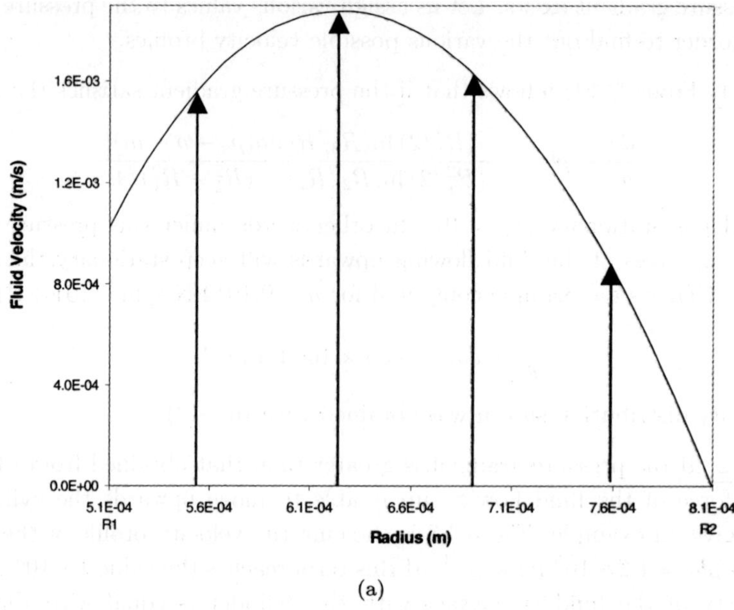

(a)

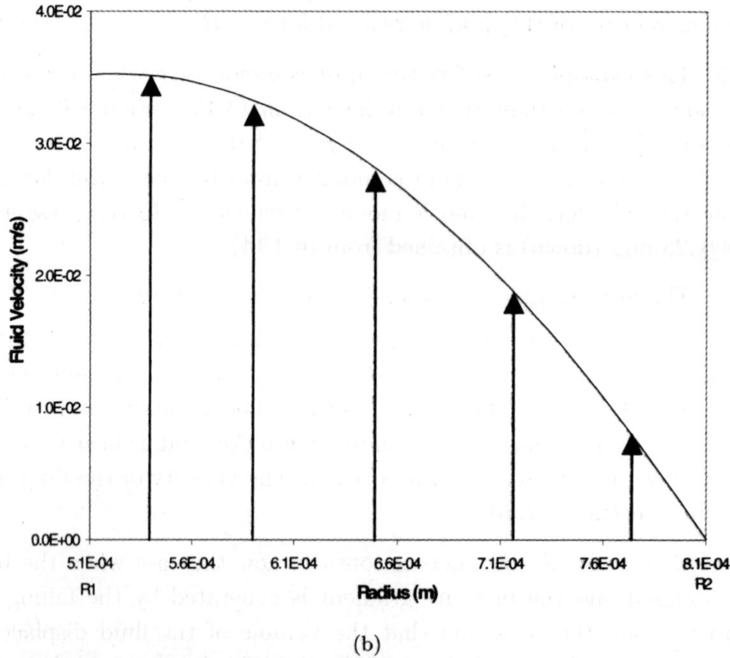

(b)

Fig. 6.4.8   Velocity profile from the fluid flowing upwards under a pressure gradient greater than (6.4.16); the fluid pushing the cylinder upwards: (a) the pressure gradient slightly bigger than (6.4.17); (b) pressure gradient bigger than (6.4.16) when the velocity of the cylinder is equal with the maximum velocity of the fluid (and a small plateau exists).

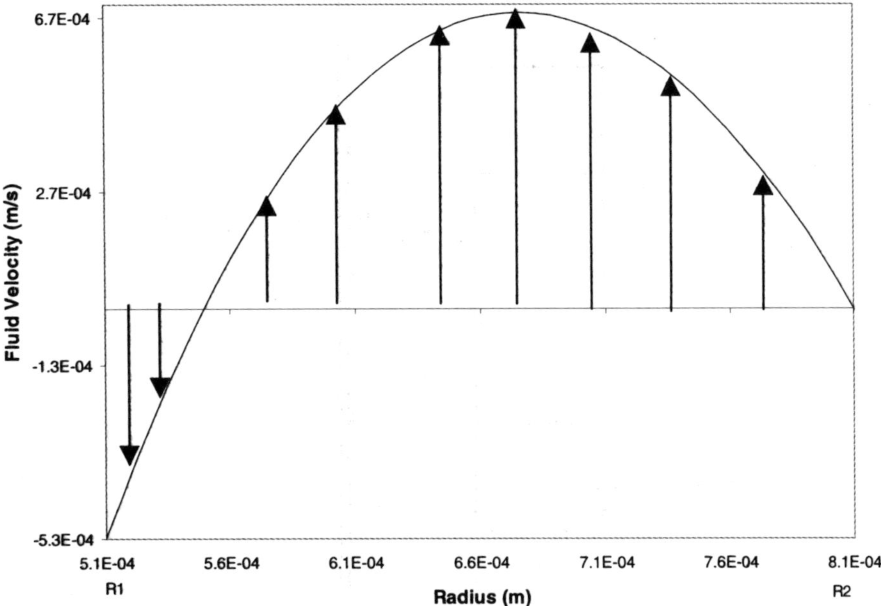

Fig. 6.4.9   Under a small pressure gradient, part of the fluid is moving up, but part is moving down with the falling cylinder.

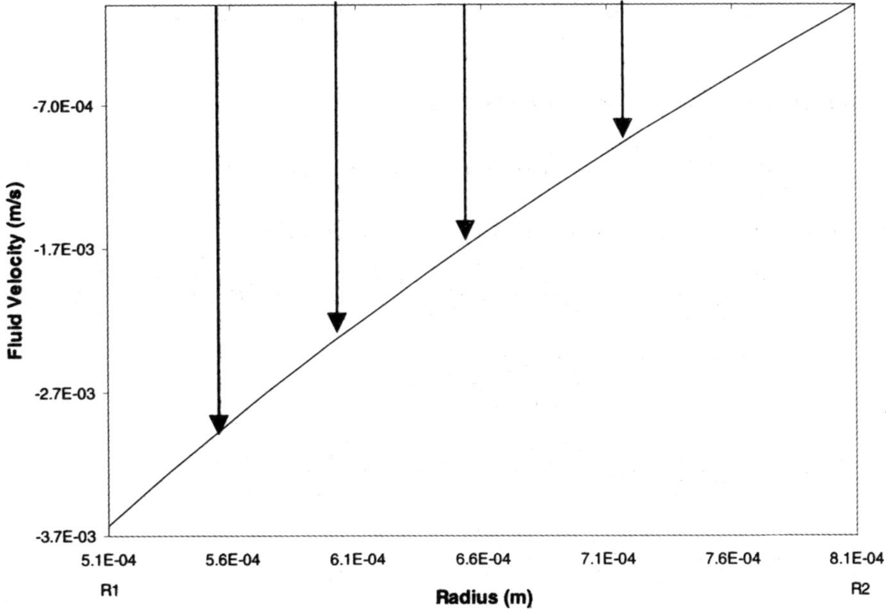

Fig. 6.4.10   The velocity profile in the fluid generated by the falling cylinder (pressure gradient is zero).

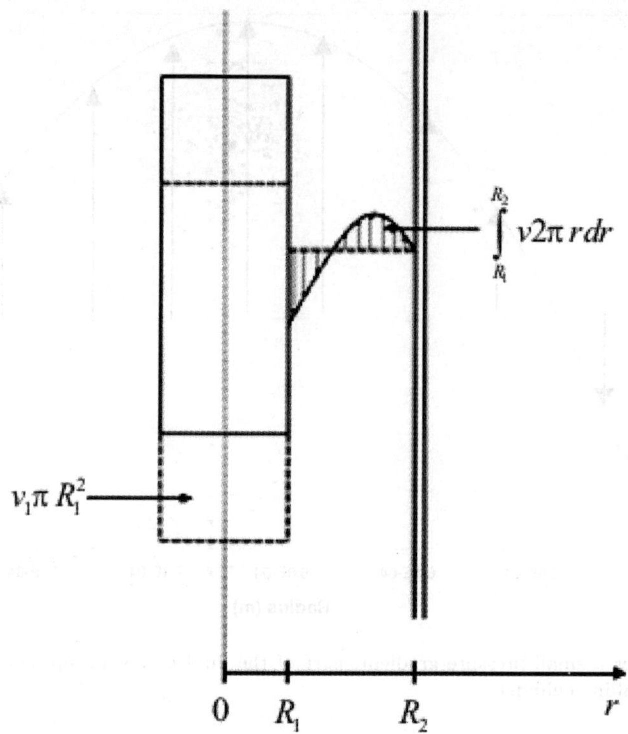

Fig. 6.4.11   Schematic of the fluid flow in the case of closed tube.

Introducing here $v(=f)$ from (6.4.11) we get

$$\frac{2}{\eta}\int_{R_1}^{R_2}\left[-\left(\frac{d\sigma}{dz}+\rho b_z\right)\frac{r^3}{4}+C_1 r\ln r+C_2 r\right]dr=v_1 R_1^2 \tag{6.4.18}$$

with the values of C_1 and C_2 from (3.14) and the notations

$$\Sigma=-\frac{R_2^4-R_1^4}{16}+\frac{R_2^2-R_1^2}{4\ln(R_2/R_1)}\left[R_2^2\left(\frac{\ln R_2}{2}-\frac{1}{4}\right)-R_1^2\left(\frac{\ln R_1}{2}-\frac{1}{4}\right)\right]$$

$$+\frac{R_2^2}{4}\frac{R_2^2-R_1^2}{2}-\frac{(R_2^2-R_1^2)^2}{8\ln(R_2/R_1)}\ln R_2$$

$$V=-\frac{1}{\ln(R_2/R_1)}\left[R_2^2\left(\frac{\ln R_2}{2}-\frac{1}{4}\right)-R_1^2\left(\frac{\ln R_1}{2}-\frac{1}{4}\right)\right]$$

$$+\frac{\ln R_2}{\ln(R_2/R_1)}\frac{R_2^2-R_1^2}{2}+\frac{R_1^2}{2} \tag{6.4.19}$$

we finally obtain

$$\frac{d\sigma}{dz}+\rho b_z=\frac{V}{\Sigma}\eta|v_1| \tag{6.4.20}$$

which established another relationship between ηv_1 and $(d\sigma/dz) + \rho b_z$, obtained from a kinematic assumption.

However, the term ηv_1 as obtained from (6.4.20) and from (6.4.15) must be equal. From this condition follows

$$\frac{d\sigma}{dz} + \rho b_z = \frac{(R_1^2/2)\ln(R_2/R_1)[g(\rho_c - \rho + m)]}{V/\Sigma - (R_1^2/2)\ln(R_2/R_1) + (R_2^2 - R_1^2)/4}. \tag{6.4.21}$$

This relation determines $(d\sigma/dz) + \rho b_z$. The magnetic force is involved as a "local density", i.e., if pushing down it is equivalent with a heavier cylinder. After calibration, this observation will be used to speed up some tests with very viscous fluids, and using the very same magnetic cylinder. Introducing this value into (6.4.20) we obtain a formula for the determination of the viscosity coefficient η:

$$\eta = \frac{\Sigma}{V|v_1|}\left(\frac{d\sigma}{dz} + \rho b_z\right). \tag{6.4.22}$$

We can consider the expression

$$\alpha = \frac{\Sigma}{V} \frac{(R_1^2/2)g\ln(R_2/R_1)}{\Sigma/V - (R_1^2/2)\ln(R_2/R_1) + (R_2^2 - R_1^2)/4}\beta \tag{6.4.23}$$

to be a viscometer constant. We have introduced β as a correction parameter, mainly because the cylindrical magnet has sometimes no perfect cylindrical shape, and also to take into account the end effects. Thus the formula for the determination of the viscosity coefficient (6.4.22) writes

$$\eta = \alpha\frac{\rho_c - \rho + m}{|v_1|}. \tag{6.4.24}$$

Shearing stress distribution. The shearing stress distribution in the layer of fluid is obtained from (6.4.10) with (6.4.12), as

$$T_{rz} = -\left(\frac{d\sigma}{dz} + \rho b_z\right)\frac{r}{2} + \frac{\eta v_1 + ((d\sigma/dz) + \rho b_z)(R_2^2 - R_1^2)/4}{\ln(R_2/R_1)}\frac{1}{r}. \tag{6.4.25}$$

The stress becomes zero there where the fluid velocity is maximum; $T_{rz} = 0$ or $v = v_{\max}$ for $r = r_0$. That is obtained from

$$-\left(\frac{d\sigma}{dz} + \rho b_z\right)\frac{r_0}{2} + C_1\frac{1}{r_0} = 0$$

i.e.,

$$r_0 = \sqrt{\frac{2C_1}{(d\sigma/dz) + \rho b_z}} \tag{6.4.26}$$

which determines r_0. From the numerical examples given in this paper one can see that r_0 depends strongly on the pressure gradient.

Example 5. In order to check the prediction of the above formulae, we have performed experiments with a variety of standardized fluids available in our laboratory.

Table 6.4.1   Summary statistics of all samples computed both experimentally and theoretically.

Fluid	Viscosity as measured on the Brookfield (cp)	Number of Falling Cylinder samples	Average Falling Cylinder Velocity (m/s)	Average Viscosity as determined from the Theory (cp)	Standard deviation of Theoretical Value	Accuracy % (theoretical value-Brookfield value)/Brookfield value)	Variability % (standard deviation/ Measured value)
1	1005	16	5.25E-04	1067	87	6.169154229	8.153701968
2	3085	17	1.71E-04	3274	241	6.126418152	7.361026268
3	9549	37	6.05E-05	9442	1564	−1.120536182	16.56428723

The tests were done with a custom made falling cylinder viscometer. The viscometer data are:

$$R_1 = 0.000505 \text{ m (cylinder radius)}$$

$$R_2 = 0.000805 \text{ m (container inner radius)}$$

$$\rho_c = 7228 \text{ kg/m}^3 \text{ (cylinder density)}$$

$$\rho_f = 1000 \text{ kg/m}^3 \text{ (fluid density)}.$$

A sample of standard 1000 cP calibration fluid was measured to verify its viscosity using a cone-plate viscometer (Brookfield, MA). We have tested 17 samples of this calibration fluid and determined a coefficient of viscosity $\mu = 1067 \pm 87$ cP compared with 1005 cP as determined by the cone-plate viscometer. The experimental viscosity coefficient is 6.17% higher than the cone-plate value. Though the matching is quite good, the existing difference may be due either to the disregarding in the theory of the end effects, or/and to the non-perfect cylindrical shape of the falling cylinder. However, for all practical measurements of viscosities, the accuracy is quite good. The data for all sample fluids is summarized in Table 6.4.1.

In order to make another check of the prediction of the above formulae, we use the experimental data given by Park and Irvine [1997] for the Canon S-60 standard

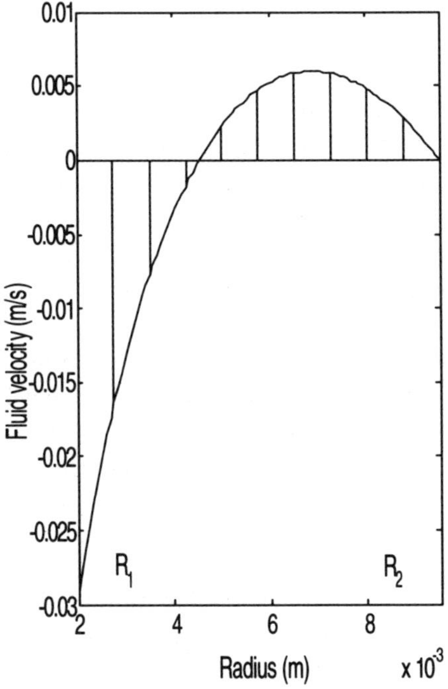

Fig. 6.4.12   Velocity profile obtained for $v_1 = 0.02903$ m/s.

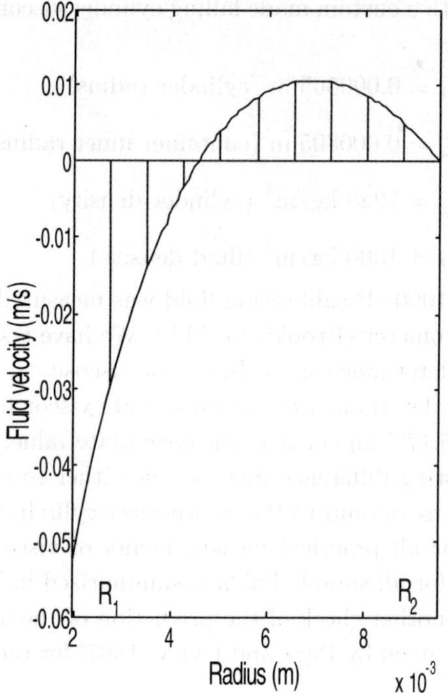

Fig. 6.4.13   Velocity profile for $v_1 = 0.05251$ m/s.

fluid. The tests done for the falling needle viscometer are: $R_1 = 1.99 \times 10^{-3}$ m (needle diameter); $R_2 = 9.525 \times 10^{-3}$ m (container diameter); $L = 10.2$ cm (needle length); $\rho = 0.8589 \times 10^3$ kg/m^3 (fluid density). For the needle density $\rho_c = 1.1019 \times 10^3$ kg/m^3 the terminal falling velocity recorded was $v_1 = 0.02903$ m/s. We have determined a coefficient of viscosity $\eta = 115$ cP as compared with 101.4 cP given in the Cannon specifications. It is 13% higher. The flow of the fluid is shown in Fig. 6.4.12.

For the needle density $\rho_c = 1.2998$ kg/m^3 the terminal falling velocity was found $v_1 = 0.05251$ m/s. The determined viscosity coefficients is $\eta = 116$ cP to be compared with 101.4 cP. The velocity profile is shown in Fig. 6.4.13. Thus the determined viscosity coefficients are about 14% higher than the Cannon specification value.

Conclusion.

A simple theory is presented in which the solution of the flow of a viscous fluid in a falling cylinder viscometer is expressed in a closed form. The formula giving the viscosity of the fluid is also obtained in finite form. Since all the formulae are obtained in finite forms, one can easily study the influence of all parameters involved [as densities (maybe variable), geometry of cylinder-tube, pressure gradient, magnetic force, etc.] on the cylinder-fluid motion. The comparison of the viscosity coefficients, as determined by our experiments with the falling cylinder micro-rheometer and using the present theory, and as determined with a classical cone-plate viscometers is quite good.

It is important to note that the formulae for velocity, stress, viscosity, etc. contain a term that describes the influence of a magnetic field on the motion of the cylinder and thus on the flow of the fluid. However, the influence and use of a magnetic field in viscometer studies will be treated in a future paper. An extension of the analysis to nonhomogeneous fluids, viscoplastic materials, and to the case when the fluid slides along the tube and/or cylinder is to be considered in the future.

Further developments. An extension of the theory for Non-Newtonian fluids was done by Tigoiu and Cipu [2005]. They have in mind that some biological fluids are of this kind. The velocity, in physical cylindrical components is:

$$v_r = 0, \quad v_\theta = 0, \quad v_z = V_1 f(r), \quad r \in [R_1, R_2]. \tag{6.4.27}$$

The corresponding boundary conditions on both the falling cylinder and the pipe are

$$v_z(R_1) = -V_1, \quad v_z(R_2) = 0. \tag{6.4.28}$$

It is well known that the velocity field (6.4.27) is associated to a viscometric motion and therefore Cauchy's stress tensor is given by

$$\mathbf{T}(\mathbf{x}, t) = -p\mathbf{I} + \tilde{\mathbf{T}}(\mathbf{A}_1, \mathbf{A}_2). \tag{6.4.29}$$

Here p is the pressure field, $\mathbf{I}$ is the identity tensor, $\mathbf{A}_1, \mathbf{A}_2$ are the first two Rivlin–Ericksen tensors. The effective stress tensor $\tilde{\mathbf{T}}$ should be given in explicit forms in next sections. After some calculi we obtain from (6.4.27)

$$\begin{aligned}
(\mathbf{A}_1)_{rz} &= V_1 f', \\
(\mathbf{A}_2)_{rr} &= 2V_1^2 f'^2, \\
(\mathbf{A}_1\mathbf{A}_2 + \mathbf{A}_2\mathbf{A}_1)_{rz} &= 2V_1^3 f'^3, \\
\mathbf{A}_3 &= \mathbf{0}
\end{aligned} \tag{6.4.30}$$

and all other components of $\mathbf{A}_1, \mathbf{A}_2$ are null.

We pass to non-dimensional variables and functions by

$$z = H\bar{z}, \quad r = R_1\bar{r}, \quad f = \bar{f}, \quad p = p_0\bar{p}, \quad \sigma_z = \rho_1 g \bar{\sigma}_z \quad \text{and} \quad \tilde{T} = T_0\bar{T} \tag{6.4.31}$$

where we denote $T_0 \equiv \eta_0(V_1/R_1)$. Here η_0 is a characteristic shear viscosity. We remark that the velocity field (6.4.28) satisfies the continuity equation and then the flow problem can be written, after using (6.4.31), as

$$h'(r) = \frac{1}{Re}\frac{1}{r}\frac{d}{dr}\left(r\tilde{T}_{rr}(r)\right),$$

$$\frac{1}{r}\frac{d}{dr}\left(rT_{rz}(r)\right) = Re\,\sigma_z\,, \tag{6.4.32}$$

$$f(1) = -1\,, \quad f\left(\frac{R_2}{R_1}\right) = 0\,,$$

where $Re \equiv (\rho_1 g R_1^2/\eta_0 V_1)$ is a modified Reynolds number, described by means of characteristic quantities of our problem (ρ_1 being the fluid mass density) and where we have suppressed the overlines. For this end we remark that $(\partial p/\partial\theta) = 0$. At this level of generality, the arguments are given by (6.4.29), (6.4.30) and Wang's representation theorem for isotropic functions. Consequently $p = C_0 z + h(r)$. In (6.4.32) we have denoted $\sigma_z = (p_0/\rho_1 g H)(\partial p/\partial z) + 1$ the "modified pressure gradient". We remark that equation (6.4.32)$_1$ leads to the determination of the pressure field, once the effective stress tensor components are known. The above mentioned components will be determined after solving the boundary value problem (6.4.32)$_2$–(6.4.32)$_3$. For this, a first integral is simply obtained from (6.4.32) and leads to the determination of the shear stress component which, in dimensional variables, is

$$\tilde{T}_{rz}(r) = \frac{1}{2}\sigma_z r + \frac{V_1 C_1}{r}\eta_0 = \frac{1}{2}\left(\frac{\partial p}{\partial z} + \rho_1 g\right)r + \frac{C_1}{r}\eta_0 V_1\,. \tag{6.4.33}$$

The dimensional constant C_1 will be independently determined for each investigated model.

We remark here that the necessary condition of having a constant velocity (for the falling cylinder) is given by the global equilibrium of forces acting on the cylinder (gravitational force, buoyancy force, shearing force and magnetic force). Some elementary calculi lead to the following formula (ρ_c being the mass density of the falling cylinder)

$$\tilde{T}_{rz}(R_1) = \frac{1}{2}(\rho_c - \rho_1 + f_m)\,gR_1\,. \tag{6.4.34}$$

Employing (6.4.33) and (6.4.34) we obtain

$$\eta_0 V_1 = \frac{1}{2C_1}(\rho_c - \rho_1 + f_m)\,g - \sigma_z\}\,R_1^2\,. \tag{6.4.35}$$

This formula is a compatibility condition between the geometry, material and flow parameter (connecting V_1, f_m, η_0 and σ_z) and which leads to the determination of the necessary magnetic field density f_m.

We remind, that the shear viscosity η on viscometric flows is given by

$$\eta(k) = \frac{\tau(k)}{k}\,, \tag{6.4.36}$$

where k stands for the shear rate and $\tau(k)$ for the shear stress. Formula (6.4.36) is the support for the T_0 expression in (6.4.31).

The Third Grade Fluid Case. For the third grade fluid we employ the effective stress tensor $\tilde{\mathbf{T}}$ and constitutive restrictions as obtained in Tigoiu [1987]

$$\tilde{\mathbf{T}}(\mathbf{A}_1, \mathbf{A}_2, \mathbf{A}_3) = \mu \mathbf{A}_1 + \alpha_1(\mathbf{A}_2 - \mathbf{A}_1^2) + \beta_1 \mathbf{A}_3$$

$$+ \beta_2(\mathbf{A}_1 \mathbf{A}_2 + \mathbf{A}_2 \mathbf{A}_1) + \beta_3(\mathrm{tr}\mathbf{A}_1^2)\mathbf{A}_1 , \qquad (6.4.37)$$

$$\mu \geq 0, \quad \beta_1 < 0, \quad \beta_1 + 2(\beta_2 + \beta_3) \geq 0,$$

where μ, α_1 and β_i, $i = 1, 2, 3$ are constant constitutive moduli.

Use of (6.4.37) in flow equation $(6.4.32)_2$ leads, after a first integration, to the following non-dimensional problem

$$f' + \beta f'^3 = \frac{1}{2} Re\sigma_z r + C_1 \frac{1}{r}, \quad f(1) = -1, \quad f(R_2/R_1) = 0, \qquad (6.4.38)$$

where the left term in equation $(6.4.37)_1$ is the non-dimensional shear component of Cauchy's stress tensor.

We see that this problem has a unique solution

$$f'(r) = \frac{1}{(2\beta)^{1/3}} \left\{ \sqrt[3]{C + \sqrt{C^2 + \frac{4}{27\beta}}} + \sqrt[3]{C - \sqrt{C^2 + \frac{4}{27\beta}}} \right\}, \qquad (6.4.39)$$

where

$$C \equiv C(r) \equiv ar + \frac{C_1}{r}, \quad a \equiv \frac{1}{2} Re\sigma_z \quad \text{and} \quad \beta \equiv \frac{2(\beta_2 + \beta_3)}{\mu R_1^2} V_1^2 .$$

We remark here that, the normal force acting on the outer cylinder is given, in non-dimensional form, by

$$T_{rr}(R_2/R_1, z) = -Re\, p(R_2/R_1, z) + \alpha f'^2(R_2/R_1) . \qquad (6.4.40)$$

The explicit formula for T_{rr} is obtained if we introduce $p(r, z) = C_0 z + h(r)$ where $h(r)$ is obtained from

$$h'(r) = \frac{\alpha}{Re} \frac{1}{r} \frac{\partial}{\partial r}(r f'^2(r)) \qquad (6.4.41)$$

and (6.4.39). In formulae (6.4.40), (6.4.41) α stands for the non-dimensional constitutive moduli α_1, $\alpha \equiv (\alpha_1 V_1/\mu R_1)$.

The solution $f(r)$ is finally obtained from (6.4.39) + $(6.4.38)_2$, after applying a shooting method (for determination of the constant C_1), by numerical integration. The results are plotted.

The Second Order Fluid Case. We consider now a falling cylinder viscometer filled with a second order fluid with the shear viscosity depending on the shear rate. The corresponding constitutive law for the effective stress is given by

$$\tilde{\mathbf{T}}(\mathbf{A}_1, \mathbf{A}_2) = \eta(\mathrm{tr}\, \mathbf{A}_1^2)\mathbf{A}_1 + \alpha_1(\mathbf{A}_2 - \mathbf{A}_1^2) , \qquad (6.4.42)$$

where we generally suppose that η is a positive, bounded and decreasing C^2 function (defined on a compact set of positive numbers). For the numerical application we consider

$$\eta(\operatorname{tr} A_1^2) = \eta_0 \exp\left(-A(\ln 10)\left\{\left[\frac{1}{2}\log(f'^2) + 1.7897\right]^2\right\}\right).$$

Here $\eta_0 = 10^{2.8}$ Pa s is supposed to be the zero shear rate viscosity.

Equation $(6.4.32)_1$ is now given in dimensional form by

$$h'(r) = V_1^2 \alpha_1 \frac{1}{r}\frac{\partial}{\partial r}(r f'^2(r)). \tag{6.4.43}$$

After obtaining f from the corresponding boundary value problem for $(6.4.32)_2$–$(6.4.32)_3$, the pressure field will be determined from (6.4.43).

We consider, as in the previous section, the non-dimensional flow problem

$$\frac{\partial}{\partial r}[r\eta(\operatorname{tr} A_1^2)f'(r)] = \frac{\bar{\sigma}_z}{Wi}r, \quad f(1) = -1; \quad f(R_2/R_1) = 0. \tag{6.4.44}$$

In (6.4.44) we denote with Wi the corresponding Weissenberg number given by $Wi = (V_1\eta_0/\rho_1 g R_1^2)$. Similar to the third grade fluid case, a first integral is obtained from $(6.4.44)_1$ as:

$$T_{rz}(r) \equiv \eta(\operatorname{tr} A_1^2)f'(r) = \frac{\sigma_z}{Wi}r + \frac{C_1}{r}. \tag{6.4.45}$$

With the above mentioned properties for η, we can easily see that the equation

$$\eta(X^2)X - g(r) = 0 \tag{6.4.46}$$

has a unique solution $X = \bar{X}(r)$. Consequently we can uniquely express from (6.4.45) the derivative of f as a continuously differentiable function of r. Finally it results that the considered problem (6.4.44) has a unique twice continuously differentiable solution. To obtain the final solution we will numerically solve the problem (6.4.44) for the mentioned solution of a polyisobutylene in decline.

Closed Domain Case. In this section we consider the case of a third grade fluid model. However, in the case of a linear viscous fluid the velocity field is analytically obtained. In almost all other cases we can not expect any analytical solutions. In order to put into evidence end effects (i.e., in a closed pipe), we simulate such effects by a kind of "volumetric flow rate" equilibrium. The result leads to the determination of a modified pressure gradient σ_z necessary for the above mentioned equilibrium. With this σ_z the characteristics of the velocity field (as solution of the new boundary value problem) are different from those of the velocity field which has been obtained in previous section. In order to do this, we write the balance of the volume of fluid (per unit time) pushed downwards by the falling cylinder with the volume of fluid (per unit time) flowing upwards and downwards between the pipe and the cylinder. Therefore we have

$$\pi R_1^2 V_1 = 2\pi V_1 \int_{R_1}^{R_2} r f(r)\, dr.$$

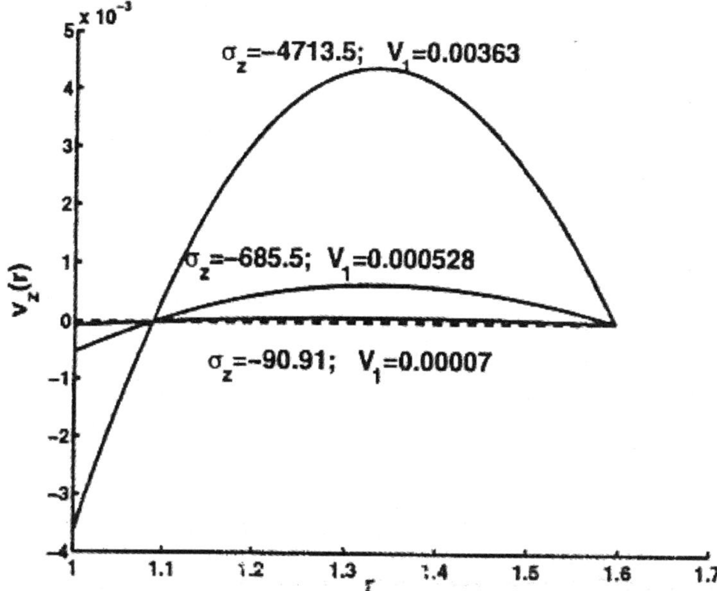

Fig. 6.4.14   Velocity fields for a closed domain for different values of computed apparent modified pressure gradient σ_z.

For a third grade fluid, the obtained values for σ_z (corresponding to $V_1 = 0.00007$ m s^{-1}, 0.000528 m s^{-1} and 0.00363 m s^{-1}, respectively) are $|\sigma_z| = 90.91$, 685.5 and 4713.5, respectively. The corresponding graphics for the velocity fields are shown in Fig. 6.4.14. In all graphics it is put into evidence a domain in neighborhood of the pipe in which an inverse flow occurs. The three domains have practically similar dimensions, that is the point of cut off with the axe $v_z = 0$ is almost the same in the three cases. However, the dimension of the domain is different from the case of a linear viscous fluid.

Bibliography

Aleksandrov S. E., Goldshtein R. V. and Lyamina E. A., 2003, Developing the concept of the strain rate intensity factor in plasticity theory, *Doklady Physics* **48**, 3, 131–133.

Alonso E. E., Gens A. and Lloret A., 1953, The landslide of Cortes de Pallas, Spain, *Géotechnique* **43**, 507–521.

Ashare E., Bird R. B. and Lescarboura J. A., 1965, Falling cylinder viscometer for non Newtonian fluids, *AIChE Jour.* **11**, 5, 910–916.

Avitzur B., 1968, *Metal Forming: Processes and Analysis*, Mc Graw Hill, New York.

Bertini T., Cugusi F., D'Elia B. and Rossi-Doria M., 1984, Climatic conditions and slow movements of colluvial covers in Central Italy, *Proc. IV Int. Symp. on Landslaides*, Toronto, Vol. I, 367–376.

Bird B. R., Armstrong R. C. and Hassager O., 1987, *Dynamics of Polymeric Liquids*, Wiley, New York.

Borisov V. B., 1998, The equation for a falling-cylinder viscometer, *High Temperature* **36**, 2, 293–298.

Bridgman P. W., 1926, The effect of pressure on the viscosity of forty-three pure liquids, *Proc. Amer. Acad. Arts Sci.* **61**, 56, 57–99.

Buckingham E., 1921, On plastic flow through capillary tubes, *Proc Amer. Soc. Testing Materials* **117**, 1154–1161.

Cazacu O. and Cristescu N., 2000, Constitutive model and analysis of creep flow of natural slopes, *Rev. italiana di geotecnica* **3**, Luglio-Settembre, 44–54.

Chan R. K. Y. and Jackson D. A., 1985, An automated falling-cylinder high pressure Laser-Doppler viscometer, *J. Phys. E: Sci. Instrum.* **18**, 510–515.

Chee K. K., Sato K. and Rudin A., 1976, Falling coaxial cylinder viscometer for polymer solutions, *Journal of Applied Polymer Science* **20**, 1467–1474.

Chen M. C. S., Lescarboura J. A. and Swift G. W., 1968, The effect of eccentricity on the terminal velocity of the cylinder in a falling cylinder viscometer, *AIChE Journal* **14**, No. 1, 123–127.

Chen M. C. S. and Swift G. W., 1972, Analysis of entrance and exit effects in a falling cylinder viscometer, *AIChE Journal* **18**, No. 1, 146–149.

Cristescu N., 1975, Plastic flow through conical converging dies using a viscoplastic constitutive equation, *Int. J. Mech. Sci.* **17**, 425–433.

Cristescu N., 1976, Drawing through conical dies. An analysis compared with experiments, *Int. J. Mech. Sci.* **18**, 1, 45–49.

Cristescu N., 1991, Nonassociated elastic/viscoplastic constitutive equations for sand, *Int. J. Plasticity* **7**, 41–64.

Cristescu N. D., 2005, *Theory of Falling Cylinder Viscometers* (in print).

Cristescu N. D., Conrad B. P. and Tran-Son-Tay R., 2002, A closed form solution for falling cylinder viscometers, *Int. J. Engn. Sci.* **40**, 605–620.

Cristescu N. D., Cazacu O. and Cristescu C., 2002, A model for slow motion of natural slopes, *Canadian Geotech. Journal* **39**, 4, 924–937.

Dandridge A. and Jackson D. A., 1981, Measurements of viscosity under pressure: a new method, *J. Phys. D: Appl. Phys.* **14**, 829–831.

Desai C. S., Samtani N. C. and Vulliet L., 1995, Constitutive modeling and analysis of creeping slopes, *J. of Geotehnical Eng.*, *ASCE* **121**, 43–56.

Mc Duffie G. E. and Barr T., 1969, Pressure viscometer for viscosities between 1 and 104 P, *The Review of Scientific Instruments* **40**, No. 5, 653–655.

Eichstadt F. J. and Swift G. W., 1966, Theoretical analysis of the falling cylinder viscometer for power law and Bingham plastic fluids *AIChE Journal* **6**, 1179–1183.

Forlati F., Gioda G. and Scavia C., 2001, Finite element analysis of a deep-seated slope deformation, *Rock Mech. and Rock Engineering* **34**, 2, 135–159.

Fu M. and Loo Z. J., 1995, *J. Materials Proc. Tech.* **55**, 442–447.

Gui F. and Irvine T. F. Jr., 1994, Theoretical and experimental study of the falling cylinder viscometer, *Int. J. Heat Mass Transfer.* **37**, 1, 41–50.

Gui F. and Irvine T. F. Jr., 1996, An absolute falling tube viscometer, *Experimental Thermal and Fluid Science* **12**, 325–337.

Hettler A., Gudehus G. and Vardoulakis I., 1984, Stress–strain behaviour of sand in triaxial tests, in *Results of the International Workshop on Constitutive Relations for Soils*, Grenoble, France, September 1982, Gudehus G., Darve F. and Vardulakis I. (eds.), A.A. Balkema, Rotterdam, The Nederlands, 55–66.

Hild P., Ionescu I. R., Lachand-Robert T. and Roşca I., 2002, The blocking of an inhomogeneous Bingham fluid. Applications to landslides, *Mat. Modelling and Num. Anal.* **36**, 6, 1013–1026.

Huang E. T. S., Swift G. W. and Kurata F., 1966, Viscosities of methane and propane at low temperatures and high pressures, *AIChE.* **12**, 5, 932–936.

Irving J. B. and Barlow A. J., 1971, An automatic high pressure viscometer, *Journal of Physics E: Scientific Instruments* **4**, 232–236.

Kiran E. and Sen Y. L., 1992, High-pressure viscosity and density of n-alkanes, *International Journal of Thermophysics* **13**, No. 3, 411–442.

Kiran E. and Sen Y. L., 1995, Viscosity of polymer solutions in near-critical and supercritical fluids. Ch. 9 in *Viscosity of Polymer Solutions in Fluids*, Kiran E. and Brennecke J. F. (eds.), ACS Symposium Series 514, *Am. Ch. Soc.* 1993, 104–120.

Kiran E. and Gokmenoglu Z., 1995, High-pressure viscosity and density of polyethylene solutions in n-pentane, *Journal of Applied Polymer Science* **58**, 2307–2324.

Levanov A. N., Kolmogorov V. L., Burkin S. P., Katak B. R., Ashpur Iu. V. and Spasskii Iu. M., 1976, *Contact Friction in the Metal Working Processes*, Metallurgia Publisher, Moscow (in Russian).

Lohrenz J., Swift G. W. and Kurata F., 1960, An experimentally verified theoretical study of the falling cylinder viscometer, *AIChE Journ.* **6**, 4, 547–550.

Park N. A. and Irvine T. F. Jr., 1995, Falling cylinder viscometer end correction factor, *Rev. Sci. Instrum.* **66**, 3982–3984.

Park N. A. and Irvine T. F. Jr., 1997, Liquid density measurements using the falling needle viscometer, *Int. Comm. Heat Mass Transfer.* **42**, 3, 303–312.

Paterson S. R. and Miller R. B., 1998, Stoped blocks in plutons: paleo-plumb bobs, viscometers, or chronometers? *J. of Structural Geology* **20**, 9/10, 1261–1272.

Petley D. N. and Allison R. J., 1997, The mechanics of deep-seated landslides, *Earth Surface Processes and Landforms* **22**, 747–758.

Phan-Thien N., Jin H. and Zheng R., 1993, On the flow past a needle in a cylindrical tube, *Journal of Non-Newtonian Fluid Mechanics* **47**, 137–155.

Pochettino A., 1914, Su le proprietà dei corpi plastici, *Il Nuovo Cimento* **8**, 6th Ser., 77–108.

Reiner M. and Riwlin R., 1927, Über die Strömung einer elastichen Flüssigkeit im Couette Apparat, *Kolloid. Z.* **43**, 1–5.

Reiner M., 1926, Über die Strömung einer elastichen Flüssigkeit durch eine Kapillare, *Kolloid Z.* **39**, 80–87.

Reiner M., 1960, *Deformation, Strain and Flow*, Wiley-Interscience, New York.

Samtani N. C., Desai C. S. and Vulliet L., 1996, An interface model to describe viscoplastic behavior, *Int. Jour. of Numerical and Analyt. Meth. in Geomech.* **20**, 4, 231–252.

Swift G. W., Lohrenz J. and Kurata F., 1966, Liquid viscosities above the normal boiling point for methane, ethane, propane, and n-butane, *AIChE Jour.* **6**, 3, 415–419.

Talmon A. M. and Huisman M., 2005, Fall velocity of particles in shear flow of drilling fluids, *Tunnelling and Underground Space Technology* **20**, 2, 193–201.

Tanaka Y., Ohta K., Kybota H. and Makita T., 1988, Viscosity of aqueous solutious of 1, 2-ethanediol and 1,2-propanediol under high pressures, *International Journal of Thermophysics* **9**, 4, 511–523.

Tanaka Y., Xiao Y. F., Matsuo S. and Makita T., 1994, Density, viscosity and dielectric constant of HCFC-225ca and HCFC-225cb at temperatures from 293 to 323 K and pressures up to 80 Mpa, *Fluid Phase Equilibria* **97**, 155–165.

Tigoiu V., 1987, Wave propagation and thermodynamics for third grade fluids, *St. Cerc. Mat.* **29**, 4, 279–347.

Tigoiu V. and Cipu C., 2005, Non-Newtonian fluid flows in a falling cylinder viscometer, trends in applications of mathematics to mechanics, Wang Y. and Hutter K. (eds.), *Proc. Of the XIVth Int. Simp. on Trends in Appl. of Math. to Mech.*, Seeheim, Germany, 22–28 Aug., 2004, Shaker Verlag.

Tran-Son-Tay R., Beaty B. B., Acker D. N. and Horchmuth R. M., 1988, *Rev. Sci. Instrum.* **59**, 1399–1404.

Wehbeh E. G., Ui T. J. and Hussey R. G., 1993, End effects for the falling cylinder viscometer, *Phys. Fluids A* **5**, 1, 25–33.

Chapter 7

Axi-Symmetrical Problems

7.1 Introduction

In previous sections we have considered cases when the boundary and initial conditions conform to the assumption that only a single component of stress and a single component of velocity occur. Sometimes, in these cases, a constitutive equation is not used at all and the procedure is somehow similar to that used in some static problems, i.e., the equation of motion and continuity condition are only combined with a yield condition but not a constitutive equation. Using this procedure a sufficient number of equations is obtained for the number of unknown quantities required. Sometimes else, a constitutive equation is used, but this is in fact a one-dimensional stress–strain relation, and therefore as a rule, the methods used do not greatly differ from those used for the study of the propagation of longitudinal waves in thin bars. In all the cases examined the propagation of a single type of wave has been considered if the equation of motion is of the hyperbolic type. Otherwise, propagation is assumed to occur by diffusion according to an equation of parabolic type.

First let as present shortly the static, plane stress, problem (Sokolovski [1969]) when at the orifice one is given a pressure. For a Mises type of yield condition

$$\frac{1}{3}(\sigma_x^2 - \sigma_x\sigma_y + \sigma_y^2) + \sigma_{xy}^2 = k^2 = \frac{1}{3}\sigma_Y^2$$

the problem is *hyperbolic* close to the orifice, but *elliptic* at farther distances Fig. 7.1.1. The characteristics are logarithmic spiral lines. The figure corresponds to $p = \sqrt{3}k$ when $b \approx 5.13a$. On this circle the problem is parabolic. For greater distances the problem is elliptic.

For the Tresca condition, written in cylindrical coordinates

$$\sigma_\theta + \sigma_r = 2k$$

the characteristics are logarithmic spirals up to $b = a\exp(p/2k)$, where p is the applied pressure at the orifice. The Fig. 7.1.2 is obtained for $p = 2k$ when the logarithmic spirals are until the circle $r = b$ is reached. Further on the problem we have a single family of straight lines, the problem being parabolic.

315

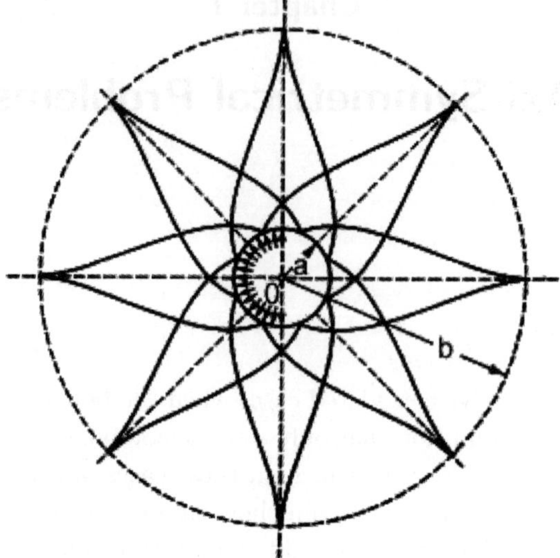

Fig. 7.1.1   The characteristics for the plane stress problem satisfying a Mises condition.

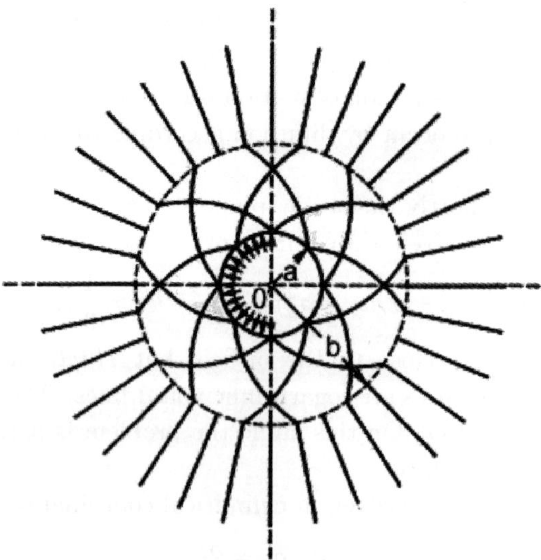

Fig. 7.1.2   The characteristics for the Tresca condition.

The simple problem of enlargement of circular holes in thin infinite plate taking into account variations in geometry, material properties, and loading conditions, was considered by several authors. Upadhyay and Stokes [1977] have used for that a Ramberg–Osgood constitutive equation:

$$\dot{T}_{ij} = 2G\dot{e}_{ij} - \frac{2n+1}{2}\left(\frac{II_{T'}}{k^2}\right)^n \left(\frac{\dot{II}_{T'}}{II_{T'}}\right)T_{ij}$$

combined with a Mises yield condition.

As remarked by Bernstein and Shokooh [1980], when subjected to creep test, aluminum exhibits characteristics which are analogous to those of thermorheologically simple material in the sense that, on a plot of logarithm of creep function versus logarithm of time, the curves corresponding to different stresses have the same shape so that they all could be shifted to form a master curve. Based on this observation and in analogy with the idea of time-temperature superposition, a theory for viscoelastic materials is developed in which the natural time appearing explicitly in the argument of relaxation or creep function is replaced by a suitable scalar-valued function of stress tensor. A comparison between the Perzyna viscoplastic model and the Consistency viscoplastic model is due to Heeres *et al.* [2002].

If, in the problem considered, there occur two or three components of velocity and therefore several components of stress and strain, and the problem is dynamic, one is obliged to consider a complete constitutive equation. Several types of constitutive equations are used in theory of plasticity. This set of constitutive equations can be divided into several group, each group of constitutive equations is characterized by certain dynamic properties. In problems in which several components of velocity, stress and strain are involved, several kind of plastic waves are present. Thus, depending on the theory of plasticity used, the plastic waves are coupled (Cristescu [1956], [1967], Nowachi [1974]).

The coupling of plastic waves was first discussed many years ago. Some authors have used for this purpose constitutive equations written in finite form. First were discussed axi-symmetrical problems. That will be discussed also below.

In order to analyze such problems, without overcomplicating the writing, we analyze a case in which the whole problem depends only on a single spatial coordinate, although many components of stress and strain are involved.

7.2   Enlargement of a Circular Orifice

The simplest possible case is the one in which two velocity components are involved. It is assumed that along a cylindrical hole of radius $r = R$ the radius is enlarged and rotated. Thus the radius is enlarged (velocity u) and rotated (velocity v) (see Fig. 7.2.1). It is question of sudden motion, so that the inertia forces are involved. From all the stresses involved we assume that σ_r, σ_θ, and $\sigma_{r\theta}$ are different of zero, but $\sigma_z = \sigma_{rz} = \sigma_{\theta z} = 0$. There are two velocity components distinct of zero; the radial one u, and the circumferential one v.

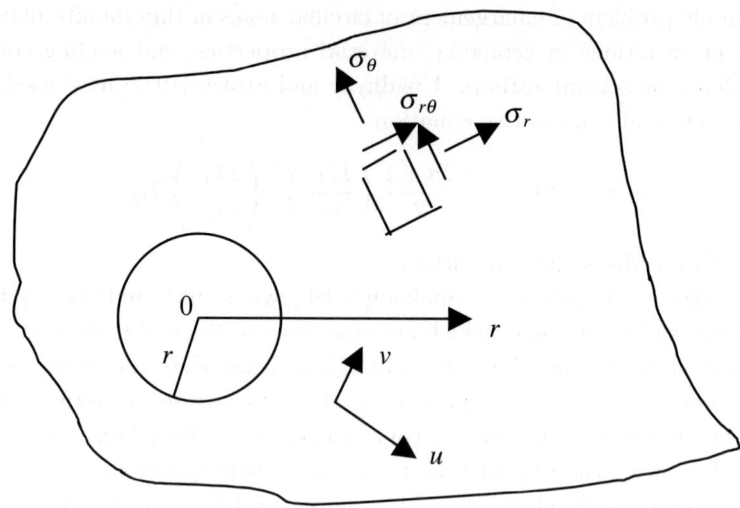

Fig. 7.2.1   Enlargement of a circular hole in a plate.

We apply the Hencky–Nadai constitutive equation. For our case we have:

$$\sigma_r' = \frac{S}{E}\varepsilon_r' \quad \text{and} \quad \sigma_\theta' = \frac{S}{E}\varepsilon_\theta'.$$

The strain components are:

$$\varepsilon_r = \frac{\partial u}{\partial r}, \quad \varepsilon_\theta = \frac{u}{r}, \quad \varepsilon_{r\theta} = \frac{1}{2}\left(\frac{\partial v}{\partial r} - \frac{v}{r}\right). \tag{7.2.1}$$

If we assume incompressibility we have:

$$\varepsilon_r + \varepsilon_\theta + \varepsilon_z = 0, \tag{7.2.2}$$

from where we obtain ε_z if the other two components are known.

The Hencky–Nadai constitutive equation is for our case:

$$\sigma_r = \frac{S}{E}\left(2\frac{\partial u}{\partial r} + \frac{u}{r}\right), \quad \sigma_{r\theta} = \frac{S}{2E}\left(\frac{\partial v}{\partial r} - \frac{v}{r}\right), \tag{7.2.3}$$

where

$$E^2 = II_\varepsilon = \varepsilon_r^2 + \varepsilon_\theta^2 + \varepsilon_r\varepsilon_\theta + \varepsilon_{r\theta}^2, \tag{7.2.4}$$

and the work-hardening law is written:

$$S = F(E). \tag{7.2.5}$$

The equation of motion, with X_r and X_θ the two body forces per unit volume are:

$$\rho\frac{\partial^2 u}{\partial t^2} - \frac{\partial \sigma_r}{\partial r} = \frac{\sigma_r - \sigma_\theta}{r} + X_r,$$

$$\rho\frac{\partial^2 v}{\partial t^2} - \frac{\partial \sigma_{r\theta}}{\partial r} = \frac{2\sigma_{r\theta}}{r} + X_\theta. \tag{7.2.6}$$

Introducing here (7.2.3) we get:

$$\rho\frac{\partial^2 u}{\partial t^2} - L\frac{\partial^2 u}{\partial r^2} - P\frac{\partial^2 v}{\partial r^2} = \Phi\,,$$

$$\rho\frac{\partial^2 v}{\partial t^2} - P\frac{\partial^2 u}{\partial r^2} - M\frac{\partial^2 v}{\partial r^2} = \Psi\,,$$

(7.2.7)

where

$$L = \frac{2S}{E} + \frac{ES' - S}{2E^3}(2\varepsilon_r + \varepsilon_\theta)^2\,,$$

$$M = \frac{S}{2E} + \frac{ES' - S}{2E^3}\varepsilon_{r\theta}^2\,,$$

$$P = \frac{ES' - S}{2E^3}(2\varepsilon_r + \varepsilon_\theta)\varepsilon_{r\theta}\,,$$

(7.2.8)

and

$$\Phi = \frac{S}{E}\frac{\varepsilon_r - \varepsilon_\theta}{r} + \frac{ES' - S}{2E^3 r}(2\varepsilon_r + \varepsilon_\theta)[(\varepsilon_r - \varepsilon_\theta)(2\varepsilon_\theta + \varepsilon_r) - 2\varepsilon_{r\theta}^2] + X_r\,,$$

$$\Psi = \frac{2S}{Er}\varepsilon_{r\theta} + \frac{ES' - S}{2E^3 r}\varepsilon_{r\theta}[(\varepsilon_r - \varepsilon_\theta)(2\varepsilon_\theta + \varepsilon_r) - 2\varepsilon_{r\theta}^2] + X_\theta\,.$$

There are two waves propagating with the velocities:

$$\left.\begin{array}{c} c_1^2 \\ c_2^2 \end{array}\right\} = \frac{M + L \pm \sqrt{(M-L)^2 + 4P^2}}{2\rho}\,.$$

(7.2.9)

Both are real (the expressions at the right hand side are positive). Also,

$$c_1^2 > c_2^2\,.$$

(7.2.10)

The differential relations satisfied on the characteristic lines are:

$$\{\Phi(M - \rho c^2) - \Psi P\}\,dt - \rho(M - \rho c^2)(c\,du_r - du_t) - \rho P(c\,dv_r - dv_t) = 0$$

(7.2.11)

where c is either c_1 or c_2, and we denote $u_r = (\partial u/\partial r)$, etc.

Let us write now the <u>jump conditions</u>. First the kinematics jump conditions are

$$\left[\frac{\partial^2 u}{\partial t^2}\right]dt + \left[\frac{\partial^2 u}{\partial r\partial t}\right]dr = 0\,,$$

$$\left[\frac{\partial^2 u}{\partial r\partial t}\right]dt + \left[\frac{\partial^2 u}{\partial r^2}\right]dr = 0\,.$$

From here we obtain easy the kinematics jump conditions, for u and v are similar:

$$\left[\frac{\partial^2 u}{\partial t^2}\right](dt)^2 - \left[\frac{\partial^2 u}{\partial r^2}\right](dr)^2 = 0\,,$$

$$\left[\frac{\partial^2 v}{\partial t^2}\right](dt)^2 - \left[\frac{\partial^2 v}{\partial r^2}\right](dr)^2 = 0\,.$$

(7.2.12)

The dynamic jump conditions are [from (7.2.7)]:

$$\rho\left[\frac{\partial^2 u}{\partial t^2}\right] - L\left[\frac{\partial^2 u}{\partial r^2}\right] - P\left[\frac{\partial^2 v}{\partial r^2}\right] = 0,$$

$$\rho\left[\frac{\partial^2 v}{\partial t^2}\right] - P\left[\frac{\partial^2 u}{\partial r^2}\right] - M\left[\frac{\partial^2 v}{\partial r^2}\right] = 0. \tag{7.2.13}$$

From (7.2.12) and (7.2.13) we get easily

$$\left[\frac{\partial^2 u}{\partial r^2}\right]\rho c^2 - L\left[\frac{\partial^2 u}{\partial r^2}\right] - P\left[\frac{\partial^2 v}{\partial r^2}\right] = 0,$$

$$\left[\frac{\partial^2 v}{\partial r^2}\right]\rho c^2 - P\left[\frac{\partial^2 u}{\partial r^2}\right] - M\left[\frac{\partial^2 v}{\partial r^2}\right] = 0, \tag{7.2.14}$$

or we can write also:

$$(\rho c^2 - L)\left[\frac{\partial^2 u}{\partial r^2}\right] - P\left[\frac{\partial^2 v}{\partial r^2}\right] = 0,$$

$$(\rho c^2 - M)\left[\frac{\partial^2 v}{\partial r^2}\right] - P\left[\frac{\partial^2 u}{\partial r^2}\right] = 0. \tag{7.2.15}$$

These relations are valid for both waves; we have to precise only the velocity. It is easy to show by direct computation that

$$\frac{P}{\rho c_1^2 - L} = -\frac{\rho c_2^2 - L}{P}. \tag{7.2.16}$$

From (7.2.16) and (7.2.15) we get

$$\frac{[\partial^2 u/\partial r^2]_1}{[\partial^2 v/\partial r^2]_1} = \frac{P}{\rho c_1^2 - L} = -\frac{[\partial^2 v/\partial r^2]_2}{[\partial^2 u/\partial r^2]_2} = -\frac{\rho c_2^2 - L}{P}, \tag{7.2.17}$$

or,

$$\left[\frac{\partial^2 u}{\partial r^2}\right]_1\left[\frac{\partial^2 u}{\partial r^2}\right]_2 + \left[\frac{\partial^2 v}{\partial r^2}\right]_1\left[\frac{\partial^2 v}{\partial r^2}\right]_2 = 0. \tag{7.2.18}$$

Thus at each point the jumps are orthogonal.

Both waves are coupled, i.e., they are both dilatational and shearing [see (7.2.11), (7.2.15), (7.2.17)]. Both involve the stresses σ_r and $\sigma_{r\theta}$ and also both components of the velocity u and v.

From (7.2.15), if the ratio between jumps is eliminated, it follows

$$\frac{P}{L - \rho c^2} = \frac{M - \rho c^2}{P}. \tag{7.2.19}$$

If $M > L$ from (7.2.19) we have $(P/(\rho c^2 - L)) > 1$, or from (7.2.17) we obtain $[\partial^2 u/\partial r^2]_1 > [\partial^2 v/\partial r^2]_1$ that is the wave "one" is mainly a dilatational one. If $M < L$ from (7.2.19) follows $(P/(\rho c^2 - L)) < 1$, and from (7.2.17) we obtain $[\partial^2 u/\partial r^2]_1 < [\partial^2 v/\partial r^2]_1$ that is the wave "one" is primarily a shearing wave. If $M \cong L$, then both waves are equally dilatant and shearing.

The coupling coefficient is P, that is if $P = 0$ an uncoupling takes place. That can happen if:

- $\varepsilon_{r\theta} = 0$ or $v = 0$. The wave "one" is pure dilatational and the only propagable, with the velocity:

$$c_1^* = \frac{1}{\rho}\left\{\frac{3S}{2E^3}\varepsilon_\theta^2 + \frac{S'}{2E^2}\right\}$$

which becomes smaller i.e., $c_1^* < c_1$.

- $u = 0$, i.e., $\varepsilon_r = \varepsilon_\theta = 0$; the only wave which exists is the wave "two" which becomes pure shearing. For the velocity of propagation we have:

$$c_2^* = \frac{S'}{2\rho\varepsilon_{r\theta}^2}$$

and we have $c_2^* > c_1$. Therefore we have $c_2 < c_2^* < c_1^* < c_1$.

- The third case is the elastic body, i.e., $ES' - S = 0$.

The coupling factor is P. If it is present one has coupling. And we have to integrate two equations separately, as if they are not influencing each other. If it is question what is coupling the waves it is the work hardening condition (7.2.5).

7.3 Thin Wall Tube

Let us consider now the problem of a thin wall tube, loaded at one of its end by tension and torque. Thus we have two loadings, and we expect a combined load. The boundary conditions are:

$$\text{at } x = 0 \text{ a combined loading}$$

$$\text{at } x = l_0 \text{ the end is fixed}.$$

We are using cylindrical coordinates x, r, and θ, and the displacements are u, v and $w = 0$. Thus the velocity components are u_t, v_t, and 0. The stress components

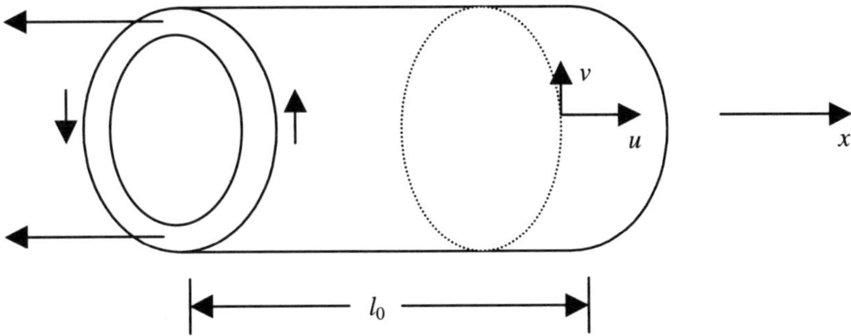

Fig. 7.3.1   Thin wall tube subjected to tension and torque.

are σ_r, $\sigma_{r\theta}$, and σ_{rx} are assumed to be small, at least compared with $\sigma_{xx} = \sigma$ and $\sigma_{\theta x} = \tau$, which are dominant. We assume also axial-symmetry thus $(\partial/\partial\theta) = 0$.

The equations of motion are:

$$\frac{\partial\sigma}{\partial x} + F_x = \rho\frac{\partial u_t}{\partial t}\,,$$

$$\frac{\partial\tau}{\partial x} + F_\theta = \rho\frac{\partial v_t}{\partial t}\,, \tag{7.3.1}$$

where F_x and F_θ are the corresponding body forces, and ρ is the constant density.

The strains are:

$$\varepsilon_{rr} = \frac{\partial w}{\partial r} = 0\,, \qquad \varepsilon_{\theta\theta} = \frac{w}{r} = 0\,, \qquad \varepsilon_{xx} = \frac{\partial u}{\partial x} = \varepsilon\,,$$

$$\varepsilon_{\theta x} = \frac{1}{2}\frac{\partial v}{\partial x} = \gamma\,, \quad \varepsilon_{rx} = \frac{1}{2}\left(\frac{\partial u}{\partial r} + \frac{\partial w}{\partial x}\right) = 0\,, \quad \varepsilon_{r\theta} = \frac{1}{2}\left(\frac{\partial v}{\partial r} - \frac{v}{r}\right) = 0\,. \tag{7.3.2}$$

We use also the noticing

$$II_{\sigma'} = \frac{1}{2}\boldsymbol{\sigma}' : \boldsymbol{\sigma}' = \frac{\sigma^2}{3} + \tau^2\,. \tag{7.3.3}$$

We introduce now the constitutive equations. We introduce one in very general form and we particularize it later on. Thus, assuming small strains,

$$\dot{\boldsymbol{\varepsilon}}' = \dot{\boldsymbol{\varepsilon}}^{E'} + \dot{\boldsymbol{\varepsilon}}^{P'}\,. \tag{7.3.4}$$

For the elastic we assume the Hooke's law:

$$2G\dot{\boldsymbol{\varepsilon}}^{E'} = \dot{\boldsymbol{\sigma}}'\,,$$

$$\dot{\sigma}_m = 3K\dot{\varepsilon}_m\,. \tag{7.3.5}$$

The plastic part of the strain rate, satisfy a general constitutive equation of the form:

$$\dot{\boldsymbol{\varepsilon}}^P = \mathbf{A}\dot{\boldsymbol{\sigma}}' + \mathbf{B}$$

where $\mathbf{A}$ is a fourth order tensor and $\mathbf{B}$ a second order tensor. For our case we have

$$\dot{\varepsilon}^P_{xx} = \varphi_{11}\dot{\sigma}'_{xx} + \varphi_{12}\dot{\sigma}'_{\theta x} + \psi_1\,,$$

$$\dot{\varepsilon}^P_{\theta x} = \varphi_{21}\dot{\sigma}'_{xx} + \varphi_{22}\dot{\sigma}'_{\theta x} + \psi_2\,.$$

Introducing here the Hooke's law, these equations becomes:

$$\frac{\partial u_t}{\partial x} = \left(\frac{2}{3}\varphi_{11} + \frac{1}{E}\right)\dot{\sigma} + \varphi_{12}\dot{\tau} + \psi_1\,,$$

$$\frac{\partial v_t}{\partial x} = \frac{4}{3}\varphi_{21}\dot{\sigma} + \left(2\varphi_{22} + \frac{1}{G}\right)\dot{\tau} + 2\psi_2\,, \tag{7.3.6}$$

where

$$E = \frac{9KG}{3K + G}, \quad \text{and} \quad \sigma_m = \frac{\sigma}{3}.$$

In order to simplify the problem, we introduce the new notation:

$$\alpha_{11} = \frac{1}{3}\left(2\varphi_{11} + \frac{1}{G} + \frac{1}{3K}\right), \quad \alpha_{12} = \varphi_{12}, \quad \beta_1 = \psi_1,$$

$$\alpha_{21} = \frac{4}{3}\varphi_{21}, \quad \alpha_{22} = 2\varphi_{22} + \frac{1}{G}, \quad \beta_2 = 2\psi_2,$$

with whom the constitutive equation becomes:

$$\frac{\partial u_t}{\partial x} = \alpha_{11}\dot{\sigma} + \alpha_{12}\dot{\tau} + \beta_1,$$

$$\frac{\partial v_t}{\partial x} = \alpha_{21}\dot{\sigma} + \alpha_{22}\dot{\tau} + \beta_2. \tag{7.3.7}$$

This is the general form of the constitutive equation. Until now we have not said what is $\alpha_{11}, \ldots\ldots$ etc. In order to make a discussion we consider now several particular cases.

P1. <u>Elastic</u>. In the elastic case, we have:

$$\varphi_{11} = \varphi_{12} = \varphi_{21} = \varphi_{22} = \psi_1 = \psi_2 = 0.$$

The constitutive equation becomes:

$$E\frac{\partial u_t}{\partial x} = \dot{\sigma}, \quad G\frac{\partial v_t}{\partial x} = \dot{\tau}.$$

The two equations are completely separated and we have two kinds of waves which propagate independently.

P2. <u>Rate semi linear</u> (the constitutive equation is of Malvern type). For such equations we have:

$$\varphi_{11} = \varphi_{12} = \varphi_{21} = \varphi_{22} = 0.$$

The constitutive equation can be written:

$$\dot{\varepsilon} = \frac{\dot{\sigma}}{E} + \frac{1}{3\eta}\left(1 - \frac{k}{\sqrt{II_{\sigma'}}}\right)\sigma,$$

$$\dot{\gamma} = \frac{\dot{\tau}}{2G} + \frac{1}{2\eta}\left(1 - \frac{k}{\sqrt{II_{\sigma'}}}\right)\tau, \tag{7.3.8}$$

where k is the yield stress, assumed constant or not, and η is a viscosity coefficient, assumed also possibly variable.

P3. <u>Piece-wise work-hardening</u> (separated for the two components).

In this case we have

$$\varphi_{12} = \varphi_{21} = 0.$$

The constitutive equations are:

$$\frac{\partial u_t}{\partial x} = \left(\frac{2}{3}\varphi_{11} + \frac{1}{E}\right)\dot{\sigma} + \psi_1 \,,$$

$$\frac{\partial v_t}{\partial x} = \left(2\varphi_{22} + \frac{1}{G}\right)\dot{\tau} + \psi_2 \,. \tag{7.3.9}$$

P4. <u>The Prandtl–Reuss</u> constitutive equation. In this case we have:

$$\psi_1 = \psi_2 = 0 \,.$$

The constitutive equation is in general:

$$2G\dot{\varepsilon}'_{ij} = \dot{\sigma}'_{ij} + \frac{2G}{H'}\frac{\sigma'_{ij}\dot{\sigma}'_{ij}}{\sigma'_{mn}\sigma'_{mn}}\sigma'_{ij}\,, \quad \text{and}$$

$$\sigma'_{ij}\sigma'_{ij} = H\left(\int \sigma'_{kl}d\varepsilon^{P'}_{kl}\right).$$

This constitutive equation particularized for the problem is:

$$\dot{\varepsilon} = \left(\frac{1}{E} + \frac{4}{27}\frac{1}{H'II_{\sigma'}}\sigma^2\right)\dot{\sigma} + \frac{2}{3}\frac{1}{H'II_{\sigma'}}\sigma\tau\dot{\tau}\,,$$

$$\dot{\gamma} = \frac{2}{9}\frac{1}{H'II_{\sigma'}}\sigma\tau\dot{\sigma} + \left(\frac{1}{2G} + \frac{1}{H'II_{\sigma'}}\tau^2\right)\dot{\tau}\,. \tag{7.3.10}$$

P5. <u>Rate quasi-linear.</u> In this case, assuming that the strains are small, we have to start with:

$$\dot{\varepsilon}'_{ij} = \dot{\varepsilon}^{E'}_{ij} + \dot{\varepsilon}^{P'}_{ij} + \dot{\varepsilon}^{VP'}_{ij}\,,$$

$$\dot{\varepsilon}'_{ij} = \frac{\dot{\sigma}'_{ij}}{2G} + \frac{1}{2\eta}\left(1 - \frac{k}{\sqrt{II_{\sigma'}}}\right)\sigma'_{ij} + \frac{F(II_{\sigma'})}{2II_{\sigma'}}\sigma'_{kl}\dot{\sigma}'_{kl}\sigma_{ij}\,,$$

and for the work-hardening:

$$\sigma'_{ij}\dot{\varepsilon}^{P'}_{ij} = F(II_{\sigma'})\sigma'_{kl}\dot{\sigma}'_{kl}\,.$$

For the problem considered this law is reduced to

$$\dot{\varepsilon} = \left(\frac{1}{E} + \frac{4}{27}\frac{F(II_{\sigma'})}{II_{\sigma'}}\sigma^2\right)\dot{\sigma} + \frac{2}{3}\frac{F(II_{\sigma'})}{II_{\sigma'}}\sigma\tau\dot{\tau} + \frac{1}{3\eta}\left(1 - \frac{k}{\sqrt{II_{\sigma'}}}\right)\sigma\,,$$

$$\dot{\gamma} = \frac{2}{9}\frac{F(II_{\sigma'})}{II_{\sigma'}}\sigma\tau\dot{\sigma} + \left(\frac{1}{2G} + \frac{F(II_{\sigma'})}{II_{\sigma'}}\tau^2\right)\dot{\tau} + \frac{1}{2\eta}\left(1 - \frac{k}{\sqrt{II_{\sigma'}}}\right)\tau\,. \tag{7.3.11}$$

From here for $F = 0$ we obtain the rate semi linear constitutive equation, while if $\eta \to \infty$ we obtain the classic plasticity theory.

The velocity of propagation for P5 and P4 are variable and equal to:

$$\left.\begin{array}{c} c^2_{TL} \\ c^2_{LT} \end{array}\right\} = \frac{\alpha_{11} + \alpha_{22} \pm \sqrt{(\alpha_{11} - \alpha_{22})^2 + 4\alpha_{12}\alpha_{21}}}{2\rho(\alpha_{11}\alpha_{22} - \alpha_{12}\alpha_{21})} \tag{7.3.12}$$

and $c_{TL} > c_{LT}$. For the particular case P3 we have:

$$c_{L3}^2 = \frac{1}{\rho\alpha_{11}}, \quad \text{and} \quad c_{T3}^2 = \frac{1}{\rho\alpha_{22}},$$

and these velocities are also variable. For the particular cases P1 and P2 the velocities are constant and equal to:

$$c_{L1}^2 = c_{L2}^2 = \frac{E}{\rho}, \quad c_{T1}^2 = c_{T2}^2 = \frac{G}{\rho}.$$

Let us consider two particular cases. If everywhere we have $v_t = 0$, then considering again the initial equations (with $\varphi_{12} = 0$) we have a single kind of longitudinal wave which propagates with the velocity

$$c_{L0}^2 = \frac{1}{\rho}\frac{3E}{2E\varphi_{11} + 3},$$

which for the special cases P1 and P2 reduce to the elastic bar velocity.

In a similar case if $u_t = 0$ the only existing wave is the one propagating with the velocity

$$c_{T0}^2 = \frac{1}{\rho}\frac{G}{2G\varphi_{22} + 1}$$

which are shearing waves and the formula is reduced to that from elastic waves in the case $\varphi_{22} = 0$.

Let us consider now the differential relations along the characteristics. We have for the general case:

$$\mp\rho c(1 - \rho\alpha_{22}c^2)\,du_t \mp \rho^2\alpha_{12}c^3 dv_t + (1 - \rho\alpha_{22}c^2)\,d\sigma + \rho\alpha_{12}c^2 d\tau$$

$$+ \{\rho\beta_1 c^2(1 - \rho\alpha_{22}c^2) + \beta_2\alpha_{12}\rho^2 c^4 \mp (1 - \rho\alpha_{22}c^2)\,F_x \mp \rho c^3\alpha_{12}F_\theta\}\,dt = 0.$$

$$(7.3.13)$$

Here we have to replace c by one of the four expressions (7.3.12).

For the particular case P3 the differential relations are:

$$\mp\rho c_{L3}\,du_t + d\sigma + (\rho\beta_1 c_{L3}^2 \mp c_{L3}F_x)\,dt = 0,$$

$$\mp\rho c_{T3}\,dv_t + d\tau + (\rho\beta_2 c_{T3}^2 \mp c_{T3}F_\theta)\,dt = 0.$$

$$(7.3.14)$$

For the particular case P2 they are:

$$\mp\sqrt{\rho E}\,du_t + d\sigma + \left(\beta_1 E \mp \sqrt{\frac{E}{\rho}}F_x\right)dt = 0,$$

$$\mp\sqrt{\rho G}\,dv_t + d\tau + \left(\beta_2 G \mp \sqrt{\frac{G}{\rho}}F_\theta\right)dt = 0.$$

$$(7.3.15)$$

For the particular case P1 they are:

$$\mp\sqrt{\rho E}\,du_t + d\sigma \mp \sqrt{\frac{E}{\rho}}F_x dt = 0\,,$$

$$\mp\sqrt{\rho G}\,dv_t + d\tau \mp \sqrt{\frac{G}{\rho}}F_\theta dt = 0\,. \qquad (7.3.16)$$

Let us write now the coupling jump conditions. We have from the constitutive equations (the brackets stand for the "jumps" of the function considered):

$$\left[\frac{\partial u_t}{\partial x}\right] = \alpha_{11}\left[\frac{\partial \sigma}{\partial t}\right] + \alpha_{12}\left[\frac{\partial \tau}{\partial t}\right],$$

$$\left[\frac{\partial v_t}{\partial x}\right] = \alpha_{21}\left[\frac{\partial \sigma}{\partial t}\right] + \alpha_{22}\left[\frac{\partial \tau}{\partial t}\right].$$

From the equation of motion we get:

$$\rho\left[\frac{\partial u_t}{\partial t}\right] = \left[\frac{\partial \sigma}{\partial x}\right],\quad \rho\left[\frac{\partial v_t}{\partial t}\right] = \left[\frac{\partial \tau}{\partial x}\right].$$

Finally, we have also the relation

$$\left[\frac{\partial \psi}{\partial t}\right] = -c\left[\frac{\partial \psi}{\partial x}\right],\quad \text{for } \psi = u_t, v_t, \sigma, \tau\,.$$

From here we obtain:

$$\left[\frac{\partial u_t}{\partial x}\right](1 - \rho\alpha_{11}c^2) = \rho\alpha_{12}c^2\left[\frac{\partial v_t}{\partial x}\right],$$

$$\left[\frac{\partial v_t}{\partial x}\right](1 - \rho\alpha_{22}c^2) = \rho\alpha_{21}c^2\left[\frac{\partial u_t}{\partial x}\right].$$

From all those relations we conclude that α_{12} and α_{21} are the coefficients that couple the wave. They are introduced by the yield condition, which are relating all stresses in a single condition. Thus we have:

	instantaneous response	non-instantaneous response
P5 $\quad \alpha_{12} \neq 0 \leftarrow$ coupling		$\leftarrow$ coupling
P4 $\quad \alpha_{21} \neq 0 \leftarrow$ coupling		$\leftarrow$ no coupling
P3 $\quad\quad\quad\quad\quad \leftarrow$ partially coupling or uncoupling		$\leftarrow$ coupling
P2 $\quad \alpha_{12} = 0 \leftarrow$ uncoupled		$\leftarrow$ coupling
P1 $\quad \alpha_{21} = 0 \leftarrow$ uncoupled		$\leftarrow$ no coupling

A combined tension–torsion impact testing apparatus and an experimental study in the incremental wave propagation is due to Tanimura [1978]. The apparatus can realize a tension–torsion loading. To obtain a practical form of the general equation, the methods to examine the existence of the instantaneous plastic property and to obtain directly the coefficients of the general equation, through the experiments in which the incremental torsional impact or the incremental combined impact is applied to the tubular specimens, have been presented. By a special mechanism for clamping and quick releasing of the input bar, a combined impulse could be generated easily with a short rise time. Through an experiment in which the torsional incremental impact was applied to the tubular specimens of commercial pure aluminum and copper, it could be observed that the incremental wave velocity, just after passing through the partial unloading, was clearly smaller than the elastic wave velocity.

A general study is given by Ting [1970] of plane and cylindrical wave propagation of combined stress in an elastic-plastic medium. The relations between the stresses on both sides of an elastic-plastic boundary are derived. Also presented are the restrictions on the speed of an elastic-plastic boundary. The combined longitudinal and torsional plastic waves in thin-walled tube of rate-independent isotropic work-hardening material are used by Ting [1973] to illustrate the problems involved when the speeds are equal. Plastic wave speeds in materials whose elastic response is linear and isotropic while the plastic flow is incompressible and isotropically work-hardening are obtained also by Ting [1977]. One of the three plastic wave speeds is identical to the elastic shear wave speed regardless of the form of the yield condition. The other two plastic wave speeds are determined for materials obeying the von Mises yield condition. See also Mandel [1974].

The propagation of waves in thin tubes, subjected to longitudinal and torsional loading, was also considered by Myers and Eisenberg [1974], [1975]. They find that the fast and slow wave speeds are

$$c_f = \left[\frac{1}{2a\rho}\{b + (b^2 - 4aA_4)^{1/2}\}\right]^{1/2}, \quad c_s = \left[\frac{1}{2a\rho}\{b - (b^2 + 4aA_4)^{1/2}\}\right]^{1/2}$$

for a constitutive equation

$$\dot{\varepsilon}_{ij}^P = \frac{3}{2}[\phi(\bar{s}, \Delta)\dot{\bar{s}}]\frac{s_{ij}}{\bar{s}}$$

$$\bar{s} = \sqrt{\frac{3}{2}s_{ij}s_{ij}}, \quad \Delta = \sqrt{\frac{2}{3}}\int \sqrt{\dot{\varepsilon}_{ij}^P \dot{\varepsilon}_{ij}^P}\, dt + \frac{\bar{s}}{E}.$$

All the other coefficients are variable. Integration along characteristics is also done, for solutions with or without radial inertia.

The thin-walled tube was considered also by Yokoyama [2001]. He has considered various models to describe the propagation only of simple torsional waves in tubes. These were elasto-plastic, elasto-viscoplastic and elasto-viscoplastic-plastic with linear work-hardening. The constitutive equation is of the form

$$\dot{\gamma} = \begin{cases} \left(\dfrac{1}{G} + \dfrac{1}{H_h'}\right)\dot{\tau} + \dfrac{k}{G}\langle\tau - g(\gamma_p)\rangle & \dot{\tau} > 0, \\[2ex] \dfrac{\dot{\tau}}{G} + \dfrac{k}{G}\langle\tau - g(\gamma_p)\rangle & \dot{\tau} < 0. \end{cases}$$

with $H_h' = dh(\gamma_p)/d\gamma_p$ and the function $h(\gamma_p)$ denotes the limiting maximum dynamic shear stress-plastic shear strain curve. When $\dot{\tau} \to \infty$ from the first equation we have

$$\frac{d\tau}{d\gamma} = \frac{1}{1/G + 1/H_h'} = \begin{cases} G & (\gamma < \gamma_Y) \\ H_t & (\gamma \geq \gamma_Y) \end{cases}$$

which is the instantaneous response curve. The integration is numerical using finite element formulation. Then one compares the velocity of propagation of the shearing strain with the tests of Yew and Richardson [1969]. One is not finding a good coincidence, since none of the considered theories, considered by the author, could predict that plastic shear strains along the axis of the thin-walled tube decrease with increasing distance from the impacted end. However one is finding that the rate dependent models give an overall better agreement with the experimental data than the rate-independent models. The smaller plastic strains are propagated with higher velocity in the torsional wave-propagation experiments than predicted by the rate-independent plastic wave theory, whereas the situation is reversed for larger plastic strains.

Assuming a one-dimensional rate independent theory of combined longitudinal and torsional plastic wave propagation in thin-walled tube, restrictions are obtained on the possible speeds of elastic-plastic boundaries by Clifton [1968].

A laser-induced deformation modes in thin metal targets is considered by O'Keefe *et al.* [1973]. An analogy with dilatational and transverse plastic waves in a thin membrane is considered.

Some experimental data are due to Hsu and Clifton [1974a] for longitudinal plastic waves propagating in thin-walled tubes of alpha-titanium. Strain-time profiles recorded in these experiments show evidence of (i) stress levels considerably above quasistatic values at the same strain, (ii) decay of the amplitude of the elastic precursor, and (iii) variation with distance of propagation of the speed at which a given level of strain propagates. These features of the strain-time profiles are interpreted as indicating that a strain-rate dependent theory is necessary to describe the observed wave phenomena. Numerical solutions based on such a theory agree reasonably well with experimental results. In another paper Hsu and Clifton [1974b] present experiments in which combined longitudinal and torsional plastic waves are generated in thin-walled tubes of alpha-titanium by subjecting pre-torque tubes to longitudinal impact. Longitudinal and torsional strain-time profiles are recorded at several stations along the specimen. These strain-time profiles exhibit features which cannot be explained within the framework of a strain-rate independent theory. The latter theory requires the wave generated under the loading employed to consist of, successively, a fast simple wave, an intermediate constant state region, a slow simple wave, and a final constant-state region. The experiments, done by the authors, show no evidence of an intermediate constant-state region; furthermore, a final constant-state region is not observed even though the latest times of observation are greater than the time at which, according to a strain-rate independent theory, a constant strain-rate region is expected. Straightforward generalization of the rate-dependent theory to allow for combined stress state leads to a theory which predicts the observed strain-time profiles with good accuracy. The theory applied by the authors is

$$\dot{\varepsilon} = \frac{\dot{\sigma}}{E} + \langle \Phi(\bar{\tau}, \bar{\gamma}^p) \rangle \frac{\partial \bar{\tau}}{\partial \sigma}, \quad \dot{\gamma} = \frac{\dot{\tau}}{G} + \langle \Phi(\bar{\tau}, \bar{\gamma}^p) \rangle \frac{\partial \bar{\tau}}{\partial \tau},$$

which are giving such results.

In another paper Abou-Sayed and Clifton [1976] consider the oblique impact of fused-silica target plates by A1 6061-T6 projectile plates. The equations of motion are

$$\frac{\partial \sigma}{\partial X} = \rho_0 \frac{\partial u}{\partial t}, \quad \frac{\partial \tau}{\partial X} = \rho_0 \frac{\partial v}{\partial t}.$$

The present analysis of the propagating waves in the fused-silica plates indicates that the main features of the normal velocity profiles at the free surface will be the same as for the case of uniaxial strain. The predicted velocity-time profile and the rise time for the normal velocity agree closely with those measured experimentally. The predicted effect of the combined loading on the normal component of the particle velocity is a shift in the final plateau value due to the interaction between the longitudinal wave reflected from the free surface and the on-coming shear wave. The effect of this interaction propagates back towards the free surface at the longitudinal wave speed so that the disturbance reaches the rear surface, before the arrival of the shear wave front. The shear wave front propagating in the shear-strain-free compressed region of the specimen remains non-dispersive. The initial jump in the transverse acceleration maintains its strength as it propagates inside the body.

A numerical solution is presented by Abou-Sayed and Clifton [1977a] for the case of symmetric impact of two skewed plates, modeled to represent 6061-T6 Aluminum. One is considering a Perzina [1966] type of constitutive equation

$$\dot{\varepsilon}_{ij} = \frac{1+v}{E}\dot{\sigma}_{ij} - \frac{v}{E}\delta_{ij}\dot{\sigma}_{kk} + \langle \Phi(\bar{\tau}, \bar{\gamma}^p) \rangle \frac{\partial f}{\partial \sigma_{ij}}.$$

The main features of the solution are, except near the impact face, the same as in previous solutions based on a rate-independent theory. Free-surface velocity-time profiles are obtained for the target rear surface.

Results are reported by Abou-Sayed and Clifton [1977b] for experiments employing two types of oblique-plate-impact configurations. One, a symmetric configuration, uses 6061-T6 Aluminum plates for both targets and projectiles. The second, an asymmetric configuration, uses a 6061-T6 Aluminum projectile and a fused silica target. The experimental results are compared to analytical solutions based on an elastic-viscoplastic model for the aluminum alloy and a hyperelastic model for fused silica. The normal velocity-time profile for the symmetric configuration shows close agreement with the predicted one.

The plastic deformation of circular membranes of strain-rate dependent, work-hardening metal is analyzed based on the strain increment theory by

Tobe *et al.* [1979]. Numerical solutions for aluminum membranes are obtained under various types of loading as well as impulsive pressure, pulses caused by underwater wire explosions. When the duration of the loading pulse is less than one-third of the time required to finish the deformation of the membrane, the deflection of a deformed membrane depends almost only on the impulse of the loading pulse and the instantaneous profile of the membrane is found to be trapezoidal during deformation. The relation between impulse of loading and deflection is calculated and illustrated. The neglect of strain-rate dependency of material leads to a notable underestimation of the forming limit.

In a paper written by Kim and Clifton [1980] one is using the plate impact experiment to study the material behavior at high strain rates. The results show that the transverse velocity profile is clearly more sensitive to the constitutive relations than is the longitudinal velocity profile. In addition, the transverse velocity profile gives the relatively long time history of the flow characteristics of solids, whereas the longitudinal one does not because the plastic strain rate decreases as the stress state becomes more nearly that of hydrostatic pressure. Improved agreement between theoretical predictions and results of pressure shear impact experiments appears to require improved constitutive models for plastic flow. Comparisons presented here suggest that models are required which characterize plastic flow characteristics accurately along loading trajectories with sharp changes in direction. The computed transverse velocity-time profiles have regions of steeper slope than observed in the experiments. This discrepancy appears to be mainly due to the inadequacy of the assumption of isotropic hardening and the yield function.

Comparison of the various strain-time profiles reveals qualitative agreement in the main features, but with several characteristic differences (Güldenpfennig and Clifton [1980]). First, the solution based on self-consistent slip models does not show a constant state region between the fast and slow simple waves. Such intermediate constant states regions are predicted by theories based on smooth yield surfaces, but are not observed in experiments. Second, the decrease in shear strain in early part of the wave profile is greater for the self-consistent slip models than for the experiments or for the predictions of an isotropic work-hardening, smooth yield surface model. Third, for small strains, the slip-models with independent and latent hardening predict the experimental results appreciably better than either of the other models. Fourth, all models predict the late-arriving, large amplitude strains would arrive earlier than observed in the experiments.

In a paper written by Chhabilidas and Swegle [1980] an experimental technique is described which uses anisotropic crystals to generate dynamic pressure-shear loading in materials. The technique has been successfully used to detect a 0.2 GPa shear wave in 6061-T6 aluminum at 0.7 GPa longitudinal stress.

A viscoplastic constitutive equation of Bingham type was used by Tayal and Natarajan [1981] together with finite element method to analyze the extrusion of rate sensitive aluminum through a conical die, extrusion of superplastic alloy through a tube of uniform diameter under complete sticking conditions and extrusion of a Pb-Sn eutectic alloy through a conical die. The results are compared with experimentally measured values. A good agreement is found.

Gilat and Clifton [1985] compared experimental results with predictions based on elastic/viscoplastic models of the form

$$\dot{\varepsilon}_{ij}^{vp} = \Phi(\bar{\tau}, \bar{\gamma}^p)\frac{\partial f}{\partial \sigma_{ij}}\,.$$

The effective shear stress $\bar{\tau}$ was taken to have the value of the plastic potential f. Two flow potentials were used:

$$f(\sigma_{ij}) = \left(\frac{1}{2}\sigma'_{ij}\sigma'_{ij}\right)^{1/2},$$

$$f(\sigma_{ij}, \alpha_{ij}) = \left[\frac{1}{2}(\sigma'_{ij} - \alpha_{ij})(\sigma'_{ij} - \alpha_{ij})\right]^{1/2},$$

where α_{ij} denotes the coordinates of the center of a yield surface that translates in the direction according to $\dot{\alpha}_{ij} = \dot{\mu}(\bar{\gamma}^p)n_{ij}$ where $n_{ij} = \partial f/\partial \sigma_{ij} = (\sigma'_{ij} - \alpha_{ij})/2\bar{\tau}$ is determined by unloading at various stages of, for example, a simple shear experiment in order to determine the extent of the elastic region. The plastic strain rate function $\Phi(\bar{\tau}, \bar{\gamma}^p)$ characterizes the material's response in simple shear. Its value during such an experiment is equal to the current value of the plastic shear strain rate. The yield surfaces are

$$\frac{1}{2}(\sigma'_{ij}\sigma'_{ij}) = k^2(\bar{\gamma}^p) \qquad \text{for isotropic hardening},$$

$$\frac{1}{2}(\sigma'_{ij} - \alpha_{ij})(\sigma'_{ij} - \alpha_{ij}) = k_0^2 \quad \text{for kinematic hardening},$$

where $k(\bar{\gamma}^p)$ is the quasi-static flow stress in pure shear at a plastic shear strain $\bar{\gamma}^p$, and k_0 is the initial yield stress in shear. For these models the transverse velocity predicted at the free surface is less for kinematic hardening than for isotropic hardening; such a difference was observed in all symmetric impact experiments. Better determination of $\Phi(\bar{\tau}, \bar{\gamma}^p)$ especially for the small plastic strains that occur in symmetric impact experiments, would be helpful. The flow stress increases strongly with increasing plastic strain rate at strain rates above 10^4 s^{-1}. Hydrostatic pressure up to 2 GPa appears to have no significant effect on plastic flow in 6061-T6 aluminum and at most a minor effect on plastic flow in alpha titanium.

In plate impact experiments, the elastic precursor is attenuated as it propagates through the shocked specimen due to plastic straining at the shock front. However, the plastic strain rates (Meir and Clifton [1986]) required to explain the observed precursor decay are much higher than the strain rates that are predicted for known initial dislocation densities in the unshocked specimen, regarded as homogeneous. When dislocation generation at the surfaces is included in the computations, the calculated precursor amplitudes are comparable with measurements only for thin

(3 mm) specimens. When subgrain boundaries are included as additional sources for dislocation generation, the computed velocity-time profiles at 3 and 6.6 mm are in good agreement with measured profiles.

An analysis is presented by Klopp and Clifton [1990] concerning the effect of the inclination of the waves due to the tilt, on the calculation of stress, strain, and strain rate in the specimen. The errors seem to be small.

Frutschy and Clifton [1997] present some modified pressure-shear plate impact experiment to test materials at high temperatures (up to 700°C). The high strain rate experiment is up to 10^6 s^{-1}. In another paper Frutschy and Clifton [1998] present the dynamic response of copper at strain rates of 10^5–10^6 s^{-1} and temperatures up to 700°C. They show that the flow stress increases with increasing strain rate, and decreases with increasing temperature. They compare the experiment with some popular models, but they show that the models do not predict the softening that is observed at large strains.

7.4 Wire Drawing

7.4.1 *Introduction*

The processes of metal extrusion or drawing have been considered within the framework of classical time-independent plasticity theory (see Avitzur [1968]). The theory considered here is to describe the influence of the speed of the process (of the order of 100 m/s) on all the other involved parameters (Cristescu [1975], [1976]). We consider mainly wires which are very thin, less than 0.5 mm in diameter. Let us assume that the main mechanical properties of the material can be described with a viscoplastic constitutive equation. Since the elastic part of the strain will be neglected the simplest possible model is a Bingham-type constitutive equation of the form

$$D_{ij} = \frac{1}{2\eta} \left\langle 1 - \frac{\sigma_m}{\sqrt{3}\sqrt{II_{\sigma'}}} \right\rangle \sigma'_{ij} \tag{7.4.1}$$

where σ_m is the mean yield stress which depends on reduction and is an approximation of the isotropic work-hardening law $\bar{\sigma} = f(\bar{\varepsilon})$ of the material, and $\langle\,\rangle$ is the positive part. Thus the model is viscoplastic rigid.

A second assumption is that the circular conical die remains rigid during plastic flow and that in the domain where plastic flow takes place (domain II) the process is axi-symmetric. Assuming volume incompressibility the following velocity field components in spherical co-ordinate r, θ, φ can be obtained

$$v = v_r = -v_f r_f^2 \left(\frac{\cos\theta}{r^2} \right), \qquad v_\theta = v_\varphi = 0, \tag{7.4.2}$$

in the domain II defined by $r_0 \leq r \leq r_f$, $0 \leq \theta \leq \alpha$, $0 \leq \varphi \leq 2\pi$. According to (7.4.2) the material particles are moving radially towards the apex 0. Since in the domain I (for $r > r_0$) the whole rod is moving as a rigid body with the absolute

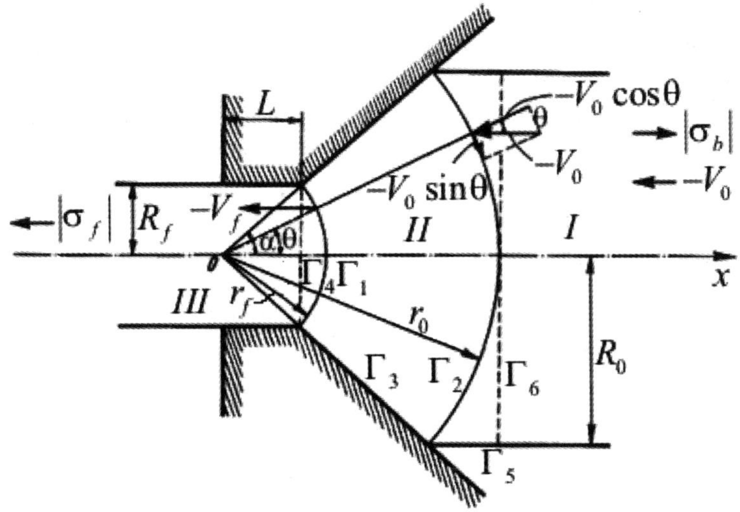

Fig. 7.4.1   Geometry of wire drawing.

velocity v_0 and since in the domain III the rod is moving again as a rigid body with the absolute velocity v_f, the spherical surfaces $r = r_0$ and $r = r_f$ are discontinuity surfaces for the velocity field.

Along the surface Γ_2 $(r = r_0)$ the tangential discontinuity of the velocity field is obtained from (7.4.2) as

$$\Delta v = -v_0 \sin\theta \text{ along } r = r_0, \quad 0 \le \theta \le \alpha \tag{7.4.3}$$

while along the surface Γ_1 $(r = r_f)$ the tangential discontinuity of the velocity is

$$\Delta v = -v_f \sin\theta \text{ along } r = r_f, \quad 0 \le \theta \le \alpha. \tag{7.4.4}$$

Another evident formulae is

$$v_0 R_0^2 = v_f R_f^2, \quad r = \frac{R}{\sin\alpha}. \tag{7.4.5}$$

From (7.4.2) the following components of the rate of strain result, in spherical coordinates,

$$D_{rr} = \frac{\partial v}{\partial r} = 2v_f r_f^2 \frac{\cos\theta}{r^3}, \qquad D_{\theta\theta} = \frac{v}{r} = -v_f r_f^2 \frac{\cos\theta}{r^3}, \qquad D_{\varphi\varphi} = \frac{v}{r},$$

$$D_{r\theta} = \frac{1}{2r}\frac{\partial v}{\partial \theta} = \frac{1}{2}v_f r_f^2 \frac{\sin\theta}{r^3}, \qquad D_{r\varphi} = D_{\theta\varphi} = 0 \tag{7.4.6}$$

and obviously

$$D_{rr} + D_{\theta\theta} + D_{\varphi\varphi} = 0. \tag{7.4.7}$$

Therefore, according to the kinematical velocity field chosen, plastic deformation takes place in region II only, while regions I and III remain rigid. Dissipation is calculated in region II, and dissipation due to the velocity discontinuity along Γ_1 and Γ_2, while along Γ_3 the dissipation is produced by the friction existing there. Γ_5 is stress free. The surface Γ_6 $(x = r_0)$ at the entrance and Γ_4 $(x = r_0 \cos\alpha)$ at the exit of the die are used for the computation of the balance of forces. The semi-angle of the die is denoted by α, the drawing stress by σ_f and a possible back-stress by σ_b.

7.4.2 Basic Equations

For the problem under consideration, taking into account the spherical symmetry and (7.4.6), the constitutive equation can be written as

$$D_{rr} = \frac{1}{2\eta}\left(1 - \frac{\sigma_m}{\sqrt{3}\sqrt{II_{\sigma'}}}\right)\frac{2}{3}(\sigma_{rr} - \sigma_{\theta\theta}),$$

$$D_{\theta\theta} = \frac{1}{2\eta}\left(1 - \frac{\sigma_m}{\sqrt{3}\sqrt{II_{\sigma'}}}\right)\frac{1}{3}(\sigma_{\theta\theta} - \sigma_{rr}),$$

$$D_{r\theta} = \frac{1}{2\eta}\left(1 - \frac{\sigma_m}{\sqrt{3}\sqrt{II_{\sigma'}}}\right)\sigma_{r\theta}. \tag{7.4.8}$$

The second invariant of the stress deviator is

$$II_{\sigma'} = \frac{1}{3}(\sigma_{rr} - \sigma_{\theta\theta})^2 + \sigma_{r\theta}^2. \tag{7.4.9}$$

Using (7.4.6) and (7.4.8) the expression for this invariant can be written as

$$\sqrt{II_{\sigma'}} = \frac{\sigma_m}{\sqrt{3}} + \frac{v_f r_f^2}{r^3}\eta\sqrt{11\cos^2\theta + 1}. \tag{7.4.10}$$

Another group of equations are the equilibrium equations. Since we have $\sigma_{r\varphi} = \sigma_{\theta\varphi} = 0$, $\sigma_{\theta\theta} = \sigma_{\varphi\varphi}$ and since the problem is axially symmetric with respect to φ, the set of equilibrium equations reduces to two

$$\frac{\partial\sigma_{rr}}{\partial r} + \frac{1}{r}\frac{\partial\sigma_{r\theta}}{\partial\theta} + \frac{2(\sigma_{rr} - \sigma_{\theta\theta}) + \sigma_{r\theta}\cot g\theta}{r} = 0,$$

$$\frac{\partial\sigma_{r\theta}}{\partial r} + \frac{1}{r}\frac{\partial\sigma_{\theta\theta}}{\partial\theta} + \frac{3\sigma_{r\theta}}{r} = 0. \tag{7.4.11}$$

From (7.4.6), (7.4.8)$_3$ and (7.4.10) we get

$$\sigma_{r\theta} = \frac{\sin\theta}{\sqrt{11\cos^2\theta + 1}}\frac{\sigma_m}{\sqrt{3}} + \eta v_f r_f^2\frac{\sin\theta}{r^3} \tag{7.4.12}$$

which satisfies the requirement that $\sigma_{r\theta} = 0$ for $\theta = 0$. Introducing (7.4.12) in (7.4.11)$_2$ a differential equation is obtained, which after integration yields

$$\sigma_{\theta\theta} = \frac{\sqrt{3}\sigma_m}{\sqrt{11}}\ln\left[\cos\theta + \sqrt{\frac{1}{11} + \cos^2\theta}\right] + C(r) \tag{7.4.13}$$

where C is a undetermined function depending on r alone.

Any one from the two equations (7.4.8) taken together with (7.4.6), (7.4.10) and (7.4.13) now yields

$$\sigma_{rr} = \frac{\sqrt{3}\sigma_m}{\sqrt{11}} \ln\left[\cos\theta + \sqrt{\cos^2\theta + \frac{1}{11}}\right] + C(r)$$

$$+ 2\sqrt{3}\sigma_m \frac{\cos\theta}{\sqrt{11\cos^2\theta + 1}} + 6v_f r_f^2 \eta \frac{\cos\theta}{r^3}. \tag{7.4.14}$$

To determine the function C for $\theta = 0$, we obtain

$$C(r) = -\frac{4}{3} v_f r_f^2 \eta \frac{1}{r^3} - \frac{\sigma_m}{\sqrt{3}} \frac{168}{12^{3/2}} \ln r + C_1 \tag{7.4.15}$$

where C_1 is an integration constant, which is determined from the condition

$$r = r_0, \quad \theta = 0 \quad \text{we have } \sigma_{rr} = \sigma_{xb}.$$

Thus it is found for C_1,

$$C_1 = \sigma_{xb} - \frac{\sqrt{3}\sigma_m}{\sqrt{11}} \ln\left[1 + \sqrt{1 + \frac{1}{11}}\right] - \frac{14}{3} \frac{v_f r_f^2 \eta}{r_0^3} + \frac{\sigma_m}{\sqrt{3}} 4.04145 \ln r_0 - \sigma_m,$$

and for $\sigma_{\theta\theta}$ we get

$$\sigma_{\theta\theta} = \frac{\sqrt{3}\sigma_m}{\sqrt{11}} \ln\left[\cos\theta + \sqrt{\cos^2\theta + \frac{1}{11}}\right] - \frac{v_f r_f^2 \eta}{3}\left(\frac{4}{r^3} + \frac{14}{r_0^3}\right)$$

$$+ \frac{\sigma_m}{\sqrt{3}} 4.04145 \ln\frac{r_0}{r} + \sigma_{xb} - \frac{\sqrt{3}\sigma_m}{\sqrt{11}} \ln\left[1 + \sqrt{1 + \frac{1}{11}}\right] - \sigma_m \tag{7.4.16}$$

and a similar procedure can be used to determine σ_{rr}. This is an approximate value since we have not taken into account the friction.

We have to compute now the stress power per unit volume

$$\sigma_{ij} D_{ij} = \frac{2\sigma_m}{\sqrt{3}} \sqrt{II_D} + 4\eta II_D, \tag{7.4.17}$$

where

$$II_D = \frac{1}{2} D_{ij} D_{ij} = \frac{1}{2}(D_{rr}^2 + D_{\theta\theta}^2 + D_{\varphi\varphi}^2) + D_{r\theta}^2 + D_{\theta\varphi}^2 + D_{\varphi r}^2. \tag{7.4.18}$$

Using (7.4.6) in (7.4.17) and (7.4.18) we get

$$\sigma_{ij} D_{ij} = \frac{2\sigma_m}{\sqrt{3}} \frac{v_f r_f^2}{2r^3} \sqrt{11\cos^2\theta + 1} + \frac{v_f^2 r_f^4 \eta}{r^6}(11\cos^2\theta + 1).$$

After integrating over the volume of zone II, and using (7.4.5):

$$\dot{W} = 2\pi \sigma_m v_f R_f^2 \left(\ln\frac{R_0}{R_f}\right) f(\alpha) + \frac{2}{3}\pi v_f^2 \frac{R_f}{\sin\alpha}\eta$$

$$\times \left[1 - \left(\frac{R_f}{R_0}\right)^3\right]\left[1 - \cos\alpha + \frac{11}{3}(1 - \cos^3\alpha)\right], \tag{7.4.19}$$

where

$$-f(\alpha) = \frac{\sqrt{11}}{2\sqrt{3}} \frac{1}{\sin^2 \alpha} \left\{ (\cos \alpha)\sqrt{\cos^2 \alpha + \frac{1}{11}} + \frac{1}{11} \ln \left[\cos \alpha + \sqrt{\cos^2 \alpha + \frac{1}{11}} \right] \right.$$
$$\left. - \sqrt{\frac{12}{11}} - \frac{1}{11} \ln \left[1 + \sqrt{\frac{12}{11}} \right] \right\}. \tag{7.4.20}$$

The first term is the time independent while the second term in (7.4.19) corresponds to the time dependent term.

7.4.3 *Friction Laws*

We first consider a Coulomb friction law

$$\tau = \mu \sigma_{\theta\theta} \,. \tag{7.4.21}$$

Here μ is the constant friction coefficient and $\sigma_{\theta\theta}$ is the stress normal to the die surface. Since the velocity along the surface is

$$v = -v_f r_f^2 \frac{\cos \alpha}{r^2}$$

the rate of the work done by the stress vector on conical die surface Γ_3 ($r_0 \geq r \geq r_f, \theta = \alpha$) is

$$\dot{W}_{\Gamma_3} = -\mu \int_{r_0}^{r_f} \sigma_{\theta\theta} v_f r_f^2 \frac{\cos \alpha}{r^2} 2\pi r \sin \alpha \, dr \,.$$

Introducing here (7.4.16) and after some algebra, we get

$$\dot{W}_{\Gamma_3} = -2\mu\pi v_f R_f^2 \cot \alpha \left\{ \frac{4}{9} v_f \eta \frac{\sin \alpha}{R_f} \left[1 - \left(\frac{R_f}{R_0} \right)^3 \right] - \frac{\sigma_m}{\sqrt{3}} 2.0207 \left(\ln \frac{R_0}{R_f} \right)^2 \right.$$
$$+ \ln \frac{R_f}{R_0} \left[\frac{\sqrt{3}\sigma_m}{\sqrt{11}} \ln \left(\cos \alpha + \sqrt{\cos^2 \alpha + \frac{1}{11}} \right) - \frac{14}{3} v_f \eta \frac{\sin \alpha}{R_0} \left(\frac{R_f}{R_0} \right)^2 + \sigma_{xb} \right.$$
$$\left. \left. - \frac{\sqrt{3}\sigma_m}{\sqrt{11}} \ln \left(1 + \sqrt{1 + \frac{1}{11}} \right) - \sigma_m \right] \right\}. \tag{7.4.22}$$

Another law in metal plasticity is that the modulus of shearing stress due to friction is proportional to the yield stress $\tau_Y = \sigma_Y / \sqrt{3}$. For viscoplastic materials this law is generalized as

$$\tau = m\sqrt{II_{\sigma'}} \tag{7.4.23}$$

where m is a constant friction coefficient and $0 \leq m \leq 1$. $m = 0$ means no friction, while $m = 1$ means adherence on the wall. Using (7.4.10) (7.4.23) can be written

$$\tau = m \left[\frac{\sigma_m}{\sqrt{3}} + \frac{v_f r_f^2}{r^3} \eta \sqrt{11 \cos^2 \alpha + 1} \right]. \tag{7.4.24}$$

With this law the rate of work done on the conical die surface is

$$\dot{W}_{\Gamma_3} = 2\pi m v_f R_f^2 \frac{\cos\alpha}{\sin\alpha} \left\{ \frac{\sigma_m}{\sqrt{3}} \ln \frac{R_f}{R_0} + \frac{v_f \eta \sin\alpha}{3R_f} \sqrt{11\cos^2\alpha + 1} \left[\left(\frac{R_f}{R_0}\right)^3 - 1 \right] \right\}.$$

$$(7.4.25)$$

Let us observe that both expressions are variable depending on v_f, η and α. The dependency on r is distinct; in the law (7.4.21) it decreases with decreasing r, while in the law (7.4.23) it increases with increasing r. We assume all the other parameters to stay constants.

7.4.4 *Drawing Stress*

We apply now the theorem of powder expanded, extended to the volume V of the viscoplastic domain. We have

$$\int_V \sigma_{ij} D_{ij} dV + \int_{\Gamma_1} \sigma_i [v_i] \, dS + \int_{\Gamma_2} \sigma_i [v_i] \, dS = \int_{\partial V} \sigma_i v_i \, dS , \qquad (7.4.26)$$

where σ_i is the component of the stress vector on Γ_1, Γ_2 or ∂V while $[v]$ is the jump of the particle velocity at the crossing of the surface Γ_1 or of the surface Γ_2.

The term on the right-hand side of (7.4.26) represents the rate of work done by the stress vector on the surface bounding the volume V. This term can be decomposed into five parts. On the plane Γ_6 where $x = r_0$ we have

$$\dot{W}_{\Gamma_6} = -\pi R_0^2 v_0 \sigma_{xb} , \qquad (7.4.27)$$

while on Γ_4 where $x = r_f$,

$$\dot{W}_{\Gamma_4} = \pi R_f^2 v_f \sigma_{xf} . \qquad (7.4.28)$$

Along Γ_5 the rate of work is zero since this surface is stress free. Along the surfaces where a friction law exists the integral is already computed.

Let us consider now the last term from right. The velocity discontinuity is $[v] = v_0 \sin\theta$. The surface element is

$$dA = r_j^2 \sin\theta \, d\theta \, d\varphi .$$

Thus

$$\int_{\Gamma_2} \sigma_i [v_i] \, dA = 2\pi v_0 r_0^2 \int_0^\alpha \sigma_i \tau_i \sin^2\theta \, d\theta \qquad (7.4.29)$$

where τ_i is the vector tangent to Γ_2 in the meridian plane. From the global condition of equilibrium of the forces which act on the domain I of the bar we have

$$-\pi \sigma_0 R_0^2 + \sigma_0^* \int_{\Gamma_2} dA = 0$$

we determine

$$\sigma_0^* = \frac{\sigma_0 R_0^2}{2r_0^2(1 - \cos\alpha)} = \frac{\sigma_0}{2}(1 + \cos\alpha) .$$

Introducing this value in (7.4.29) we get

$$\int_{\Gamma_2} \sigma_i[v_i]\,dA = -(1+\cos\alpha)\pi v_0 r_0^2 \sigma_0 \int_0^\alpha \sin^3\theta\,d\theta$$

or

$$\int_{\Gamma_2} \sigma_i[v_i]\,dA = -\frac{-\cos\alpha + 1/3\cos^3\alpha + 2/3}{1-\cos\alpha}\pi v_0 R_0^2 \sigma_0\,. \tag{7.4.30}$$

Similar for the surface Γ_4 we get

$$\int_{\Gamma_4} \sigma_i[v_i]\,dA = \pi v_f r_f^2 \sigma_f(1+\cos\alpha)\left[(1-\cos\alpha)+\frac{1}{3}(1-\cos^3\alpha)\right]. \tag{7.4.31}$$

Combining all these formulae, for the second friction law, we have

$$\frac{\sigma_f}{\sigma_m} = \frac{\sin^2\alpha}{2\cos\alpha - \cos^2\alpha - (2/3)\cos^3\alpha - 1/3}\left\{\frac{\sigma_b}{\sigma_m}\left[1+2\frac{-\cos^2\alpha + (1/3)\cos^3\alpha + 2/3}{\sin^2\alpha}\right]\right.$$

$$+2f(\alpha)\ln\frac{R_0}{R_f} + \frac{2}{3}\frac{\eta v_f}{\sigma_m R_f}\frac{1}{\sin\alpha}\left[1-\left(\frac{R_f}{R_0}\right)^3\right]\left[1-\cos\alpha+\frac{11}{3}(1-\cos^3\alpha)\right]$$

$$+\frac{2}{\sqrt{3}}mctg\alpha\left\{\ln\frac{R_0}{R_f} + \frac{\eta v_f}{\sigma_m R_f}\frac{\sin\alpha}{\sqrt{3}}\sqrt{11\cos^2\alpha + 1}\left[1-\left(\frac{R_f}{R_0}\right)^3\right]\right\}\right\}$$

$$+\frac{2}{3}\frac{L}{R_f}m\frac{\sigma_Y}{\sigma_m} \tag{7.4.32}$$

where

$$-f(\alpha) = \frac{\sqrt{11}}{2\sqrt{3}}\frac{1}{\sin^2\alpha}\left\{(\cos\alpha)\sqrt{\cos^2\alpha + \frac{1}{11}} + \frac{1}{11}\ln\left(\cos\alpha + \sqrt{\cos^2\alpha + \frac{1}{11}}\right)\right.$$

$$\left. -\sqrt{\frac{12}{11}} - \frac{1}{11}\ln\left(1 + \sqrt{\frac{12}{11}}\right)\right\}.$$

and

$$N = \frac{\eta v_f}{\sigma_m R_f} \tag{7.4.33}$$

is a "speed effectiveness parameter" expressing the influence of the speed on the energy dissipated in the domain of viscoplastic deformation. Further the term containing the factor m expresses the energy dissipation due to friction; in this term a component due to the speed influence is also present. The final term in Eq. (7.4.32) is due to the friction along the die land of length L. In the same equation the dissipation term due to the presence of discontinuity surfaces Γ_1 and Γ_2 is also involved, but not additively.

Since the rates of deformation involved in viscoplastic deformation in region II are very high, the resultant heating was also considered (Cristescu [1980]), in order to be able to choose the values of the various constants from the constitutive equation correctly. It was assumed that the heating is <u>adiabatic</u>, that is

$$\rho c \frac{dT}{dt}\Big|_{med} = \int_V \sigma_{ij} D_{ij} dV \tag{7.4.34}$$

where c is the specific heat (assumed constant), T is the temperature and $dT/dt|_{med}$ is the mean temperature gradient over the volume V, defined by

$$\frac{dT}{dt}\Big|_{med} = \frac{1}{V} \int_V \frac{dT}{dt} dV.$$

On the right-hand side of (7.4.34) is the stress power over the volume V, which is given by the above formula. This equation becomes

$$\frac{dT}{dt}\Big|_{med} = \frac{\pi v_f R_f^2 \sigma_m}{V \rho c} \left\{ 2f(\alpha) \ln \frac{R_0}{R_f} + \frac{2}{3}\frac{N}{\sin \alpha} \left[1 - \left(\frac{R_f}{R_0}\right)^3 \right] \times \left[1 - \cos \alpha + \frac{11}{3}(1 - \cos^3 \alpha) \right] \right\}.$$

This equation was used to estimate the mean rise of temperature due to fast viscoplastic deformation. For steel wire and $R_0 = 0.5$ mm, $v_f = 1$ m/s, $\alpha = 6°$, $\eta = 0.34$ Nmm^{-2} s, $\rho c = 3.60$ Nmm^{-2} (°C)$^{-1}$, $r\% = 20\%$, an approximate mean temperature rise of $\Delta T_{med} = 84.0°$C (28.2°C due to plastic deformation and 55.8°C due to the speed effect), is obtained, estimating that the particles cross region II in a time interval less than 5×10^{-4} s. For $v_f = 50$ m/s and $\eta = 0.023$ Nmm^{-2} s, the mean temperature rise due to viscolastic deformation is $\Delta T_{med} = 121°$C (29.6°C due to plastic deformation and the remaining due to the speed effect).

Another estimation of temperature rise was done by integrating numerically the equation (7.4.34). This is no more done since the temperature rise is less than 150°C at the most and thus do not change the constants of the materials significantly.

7.4.5 *Comparison with Experimental Data*

We have chosen the experimental data by Wistreich [1955] for copper wires with $R_f = 1.27$ mm and mean yield stress of $\sigma_Y = 349.7$ MN/m^2 have been drawn at a speed of $v_f = 33$ mm/sec, the friction coefficient is $\mu = 0.025$. The viscosity coefficient for copper was taken from Lindholm and Bessey [1969] $\eta = 2.78$ MN/m^{-2} from where $N = 0.2069$.

For these values of the constants, the results shown in Fig. 7.4.2 were obtained. For various reductions in area defined by

$$r\% = 100 \left[1 - \left(\frac{R_f}{R_0}\right)^2 \right]$$

the variation of the drawing stress with semi-cone angle α is plotted. The discrepancy with experimental data existing for α relatively big and/or r small is due to the bulging phenomenon.

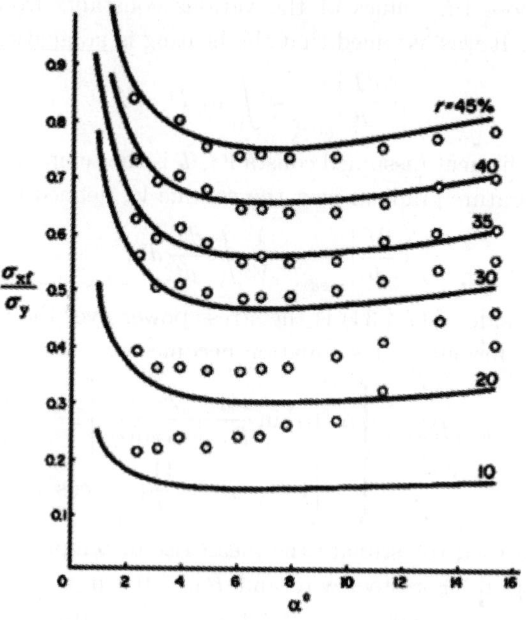

Fig. 7.4.2 Variation of drawing stress with semi-cone angle for various reductions in area: comparison between theory and experiment.

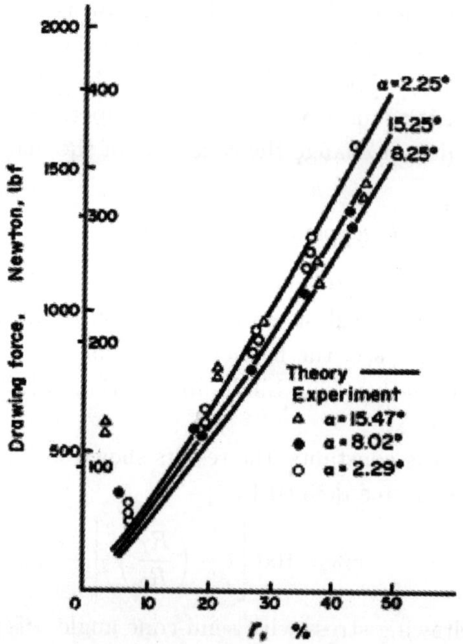

Fig. 7.4.3 Total drawing force for various reductions and area and three semi-cone angles.

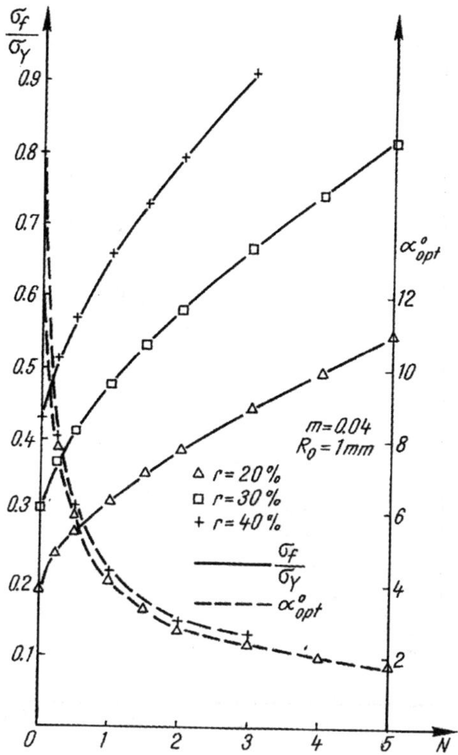

Fig. 7.4.4 Variation of relative drawing stress and optimum die angle with the speed effectiveness parameter N, for different degrees of reduction.

Another comparison with experimental data has been done for the total drawing force as measured by Wistreich for semi-cone die angles equal to: 2.29°, 8.02° and 15.47°. These experimental results are shown in Fig. 7.4.3 together with the theory. For mild steel Manjoine (see Lindholm and Bessey [1969]) and Cristescu [1977] give $D \approx 10^{-1}$–1 s^{-1}, $\eta \approx 10^4$ kNm^{-2} s, for $D \approx 1$–10 s^{-1}, $\eta \approx 1700$ kNm^{-2} s, for $D \approx 10$–10^2 s^{-1}, $\eta \approx 200$ kNm^{-2} s, and for $D \approx 10^2$–10^3 s^{-1}, $\eta \approx 24$ kNm^{-2} s. For higher rates of deformation Campbell [1973] suggests $\eta \approx 2$ kNm^{-2} s and it is varying very little. Let us observe that an increase of v_f does not imply an increase on N, since η may decrease significantly with the increase of v_f. In the example given below, $\sigma_b = 0$ and $L = 0$, the consideration of these two parameters can be done very easily.

In Fig. 7.4.4 are given for three reductions, the optimum relative drawing stress for different values of N (full lines). From this figure it can be seen that the drawing stress increases with reduction and become large for very thin wires.

A comparison with the experiments was done also by Cristescu [1977] for copper of diameter wires $2R_0 = 0.94$ mm and steel wires of $2R_0 = 1.02$ mm, tested slowly.

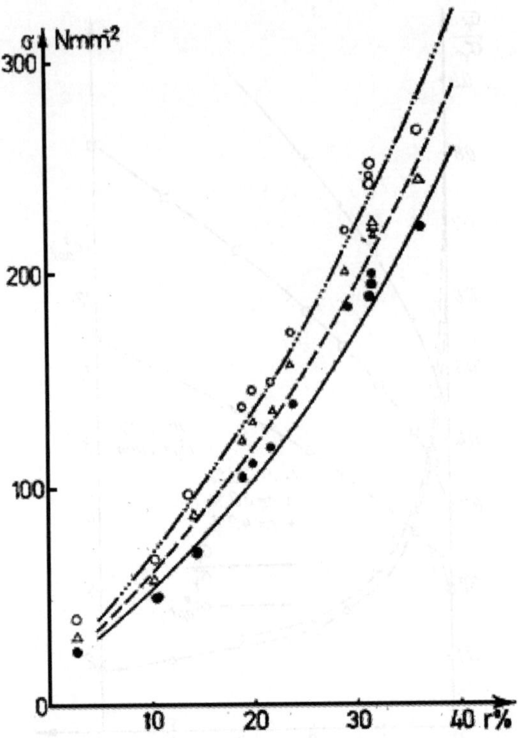

Fig. 7.4.5   Drawing forces for copper wires.

For the die semi-angle $\alpha = 8°$, the length of the land $L = 0.5$ mm, friction coefficient on the land $m = 0.08$, friction on the conical surface $\mu = 0.09$, and viscosity of the copper $\eta = 2.94$ Nmm^{-2} s. Figure 7.4.5 shows some results for $\alpha = 8°$ (smaller figures) and $\alpha = 7°$. That is because $\alpha = 8°$ is closer to the optimum die angle. The lines are the theoretical ones. The figure corresponds to $L = 0.5$ mm, $\mu = 0.06$, $m = 0.08$, and $\eta = 2.94$ Nmm^{-2} s.

Similar results for steel wires.

7.4.6   *Other Papers*

A visio-plastic method, for big deformations, was developed by Dahan and Le Nevez [1983]. First some extrusion pieces are obtained and measured. Then by a finite difference one is obtaining for a rigid work-hardening and isotropic material, the stress deviator and the strain.

Also a Bingham model was used by Ionescu and Vernescu [1988], to describe numerically the wire drawing. The friction conditions are of two types: a Coulomb friction law and a viscoplastic one. The drawing speeds is low, with the viscosity coefficient 2.94 Ns/mm^2.

The same theory was presented in a book devoted to "Mathematical Models in Working Metals" by Camenschi and Şandru [2003]. The book contains much more than drawing of bars but this theory is presented. In reality the book presents some papers published since 1978. For instance the drawing of bars with asymptotic developments was firs published in Camenschi *et al.* [1979] for a friction law of the form (7.4.23) and in Camenschi *et al.* [1983] for a Coulomb friction law.

Concerning the drawing theory one is assuming $N < 1$, that is slow processes. The solution is obtained for the incompressibility condition by introducing the flow function $\psi = \psi(r,\theta) = R_2^2 v_f \psi^0(r^0,\theta)$, and

$$v_r^0 = -\frac{1}{r^{02}\sin\theta}\frac{\partial\psi^0}{\partial\theta}, \quad v_\theta^0 = -\frac{1}{r^0\sin\theta}\frac{\partial\psi^0}{\partial r^0}$$

where the upper index ^{0} is for dimensionless magnitudes. For asymptotic developments one is writing

$$\psi^0(r^0,\theta) = \psi_0^0(\theta) + N\psi_1^0(r^0,\theta) + O(N^2),$$

$$p^0(r^0,\theta) = p_0^0(r^0,\theta) + Np_1^0(r^0,\theta) + O(N^2).$$

and the solution is obtained in dimension quantities with a formula for the stress as well. One is obtaining the drawing force as

$$\frac{|\sigma_f|}{\sigma_Y} = \frac{2}{\sqrt{3}N}\left[\left(1 - \frac{R_2^3}{R_1^3}\right)F(\alpha,m) + N\ln\left(\frac{R_1}{R_2}\right)^3 G(\alpha,m)\right]$$

where one have used the friction law (7.4.23) and

$$\gamma = \frac{m}{\sqrt{1-m^2}},$$

$$I_k = \int_0^\alpha \frac{\sin^{2k-1}\theta\, d\theta}{\sqrt{1-\sin^2\theta(\lambda_1 - \lambda_2\sin^2\theta)}}, \quad k = 1,2,3$$

$$\lambda_1 = \frac{3\gamma(\sqrt{3}\sin 2\alpha - 3\gamma\cos 2\alpha + 2\gamma)}{\sin^2\alpha(\sqrt{3}\cos\alpha + 3\gamma\sin\alpha)^2},$$

$$\lambda_2 = \frac{6\gamma^2}{\sin^2\alpha(\sqrt{3}\cos\alpha + 2\gamma\sin\alpha)^2},$$

$$F(\alpha,m) = \frac{(\cos^3\alpha/2)(2\gamma + \sqrt{3}\sin 2\alpha)}{\gamma(\sin\alpha/2)(1 + 2\cos\alpha) + \sqrt{3}\cos\alpha\cos(\alpha/2)},$$

$$G(\alpha,m) = \frac{(m/3)\cos\alpha\sin^2\alpha - \sin\alpha(\gamma\sin\alpha - (2/\sqrt{3})\cos\alpha)I_1}{(1-\cos\alpha)[\sin 2\alpha - (\gamma/\sqrt{3})(1-\cos\alpha)(1+2\cos\alpha)]}$$

$$- \frac{(\gamma(\sqrt{3}\sin 2\alpha + 6\gamma\cos 2\alpha - 4\gamma)I_2 + 4\gamma^2 I_3)/(\sqrt{3}\sin 2\alpha + 6\gamma\sin^2\alpha)}{(1-\cos\alpha)[\sin 2\alpha - (\gamma/\sqrt{3})(1-\cos\alpha)(1+2\cos\alpha)]}.$$

The authors have determined also several optimum angles for the drawing, depending on friction, reduction and N. They have also considered a Coulomb friction law in Camenschi *et al.* [1983].

The same authors, Şandru and Camenschi [1979] have considered the rolling of a material, by assuming that the rolls can be approximated by two straight lines. They apply an asymptotic series development.

The kinetic and dynamic effects, on the upper bound loads in metal forming processes, are due to Tirosh and Kobayashi [1976]. It is question of forging, extrusion and piercing. The same kind of approach was applied for rolling of viscoplastic materials (Tirosh *et al.* [1985]). Since the arc of contact area is practically small ($\alpha \approx 5$ deg) and the thickness of the sheet is small with respect to the roller diameter, the authors replace the arc of the contact area by a straight line. The strains are

$$D_{rr} = \frac{1}{2\eta} \left(1 - \frac{\sigma_m}{\sqrt{3}\sqrt{II_{\sigma'}}} \right) \frac{\sigma_{rr} - \sigma_{\theta\theta}}{2} ,$$

$$D_{r\theta} = \frac{1}{2\eta} \left(1 - \frac{\sigma_m}{\sqrt{3}\sqrt{II_{\sigma'}}} \right) \tau_{r\theta} ,$$

and repeat the theorem of powder expansion and also with the same friction laws.

The comparison with the tests seems good. Comparisons with experiments and with FEM are done. For instance in Fig. 7.4.6 a comparison with the data by Shida and Awazuhara [1973] and the FEM method by Li and Kobayashi [1982] are done.

In another paper by Iddan and Tirosh [1996] one renounces at this assumption.

The dynamic fracture by spallation in metals was considered by Gilman and Tuler [1970]. The stress state associated with plane strain is described, and the effect of the initial conditions and the time dependent conditions of the material on spallation is discussed.

Fu and Luo [1995] consider the rigid-viscoplastic FEM to analyze the viscoplastic forming process. The general rigid-viscolastic FEM formulation is given and specific formulations of isothermal forging processes and superplastics forming are listed also. As an application, the combined extension process of pure lead, which is strain-rate sensitive at room temperature, is analyzed. The simulated results reveal the variation of the forming load with the stroke and its dependence on conditions. On the basis of the metal flow patterns defined by the rigid-viscoplastic finite-element method, the change of the position of the neutral layer is given and it is found that the occurrence of folding at the flange may be attributed mainly to an abnormal flow pattern. Moreover, the calculated results bring to light the rule of deformation distribution and its dependence on strain rate.

Alexandrov and Alexandrova [2000], assuming a rigid viscoplastic material model, show that the velocity fields adjacent to the surfaces of maximum friction must satisfy sticking conditions. This means that the stress boundary conditions, the maximum friction law, may be replaced by the velocity boundary condition.

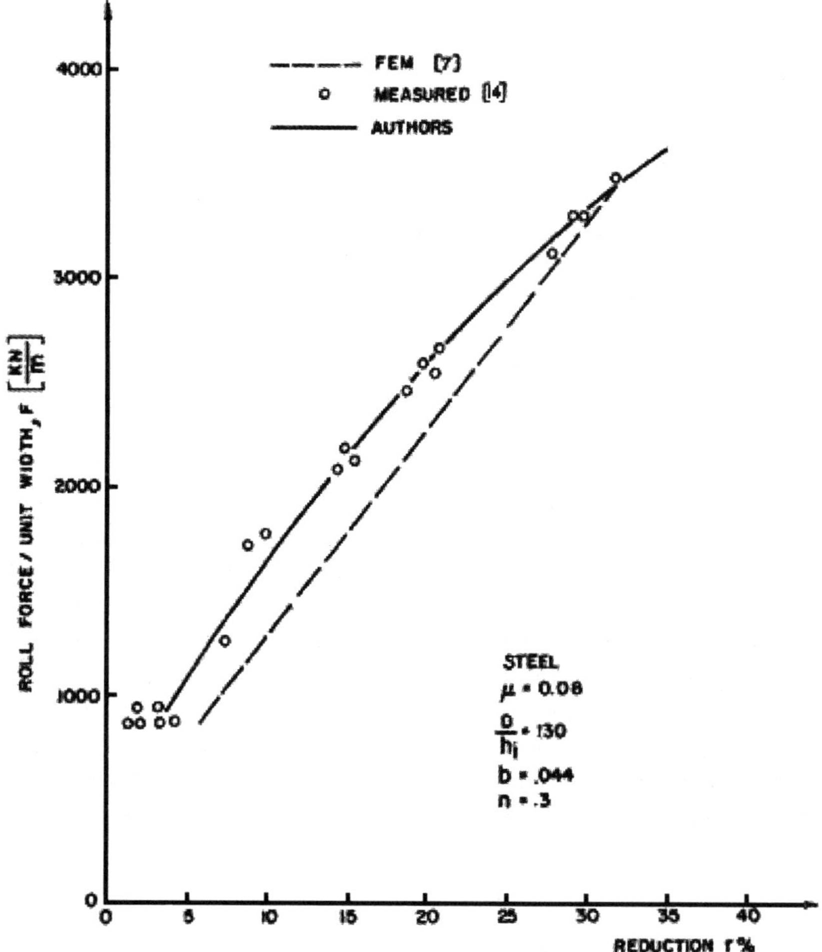

Fig. 7.4.6   Comparisons between measured, predicted, and computed roll force versus reduction for steel.

For planar flows, they show that plastic deformation in the vicinity of maximum friction surfaces is possible.

The metal forming processes of aluminum-alloy wheel forging at elevated temperature are analyzed by the finite element method by Kim *et al.* [2002]. A coupled thermo-mechanical model for the analysis of plasic deformation and heat transfer is adapted in the finite element formulation.

The rolling problem is not mentioned in the present book. We would like to mention the paper by Angelov [2004] in which a variational analysis of a rigid-plastic rolling problem was considered. The material is rigid-plastic, strain-rate sensitive and incompressible. A nonlinear friction law is considered.

7.5   Floating Plug

7.5.1   *Formulation of the Problem*

In the present text the geometry of the floating plug is studied in order to optimize the tube drawing process, i.e., in order to increase the drawing speed, to increase the reduction per pass, to reduce the consumption of energy used, etc. For this purpose the theory of fast tube drawing with floating plug is formulated and a formula is given expressing the drawing stress as function of various parameters involved in the drawing process (drawing speed, semi-cone die angle α, tube geometry, mechanical properties of the material, friction coefficients and plug geometry).

Some data, mainly of experimental character and of industrial practice can be found in Bisk and Shveikin [1963] and Perlin [1957]. From the experiments described in these books it results that when the drawing speed is increased from 20 m/min to 60 m/min the drawing force increases several times (two or even three times). The significant influence of the speed on the wire drawing process was mentioned. Since the rates of strains are higher than 10^2 s^{-1}, one has to describe theoretically this process with a viscoplastic constitutive equation (Cleja and Cristescu [1979]).

Due to the geometrical symmetry of the die, plug and of the tube the problem can be considered to be axisymmetrical. For this reason the geometry of the deformation process was represented in Fig. 7.5.1 in a symmetry plane. The tube of initial thickness $R_1 - R_3$ enter in the die with the speed v_0 and leaves it with the speed v_f. The thickness of the tube at the exit is $R_2 - R_4$. Three domains can be distinguished: in domains I and III the material is assumed rigid, while in domain II the viscoplastic deformation takes place and a Bingham type of constitutive equation

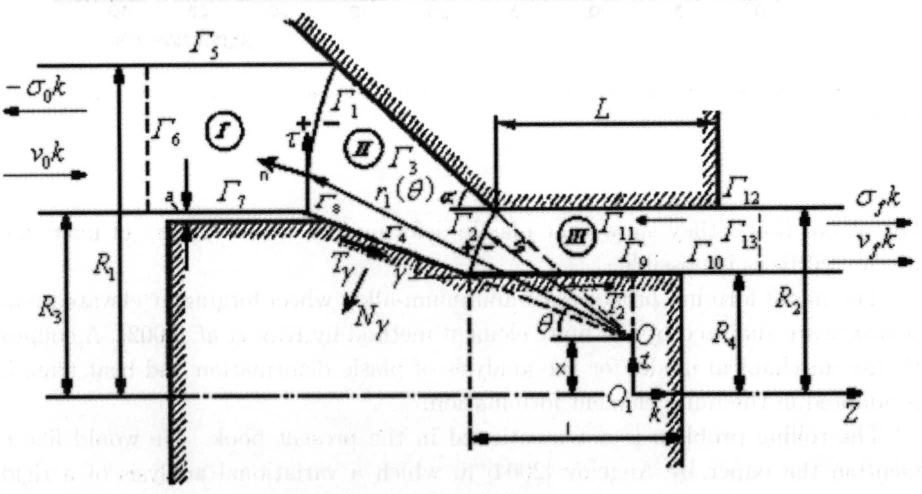

Fig. 7.5.1   Geometry of floating plug.

$$D_{ij} = \frac{1}{2\eta} \left\langle 1 - \frac{\sigma_m}{\sqrt{3}\sqrt{II_{\sigma'}}} \right\rangle \sigma'_{ij} \tag{7.5.1}$$

is satisfied. Here again σ_m is the "mean" yield stress, i.e., the area under the universal stress–strain curve divided by the equivalent strain, while

$$\langle Z \rangle = \begin{cases} Z & \text{if } Z > 0 \\ 0 & \text{if } Z \le 0. \end{cases} \tag{7.5.2}$$

It is assumed that the domains previously mentioned are separated by the discontinuity surfaces Γ_1 and Γ_2. Generally it is assumed that one of the two curves Γ_1 or Γ_2 has a prescribed shape and the second one is then determined accordingly. In what follows Γ_2 is taken as an arc of a circle of radius r_2.

It will be assumed that in the domain II the material flows radially towards the pole 0 situated at the distance x from the symmetry axis; in what follows x is to be determined.

Here we will determine the semi-angle γ of the floating plug. From the point of view of the optimization of the drawing process, this angle is a fundamental parameter of the geometry of the floating plug.

Let V be the domain filled up with the viscoplastic material and $\Gamma = \Gamma_\sigma \cup \Gamma_v = \partial V$, where Γ_σ is the portion of the boundary where the stress vector is prescribed and Γ_v the one on which the velocity field is prescribed. Let $\mathbf{v}$ be a kinematic admissible field (i.e., $\operatorname{div} \mathbf{v} = 0$ in V and $\mathbf{v}|_{\Gamma_v} = \mathbf{v}_0$). We denote by D the set of kinematic admissible fields on V. Then the functional

$$J(\mathbf{v}) = \int_V \left(\frac{1}{2}\sigma_{ij}D_{ij} + \frac{\sigma_m}{\sqrt{3}}\sqrt{II_D} \right) dV + \sum_{j=1}^{2} \int_{\Gamma_j} [\mathbf{t} \cdot \mathbf{v}] \, d\sigma$$

$$- \int_V \mathbf{X} \cdot \mathbf{v} \, dV - \int_{\Gamma_\sigma} \mathbf{t} \cdot \mathbf{v} \, d\sigma \tag{7.5.3}$$

is minimized. Here $[f] = f^+ - f^-$ is the jump of f at the crossing of the considered surface.

Let $D_\gamma \subset D_0$ be a family of kinematic admissible fields depending on a parameter γ. The kinematic field which corresponds to that value of γ for which $J(\mathbf{v})$ on D_γ is minimum will be determined.

7.5.2 *Kinematics of the Deformation Process*

From

$$r_2 \sin\alpha = R_2 - x, \quad r_2 \sin\gamma = R_4 - x \tag{7.5.4}$$

the following expression for x

$$x = \frac{R_2 \sin\gamma - R_4 \sin\alpha}{\sin\gamma - \sin\alpha} \tag{7.5.5}$$

and for r_2

$$r_2 = \frac{R_2 - x}{\sin \alpha} = \frac{R_2 - R_4}{\sin \alpha - \sin \gamma} \qquad (7.5.6)$$

are obtained.

The relation between the velocities v_0 at the entrance and v_f at the exit is obtained as

$$v_0 = \frac{R_2^2 - R_4^2}{R_1^2 - R_3^2} v_f \,. \qquad (7.5.7)$$

Writing now the condition of equal rates of flow for a circular annulus comprised at the entrance between radii $r_1(\theta)\sin\theta + x$ and R_3 and at the exit between $r_2\sin\theta + x$ and R_4, the equation of the discontinuity surface Γ_1

$$r_1(\theta) = \frac{1}{\sin\theta}\left\{\sqrt{[(r^2\sin\theta + x)^2 - R_4^2]\frac{R_1^2 - R_3^2}{R_2^2 - R_4^2} + R_3^2} - x\right\} \qquad (7.5.8)$$

is obtained. Here $r_1(\theta)$ can be either a decreasing or increasing function of θ.

In order to determine a kinematic admissible velocity field we introduce the curvilinear coordinates θ, r, φ by

$$X = (r\sin\theta + x)\cos\varphi\,, \quad Y = (r\sin\theta + x)\sin\varphi\,, \quad Z = -r\cos\theta \qquad (7.5.9)$$

where X, Y, Z are the Cartesian coordinates of a point P belonging to the domain II, the OZ axis is the symmetry axis, and $\theta \in [\gamma, \alpha]$, $r \in [r_2, r_1[\theta]]$, $\varphi \in [0, 2\pi]$ (see Fig. 7.5.1 which corresponds to $\varphi = 0$). Denoting by v_r, v_θ and v_φ the components of the particle velocity in the curvilinear coordinates chosen, and taking into account that due to symmetry $v_\theta = v_\varphi = 0$, from the incompressibility condition, we get

$$v_r = \frac{C(\theta)}{r(r\sin\theta + x)}$$

where the function $C(\theta)$ is determined from the continuity of the normal component of the velocity at the crossing of the surface Γ_2, i.e., for $r = r_2 : v_r = v_f\cos\theta$. The following velocity field is thus found

$$v_r = v_f \frac{r_2(r_2\sin\theta + x)}{r(r\sin\theta + x)}\cos\theta\,, \quad \theta \in (\gamma, \alpha)\,, \quad r \in (r_2, r_1(\theta))$$

$$v_\theta = v_\varphi = 0\,. \qquad (7.5.10)$$

The components of the rate of strain are

$$D_{rr} = -v_f r_2(r_2\sin\theta + x)\cos\theta\frac{2r\sin\theta + x}{[r(r\sin\theta + x)]^2}\,,$$

$$D_{\theta\theta} = v_f r_2 \frac{r_2\sin\theta + x}{r^2(r\sin\theta + x)}\cos\theta\,,$$

$$D_{\varphi\varphi} = v_f r_2 \frac{r_2\sin\theta + x}{r(r\sin\theta + x)^2}\cos\theta\sin\theta\,,$$

$$D_{\theta r} = -v_f r_2 \frac{r(r_2\sin\theta + x) + x^2\sin\theta - xr_2\cos2\theta}{2r^2(r\sin\theta + x)^2}\,,$$

$$D_{\theta\varphi} = D_{\varphi r} = 0\,. \qquad (7.5.11)$$

With these formulas the second invariant of the rate of strain tensor is

$$2II_D = D_{ij}D_{ij} = \left[\frac{v_f r_2}{r^2(r\sin\theta + x)^2}\right]^2 (Ar^2 + Br + C),\tag{7.5.12}$$

where

$$A = 6\cos^2\theta\sin^2\theta(r_2\sin\theta + x)^2 + \frac{1}{2}(r_2\sin^3\theta + x)^2,$$

$$B = 6x\sin\theta\cos^2\theta(r_2\sin\theta + x)^2 + x(r_2\sin^3\theta + x)(x\sin\theta - r_2\cos 2\theta),$$

$$C = 2x^2\cos^2\theta(r_2\sin\theta + x)^2 + \frac{x^2}{2}[x\sin\theta - r_2\cos 2\theta]^2.\tag{7.5.13}$$

With the help of the invariant II_ε for the Bingham model the stress power can be expressed as

$$\sigma_{ij}D_{ij} = \frac{2\sigma_m}{\sqrt{3}}\sqrt{II_D} + 4II_D,\tag{7.5.14}$$

where σ_{ij} are the stress components.

For the whole volume V of the domain II defined by $\theta \in (\gamma, \alpha)$, $r \in (r_2, r_1(\theta))$, $\varphi \in [0, 2\pi]$, using (7.5.12)–(7.5.14) we get

$$\int_V \sigma_{ij}D_{ij}\,dv = \frac{4\pi\sigma_m}{\sqrt{6}}v_f r_2 I_1(\gamma, \alpha; r_2, r_1(\theta)) + 4\pi\eta(v_f r_2)^2 I_2(\gamma, \alpha; r_2, r_1(\theta)),$$

$$\tag{7.5.15}$$

where the following notations

$$I_1(\gamma, \alpha) = \int_\gamma^\alpha \int_{r_2}^{r_1(\theta)} \frac{\sqrt{Ar^2 + Br + C}}{r(r\sin\theta + x)}\,dr d\theta\tag{7.5.16}$$

and

$$I_2(\gamma, \alpha) = \int_\gamma^\alpha \int_{r_2}^{r_1(\theta)} \frac{Ar^2 + Br + c}{r^3(r\sin\theta + x)^3}\,dr d\theta\tag{7.5.17}$$

were used, together with the Jacobian of (7.5.9).

7.5.3 *Friction Forces*

Friction forces are involved at the interfaces between the tube and die and between the tube and the plug.

On the conical surface of the plug it will be assumed that the friction between the rigid surface of the plug and the viscoplastic body, can be described by

$$|t_\tau| = m\sqrt{II_{\sigma'}} = k + 2\eta\sqrt{II_D}.\tag{7.5.18}$$

The friction force T_2 between the tube and the plug land at exit, Γ_9 (see Fig. 7.5.1) has the direction k and the modulus

$$|\mathbf{T}_2| = 2\pi m_2 \tau_2 l R_4\tag{7.5.19}$$

where m_2 is the friction factor, τ_2 is the actual shearing yield stress of the tube at the exit while l is the length of the plug land at the exit.

The resultant of the normal forces on the conical portion of the floating plug ($\theta = \gamma$) has the modulus

$$N_\gamma = 2\pi \int_{r_2}^{r_1(\gamma)-\varepsilon} T_{\theta\theta}|_{\theta=\gamma}(r\sin\gamma + x)\,dr\,, \qquad (7.5.20)$$

$T_{\theta\theta}$ being the normal component on the conical portion of the plug. The contact between the tube and the plug, on the conical portion, is expressed by a relationship between r_2 and $r_1(\gamma) - \varepsilon$ with $a = \varepsilon\sin\gamma$, representing the clearance between the maximum diameter of the plug and the inside diameter of the tube at entrance.

The magnitude of $T_{\theta\theta}$ will be obtained assuming that the friction law (7.5.18) is giving the same tangential stress with the one given by a law of Coulomb type $|t_\tau| = \mu|T_{\theta\theta}|$, μ being the corresponding friction factor. Thus

$$\mu T_{\theta\theta} = m_4 \sqrt{II_{\sigma'}}\,. \qquad (7.5.21)$$

This suggestion is done by the computations done at the bar.

On the conical portion of the plug the modulus of the friction force is

$$T_\gamma = 2\pi m_4 \int_{r_2}^{r_1(\gamma)-\varepsilon} \sqrt{II_{\sigma'}}|_{\theta=\gamma}(r\sin\gamma + x)\,dr\,. \qquad (7.5.22)$$

The equilibrium condition of all the forces which act on the floating plug is

$$\mathbf{N}_\gamma + \mathbf{T}_2 + \mathbf{T}_\gamma = 0 \qquad (7.5.23)$$

since the cylindrical portion of the plug, of radius R_3, is not in contact with the interior of the tube.

Projecting now (7.5.18) on the symmetry axes of the plug and using (7.5.19), (7.5.20) and (7.5.22), we obtain the following equation which determine

$$l = \frac{m_4\tau_2^{-1}}{m_2 R_4}\left(\frac{\sin\gamma}{\mu} - \cos\gamma\right) I_3(\gamma; r_2, r_1(\gamma) - \varepsilon) \qquad (7.5.24)$$

with the notation

$$I_3(\gamma; r_2, r_1(\gamma) - \varepsilon) = \int_{r_2}^{r_1(\gamma)-\varepsilon} \sqrt{II_{\sigma'}}|_{\theta=\gamma}(r\sin\gamma + x)\,dr$$

$$= \frac{1}{2}k\sin\gamma[(r_1(\gamma) - \varepsilon)^2 - r_2^2] + xk(r_1(\gamma) - \varepsilon - r_2)$$

$$+ 2\eta \int_{r_2}^{r_1(\gamma)-\varepsilon} \sqrt{II_D}|_{\theta=\gamma}(r\sin\gamma + x)\,dr\,. \qquad (7.5.25)$$

In the next paragraph we first determine the value of γ which minimize the functional $J(\mathbf{v})$ and which is used to determine the length l of the cylindrical part of the plug.

7.5.4 *Determination of the Shape of the Floating Plug*

For the problem considered here the functional $J(\mathbf{v})$ on $U_{ad}(\gamma)$ becomes a function of γ

$$j(\gamma) = 2\pi v_f \left\{ \frac{1}{2}\sigma_0(R_2^2 - R_4^2) - \frac{\pi(R_1^2 - R_3^2)}{A(S_1)}I_5 + \sqrt{2}kr_2I_1(\gamma, \alpha) \right.$$

$$+ km_3 r_2 R_2 \ln \frac{r_1(\alpha)}{r_2}\cos\alpha + km_4 r_2 R_4 \ln \frac{r_1(\gamma) - \varepsilon}{r_2}\cos\gamma$$

$$+ \eta v_f r_2^2 \left[\sqrt{2}m_3 R_2 I_4(\alpha, r_2, r_1(\alpha))\cos\alpha \right.$$

$$\left. + \sqrt{2}m_4 R_4 I_4(\gamma, r_2, r_1(\gamma) - \varepsilon) + I_2(\gamma\alpha) \right]$$

$$+ m_2\tau_2 l R_4 + m_1\tau_1 L R_2 \tag{7.5.26}$$

with the notations I_j introduced by the formulae (7.5.16) and (7.5.17) and $A(S_1)$ is the area of the surface S_1.

Proof. For the computation of $J(\mathbf{v})$ on $U_{ad}(\gamma)$ we use the formula

$$\operatorname{div}\boldsymbol{\sigma} + \rho\mathbf{b} = 0$$

and we observe that the integrals on the volume are computed by (7.5.15)–(7.5.17), with the remark that I_2 is now multiply with $1/2$. The surfaces S_j involved, correspond to the values $j = 3, 5\text{–}12$, while the surfaces S_k, correspond to the value $k = 1, 4$. S_2 is not involved in the computation of $J(\mathbf{v})$ since on this surface the stress vector is not known.

On the surface S_3 — since Γ_3 is a flow line — the speed has the tangential component equal to v_r, and the modulus of the tangential vector of the stress is given by (7.5.18). Thus

$$\int_{S_3} \mathbf{t} \cdot \mathbf{v}\, dA = -2\pi m_3 \int_{r_2}^{r_1(\alpha)} v_r \sqrt{II_{\sigma'}}|_{\theta=\alpha}(r\sin\alpha + x)\, dr. \tag{7.5.27}$$

With (7.5.10) this relation becomes

$$\int_{S_3} \mathbf{t} \cdot \mathbf{v}\, dA = -2\pi m_3 v_f r_2 R_2 I_6(\alpha; r_2, r_1(\alpha))\cos\alpha \tag{7.5.28}$$

with the notation

$$I_6(\alpha; r_2, r_1(\alpha)) = \int_{r_2}^{r_1(\alpha)} \frac{\sqrt{II_{\sigma'}|_{\theta=\alpha}}}{r}\, dr. \tag{7.5.29}$$

Using now (7.5.18), (7.5.29) is now written

$$I_6(\alpha; r_2, r_1(\alpha)) = k\ln\frac{r_1(\alpha)}{r_2} + \sqrt{2}\eta v_f r_2 I_4(\alpha; r_2, r_1(\alpha)) \tag{7.5.30}$$

where

$$I_4(\alpha; r_2, r_1(\alpha)) = \int_{r_2}^{r_1(\alpha)} \frac{\sqrt{Ar^2 + Br + C}}{r^3(r\sin\theta + x)}\bigg|_{\theta=\alpha} dr. \tag{7.5.31}$$

In a similar way we compute the integral on S_4 which becomes

$$\int_{S_4} \mathbf{t} \cdot \mathbf{v}\, dA = -2\pi m_4 \int_{r_2}^{r_1(\alpha)-\varepsilon} v_r \sqrt{II_{\sigma'}}\big|_{\theta=\gamma}(r\sin\alpha + x)\, dr$$

$$= -2\pi m_4 v_f r_2 R_4 I_6(\gamma; r_2, r_1(\gamma) - \varepsilon)\cos\gamma \tag{7.5.32}$$

while I_6 is computed with (7.5.29)–(7.5.31), with the integral limits taken correspondingly. The surface S_4 is a surface of discontinuity for the vector field, $[\mathbf{v}]|_{S_4} = -\mathbf{v}$.

The integral S_6 is

$$\int_{S_6} \mathbf{t} \cdot \mathbf{v}\, dA = -\pi v_0 \sigma_0 (R_1^2 - R_3^2). \tag{7.5.33}$$

The integral S_1 will be computed from the term $[\mathbf{v}] \cdot \mathbf{t}$, where $\mathbf{t}$ is the stress vector which act on this surface. In the symmetry plane considered, the tangent and normal unit vectors on S_1, defined by the equation $r = r_1(\theta)$, defined by (7.5.8), are

$$\boldsymbol{\tau} = \frac{1}{\sqrt{r_1^2 + (r_1')^2}}[(-r_1'\cos\theta + r_1\sin\theta)\mathbf{k} + (r_1'\sin\theta + r_1\cos\theta)\mathbf{i}]$$

$$\mathbf{n} = \frac{1}{\sqrt{r_1^2 + (r_1')^2}}[-(r_1'\sin\theta + r_1\cos\theta)\mathbf{k} + (r_1\sin\theta - r_1'\cos\theta)\mathbf{i}]. \tag{7.5.34}$$

The tangential components of the velocity are:

$$v_\tau^+ = \mathbf{v}_0 \cdot \boldsymbol{\tau} = \frac{v_0}{\sqrt{(r_1')^2 + r_1^2}}(r_1\sin\theta - r_1'\cos\theta)$$

$$v_\tau^- = \mathbf{v}_r \cdot \boldsymbol{\tau} = (v_r\cos\theta\,\mathbf{k} - v_r\sin\theta\,\mathbf{i}) \cdot \boldsymbol{\tau} = -\frac{v_r r_1'}{\sqrt{(r_1')^2 + r_1^2}}. \tag{7.5.35}$$

The normal component of the velocity is continuous at the crossing of S_1, and is

$$v_n = \mathbf{v}_0 \cdot \mathbf{n} = -v_0\frac{r_1'\sin\theta + r_1\cos\theta}{\sqrt{(r_1')^2 + r_1^2}}. \tag{7.5.36}$$

The traction vector which act on the surface S_1 (as a part of the boundary of III, thus of normal $\mathbf{n}$) is determined from a mean formula, obtained from the equilibrium forces which act on the domain III:

$$-\pi(R_1^2 - R_3^2)\sigma_0\mathbf{k} + \sigma_0^* A(S_1)\mathbf{k} = 0. \tag{7.5.37}$$

For the computation of the area $A(S_1)$ we take care of the element of the surface:

$$dA = (r_1\sin\theta + x)\sqrt{(r_1')^2 + r_1^2}\, d\theta d\varphi \tag{7.5.38}$$

with $r = r_1(\theta)$ given by (7.5.8) and thus

$$A(S_1) = 2\pi \int_\gamma^\alpha (r_1 \sin\theta + x)\sqrt{(r_1')^2 + r_1^2}\, d\theta\,. \tag{7.5.39}$$

From (7.5.37) and (7.5.39) we obtain:

$$\sigma_0^* = \frac{\pi(R_1^2 - R_3^2)}{A(S_1)}\sigma_0\,. \tag{7.5.40}$$

The stress vector on the surface S_1 is given by the formula:

$$\mathbf{t} = \mathbf{t}(\mathbf{x}, \mathbf{n}) = -\sigma_0^*\mathbf{k}\,. \tag{7.5.41}$$

Thus from (7.5.35), (7.5.40), (7.5.41) and (7.5.34) we find:

$$\int_{S_1} [\mathbf{v}] \cdot \mathbf{t}\, dA = \frac{\pi(R_1^2 - R_3^2)}{A(S_1)}\sigma_0 2\pi \int_\gamma^\alpha \frac{v_0(r_1 \sin\theta - r_1'\cos\theta) + v_r r_1'}{\sqrt{(r_1')^2 + r_1^2}}$$

$$\times\, (-r_1'\cos\theta + r_1\sin\theta)(r_1\sin\theta + x)\, d\theta\,. \tag{7.5.42}$$

We introduce the notation:

$$I_5 = \frac{v_0}{v_f} \int_\gamma^\alpha \left[(r_1\sin\theta - r_1'\cos\theta) + \frac{v_r}{v_0}r_1'\right] \frac{(r_1\sin\theta - r_1'\cos\theta)(r_1\sin\theta + x)}{\sqrt{(r_1')^2 + r_1^2}}\, d\theta \tag{7.5.43}$$

and the relation (7.5.42) becomes:

$$\int_{S_1} \mathbf{t} \cdot [\mathbf{v}]\, dA = -\frac{\pi(R_1^2 - R_3^2)}{A(S_1)} 2\pi\sigma_0 v_f I_5\,. \tag{7.5.44}$$

On the surfaces S_9 and S_{11}, taking into account the friction forces on the plug, we have:

$$\int_{S_9} \mathbf{t} \cdot \mathbf{v}\, dA = -2\pi m_2 \tau_2 v_f l R_4\,,$$

$$\int_{S_{11}} \mathbf{t} \cdot \mathbf{v}\, dA = -2\pi m_1 \tau_1 v_f L R_2\,, \tag{7.5.45}$$

where τ_j (with $j = 1, 2$) are the plasticity limits at shearing, corresponding to the surfaces, and l the cylindrical lengths of the plug at the exit is given by (7.5.24).

The integral S_4 is computed with (7.5.32):

$$\int_{S_4} \mathbf{t} \cdot [\mathbf{v}]\, dA = -\int_{S_4} \mathbf{t} \cdot \mathbf{v}\, dA\,. \tag{7.5.46}$$

Introducing the integrals already computed we get:

$$J(\mathbf{v}) = 2\sqrt{2}\pi k v_f r_2 I_1(\gamma, \alpha) + 2\pi\eta(v_f r_2)^2 I_2(\gamma, \alpha) - 2\pi\sigma_0 v_f I_5 \frac{\pi(R_1^2 - R_3^2)}{A(S_1)}$$

$$+ 2\pi m_3 v_f r_2 R_2 \cos\alpha \left[k \ln \frac{r_1(\alpha)}{r_2} + \sqrt{2}\eta v_f r_2 I_4(\alpha, r_2, r_1(\alpha))\right]$$

$$+ 2\pi m_4 v_f r_2 R_4 \cos\gamma \left[k \ln \frac{r_1(\gamma) - \varepsilon}{r_2} + \sqrt{2}\eta v_f r_2 I_4(\gamma, r_2, r_1(\gamma) - \varepsilon)\right]$$

$$+ \pi v_0 \sigma_0 (R_1^2 - R_3^2) \tag{7.5.47}$$

which will be denoted by $j(\gamma)$ and which by minimization will determine a value which will be denoted $\gamma^* = \gamma_{opt}$; because the value of the functional $J(\mathbf{v})$ computed on real fields towards the pole O, in the plane of symmetry, (in the deforming field) becomes a function of γ.

The shape of the plug is given by the value γ_{opt} and the lengths l computed from (7.5.24), for this value of γ. With these values of γ and l we will compute the drawing force.

7.5.5   *The Drawing Force*

The drawing stress will be determined from the theorem of powder expanded applied to the domain V bounded by the boundaries S_3–S_{13} and which contains the discontinuity surfaces S_1 and S_2:

$$\int_V \boldsymbol{\sigma} \cdot \boldsymbol{D}\, dx + \sum_{k=1}^{2} \int_{S_k} \mathbf{t} \cdot [\mathbf{v}]\, dA = \int_{\partial V} \mathbf{t} \cdot \mathbf{v}\, dA, \tag{7.5.48}$$

(with $\partial V = \bigcup_{j=3}^{13} S_j$). In this formula $\dot{\boldsymbol{\varepsilon}}$ will be computed with the kinematics admissible field (7.5.10) using $\gamma = \gamma_{opt}$, while the stress field $\boldsymbol{\sigma}$ must be considered a static admissible field (i.e., satisfying the equilibrium equations and the boundary conditions). Since we were unable to find such a field, we compute the power $\boldsymbol{\sigma} \cdot \boldsymbol{D}$ with the formula (7.5.15) with $\boldsymbol{D}$ found previously. Therefore we found the approximate values for the velocity by a minimization procedure.

The integral on the volume from (7.5.48) is given by (7.5.15).

The integral on S_2 is computed with a similar procedure with the one which was used to get (7.5.43) and (7.5.44). The unit vectors on the tangent and normal at Γ_2 are $\boldsymbol{\tau} = \sin\theta\, \mathbf{k} + \cos\theta\, \mathbf{i}$ and $\mathbf{n} = -\cos\theta\, \mathbf{k} + \sin\theta\, \mathbf{i}$ and therefore

$$v_\tau^+ = 0, \quad v_r^- = v_f \mathbf{k} \cdot \boldsymbol{\tau} = v_f \sin\theta. \tag{7.5.49}$$

The stress vector on S_2 will be computed from a global condition of equilibrium of the forces which act on the rigid domain III of the tube:

$$0 = \left[-2\pi(m_1\tau_1 L R_2 + m_2\tau_2 l R_4) + \pi\sigma_f(R_2^2 - R_4^2) - \sigma_f^* \int_{S_2} dA \right] \mathbf{k}, \tag{7.5.50}$$

since we assume that $\sigma_f^* \mathbf{k}$ is a mean stress on the surface. With the expression $dA = r_2(r_2 \sin\theta + x)\, d\theta d\varphi$ of the element of the surface S_2, with (7.5.49) and (7.5.50), the integral on S_2 is computed from

$$\int_{S_2} \mathbf{t} \cdot [\mathbf{v}]\, dA = -2\pi r_2 v_f \sigma_f^* \left[r_2 \left(-\cos\theta + \frac{\cos^3\theta}{3} \right) + \frac{x}{2} \left(\theta - \frac{\sin 2\theta}{2} \right) \right] \Bigg|_\gamma^\alpha \tag{7.5.51}$$

or if we replace σ_f^* from (7.5.50)

$$\int_{S_2} \mathbf{t} \cdot [\mathbf{v}]\, dA = \frac{\pi v_f}{r_2(\cos\gamma - \cos\alpha) + x(\alpha - \gamma)} \left[r_2\left(-\cos\theta + \frac{\cos^3\theta}{3}\right) \right.$$

$$\left. + \frac{x}{2}\left(\theta - \frac{\sin 2\theta}{2}\right) \right]\Big|_\gamma^\alpha [2(\tau_1 m_1 L R_2 + m_2\tau_2 l R_4) - \sigma_f(R_2^2 - R_4^2)].$$

$$(7.5.52)$$

The integrals on the surfaces S_9 and S_{11} are computed in (7.5.45) taking into account the friction forces between the tube at exit and the plug. The integral on S_{13} is

$$\int_{S_{13}} \mathbf{t} \cdot \mathbf{v}\, dA = \pi v_f \sigma_f(R_2^2 - R_4^2). \tag{7.5.53}$$

Introducing now in (7.5.48) the formulae (7.5.15), (7.5.44), (7.5.52), (7.5.28)–(7.5.33), (7.5.45) and (7.5.53) and taking into account the integrals on S_j with $j = 5, 7, 8, 10, 12$ are zero, we obtain:

$$\sqrt{2}kr_2 I_1(\gamma, \alpha) + 2\eta v_f r_2^2 I_2(\gamma, \alpha) - \sigma_0 I_5 \frac{\pi(R_1^2 - R_3^2)}{A(S_1)}$$

$$+ \frac{m_1\tau_1 L R_2 + m_2\tau_2 l R_4 + (1/2)\sigma_f(R_2^2 - R_4^2)}{r_2(\cos\gamma - \cos\alpha) + x(\alpha - \gamma)}$$

$$\times \left[r_2\left(-\cos\theta + \frac{\cos^3\theta}{3}\right) + \frac{x}{2}\left(\theta - \frac{\sin 2\theta}{2}\right) \right]\Big|_\gamma^\alpha$$

$$= -m_3 r_2 R_2 \cos\alpha I_6(\alpha, r_2, r_1(\alpha)) - m_4 r_2 R_4 \cos\gamma I_6(\gamma, r_2, r_1(\gamma) - \varepsilon)$$

$$- \frac{v_0\sigma_0}{2v_f}(R_1^2 - R_3^2) - m_2\tau_2 l R_4 - m_1\tau_1 L R_2 + \frac{\sigma_f}{2}(R_2^2 - R_4^2). \tag{7.5.54}$$

We use now (7.5.26) and (7.5.30), and grouping the terms which contain v_f we obtain from (7.5.54) the following expression for the drawing force

$$\sigma_f = \frac{\tau_2 m_2 l R_4 + m_1\tau_1 L R_2}{R_2^2 - R_4^2}$$

$$+ \left\{ \frac{R_2^2 - R_4^2}{2[r_2(\cos\gamma - \cos\alpha) + x(\gamma - \alpha)]} \left[r_2\left(-\cos\theta + \frac{\cos^3\theta}{3}\right)\Big|_\gamma^\alpha \right. \right.$$

$$\left. \left. + \frac{x}{2}\left(\theta - \frac{\sin 2\theta}{2}\right)\Big|_\gamma^\alpha \right] + \frac{R_2^2 - R_4^2}{2} \right\}^{-1} \left\{ \frac{\min J(\mathbf{v})}{2\pi v_f} + \eta v_f r_2^2 I_2(\gamma, \alpha) \right\}\Big|_{\gamma=\gamma^*}.$$

$$(7.5.55)$$

In this formula, the terms which contains v_f will describe the speed influence on the drawing force. We note that all the integrals I_1, I_2, I_4, I_5 do not depend on v_f.

At the end, we will make precise how to make these computations, which are involved in the drawing force σ_f [the formula (7.5.55) as well as in $J(\mathbf{v})$ — the formula (7.5.47)].

Let

$$\delta = x^2 \cos^2 \theta (r_2 \sin \theta + x)^2 \left(2 \sin^2 \theta + \frac{\cos^2 \theta}{2} \right) \tag{7.5.56}$$

$$x_1 = \frac{\sqrt{A}x + \sqrt{\delta}}{\sin \theta}, \qquad x_2 = \frac{\sqrt{A}x + \sqrt{\delta}}{\sin \theta} \tag{7.5.57}$$

and the variable t:

$$t = \sqrt{Ar^2 + Br + C} - \sqrt{A}r . \tag{7.5.58}$$

Then for $I_1(\gamma, \alpha)$ we obtain the formula:

$$I_1(\gamma - \alpha)$$

$$= \int_{\gamma}^{\alpha} \left(\frac{\sqrt{C}}{x} \ln \left| \frac{t - \sqrt{C}}{t + \sqrt{C}} \right| - \frac{\sqrt{A}}{\sin \theta} \ln |2\sqrt{A}t - B| - \frac{\sqrt{\delta}}{x \sin \theta} \ln \left| \frac{t - x_1}{t - x_2} \right| \right) \Bigg|_{r_2}^{r_1(\theta)} d\theta \tag{7.5.59}$$

while for $I_2(\gamma, \alpha)$ — formula (7.5.17) — follows:

$$I_2(\gamma - \alpha) = \int_{\gamma}^{\alpha} \left[a \ln \left| \frac{r}{r \sin \theta + x} \right| - \frac{b}{r} - \frac{e}{\sin \theta (r \sin \theta + x)} \right.$$

$$\left. - \frac{1}{2} \left(\frac{c}{r^2} + \frac{f}{\sin \theta (r \sin \theta + x)^2} \right) \right] \Bigg|_{r_2}^{r_1(\theta)} d\theta \tag{7.5.60}$$

where

$$a = \frac{1}{x^3} \left(A - \frac{3B \sin \theta}{x} + \frac{6C \sin^2 \theta}{x^2} \right),$$

$$b = \frac{1}{x^3} \left(B - \frac{3C \sin \theta}{x} \right), \qquad c = \frac{C}{x^3},$$

$$e = \frac{\sin \theta}{x^2} \left(-A + \frac{2B \sin \theta}{x} - \frac{3C \sin^2 \theta}{x^2} \right),$$

$$f = \frac{\sin \theta}{x} \left(-A + \frac{B \sin \theta}{x} - \frac{C \sin^2 \theta}{x^2} \right).$$

The integral (7.5.31) for $\theta \in [\gamma, \alpha]$ is computed according to:

$$I_4(\theta, r_2, r_1(\theta)) = \left(a_1 \ln \left| \frac{t - \sqrt{C}}{t + \sqrt{C}} \right| - \frac{b_1}{t - \sqrt{C}} - \frac{b_2}{t + \sqrt{C}} - \frac{c_1}{(t - \sqrt{C})^2} \right.$$

$$\left. - \frac{c_2}{(t + \sqrt{C})^2} + e_1 \ln \left| \frac{t - x_1}{t - x_2} \right| - \frac{f_1}{t - x_1} - \frac{f_2}{t - x_2} \right) \Bigg|_{r_2}^{r_1(\theta)} \tag{7.5.61}$$

where

$$a_1 = \frac{1}{\sqrt{C}x^2}\left(A - \frac{B^2}{8C} - \frac{B\sin\theta}{x} + \frac{3C\sin\theta}{x^2}\right),$$

$$b_1 = \frac{(B - 2\sqrt{AC})\left(B - 2\sqrt{AC} - 8C\sin\theta/x\right)}{8Cx^2},$$

$$b_2 = \frac{(B + 2\sqrt{AC})(B + 2\sqrt{AC} - 8C\sin\theta/x)}{8Cx^2},$$

$$f_1 = \frac{(x_1^2 - C)\sin^2\theta}{2x^4}, \quad f_2 = \frac{(x_2^2 - C)\sin^2\theta}{2x^4},$$

$$c_1 = \frac{(B - 2\sqrt{AC})^2}{a\sqrt{C}x^2},$$

$$e_1 = \frac{\sin\theta}{x^3}\left(-\sqrt{A} + \frac{(x_1^2 - C)\sin^2\theta}{2x\sqrt{\delta}} - \frac{3\sqrt{\delta}}{x}\right),$$

$$c_2 = \frac{(B + 2\sqrt{AC})^2}{8\sqrt{C}x^2}.$$

The integral $I_3(\gamma, r_2, r_1(\gamma) - \varepsilon)$ — the formula (7.5.25) — needs the following computation:

$$\int_{r_2}^{r_1(\gamma)-\varepsilon} \sqrt{II_D}\big|_{\theta=\gamma}(r\sin\gamma + x)\,dr$$

$$= \int_{r_2}^{r_1(\gamma)-\varepsilon} \frac{\sqrt{Ar^2 + Br + C}}{r^2(r\sin\gamma + x)}\bigg|_{\theta=\gamma}\,dr$$

$$= \left[\frac{B - 2C\sin\gamma}{2x^2\sqrt{C}}\ln\left|\frac{t - \sqrt{C}}{t + \sqrt{C}}\right| - \frac{B - 2\sqrt{AC}}{2x}\frac{1}{t - \sqrt{C}}\right.$$

$$\left. - \frac{B + 2\sqrt{AC}}{2x}\frac{1}{t + \sqrt{C}} + \frac{\sqrt{\delta}}{x^2}\ln\left|\frac{t - x_1}{t - x_2}\right|\right]\Bigg|_{r_2}^{r_1(\theta)-\varepsilon}\Bigg|_{\theta=\gamma}.$$

$$(7.5.62)$$

The integral I_5, given by formula (7.5.43), which is involved in the computation of the functional is computed numerically. The same with the integrals with respect to θ, from the relations (7.5.59) and (7.5.60), one compute them numerically.

7.5.6 *Numerical Examples*

In order to show how the previously established formula can be used several numerical examples will be given. In choosing the constants involved in the constitutive equation, some experimental data obtained for mild steel was used. For this material some mechanical properties which can be described by a Bingham model were already given previously. We stress however, that the numerical examples given here have a qualitative character only.

Drawing stress. Several reductions in area were considered. By definition the reduction in area is

$$r\% = 100 \left[1 - \frac{R_2^2 - R_4^2}{R_1^2 - R_3^2}\right] \tag{7.5.63}$$

while

$$\lambda = \frac{R_1^2 - R_3^2}{R_2^2 - R_4^2} = \frac{1}{1 - r/100} \tag{7.5.64}$$

is called the coefficient of extension.

In order to give more general examples, the influence of the speed will be described with the parameter ηv_f which is called "speed factor", or by the dimensionless number

$$N = \frac{\eta v_f}{\sigma_{Y0} R_1} \tag{7.5.65}$$

which is called the "parameter of speed influence". Here σ_{Y0} is the initial limit of plasticity.

For big speeds, η decreases very little. We remind that in order to establish from one dimensional test a relation between η and the speed, we define a mean value for the strain rate

$$D_m = \frac{1}{V} \int_V \sqrt{II_D} dV = \frac{\sqrt{3}\pi v_f}{V} \left[\frac{2r_2}{\sqrt{6}} I_1(\gamma, \alpha)\right] \tag{7.5.66}$$

extended over the whole viscoplastic domain of volume V estimated by

$$V = 2\pi \left[\frac{r_1^3 - r_2^3}{3}(\cos\gamma - \cos\alpha) - \frac{r_1^2 - r_2^2}{2}x(\alpha - \gamma)\right] \tag{7.5.67}$$

where r_1 is a mean value $r_1(\text{mean}) = (1/2)(r_1(\alpha) + r_1(\gamma))$. The term from the square bracket from (7.5.66) is computed numerically. The formula (7.5.67) introduces errors in the estimation of D_m of 1%. Then $\bar{D} = 2D_m/\sqrt{3}$ was used as the mean equivalent speed of deformation.

The coefficient of viscosity in the Bingham model was chosen from the experimental data which correspond to the speed of deformation $D_1 = \bar{D}$. From the experimental data of Manjoine [1944] and Cristescu [1977] was establish the empirical formula $\eta = (0.74/\bar{D}) + 0.000204$, with η in kgf mm^{-2} s and $\bar{D}$ in s^{-1}.

Another parameter involved in the theory of tube drawing is the reduction in diameter defined by the ratio R_3/R_4. The thickness of the wall is expressed by R_3/R_1.

Thus in order to characterize the geometry of the tube drawing we use:

- the reduction in area $r\%$ or λ;
- the reduction in diameter expressed by R_3/R_4;
- the reduction of the wall thickness expressed by R_3/R_1.

In all examples given below the radius R_4 was maintained constant and all the other radius were varied in order to obtain various reductions in areas, various reduction in diameter and various wall thicknesses.

Most computations were done for the angle $\alpha = 12°$, but other cases were also considered.

Concerning the free surface a, in most of the cases it was taken $a = 0.5$ mm, and sometimes $a = 1$ mm. Generally this free surface must be small, and for tubes with thin walls and carefully prepared drawing, a can be chosen smaller than 0.5 mm.

In the following examples several values for the friction factor m were considered. In the computation of the land l according to the formula (7.5.24) the friction coefficient μ is also involved. After choosing m, the value for μ was obtained from the empirical formula $\mu = (3/4)m$, which was obtained for drawing of wires.

The optimum γ angle. We observe also that $J(\mathbf{v})$ increases faster when γ increases than when γ decreases. That suggests for the practice the idea that the floating plug is to be computed for higher speeds and that will be good for smaller speeds as well.

In Fig. 7.5.2 is given a figure showing the contribution of various terms in the minimum of the functional. The first curve is the friction floating plug-tube, the second the friction matrix-tube and the third the internal dissipation. The influence of the speed and of the friction factors on the value of the functional.

One can see a significant influence on the functional of the speed and of the friction factors (Fig. 7.5.3).

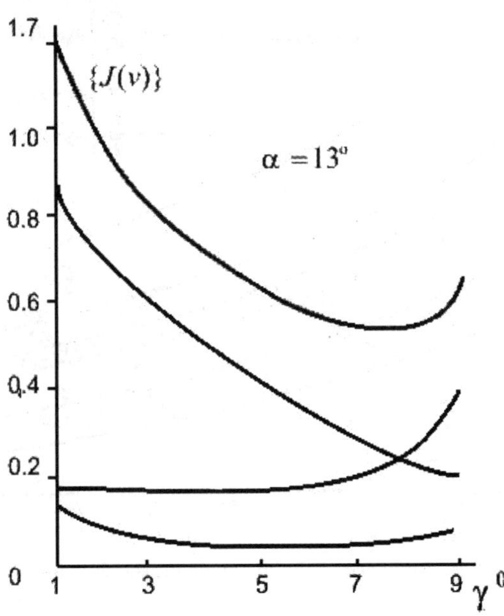

Fig. 7.5.2 Contribution of various terms contained in the functional for $\alpha = 13°$.

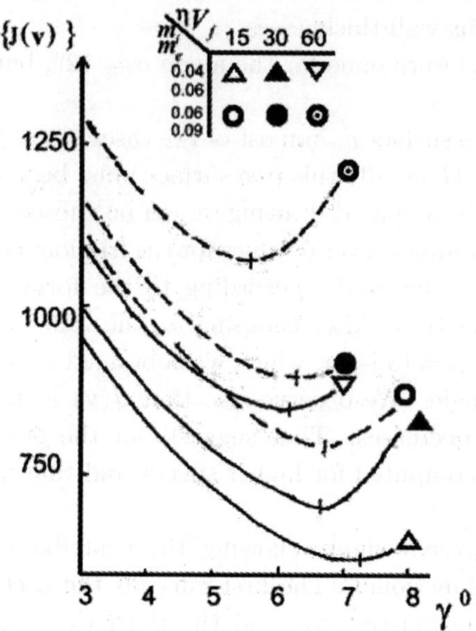

Fig. 7.5.3   Speed influence and friction influence on the functional $J(\mathbf{v})$.

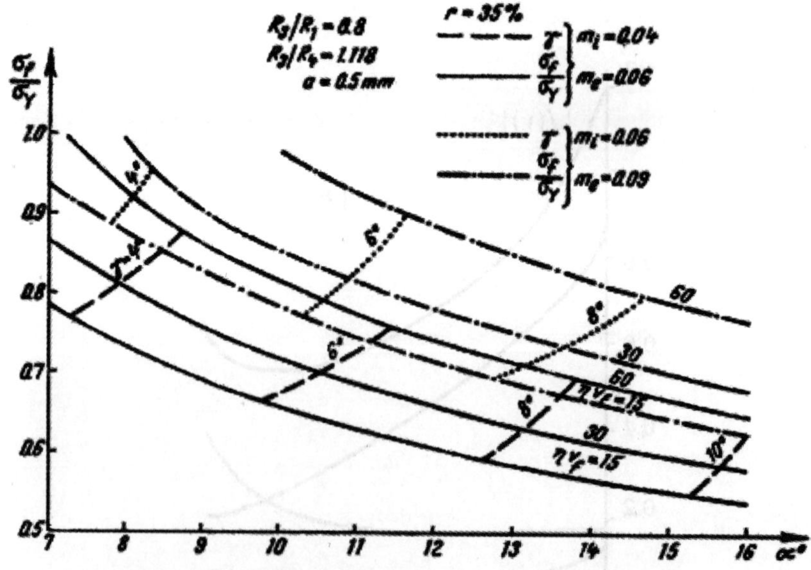

Fig. 7.5.4   Influence of the die angle on the drawing stress.

The influence of the die angle. The influence of the die angle on the drawing stress and corresponding plug angle is shown on Fig. 7.5.4 for two sets of friction factors and three sets of velocity factors. One can see that the increase of the drawing stress with speed is more significant at small die angles. On the same figure by dotted or interrupted lines are given the values of the plug angle corresponding to various frictions, speeds and die angles. When the die angle is increased, the corresponding plug angle must also be increased. The difference $\alpha - \gamma$ increases slightly when α increases. The drawing stress is decreasing when the die angle is increasing. The optimum angle γ decreases with the increase of the speed. That is more important at small speeds. For higher speeds this decrease is smaller. Our solution does not allow a too much increase of α, since in this case we have some other phenomena (Bramley and Smith [1976]).

The shape of the floating plug is defined essentially by the angle γ and the length l of the land. While γ is uniquely defined, l is not. Here the maximum value for l was computed, that is the value which would keep the floating plug in its most advanced position there where the whole conical surface of the plug is in contact with the inner surface of the tube. For practical purposes l is to be reduced with one third or even with one half of this computed length. Figure 7.5.5 gives the values of γ and l for various die angles and three drawing speeds. It is shown that l

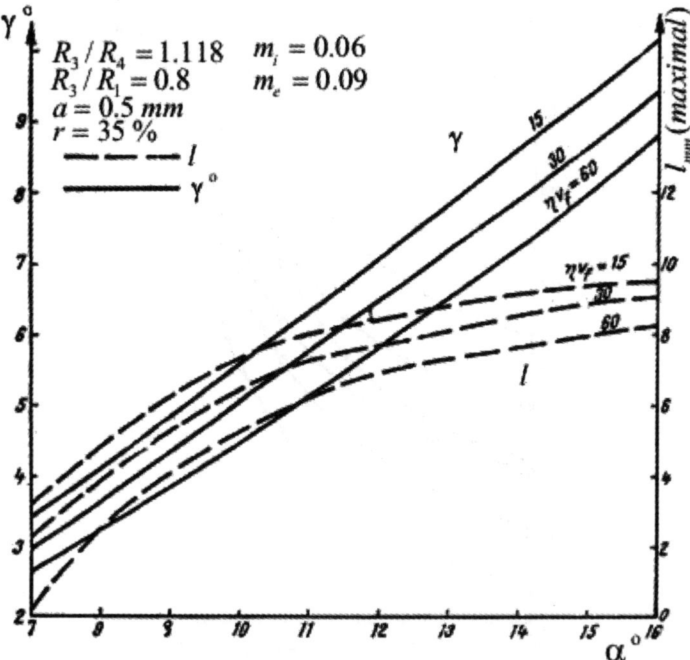

Fig. 7.5.5 Computation of γ and l, for three drawing speeds.

increases with α, but decreases when the drawing speed is increased. For a certain high speed and/or certain small α, l can decrease even to zero, i.e., the floating plug can be kept in equilibrium without any cylindrical land. A further increase of the drawing speed or a further decrease of the die angle is certainly impossible since the plug cannot be maintained in equilibrium any more. There are also some other limit drawing regimes which may occur when the drawing stress equals the current yield stress of the material; these limit drawing regimes can also be found quite easily using the main formula (7.5.55) with the condition $\sigma_f < \sigma_Y$. Thus another limitation for the drawing speed and die angle is found. It is also found that if the friction coefficient is higher, then the angle γ is only slightly reduced but l is reduced significantly.

The influence of the tube geometry on the drawing process with floating plug can be considered easily using formulas (7.5.47), (7.5.24) and (7.5.55). One of the geometrical parameters of significance for the tube drawing is the wall thickness expressed by the ratio R_3/R_1. Generally, when the wall thickness decreases stress increases (more significantly at high reductions in area and high drawing speeds) (Fig. 7.5.6), the plug angle increases and l also increases (Figs. 7.5.7 and 7.5.8).

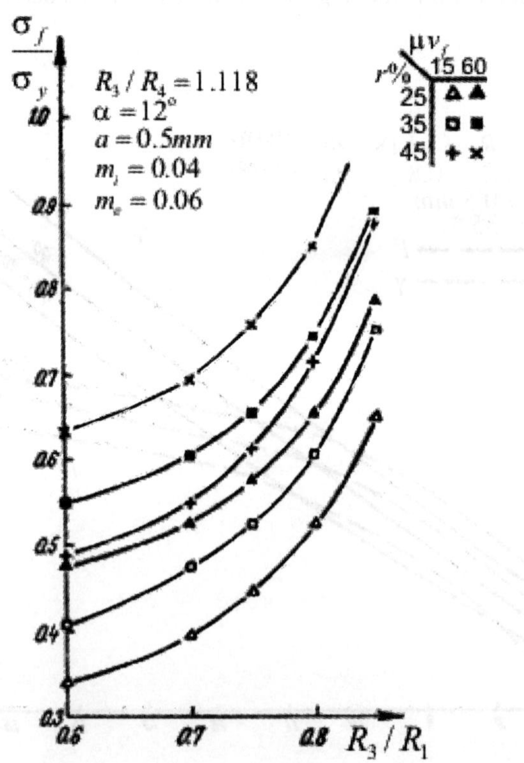

Fig. 7.5.6   Variation of wall thickness for various speeds.

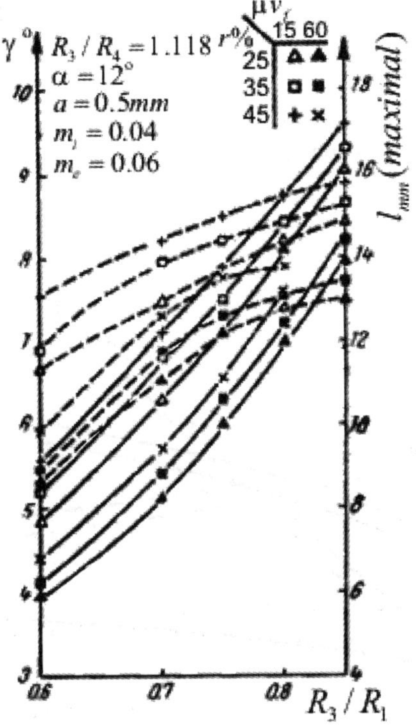

Fig. 7.5.7   Variation of plug angle and of length l.

Another important drawing parameter is the diameter reduction expressed by the ratio R_3/R_4. Generally an increase of this ratio decreases γ, increases l maximal, increase σ_f, etc. Significant diameter reductions are impossible with thin wall tubes, high reductions in area or high drawing speeds. All these limit drawing regimes can be easily computed. The clearance a, is of less importance in the whole process. Certainly an increase of a, decreases the drawing stress and therefore allows an increase of the drawing speed.

<u>Conclusions</u>. The most important conclusions are:

(1) The drawing stress for tube drawing with floating plug is obtained from formula (7.5.55). It gives the drawing stress as function of the drawing speed, die angle, tube geometry, and mechanical properties of the material, friction coefficients and plug geometry.

(2) The optimum geometry of the floating plug is obtained for various drawing conditions.

(3) The optimum semi-cone plug angle is decreasing with decreasing semi-cone die angle, increasing speed, increasing friction, increasing tube wall thickness, increasing diameter reduction.

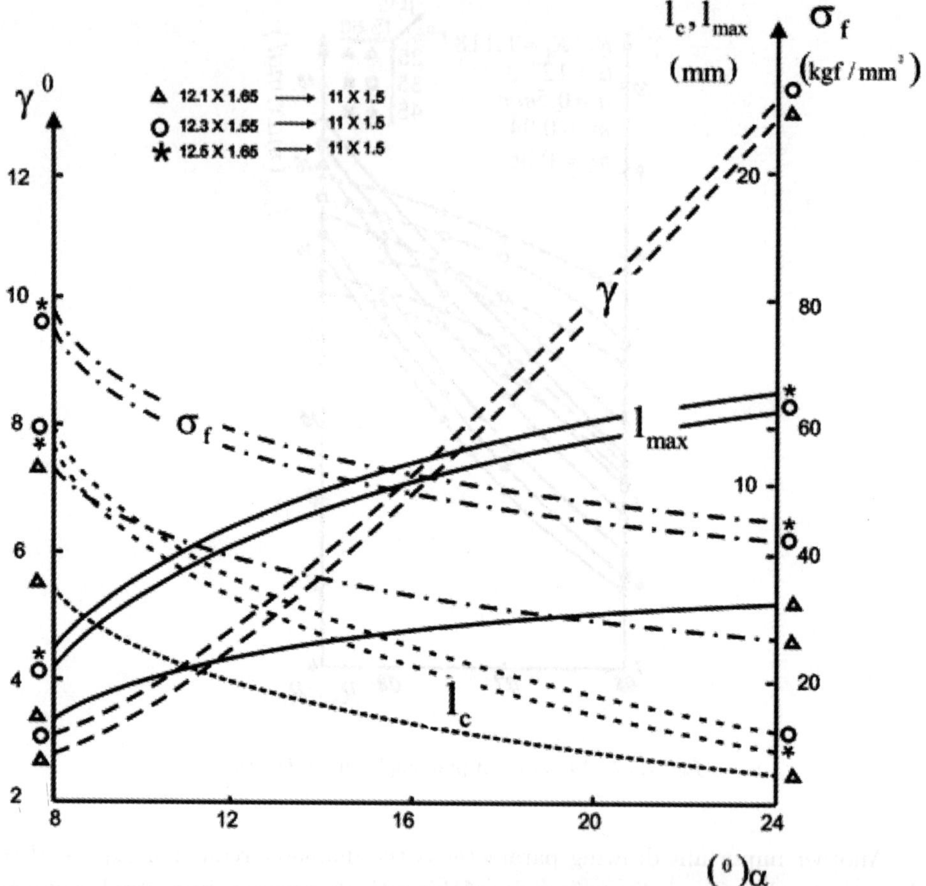

Fig. 7.5.8 Nomogram for the computation of floating plugs for drawing of small tubes. There are three programs: initial diameter × initial thickness of the wall → final diameter × the final thickness of the wall. Various angles α and various drawing stress are given.

(4) The plug land is increasing with α. Also l is increasing when the friction decreases and when the speed decreases. The plug land increases significantly with the increase of the diameter reduction, and increases when the wall thickness is decreasing.

(5) There are limit drawing regimes which cannot be surpassed; these correspond to cases when the drawing stress reaches the actual yield stress of the tube or when the plug land decreases to zero.

Another theory of drawing with floating plug is due to Şandru and Camenschi [1988] (see also Ioan and Ioan [1999]).

They are assuming again slow processes so that $N < 1$, and asymptotic developments follows. The drawing stress is

$$\frac{\sigma_f}{\sigma_Y} = \frac{2}{\sqrt{3}} \frac{R_2 L m_L}{R_2^2 - R_4^2} + \frac{2}{N} \frac{R_2^2}{\sqrt{3}(R_2^2 - R_4^2)(R_1^2 - R_3^2)}$$

$$\times \left[\frac{R_2(R_3^2 - R_2^2) + R_1(R_1^2 - R_3^2)}{R_1} F(\alpha) + \frac{R_2(R_2^2 - R_1^2)}{R_3} F(\gamma) \right]$$

$$- \frac{R_3^2(R_1^2 - R_2^2)}{\sqrt{3}(R_2^2 - R_4^2)(R_1^2 - R_3^2)} G(\alpha, \gamma) + \frac{A}{\sqrt{3}(R_1^2 - R_3^2)(R_2^2 - R_4^2)}$$

$$\times \left[R_1^2(R_3^2 - R_2^2) \ln \frac{R_1}{R_2} + R_3^2(R_2^2 - R_1^2) \ln \frac{R_3}{R_1} \right]$$

were all the constants are determined from other equations. Here

$$u(\theta) = a + bp(\theta) + q(\theta),$$

$$v(\theta) = \frac{A}{6} + Bp(\theta) + Dq(\theta) + K_1(\theta)p(\theta) + K_2(\theta)q(\theta),$$

$$K_1(\theta) = -\frac{9}{16} \int_\gamma^\alpha f(t)q(t) \sin t \, dt,$$

$$K_2(\theta) = \frac{9}{16} \int_\gamma^\alpha f(t)p(t) \sin t \, dt,$$

$$p(\theta) = \frac{1}{3} + \cos 2\theta,$$

$$q(\theta) = \left(\frac{1}{3} + \cos 2\theta \right) \ln \left(\tan \frac{\theta}{2} \right) - (1 - 3\cos\theta)(1 + \cos\theta),$$

$$F(\alpha) = [2a \sin\alpha - u'(\alpha) \cos\alpha] \sin^2 \alpha,$$

$$F(\gamma) = [2a \sin\gamma - u'(\gamma) \cos\gamma] \sin^2 \gamma,$$

$$G(\alpha, \gamma) = m_\alpha \cot\alpha + m_\gamma \cot\gamma - 4[v(\alpha) - v(\gamma)] - v'(\alpha) \cot\alpha + v'(\gamma) \cot\gamma$$

$$+ [A\cos\gamma + v'(\gamma)\sin\gamma + m_\gamma \sin\gamma] \ln \left(\frac{\tan \alpha/2}{\tan \gamma/2} \right)$$

$$- 2 \int_\gamma^\alpha \frac{u'(t)}{\sqrt{3u^2(t) + u'^2(t)/4}} \, dt + 6 \left[\ln \left(\tan \frac{\alpha}{2} \right) \right.$$

$$\times \int_\gamma^\alpha \frac{u(t) \sin t}{\sqrt{3u^2(t) + u'^2(t)/4}} \, dt - \left. \int_\gamma^\alpha \frac{u(t) \sin t \ln(\tan t/2)}{\sqrt{3u^2(t) + u'^2(t)/4}} \, dt \right].$$

Several examples are given for the determination of the drawing stress and all the other unknown quantities.

7.6   Extrusion

7.6.1   *Formulation of the Problem*

Let us consider now a mathematical model for hot tube extrusion, taking into account the influence of the working speed, when melted glass is used as a lubricant. We consider the problem of being stationary and isothermal. The working speed is considered variable during the process but for a certain magnitude of this velocity it is assumed that it is maintained constant for certain time interval so that during this time the process is considered "stationary". Therefore, the variation of the working speed is described by a set of stationary cases (Cleja-Tigoiu and Cristescu [1985]).

The extrusion problem considered with classical plasticity theory was considered by many authors. We mention here only Avitzur [1968], Blazynski [1976], Maneghin *et al.* [1980], Medvedev *et al.* [1980], Hartley [1971], etc. to mention just a few.

The cylindrical billet of annular cross-section with outer radius R_1, inner radius R_3 and length l is pushed out from a cylindrical container using a ram which is moving with the speed v_0 (Fig. 7.6.1). In the interior of the billet a cylindrical plug of radius R_3 and moving together with the ram is used to control the final inner radius of the tube. In front of the billet, between the billet and container, is placed a glass ring which at the beginning of the process melts and forms a dead zone which during the extrusion process is an extrusion of the die. All other surfaces of the billet are also glass lubricated. As is well known, glass lubrication allows higher working speeds and therefore high rates of deformation are involved (of the order of 10^2 s^{-1}). Due to the symmetry of the process the considered problem is axial symmetric and the geometry of the process is shown in Fig. 7.6.1 (not to

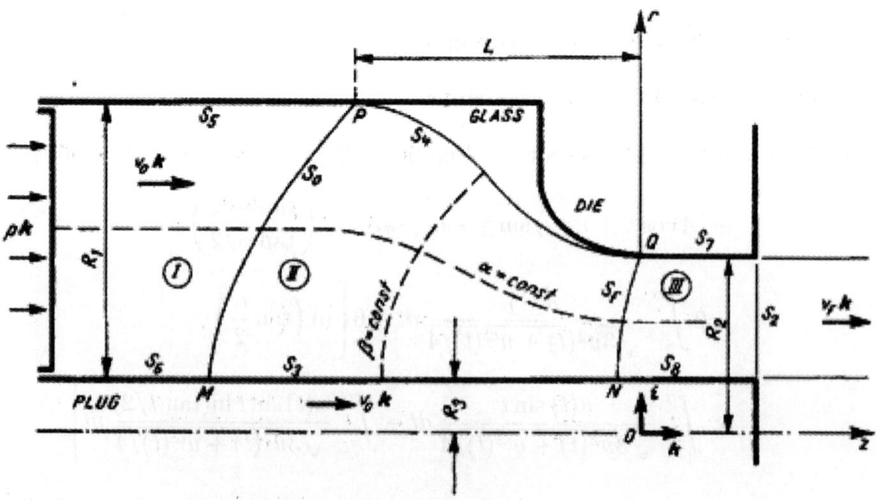

Fig. 7.6.1   Geometry of extrusion process.

scale). The mathematical problem reduces to the determination of the velocity field v and the stress field $\boldsymbol{\sigma}$, smooth in V, which must satisfy the equilibrium condition, incompressibility condition and a constitutive equation of Bingham type:

$$\mathrm{div}\,\boldsymbol{\sigma} = 0\,, \quad \mathrm{div}\,\mathbf{v} = 0\,, \quad \boldsymbol{D} = \frac{1}{2\eta}\left\langle 1 - \frac{k}{\sqrt{II_{\sigma'}}}\right\rangle \boldsymbol{\sigma}' \tag{7.6.1}$$

where the same notation as given previously are used. η is the viscosity coefficient and k is the mean yield stress in pure shear, both η and k being temperature dependent. A prime denotes "deviator". Boundary conditions, which will be given below, have also to be satisfied.

The kinematic velocity field is obtained from the following assumptions. There are two subdomains I and III, of V where the material is assumed to be in a rigid state and moving with velocity $v_0\mathbf{k}$ and $v_f\mathbf{k}$, respectively. Thus the incompressibility condition yields the following relationship between v_0 and v_f:

$$v_f(R_2^2 - R_3^2) = v_0(R_1^2 - R_3^2)\,. \tag{7.6.2}$$

Further it is assumed that in the domain II viscoplastic deformation takes place. In the interior of V the velocity field is assumed to be smooth. Thus the surfaces S_0 and S_f (the rigid-viscoplastic interfaces) are continuity surfaces for the velocity field. We assume also that the melted glass forms a dead zone, the billet-melted glass interface surface, denoted by S_4, is a flow surface (flow line tube) which can be considered to be a surface of velocity discontinuity. The shape of this surface changes during the extrusion process. Finally we assume that the plug-billet interface S_3 is also a flow surface. On these surfaces friction takes place which will be described by the law:

$$|\mathbf{t}_\tau| = m\sqrt{II_{\sigma'}} \tag{7.6.3}$$

where $\mathbf{t}_\tau$ is the friction stress and $0 \le m \le 1$ is the constant friction factor.
On the surfaces S_j $(j = 5,\dots,8)$ there is a friction between the rigid material and rigid tool surface, while the velocities on S_1 and S_2 are known and related by Eq. (7.6.2).

To solve the problem the velocity field and the ram pressure have to be determined. The velocity field is chosen from the range of kinematically admissible velocity fields as being the one for which the functional $J(\mathbf{v})$ related to the Bingham model takes the smallest value.

Starting from the equation of the flow lines in V a smooth kinematic admissible velocity field is derived which satisfies the incompressibility condition, and takes given values on S_1 and S_2. This velocity field depends on the functions g_0 and g_f which are involved in the definition of the surfaces S_0 and S_f. The velocity field which approximates the real one corresponds to the values g_0^* and g_f^* for which $J(\mathbf{v})$, considered a functional of g_0 and g_f, takes the smallest value.

The ram pressure is obtained from the theorem of power expanded written for the approximate velocity field previously determined. The approximate velocity field and the ram pressure can be determined numerically only. Since significant computing time is involved, several simplifying assumptions concerning the shape of the surfaces S_0 and S_f will be made. Thus the functional $J(\mathbf{v})$ is transformed into a real function depending on a certain number of real parameters.

The influence of the working speed on the process (ram pressure, dead zone shape, etc.) is obtained numerically for cases suggested by practical applications, assuming that the mechanical properties of the material, the lubrication conditions and the geometry of the process (i.e., R_i) are known.

7.6.2   *Kinematics of the Process*

Due to the axial symmetry of the problem, cylindrical coordinates (r, θ, z) will be used. The equation of the flow surfaces are

$$r = G(z, \alpha) \tag{7.6.4}$$

with $\alpha \in [R_3, R_2]$. The cylindrical surfaces $r = \alpha$ are the flow surfaces in domain III. Let us assume that the surfaces S_0 and S_f of equations surfaces are

$$z = \Gamma_0(r), \quad z = \Gamma_f(r) \tag{7.6.5}$$

have unique intersections with the flow surfaces $\alpha = $ const. for any $\alpha \in [R_3, R_2]$. It follows that the equations of these surfaces can be written as function of

$$r = f_0(\alpha), \quad r = f_f(\alpha) \tag{7.6.6}$$

where f_0 yields from the incompressibility condition written in the form

$$v_0[f_0^2(\alpha) - R_3^2] = v_f(\alpha^2 - R_3^2) \tag{7.6.7}$$

or

$$f_0(\alpha) = \left[\frac{R_1^2 - R_3^2}{R_2^2 - R_3^2}(\alpha^2 - R_3^2) + R_3^2 \right]^{1/2} \tag{7.6.8}$$

if Eq. (7.6.2) is also used. From Eqs. (7.6.5) and (7.6.6) the equations of the surfaces S_0 and S_f can be written as

$$z = \Gamma_0(f_0(\alpha)) \equiv g_0(\alpha), \quad z = \Gamma_f(\alpha) \equiv g_f(\alpha). \tag{7.6.9}$$

Thus the equations of the flow surface are

$$r = G(z, \alpha) \equiv \begin{cases} f_0(\alpha), & z \leq g_0(\alpha) \\ F(z, \alpha), & g_0(\alpha) \leq z \leq g_f(\alpha) \\ \alpha, & z \geq g_f(\alpha), \end{cases} \tag{7.6.10}$$

for $\alpha \in [R_3, R_2]$.

Since the velocity field is smooth in V, the derivative of the function $F(z, \alpha)$, with respect to z must satisfy the conditions

$$\lim_{z \searrow g_0(\alpha)} F_z(z, \alpha) = 0, \quad \lim_{z \nearrow g_f(\alpha)} F_z(z, \alpha) = 0. \tag{7.6.11}$$

It will be assumed also that

$$F_\alpha(z,\alpha)F_z(z,\alpha) \neq 0 \qquad (7.6.12)$$

in the interior of II and that it is finite. Since the inner cylindrical surface of the tube is $\alpha = R_3$ we have

$$F(z, R_3) = R_3 \qquad (7.6.13)$$

for any $z \in [g_0(\alpha), g_f(\alpha)]$.

From (7.6.12) we conclude that through any point P of II passes a unique flow line. Therefore, the family $\beta = $ const. of orthogonal lines to the flow lines exists; their equation in the plane (z, r) is

$$\Phi(z,r) = \beta, \qquad (7.6.14)$$

with β a real number belonging to a certain interval. The orthogonally condition of the curves (7.6.10) and (7.6.14) is

$$\Phi_z = F_z \Phi_r . \qquad (7.6.15)$$

Since $\Phi_r(z,\alpha) \neq 0$ in II, (7.6.14) can be solved, with respect to r.

From (7.6.10) and (7.6.14) written as

$$f_1(z,r;\alpha,\beta) = r - F(z,\alpha) = 0,$$
$$f_2(z,r;\alpha,\beta) = \Phi(z,\alpha) - \beta \qquad (7.6.16)$$

and using the theorem of implicit functions we get the nonsingular mapping

$$r = u(\alpha,\beta), \quad z = v(\alpha,\beta) \qquad (7.6.17)$$

because

$$\Delta = \Phi_r \frac{1 + F_z^2}{F_z} \qquad (7.6.18)$$

is finite and non-zero in the interior of II. Further the formulae

$$\frac{\partial r}{\partial \alpha} = \frac{F_\alpha}{1 + F_z^2}, \qquad \frac{\partial z}{\partial \alpha} = -\frac{F_\alpha F_z}{1 + F_z^2},$$

$$\frac{\partial r}{\partial \beta} = \frac{F_z^2}{\Phi_r(1 + F_z^2)}, \qquad \frac{\partial z}{\partial \beta} = \frac{F_z}{\Phi_r(1 + F_z^2)}, \qquad (7.6.19)$$

yield from Eqs. (7.6.16) and (7.6.17). Thus

$$x = u(\alpha,\beta)\cos\theta, \quad y = u(\alpha,\beta)\sin\theta, \quad z = v(\alpha,\beta) \qquad (7.6.20)$$

define a nonsingular mapping from $(x,y,z) \rightarrow (q^1 = \alpha, q^2 = \theta, q^3 = \beta)$, since the Jacobian

$$\sqrt{g} \equiv r \left| \frac{\partial z}{\partial \alpha} \frac{\partial r}{\partial \beta} - \frac{\partial z}{\partial \beta} \frac{\partial r}{\partial \alpha} \right| \qquad (7.6.21)$$

computed with Eq. (7.6.19)

$$\sqrt{g} = \frac{F(z,\alpha)}{1 + F_z^2} \left| \frac{F_\alpha F_z}{\Phi_r} \right| \qquad (7.6.22)$$

is non-zero.

In the new system of coordinates the incompressibility condition is

$$\operatorname{div} \mathbf{v} = \frac{1}{\sqrt{g}} \sum_{i=1}^{3} \frac{\partial}{\partial q^i}(\sqrt{g}\,v^i) = 0 \tag{7.6.23}$$

with v^i the contravariant components of the velocity.

Since $q^1 = \alpha = \text{const.}$ are flow lines it yields

$$\mathbf{v} = v^\beta \mathbf{e}_\beta \quad \text{(no summation)} \tag{7.6.24}$$

with $\mathbf{e}_\beta$ directed along the flow lines in the flow sense.

From Eqs. (7.6.23) and (7.6.24) and the axial symmetry we get

$$\sqrt{g}\,v^\beta = C(\alpha)\,. \tag{7.6.25}$$

The function $C(\alpha)$ is determined from the condition

$$\lim_{\substack{r=F(z,\alpha) \\ z \searrow g_0(\alpha)}} \mathbf{v}(r,z) = v_0\mathbf{k}\,. \tag{7.6.26}$$

Thus the kinematic velocity field considered is

$$\mathbf{v} = C(\alpha)\frac{\mathbf{e}_\beta}{\sqrt{g}} = \frac{C(\alpha)}{\sqrt{g}}\left(\frac{\partial r}{\partial \beta}\,\mathbf{i}_r + \frac{\partial z}{\partial \beta}\,\mathbf{k}\right), \tag{7.6.27}$$

where $\mathbf{i}_r$ and $\mathbf{k}$ are unit vectors (see Fig. 7.6.1).

By inverting the relations (7.6.17) and using (7.6.21) we get

$$\frac{\partial \alpha}{\partial r} = \frac{r}{\sqrt{g}}\frac{\partial z}{\partial \beta}, \quad \frac{\partial \alpha}{\partial z} = -\frac{r}{\sqrt{g}}\frac{\partial r}{\partial \beta}\,. \tag{7.6.28}$$

From (7.6.27) and (7.6.24) the components of the kinematic velocity field in the cylindrical coordinates are

$$v_r = -\frac{1}{r}C(\alpha)\frac{\partial \alpha}{\partial z}, \quad v_\theta = 0, \quad v_z = \frac{1}{r}C(\alpha)\frac{\partial \alpha}{\partial r}, \tag{7.6.29}$$

where α is considered to be a function of z and r according to Eq. (7.6.10):

$$\alpha = H(z,r) \tag{7.6.30}$$

so that $r = F(z, H(z,r))$.

To get the explicit expression of $C(\alpha)$ involved in Eqs. (7.6.26) and (7.6.27) the following limits are necessary:
from $(7.6.19)_1$ and $(7.6.11)_1$ yields

$$\lim_{\substack{r=F(z,\alpha) \\ z \searrow g_0(\alpha)}} \frac{\partial r}{\partial \alpha} = F_\alpha(g_0(\alpha),\alpha)\,, \quad \lim_{\substack{r=F(z,\alpha) \\ z \searrow g_0(\alpha)}} \frac{\partial z}{\partial \alpha} = 0\,; \tag{7.6.31}$$

from $(7.6.19)_2$, (7.6.22) and $(7.6.11)_1$ we have

$$\lim_{\substack{r=F(z,\alpha) \\ z \searrow g_0(\alpha)}} \frac{1}{\sqrt{g}}\frac{\partial z}{\partial \beta} = \frac{\displaystyle\lim \operatorname{sign}(\Phi_r F_z)_{\substack{r=F(z,\alpha) \\ z \searrow g_0(\alpha)}}}{F(g_0(\alpha),\alpha)|F_\alpha(g_0(\alpha),\alpha)|} \tag{7.6.32}$$

and

$$\lim_{\substack{r=F(z,\alpha)\\ z\searrow g_0(\alpha)}} \frac{1}{\sqrt{g}}\frac{\partial r}{\partial \beta} = 0\,. \tag{7.6.33}$$

Thus

$$C(\alpha) = v_0 F(g_0(\alpha),\alpha)|F_\alpha(g_0(\alpha),\alpha)|\,. \tag{7.6.34}$$

Using (7.6.34) in (7.6.29) and (7.6.10) the kinematic velocity field expressed as a function of z, r is

$$v_r = -v_0 f_0(\alpha)|F_\alpha(g_0(\alpha),\alpha)|\frac{1}{r}\frac{\partial \alpha}{\partial z}\,,$$

$$v_z = v_0 f_0(\alpha)|F_\alpha(g_0(\alpha),\alpha)|\frac{\partial \alpha}{\partial r}\,, \tag{7.6.35}$$

$$v_\theta = 0\,,$$

with $\alpha = H(z,r)$.

By differentiation of Eq. (7.6.10)

$$\frac{\partial \alpha}{\partial r} = \frac{1}{F_\alpha(z,\alpha)}\,, \quad \frac{\partial \alpha}{\partial z} = \frac{F_z(z,\alpha)}{F_\alpha(z,\alpha)}\,, \tag{7.6.36}$$

is obtained. From (7.6.36), (7.6.35) and (7.6.10), another form of the velocity field, this time as function of z and α, can be obtained

$$v_r = \frac{v_0 f_0(\alpha)}{F(z,\alpha)}\frac{|F_\alpha(g_0(\alpha),\alpha)|}{F_\alpha(z,\alpha)}F_z(z,\alpha)\,,$$

$$v_z = \frac{v_0 f_0(\alpha)}{F(z,\alpha)}\frac{|F_\alpha(g_0(\alpha),\alpha)|}{F_\alpha(z,\alpha)}\,, \tag{7.6.37}$$

$$v_\theta = 0\,.$$

The previous formulae will be written in a particular form for the case when the flow lines are sinusoidal lines, defined by

$$r = F(z,\alpha) = A(\alpha)\{\sin[a(\alpha)z + b(\alpha)] + 1\} + \alpha\,, \tag{7.6.38}$$

with

$$A(\alpha) = \frac{1}{2}[f_0(\alpha) - \alpha]\,, \tag{7.6.39}$$

where $a(\alpha)$ and $b(\alpha)$ are determined by (7.6.11) as

$$a(\alpha) = \frac{\pi}{g_f(\alpha) - g_0(\alpha)}\,, \quad b(\alpha) = \frac{\pi}{2}\frac{g_f(\alpha) - 3g_0(\alpha)}{g_f(\alpha) - g_0(\alpha)}\,. \tag{7.6.40}$$

Condition (7.6.13) is satisfied since (7.6.8) yields $A(R_3) = 0$. Formulae (7.6.38) to (7.6.40) yield

$$F(g_0(\alpha),\alpha) = f_0'(\alpha) \equiv \frac{v_f}{v_0}\frac{\alpha}{f_0(\alpha)}\,. \tag{7.6.41}$$

In this particular case the velocity field is written as

$$v_r = -v_f \frac{\alpha}{r}\frac{\partial \alpha}{\partial z}, \quad v_z = v_f \frac{\alpha}{r}\frac{\partial \alpha}{\partial r}, \quad v_\theta = 0, \tag{7.6.42}$$

with $\alpha = H(z,r)$ obtained from Eq. (7.6.39) by inversion, with respect to α.

In what follows we will need the expression of the second invariant II_D in cylindrical coordinates, which, using the incompressibility condition, becomes

$$II_D = D_{rr}^2 + D_{\theta\theta}^2 + D_{rz}^2 + D_{rr}D_{\theta\theta}. \tag{7.6.43}$$

The components of the rate of deformation tensor corresponding to the field (7.6.42) are

$$D_{rr} = -v_f \left(\frac{1}{r}\frac{\partial \alpha}{\partial r} - \frac{\alpha}{r^2} \right)\frac{\partial \alpha}{\partial z} + \frac{\alpha}{r}\frac{\partial^2 \alpha}{\partial r \partial z},$$

$$D_{\theta\theta} = -v_f \frac{\alpha}{r^2}\frac{\partial \alpha}{\partial z},$$

$$D_{rz} = -\frac{1}{2}\frac{v_f}{r}\left\{ \left(\frac{\partial \alpha}{\partial z}\right)^2 + \alpha\frac{\partial^2 \alpha}{\partial z^2} - \left(\frac{\partial \alpha}{\partial r}\right)^2 + \frac{\alpha}{r}\frac{\partial \alpha}{\partial r} - \alpha\frac{\partial^2 \alpha}{\partial r^2} \right\}. \tag{7.6.44}$$

In forming the partial derivatives of the function $\alpha = H(z,r)$, (7.6.38) is used. Thus the second invariant II_D can be considered either as a function of z and r if $\alpha = H(z,r)$ is used, or a function of z and α if r is represented by $F(z,\alpha)$. The latter version will be used further on.

7.6.3   *The Approximate Velocity Field*

The functional attached to the Bingham model, for the case considered here is

$$J(\mathbf{v}) = 2\eta \int_V II_D \, dV + 2k \int_V \sqrt{II_D} \, dV - \sum_j \int_{S_j} \mathbf{t} \cdot \mathbf{v} \, d\sigma + \sum_i \int_{S_i} \mathbf{t} \cdot [\mathbf{v}] \, d\sigma, \tag{7.6.45}$$

where S_j with $j = 2$ and 5–8 are the surfaces on which the velocity is prescribed and S_i with $i = 3, 4$ are the discontinuity surfaces for the velocity field and on which it is assumed that the stress vector t is known. The approximate velocity field for the extrusion problem is determined by minimizing on the kinematical velocity field (7.6.42) using (7.6.38) to form the appropriate expression of $J(\mathbf{v})$ [in the sense that in this functional on surfaces S_4 and S_3 the stress vector is replaced by Eq. (7.6.3)]. On the family of kinematic admissible velocity fields J becomes a functional of g_0 and g_f. The values g_0^* and g_f^* for which J takes the smallest value define the approximate solution of the problem. The geometry of the viscoplastic domain is known if the surfaces S_4, S_0 and S_f are known.

The volume integrals from (7.6.45) are calculated with (7.6.43) and (7.6.44) by assuming that II_D depends on α and z alone. Passing from the variables (z, r, θ) to (z, α, θ) with $r = F(z, \alpha)$, we obtain for instance for the first integral

$$\int_V II_D \, dV = 2\pi \int_{R_3}^{R_2} \int_{g_0(\alpha)}^{g_f(\alpha)} II_D(z,\alpha)F(z,\alpha)|F_\alpha(z,\alpha)| \, dzd\alpha \,. \tag{7.6.46}$$

On the boundaries S_3 and S_4 the tangential component of the stress vector, according to (7.6.3) is given by

$$|\mathbf{t}_\tau| = m_i \sqrt{II_{\sigma'}} = m_i 2\eta(\sqrt{II_D} + k) \,, \tag{7.6.47}$$

where $m_i (i = 3, 4)$ is the constant friction factor on the corresponding surface.

The equation of the surface S_4 defined by $\alpha = R_2$ is

$$x = F(z, R_2)\cos\theta \,,$$

$$y = F(z, R_2)\sin\theta \,, \tag{7.6.48}$$

$$z = z \,,$$

with $\theta \in [0, 2\pi]$ and $z \in [g_0(R_2), g_f(R_2)]$. The element of area is

$$d\sigma = F(z, R_2)\sqrt{1 + F_z^2(z, R_2)} \, d\theta dz \,. \tag{7.6.49}$$

The velocity field is tangent to the surface S_4 and from (7.6.42) yields

$$|\mathbf{v}_\tau| = v_f \frac{R_2}{F(z, R_2)} \sqrt{\left(\frac{\partial\alpha}{\partial r}\right)^2 + \left(\frac{\partial\alpha}{\partial z}\right)^2} \,, \tag{7.6.50}$$

where $\partial\alpha/\partial r$ and $\partial\alpha/\partial z$ are computed with (7.6.36). Thus from (7.6.47)–(7.6.49), the integral on S_4 is obtained with

$$\int_{S_4} \mathbf{t} \cdot \mathbf{v} \, d\sigma = -2\pi m_4 v_f R_2 \int_{g_0(R_2)}^{g_f(R_2)} (2\eta\sqrt{II_D(z, R_2)} + k)\frac{1 + F_z^2(z, R_2)}{|F_\alpha(z, R_2)|} \, dz \,. \tag{7.6.51}$$

On the cylindrical surface $S_3(\alpha = R_3)$ the velocity is tangent to the surface and from (7.6.42), (7.6.36) and (7.6,13) we get the relative velocity between the billet and the plug:

$$\mathbf{v}_3 = v_f \left(\frac{1}{F_\alpha(z, R_3)} - \frac{v_0}{v_f}\right) \mathbf{k} \,. \tag{7.6.52}$$

The integral over S_3 is computed with (7.6.47) and (7.6.52) as

$$\int_{S_3} \mathbf{t} \cdot \mathbf{v} \, d\sigma = -2\pi m_3 R_3 \int_{g_0(R_3)}^{g_f(R_3)} (2\eta\sqrt{II_D(z, R_3)} + k)|v_3| \, dz \,. \tag{7.6.53}$$

The integral over S_2 is zero since this surface is stress free. On the surface S_j with $j = 5$–8 friction between rigid tool walls and billet takes place and is prescribed by the law

$$|\mathbf{t}_\tau| = m_j \tau_j \quad \text{(no summation)} \,, \tag{7.6.54}$$

where τ_j is the constant shearing yield stress of the material along the wall S_j. The law (7.6.54) is a particular version of (7.6.3). Denoting by L_j the length of the cylindrical surfaces S_j, the integrals over these surfaces are

$$\int_{S_5} \mathbf{t} \cdot \mathbf{v}\, d\sigma = -2\pi m_5 L_5 \tau_5 R_1 v_0\,,$$

$$\int_{S_7} \mathbf{t} \cdot \mathbf{v}\, d\sigma = -2\pi m_7 L_7 \tau_7 R_2 v_f\,, \tag{7.6.55}$$

$$\int_{S_8} \mathbf{t} \cdot \mathbf{v}\, d\sigma = -2\pi m_8 L_8 \tau_8 R_3 |v_f - v_0|\,,$$

while the integral on S_6 is zero.

With the expression of all these integrals, the functional which has to be minimized, with respect to g_0 and g_f becomes

$$J(\mathbf{v}) = 4\pi \int_{R_3}^{R_2} \int_{g_0(\alpha)}^{g_f(\alpha)} (\eta II_D + k\sqrt{II_D})F(z,\alpha)|F_\alpha(z,\alpha)|\, dz\, d\alpha$$

$$+\, 2\pi m_5 L_5 \tau_5 R_1 v_0 + 2\pi m_7 L_7 \tau_7 R_2 v_f + 2\pi m_8 L_8 \tau_8 R_3 |v_f - v_0|$$

$$+\, 2\pi m_3 R_3 v_f \int_{g_0(R_3)}^{g_f(R_3)} (2\eta\sqrt{II_D(z,R_3)} + k)\left|\frac{1}{F_\alpha(z,R_3)} - \frac{v_0}{v_f}\right| dz$$

$$+\, 2\pi m_4 R_2 v_f \int_{g_0(R_2)}^{g_f(R_2)} (2\eta\sqrt{II_D(z,R_2)} + k)\frac{1 + F_z^2(z,R_2)}{|F_\alpha(z,R_2)|}\, dz\,, \tag{7.6.56}$$

where $F(z,\alpha)$ given by (7.6.38) depends on g_0 and g_f.

7.6.4 *The Extrusion Pressure*

The extrusion pressure is determined from the theorem of power expanded

$$\int_V \boldsymbol{\sigma} \cdot \mathbf{D}\, dV = \int_S \mathbf{t} \cdot \mathbf{v}\, d\sigma \tag{7.6.57}$$

with

$$S = \bigcup_{j=1}^{8} S_j$$

and in which the internal dissipation and those on various surfaces are computed for the approximate velocity field already determined.

The dissipation on the surface S_1 is

$$\int_{S_1} \mathbf{t} \cdot \mathbf{v}\, d\sigma = \pi(R_1^2 - R_3^2)v_0 p \tag{7.6.58}$$

in which p is the extrusion pressure.

For the Bingham model (7.6.1) the stress power is

$$\int_V \sigma_{ij} D_{ij}\, dV = 4\eta \int_V II_D\, dV + 2k \int_V \sqrt{II_D}\, dV \qquad (7.6.59)$$

with the volume integrals calculated according to the same procedure as for (7.6.46).

Taking into account in (7.6.57) the integral on S_1 as given by (7.6.58), the integral on S_4 and S_5 furnished by (7.6.51) to (7.6.53), the integrals on S_5, S_7 and S_8 given by (7.6.55) (the integrals on S_2 and S_6 are zero) we get the ram pressure

$$p = \frac{1}{R_1^2 - R_3^2}\frac{1}{\pi v_0}\int_V \boldsymbol{\sigma}\cdot\boldsymbol{D}\, dV + 2m_3 R_3 \int_{g_0^*(R_3)}^{g_f^*(R_3)} (2\eta\sqrt{II_D(z, R_3)} + k)|v_3|\, dz$$

$$+ 2m_4 R_2\frac{v_f}{v_0}\int_{g_0^*(R_2)}^{g_f^*(R_2)} (2\eta\sqrt{II_D(z, R_2)} + k)\frac{1 + F_z^2(z, R_2)}{|F_\alpha(z, R_2)|}\, dz$$

$$+ 2m_5 L_5\tau_5 R_1 + 2m_7 L_7\tau_7 R_2\frac{v_f}{v_0} + 2m_8 L_8\tau_8 R_3\left(\frac{v_f}{v_0} - 1\right). \qquad (7.6.60)$$

In this formula g_0^* and g_f^* are the functions which define the surfaces S_0 and S_f.

From (7.6.60) the ram pressure can be determined if either the velocity of the ram or the velocity v_f at the exit of the container is known.

The value of the extrusion pressure depends on the parameters defining the geometry (R_j, g_0^* and g_f^*), on the mechanical properties of the billet (η and k), on the friction factors (m_i) on various interface surfaces, and certainly on v_0 or v_f. The mechanical parameters and the friction conditions are strongly temperature dependent.

7.6.5 *Numerical Examples*

In order to determine numerically the ram pressure and the influence of the working velocity on this pressure, we have to find first g_0^* and g_f^*. Since significant computing time is necessary in order to find g_0^* and g_f^* by minimizing the functional J, simplifying assumptions concerning the shape of these surfaces are necessary. These simplifications, suggested by experimental evidence, must keep the formulae involved in the problem as simple as possible.

In a first particular case the surfaces S_0 and S_f are assumed to be conical, i.e., straight lines in the symmetry plane. These straight lines pass trough the points $Q(0, R_2)$ and $P(-L, R_1)$, respectively

$$r - R_1 = m_0(z - L)$$

$$r - R_2 = m_f z. \qquad (7.6.61)$$

If m_0 and m_f are finite, (7.6.61) and (7.6.6) yield

$$\frac{\alpha - R_2}{m_f} \equiv g_f(\alpha) = z$$

$$\frac{f_0(\alpha) - R_1 + m_0 L}{m_0} \equiv g_0(\alpha) = z\,. \tag{7.6.62}$$

With these expressions for g_0 and g_f the functional J becomes a function of the real parameters γ_0, γ_f and L. These parameters may vary in the intervals

$$L \in (0, l)\,, \quad \gamma_0 \in \left(\arctan\frac{R_1}{L_0}, \pi\right)\,, \quad \gamma_f \in \left(\arctan\frac{R_2}{L_0}, \pi\right)\,, \tag{7.6.63}$$

where l is the initial length of the billet, $m_0 = \tan\gamma_0$ and $m_f = \tan\gamma_f$.

Another case in which simplifications are possible is the one in which it is assumed that

$$g_f(\alpha) - g_0(\alpha) = L\,. \tag{7.6.64}$$

Equations (7.6.40), (7.6.41) and (7.6.64) then yield

$$a = \frac{\pi}{L}\,, \quad b = \frac{\pi}{2}\left[1 - \frac{2g_0(\alpha)}{L}\right]$$

in which the function g_0 and the real parameter L are arbitrary.

In another particular case we can assume that g_0 and g_f are chosen so that the surfaces S_0 and S_f are circles in the plane of symmetry: for instance circles with centers at the points $C_0(Z_0, R_0)$ and $C_f(Z_f, R_f)$ and passing through P and Q, respectively. Then we have

$$z = Z_0 - \sqrt{\rho_0^2 - [f_0(\alpha) - R_0]^2} \equiv g_0(\alpha)$$

$$z = Z_f - \sqrt{(R_2 - R_f)^2 + Z_f^2 - (\alpha - R_f)^2} \equiv g_f(\alpha)\,. \tag{7.6.65}$$

In this case the real parameters on which the function J depends are R_0, Z_0, ρ_0, R_f and Z_f.

Some combinations of the particular cases given above can be imagined.

The mechanical properties of the billet at high temperatures are described by choosing appropriate values for the viscosity coefficient η and the mean yield stress k.

In the numerical examples we have taken into account the dependence of η on the mean rate of deformation D_m defined by

$$\bar{D} = \frac{2}{\sqrt{3}}D_m = \frac{2}{\sqrt{3}}\frac{1}{V}\int_V II_D\, dV\,, \tag{7.6.66}$$

where V is the volume of the viscoplastic domain. $\bar{D}$ is computed for each particular case of extrusion considered. From the graphical plot of the function $\eta = \eta(\bar{D})$ obtained from experimental data, the corresponding value of the viscosity coefficient is obtained.

For high temperatures the value of k was determined from uniaxial experiments.

The geometry of the extrusion process will be characterized by the reduction in area

$$r\% = 100 \left(1 - \frac{R_2^2 - R_3^2}{R_1^2 - R_3^2} \right) \tag{7.6.67}$$

and the geometry of the billet.

In the numerical examples given below the data for steel AISI 304 as given by Polukhin *et al.* [1976] have been used. Thus for the temperature range of 1100–1200°C, the mean yield stress is $k = 3.5$ kgf mm^{-2} and $\sigma_{Yf} = \sqrt{3}\tau_7 = 8$ kgf mm^{-2}. For the working speed v_0 ranging between 200 mm s^{-1} and 400 mm s^{-1} and reductions of 96.14% and of 97%, $\bar{D}$ takes values between 38.12 s^{-1} and 58.40 s^{-1}. The corresponding values for η are 2.2×10^6 Poise and 1.5×10^6 Poise. For the reduction mentioned, the radii R_i are: $R_1 = 74$ mm, $R_2 = 28.5$ mm, $R_3 = 25$ mm and $R_1 = 74$ mm, $R_2 = 23.5$ mm, $R_3 = 20$ mm, respectively.

On the billet-plug and billet-container interfaces, the lubricant used is powdered glass, which at high pressure and temperature melts forming a thin film with the thickness of a few microns (5–17 μm). On the surface S_4 the interface between the melted glass and the billet, the friction factor is very small. From the experimental data by Maneghin *et al.* [1980] the friction factor at all surfaces lubricated by melted glass was taken $m = 0.02$ and in one single case $m = 0.04$.

For the particular shape of the surfaces S_0 and S_f described by formulae (7.6.62) the numerical values of the parameters γ_0, γ_f and L for which the functional $J(\mathbf{v})$ — now a function $J(L, \gamma_0, \gamma_f)$ of real variables — takes a minimal value, have been determined. This is done for various prescribed values of v_0, various geometries of the billet and product, prescribed mechanical properties of the metal and those of the lubricant, by a trial and error procedure. Then the value of L for which the function $J(L, \gamma_0, \gamma_f)$ has a minimum is found. With this value of L, the parameter γ_0 is varied and its value which minimizes J is found. We return now and with this new value of γ_0 kept constant vary again L, etc. After several iterations the parameter γ_f is varied also. The final values of the tree parameters obtained by this trial and error procedure are denoted by $(L^*, \gamma_0^*, \gamma_f^*)$.

We start with prescribed values of γ_0 and γ_f in the neighborhood of 90°.

On Fig. 7.6.2 for $v_0 = 300$ mm s^{-1} is shown the dependence of $J(L, \gamma_0^*, \gamma_f^*)$ on L for fixed values of γ_0^* and γ_f^*. The dependence of $J(L^*, \gamma_0, \gamma_f^*)$ on γ_0 is shown on Fig. 7.6.3 while that of $J(L^*, \gamma_0^*, \gamma_f)$ on γ_f is on Fig. 7.6.4. If the reduction in area is increased by a few per cent the minimum value of J increases. The values of γ_0^* and γ_f^* do not change significantly but the value of L^* increases when reduction is increased.

With an increase of the friction factor, the minimum value of J increases, L^* slightly increases but γ_0^* and γ_f^* are practically the same.

For each fixed value of v_0, geometry conditions, mechanical proprieties and lubrication conditions, using the determined values of $L^*, \gamma_0^*, \gamma_f^*$, the ram pressure is

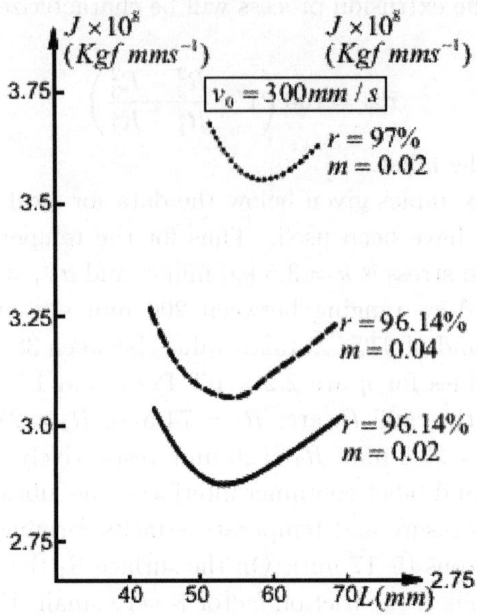

Fig. 7.6.2   Dependence of functional J on the length L for various reductions and friction factors, showing the minimum.

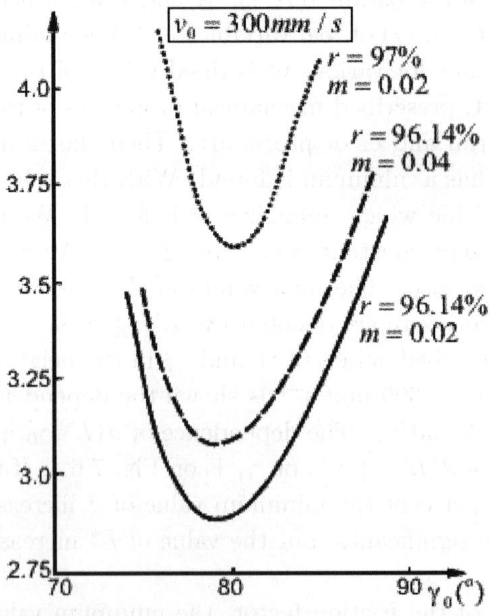

Fig. 7.6.3   Dependence of functional J on angle γ_0, for various reductions and friction factors, showing the minimum.

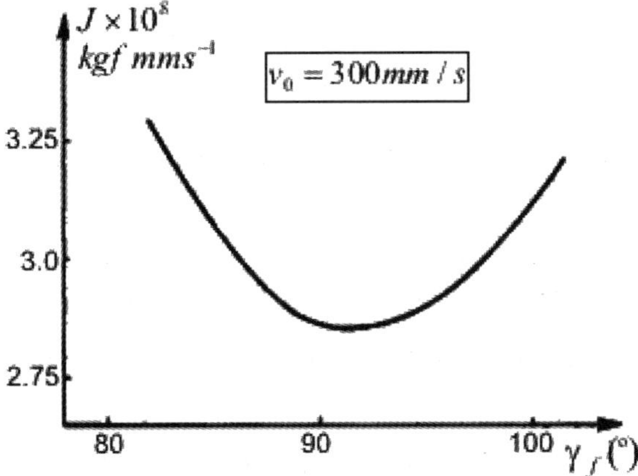

Fig. 7.6.4   Dependence of functional J on the angle γ_f; the minimum is close to $90°$.

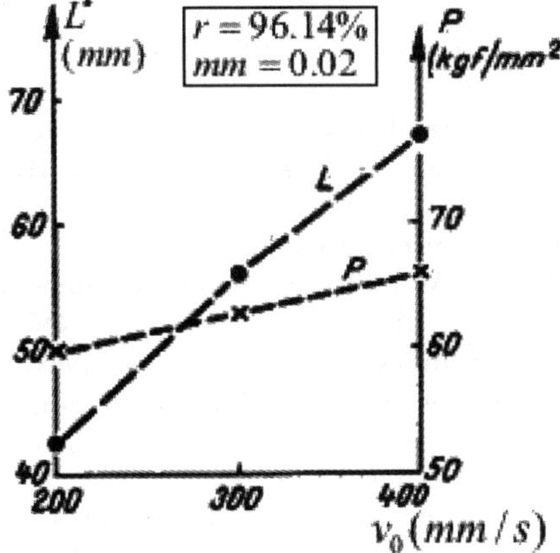

Fig. 7.6.5   The influence of working speed on the length L of the viscoplastic domain and on the extrusion pressure.

obtained from formula (7.6.60). On Fig. 7.6.5 is shown the working speed influence on the extrusion pressure: the extrusion pressure increasing with increasing v_0. The influence of v_0 on L^* is also shown on this figure: L^* increases significantly with v_0. The significant increase of L^* with v_0 is accompanied by an increase of the volume V_g of the melted glass.

It has been shown that the geometry of the viscoplastic domain, and therefore of the melted glass, is strongly dependent on the working speed. During the extrusion process, the melted glass is a self-adjusting die. A control volume of the melted glass during extrusion may ensure the optimal working conditions.

Bibliography

Abou-Sayed A. S. and Clifton R. J., 1976, Pressure shear waves in fused silica, *J. Appl. Phys.* **47**, 5, 1762–1770.

Abou-Sayed A. S. and Clifton R. J., 1977a, Analysis of combined pressure-shear waves in an elastc/viscoplastic material, *J. Appl. Mech. Trans. ASME* **44**, 79–84.

Abou-Sayed A. S. and Clifton R. J., 1977b, Pressure-shear waves in 6061-T6 aluminum due to oblique-plate-impact, *J. Appl. Mech. Trans. ASME* **44, 1**, 85–88.

Alexandrov S. and Alexandrova N., 2000, On the maximum friction law in viscoplasticity, *Mech. Time-Dependent Materias* **4**, 99–104.

Angelov T. A., 2004, Variational analysis of a rigid-plastic roling problem, *Int. J. Engng. Sci.* **42**, 1779–1792.

Avitzur B., 1968, *Metal Forming: Processes and Analysis*, McGraw-Hill, New York.

Bernstein B. and Shokooh A., 1980, *J. Rheology* **24**, 2, 189–211.

Bisk M. B. and Shveikin V. V., 1963, Tube drawing with floating plugs, *Metalurgya*, Moscow (in Russian).

Blazynski T. Z., 1976, *Metal Forming. Tool Profiles and Flow*, Macmillan, London.

Bramley A. N. and Smith D. J., 1976, *Metals Technology*, 322.

Camenschi G. and Şandru N., 2003, *Mathematical Models in Metal Working*, Bucharest, 188 pp. (in Roumanian).

Camenschi G., Cristescu N. and Şandru N., 1979, High speed wire drawing, *Archives of Mechanics* **31**, 5.

Camenschi G., Cristescu N. and Şandru N., 1983, Developments of high speed viscoplastic flow through conical converging dies, *Trans. ASME, J. Appl. Mech.* **50**.

Campbell J. D., 1973, Dynamic plasticity — macroscopic and microscopic aspects, *Materials Science and Engineering* **12**, 1, 3–21.

Chhabildas L. C. and Swegle J. W., 1980, Dynamic pressure-shear loading of materials using anisotropic crystals, *J. Appl. Phys.* **51**, 9, 4799–4807.

Cleja S. and Cristescu N., 1979, Influence of the drawing speed on the optimum shape of the floating plug, *Rev. Roum. Sci.-Mec. Appl.* **24**, 3, 357–377.

Cleja-Tigoiu S. and Cristescu N., 1985, A flow analysis of rigid-viscoplastic body through annular orifice, *Int. J. Mech. Sci.* **27**, 5, 291–301.

Clifton R. J., 1968, Elastic-plastic boundaries in combined longitudinal and torsional plastic wave propagation, *J. Appl. Mech. Trans. ASME* **35**, 4, 782–786.

Cristescu N., 1956, Some remarks concerning the case of plane axial symmetrical deformation in dynamic problems of plasticity, *Comm. Acad. Rep. Pop. Rom.* **6**, 19–28 (in Romanian).

Cristescu N., 1967, *Dynamic Plasticity*, North-Holland Pub. Co., 614 pp.

Cristescu N., 1975, Plastic flow through conical converging dies, using a viscoplastic constitutive equation, *Int. J. Mech. Sci.* **17**, 425–433.

Cristescu N., 1976, Drawing through conical dies — an analysis compared with experiments, *Int. J. Mech. Sci.* **18**, 45–49.

Cristescu N., 1977, Speed influence in wire drawing, *Rev. Roum. Sci. Techn. Mec. Appl.* **22**, 3, 391–399.

Cristescu N., 1980, On the optimum die angle in fast wire drawing, *J. Mech. Working Tech.* **3**, 275–287.

Dahan N. and Le Nevez P., 1983, Méthode de calcul des grandes déformations plastiques et endommagement dans les pièces extrudees, *Mémoire et Etudes Scientifiques Revue de Métallurgie* **80**, 10, 557–566.

Frutschy K. J. and Clifton R. J., 1997, High-temperature pressure-shear plate impact experiments using pure tungsten carbide impactors, *Experimental Mechanics* **38**, 2, 116–125.

Frutschy K. J. and Clifton R. J., 1998, High-temperature pressure-shear plate impact experiments on OFHC copper, *J. Mech. Phys. Solids* **46**, 19, 1723–1743.

Fu M. and Luo Z. J., 1995, The simulation of the viscoplastic forming process by the finite element method, *J. Materials Process. Tech.* **55**, 442–447.

Gilat A. and Clifton R. J., 1985, Pressure-shear waves in 6061-T6 aluminum and alpha titanium, *J. Mech. Phys. Solids* **33**, 3, 263–284.

Gilman J. J. and Tuler F. R., 1970, Dynamic fracture by spallation in metals, *Int. J. Fracture Mech.* **6**, 2, 169–182.

Güldenpfennig J. and Clifton R. J., 1980, Plastic waves of combined stress based on self consistent slip models, *J. Mech. Phys. Solids* **28**, 3–4, 201–219.

Hartley C., 1971, *Int. J. Mech. Sci.* **17**, 651–663.

Heeres O. M., Suiker A. S. J. and de Borst R., 2002, A comparison between the Perzyna viscoplastic model and the Consistency viscoplastic model, *Europ. J. Mech. A/Solids* **21**, 1–12.

Hsu J. C. C. and Clifton R. J., 1974a, Plastic waves in a rate sensitive material-I. Waves of uniaxial stress, *J. Mech. Phys. Solids* **22**, 233–253.

Hsu J. C. C. and Clifton R. J., 1974b, Plastic waves in a rate sensitive material-II. Waves of combined stress, *J. Mech. Phys. Solids* **22**, 255–266.

Ioan S. and Ioan R., 1999, A mathematical model of bimetallic tube drawing with floating plug, *Rev. Roum. Sci. Tech.-Mec. Appl.* **44**, 1, 49–72.

Iddan D. and Tirosh J., 1996, Analysis of high-speed rolling with inertia and rate effects, *J. Appl. Mech. Trans. ASME* **63**, 27–37.

Ionescu I. R. and Vernescu B., 1988, A numerical method for viscoplastic problem. An application to wire drawing, *Int. J. Engng. Sci.* **26**, 6, 627–633.

O'Keefe J. D., Skeen C. H. and York C. M., 1973, Laser-induced deformation modes in thin metal targets, *J. Appl. Phys.* **44**, 10, 4622–4626.

Kim K. S. and Clifton R. J., 1980, Pressure-shear impact of 6061-T6 aluminum, *J. Appl. Mech.* **47**, 1, 11–16.

Kim Y. H., Ryou T. K., Choi H. J. and Hwang B. B., 2002, An analysis of the forging processes for 6061 aluminum-alloy wheels, *J. Materials Process. Tech.* **123**, 279–276.

Klopp R. W. and Clifton R. J., 1990, Analysis of tilt in the high-strain-rate pressure-shear plate impact experiment, *J. Appl. Phys.* **67**, 11, 7171–7173.

Li G. J. and Kobayashi S., 1982, Rigid-plastic finite-element analysis of plane-strain roling, *ASME, J. Eng. for Industry* **104**, 1, 55–64.

Lindholm U. S. and Bessey R. L., 1969, A survey of rate dependent strength properties of metals, *Rept. AFML-TR-69-119*.

Mandel J., 1974, Introduction a la mecanique des milieux continues deformables, *PWN*, Warszawa.

Maneghin I. V., Pritomanov A. E., Shpittel T. and Knauschner A., 1980, Hot extrusion of tubes and profiles, *Metallurgya*, Moscow (in Russian).

Manjoine M. J., 1944, The influence of rate of strain and temperature on yield stresses in mild steel, *J. Appl. Mech.* **66**, A-211.

Medvedev M. I., Laskutov P. A. and Ratner A. G., 1980, *Metallurgya*, Moscow (in Russian).

Meir G. and Clifton R. J., 1986, Effects of dislocation generation at sursaces and subgrain boundaries on precursor decay in high-purity LiF., *J. Appl. Phys.* **59**, 1, 124–148.

Myers C. D. and Eisenberg M. A., 1974, The effect of radial inertia on combined plastic stress wave propagation in thin-walled tubes, *J. Appl. Mech. Trans. ASME* **41**, 3, 619–624.

Myers C. D. and Eisenberg M. A., 1975, The effect of radial inertia on the plastic stress wave speeds in thin-walled tubes under combined stress, *Acta Mechanica* **22**, 163–170.

Nowachi W. K., 1974, *Stress Waves in Non-Elasic Solids*, Pergamon Press, Oxford.

Perlin I. A., 1957, *Drawing Theory*, Metalurghizdat (in Russian).

Perzina P., 1966, Fundamental problems in viscoplasticity, *Advances in Appied Mechanics*, Academic Press, New York, **9**, 243–377.

Polukhin P. I., Gun G. Y. and Galkin A. M., 1976, Plastic deformation strength of metals and alloys, *Reference Book, Metallurgya*, Moscow (in Russian).

Şandru N. and Camenschi G., 1979, Viscoplastic flow through inclined planes with applications to the strip drawing, *Lett. Appl. Engng. Sci.* **17**, 773–784.

Şandru N. and Camenschi G., 1988, Mathematical model of the tube drawing with floating plug, *Int. J. Engng. Sci.* **26**, 6, 567–585.

Shida S. and Awazuhara H., 1973, *J. Japan Soc. Tech. Plasticity* **14**, 267–272.

Sokolovski V. V., 1969, Theory of plasticity, *Iz. Vashaia Skola* (in Russian).

Tanimura S., Igaki H., Majima H. and Tada M., 1978, Combined tension–torsion impact testing apparatus and an experimenal study in the incremental wave propagation, *Bulletin of the JSME* **21**, 160, 1455–1461.

Tayal A. K. and Natarajan R., 1981, Extrusion of rate-sensitive materials using a viscoplastic constitutive equation and the finite element method, *Int. J. Mech. Sci.* **23**, 89–98.

Ting T. C. T., 1970, Elastic-plastic boundaries in the propagation of plane and cylindrical waves of combined stress, *Quart. Appl. Math.* **27**, 4, 441–449.

Ting T. C. T., 1973, On wave propagation problems in which $c_f = c_s = c_2$ occurs, *Quart. Appl. Math.* **31**, 3, 275–286.

Ting T. C. T., 1977, Plastic wave speeds in isotropically work-hardening materials, *J. Appl. Mech. Trans. ASME* **44**, 1, 68–72.

Tirosh J. and Kobayeski S., 1976, Kinetic and dynamic effects on the upper-bound loads in metal-forming processes, *J. Appl. Mech. Trans. ASME* **43**, 2, 314–318.

Tirosh J., Iddan D. and Pawelski O., 1985, The mechanics of high-speed rolling of viscoplastic materials, *J. Appl. Mech. Trans. ASME* **52**, 309–318.

Tobe T., Kato M. and Obara H., 1979, Metal forming by underwater wire explosion 1. An analysis of plastic deformation of circular membranes under impulsive loading, *Bulletin JSME* **22**, 164, 271–278.

Upadhyay P. C. and Stokes V. K., 1977, On the dynamic expansion of a circular hole in an infinite plate, *J. Appl. Mech. Trans. ASME* **44**, 577–582.

Wistreich J. G., 1955, *Proc. Inst. Mech. Engng. London* **169**, 654.

Yew C. H. and Richardson H. A. Jr., 1969, Strain-rate effect and incremental plastic wave in copper, *Exp. Mech.* **9**, 8, 366–373.

Yokoyama T., 2001, Finite element computation of torsional plastic waves in a thin walled tube, *Archive of Appl. Mech.* **71**, 6–7, 359–370.

Chapter 8

Plastic Waves. Perforation

8.1 Introduction

Problems of plastic waves were considered by a lot of authors from variety points of view. Some restrictive relations for possible constitutive equations are due to Béda [1997]. Problems of plastic waves, problems of perforation etc. were considered; for a review of the subject see Davids [1960], Shewmon and Zackay [1961], Kolsky and Prager [1964], Cristescu [1967], Campbell [1972], Burke and Weiss [1971], Ezra [1973], Rinehart [1975], Cristescu *et al.* [2003], Zukas [2004] and the papers Backman and Goldsmith [1978], Zhu *et al.* [1992], Fomin and Kiselev [1997], Cheeseman and Bogetti [2003].

Chernyshov [1966] has shown that the structure of shock waves is practically the same as for gases. A sufficient condition is established by Lee [1975] for uniqueness of the dynamic path of an elastic-plastic body subject to prescribed kinematic boundary conditions on a part of the surface and prescribed time-dependent Lagrangian tractions on the remainder. The problems considered until now are of some importance for civil engineering, mechanical engineering, seismology, military applications, etc. The history of soil history is older than that of theory of plasticity. In the last times the theory of plasticity has developed faster, since the experimental facts are easier than in the soils or rock mechanics. Also it is more difficult to interpret the results of soil and rock mechanics. While the plastic deformation in metallic bodies is studied in more than half a century, the experiments with soils and rock mechanics are done only in the last decades. The physics is also quite complex: the soils is considered a continuous medium where solid bodies, fluids and gas are mixed together in a complex manner in what is called a three-phase medium. Sometimes one of these tree medium is absent and the body becomes a two-phase medium. If $\alpha_1, \alpha_2, \alpha_3$ denote the relative proportions by volume of gas, liquid and solid constituents of the soil, so that

$$\alpha_1 + \alpha_2 + \alpha_3 = 1 \,, \tag{8.1.1}$$

and if ρ_1, ρ_2 and ρ_3 are the corresponding densities (in g/cm^3), the density ρ of the soil considered as a three-phase medium will be

$$\rho = \alpha_1\rho_1 + \alpha_2\rho_2 + \alpha_3\rho_3 \,. \tag{8.1.2}$$

383

The density of air is $\rho_1 = 1.2 \times 10^{-3}$ g/cm^3 of water $\rho_2 = 1$ g/cm^3, while for the solid an average value $\rho_3 = 2.65$ g/cm^3 is used. One is introducing also the concept of porosity as the ratio between the volume of the pores and that of the soil itself.

Various constitutive equations have been established for soils and rocks (see Ch. 2). Sometimes Eulerian methods were used since "big" strains are involved. In hydrostatic tests the compressibility of the soil is considered. The compaction property is highly dependent on the water content of the body and is different under dynamic and static loadings. In static experiments it is the solid skeleton of the body that generally supports most of the pressure if the sample is drained, because there is sufficient time for the water to drain away from the pores during the experiment. In dynamic experiments the situation is quite different. The mixture of water and air, if the sample is drained, there is not sufficient time for the water to escape during a dynamic experiment. In dynamic loading experiments it is the water–air mixture that supports most of the applied pressure. However if the water–air mixtures, the total volume of air is larger than that of the water, then the mixture is much more compressible than the solid skeleton: in these cases the skeleton supports most of the pressure. For very high pressures however, skeleton compressibility plays an important role in the overall compressibility of the body.

8.2 Various Theories

In the previous sections we have discussed the laws of variation of the volume as well as various possible critical stress conditions. In the present section we present several old stress–strain relations which have played a role in the development of the theory. They are very similar to the classic theories of plasticity.

<u>S. S. Grigorian</u> [1959, 1960b, 1967] assumes that the Prandtl–Reuss theory can be used to describe the mechanical properties of soft soils. If v_i are the velocity components, and

$$D_{ij} = \frac{1}{2}\left(\frac{\partial v_i}{\partial x_j} + \frac{\partial v_j}{\partial x_i} \right)$$

(8.2.1)

are the components of the rate of strain, and x_i are the Cartesian coordinates, then the constitutive equations are

$$2GD'_{ij} = \frac{\tilde{d}\sigma'_{ij}}{dt} + \lambda \sigma'_{ij}\,,$$

(8.2.2)

where the derivatives are taken in the sense of Jauman

$$\frac{\tilde{d}\sigma'_{ij}}{dt} = \frac{\partial \sigma'_{ij}}{dt} + v_k \frac{\partial \sigma'_{ij}}{\partial x_k} - \Omega_{ik}\sigma'_{jk} - \Omega_{jk}\sigma'_{ik}\,,$$

(8.2.3)

and

$$2\Omega_{ij} = \frac{\partial v_i}{\partial x_j} - \frac{\partial v_j}{\partial x_i}\,.$$

(8.2.4)

The factor λ in (8.2.2) plays the same role as in classical plasticity theory. The elastic behavior of the material corresponds to $\lambda = 0$, and the plastic state to $\lambda > 0$. This factor can be expressed in terms of invariants following the usual procedure: by multiplying (8.2.2) by σ'_{ij} and summation, we obtain

$$4GW \equiv 2G\sigma'_{ij}D'_{ij} = \sigma'_{ij}\frac{\tilde{d}\sigma'_{ij}}{dt} + \lambda\sigma'_{ij}\sigma'_{ij}\,. \tag{8.2.5}$$

Since

$$II_{\sigma'} = F(p) \quad \text{and} \quad \sigma'_{ij}\frac{\tilde{d}\sigma'_{ij}}{dt} = \frac{dII_{\sigma'}}{dt}\,,$$

it follows from (8.2.5) that

$$\lambda = \frac{4GW - F'(p)(\partial p/\partial t + v_i\partial p/\partial x_i)}{2F(p)}\,. \tag{8.2.6}$$

This formula is used only if

$$4GW > F'(p)\left(\frac{\partial p}{\partial t} + v_i\frac{\partial p}{\partial x_i}\right),$$

otherwise $\lambda = 0$. To the constitutive equation one is adding the equation of motion and the continuity equation. The Prandtl–Reuss constitutive equation is proposed for dynamic problems for soils since soils being dispersed media, there is not a single reference configuration for them. The system of equation was analyzed also from the thermodynamic point of view. There are three functions (Grigorian [1960a]) to be determined from tests:

$$p = f(\rho, \rho_m, H)\,, \quad H_{\sigma'} = F(p, H)\,, \quad G = G(\rho_m, H)\,. \tag{8.2.7}$$

Here H is the humidity, i.e., the ratio between the volume of the water and that of the mineral particles, per unit mass of soil. Since in dynamic problems there is no time for water to drain away from the pores $dH/dt = 0$. If $H = H_{\max}$, that is the soil is fully saturated, in dynamic experiments such soils behave as barotropic liquids. Thus for small or moderate values of H the soil is irreversible compressible. The other functions involved in (8.2.7) are determined as usual for compressible materials.

The Coulomb flow rule was used by Chadwick *et al.* [1964], as associated flow rule. They have assumed that the soil obeys Hooke's law within the elastic range and the Coulomb's criterion and the associated flow rule during yielding under the restriction of perfectly plastic flow. According to the rule mentioned, if $f(\sigma_{ij}) = 0$ denotes the yield function and the elastic domain corresponds to $f < 0$, then the plastic part of the rate of strain is furnished by

$$\dot{\varepsilon}^p_{ij} = \dot{\lambda}\frac{\partial f}{\partial \sigma_{ij}}\,, \tag{8.2.8}$$

where $\dot{\lambda}(x,t)$ is a scalar function satisfying the condition $\lambda \geq 0$. The elastic strain rates satisfy a Hookean law. The authors consider only spherical-symmetry problems and therefore spherical polar coordinates r, θ, φ are used. The principal components of the stress are σ_r, σ_θ and σ_φ. If

$$\dot{\varepsilon}_r = \frac{\partial v}{\partial r}, \quad \dot{\varepsilon}_\theta = \dot{\varepsilon}_\varphi = \frac{v}{r} \tag{8.2.9}$$

are the components of the strain rate then the associated flow rule is

$$\dot{\varepsilon}_r^p = \dot{\lambda}\frac{\partial f}{\partial \sigma_r}, \quad \dot{\varepsilon}_\theta^p = \dot{\lambda}\frac{\partial f}{\partial \sigma_\theta}, \quad \dot{\varepsilon}_\varphi^p = \dot{\lambda}\frac{\partial f}{\partial \sigma_\varphi}, \tag{8.2.10}$$

where f is one of the six expressions involved in the Coulomb yield condition. At singular points the generalized flow rule proposed by Koiter [1953] and Prager [1953] is used.

The "plastic gas" model was considered by Rakhmatulin and Stepanova [1958]. It was used for soils in one-dimensional problems. By definition a "plastic gas" is a continuous body whose density changes during loading (increasing pressure) according to a prescribed law, while during unloading (decreasing pressure) the density remains constant and equal to the maximum density reached during loading. On repeated loading the density does not change as long as the pressure does not exceed the value reached previously. This model was proposed for the use in problems in which high pressures are involved. Since experiments have shown that for such pressures the temperature rise is negligible, temperature changes are generally disregarded.

Very many problems have been solved using the model mentioned above (Rakhmatulin *et al.* [1964]). Let us see what Sagomonian [1961] has to say about this problem. The problem is one-dimensional and cylindrically symmetrical. In cylindrical Lagrange coordinates r and θ, the equation of motion and the equation of mass conservation are

$$\rho_0 r \frac{\partial^2 u}{\partial t^2} = (r+u)\frac{\partial \sigma_r}{\partial r} + (\sigma_r - \sigma_\theta)\frac{\partial}{\partial r}(r+u),$$

$$\frac{1}{2}\frac{\partial}{\partial r}(r+u)^2 = \frac{\rho_0}{\rho}r. \tag{8.2.11}$$

Assuming that $\sigma_r = \sigma_\theta = -p$, the first equation becomes

$$\rho_0 \frac{\partial^2 u}{\partial t^2} = -\frac{r+u}{r}\frac{\partial p}{\partial r}. \tag{8.2.12}$$

The problems to be solved are the following. An infinitely long cylindrical explosive of initial radius r_0 is exploded in an infinite medium. It is assumed that the explosion products expand as a polytrophic gas whose adiabatic exponent is γ. Since in most cases the maximum pressure is reached just behind the shock wave front, Sagomonian assumes that the density of the plastic gas changes only on the front (in the transition zone). Thus, behind the shock wave front the density is a function only of r, and the motion is that of a non-homogeneous incompressible medium.

To the equation (8.2.11) one attaches the law of soil compressibility

$$\frac{1}{3}(\sigma_r + \sigma_\theta + \sigma_z) = F\left(1 - \frac{\rho_0}{\rho}\right),\tag{8.2.13}$$

and the Coulomb condition is written in the form

$$\sigma_r - \sigma_\theta = -\tau_0 + \mu(\sigma_r + \sigma_\theta)\tag{8.2.14}$$

with $\tau_0 = 2k\cos\varphi$ and $\mu = \sin\varphi$ and assuming that $\sigma_r > \sigma_\theta$. The first case considered is $\mu = 0$, i.e., the case of frictionless soils. The Eq. (8.2.11) becomes

$$\rho_0 r \frac{\partial^2 u}{\partial t^2} = (r+u)\frac{\partial \sigma_r}{\partial r} - \tau_0 \frac{\partial}{\partial r}(r+u),\tag{8.2.15}$$

or, by formal integration with respect to r,

$$\sigma_r(r,t) = \rho_0 \int_{r_0}^{r^*} \frac{r}{r+u}\frac{\partial^2 u}{\partial t^2}\,dr + \tau_0\{\ln(r+u) - \ln R\} - p_k(t).\tag{8.2.16}$$

Here r_0 is the radius of the charge hole at the moment of explosion, $p_k(t)$ is the pressure at $r = r_0$, while the radius of the hole at a certain moment is $R(t) = r_0 + u(r_0, t)$.

The relation (8.2.16) holds for all points between the hole and the shock wave front. Thus denoting by an asterisk the values of any function on this front, we have

$$\sigma_r^* = \rho_0 \int_{r_0}^{r^*} \frac{r}{r+u}\frac{\partial^2 u}{\partial t^2}\,dr + \tau_0\{\ln r^* - \ln R\} - p_k(t).\tag{8.2.17}$$

From (8.2.16) and (8.2.17), it follows that

$$\sigma_r - \sigma_r^* = -\rho_0 \int_{r_0}^{r^*} \frac{r}{r+u}\frac{\partial^2 u}{\partial t^2}\,dr - \tau_0\{\ln r^* - \ln(r+u)\}.\tag{8.2.18}$$

If, in (8.2.14) $\mu \neq 0$, then the first equation (8.2.11) becomes

$$(r+u)\frac{\partial \sigma_r}{\partial r} + v\frac{\partial}{\partial r}(r+u)\sigma_r = \rho_0 r\frac{\partial^2 u}{\partial t^2} + \frac{\tau_0}{1+\mu}\frac{\partial}{\partial r}(r+u),\tag{8.2.19}$$

with $v = 2\mu/(1+\mu)$. By multiplying both sides of (8.2.19) by $(r+u)^{v-1}$ and integrating with respect to r, we obtain

$$(r+u)^v \sigma_r(r,t) = \rho_0 \int_{r_0}^{r} (r+u)^{v-1} r\frac{\partial^2 u}{\partial t^2}\,dr$$

$$+ \frac{\tau_0}{1+\mu}\frac{1}{v}[(r+u)^v - R^v] - R^v p_k(t).\tag{8.2.20}$$

Hence, on the shock wave front we have

$$r^{*v}\sigma_r^* = \rho_0 \int_{r_0}^{r^*} (r+u)^{v-1} r\frac{\partial^2 u}{\partial t^2}\,dr + \frac{\tau_0}{1+\mu}\frac{1}{v}(r^{*v} - R^v) - R^v p_k(t).\tag{8.2.21}$$

Thus, from the last two formulae it follows that

$$(r+u)^v \sigma_r - r^{*v} \sigma_r^* = -p_0 \int_r^{r^*} (r+u)^{v-1} \frac{\partial^2 u}{\partial t^2} \, r dr$$

$$- \frac{\tau_0}{v(1+\mu)} \{ r^{*v} - (r+u)^v \} . \qquad (8.2.22)$$

For cohesionless soils ($\tau_0 = 0$) this formula becomes

$$(r+u)^v \sigma_r - r^{*v} \sigma_r^* = -\rho_0 \int_r^{r^*} (r+u)^{v-1} r \frac{\partial^2 u}{\partial t^2} \, dr . \qquad (8.2.23)$$

As has already been mentioned, it is assumed that the density change occurs only in the transition zone, and therefore, that behind the shock wave front, the density is a function only of the Lagrangiann coordinate r. Thus, from the second equation (8.2.11), by integration with respect to r, it follows that

$$(r+u)^2 = 2\psi(r) + \varphi(t)$$

with

$$\psi(r) = \int_{r_0}^r \frac{\rho_0}{\rho(\bar{r})} \bar{r} d\bar{r} . \qquad (8.2.24)$$

The arbitrary function $\varphi(t)$ is determined by the boundary conditions

$$r = r_0 : \quad \psi(r_0) = 0 , \quad \varphi(t) = \{ r_0 + u(r_0, t) \}^2 = R^2(t)$$

so that

$$(r+u)^2 = 2\psi(r) + R^2(t) \qquad (8.2.25)$$

or, on the shock wave front

$$r^{*2} = 2\psi(r^*) + R^2(t) . \qquad (8.2.26)$$

From (8.2.25), we obtain the velocity and acceleration as

$$\frac{\partial u}{\partial t} = \frac{R\dot{R}}{\sqrt{2\psi(r) + R^2}} ,$$

$$\frac{\partial^2 u}{\partial t^2} = \frac{\dot{R}^2 + R\ddot{R}}{\sqrt{2\psi(r) + R^2}} - \frac{(R\dot{R})^2}{[2\psi(r) + R^2]^{3/2}} . \qquad (8.2.27)$$

Across the shock wave front the jump conditions

$$\rho_0 D = \rho(D - \dot{u}^*) , \quad \rho_0 D \dot{u}^* = -\sigma_r^* - p_a$$

are satisfied. In these conditions, D is the velocity of the displacement of the wave, $\dot{u}^*$ the material velocity just behind the shock front, and p_a is the pressure ahead of the shock wave. We therefore obtain

$$-\sigma_r^* = \frac{\rho_0 \dot{u}^{*2}}{1 - b(r^*)} + p_a , \quad D = \frac{\dot{u}^*}{1 - b(r^*)} , \quad b(r^*) = \frac{\rho_0}{\rho(r^*)} = \left(\frac{d\psi}{dr} \frac{1}{r} \right)_{r=r^*} .$$

$$(8.2.28)$$

Introducing (8.2.25)–(8.2.28) into (8.2.18), (8.2.22) and (8.2.23), we obtain

$$-\sigma_r(r,t) = (R\ddot{R} + \dot{R}^2)\rho_0 \int_r^{r^*} \frac{r\,dr}{2\psi + R^2} - (R\dot{R})^2\rho_0 \int_r^{r^*} \frac{r\,dr}{(2\psi + R^2)^2}$$

$$+ \frac{\rho_0}{1-b}\frac{(R\dot{R})^2}{r^{*2}} + \tau_0 \ln\frac{r^*}{r+u} + p_a\,, \tag{8.2.29}$$

$$-(r+u)^v\sigma_r = (R\ddot{R} + \dot{R}^2)\rho_0 \int_r^{r^*} \frac{r\,dr}{(2\psi + R^2)^{1-(v/2)}}$$

$$- (R\dot{R})^2\rho_0 \int_r^{r^*} \frac{r\,dr}{(2\psi + R^2)^{2-(v/2)}} + \frac{\rho_0}{1-b}\frac{(R\dot{R})^2}{(r^*)^{2-v}}$$

$$+ \frac{\tau_0}{v(1+\mu)}(r^{*v} - (r+u)^v) + p_a r^{*v}\,, \tag{8.2.30}$$

$$-(r+u)^v\sigma_r = (R\ddot{R} + \dot{R}^2)\rho_0 \int_r^{r^*} \frac{r\,dr}{(2\psi + R^2)^{1-(v/2)}}$$

$$- (R\dot{R})^2\rho_0 \int_r^{r^*} \frac{r\,dr}{(2\psi + R^2)^{2-(v/2)}}$$

$$+ \frac{\rho_0}{1-b}\frac{(R\dot{R})^2}{(r^*)^{2-v}} + p_a r^{*v}\,. \tag{8.2.31}$$

From the relations (8.2.27)–(8.2.31) it follows that all the quantities required to describe the motion can be expressed by two functions, i.e., a function only of the Lagrangean coordinate $\psi(r)$ and another one function only of time $R(t)$.

A first solution is obtained, assuming that behind the shock wave front the density is a constant ρ_1 and that

$$\rho_0/\rho_1 = b_1\,. \tag{8.2.32}$$

From (8.2.24) and (8.2.6) it then follows that

$$\psi(r) = \frac{1}{2}b_1(r^2 - r_0^2)\,, \quad r^{*2} = \frac{R^2(t) - b_1 r_0^2}{1 - b_1}\,. \tag{8.2.33}$$

With this expression for $\psi(r)$, one can compute the integrals in (8.2.29)–(8.2.31) and thus express the solution in terms of the function $R(t)$ and its first two derivatives.

The case when the density is not constant behind the shock wave front has also been considered by an iterative method. In the same paper by Sagomonian [1961], solutions for the cases of polar symmetry and of the propagation of plane waves were also given.

Many other problems were solved using the previously mentioned model for the soils concerned. For instance, the penetration of sharp, rotationally symmetrical bodies into the ground was examined by Rachmatulin *et al.* [1964] in Ch. IV, where a bibliography of the problem is also given.

The propagation of spherical collapsing waves which propagate in an elastic-brittle material was studied by Aliev [1963]. The same problem for an elastic-plastic medium, which is incompressible both during loading and unloading and satisfies a Tresca yield condition during loading, was considered by Shemiankin [1961]. In another paper, Medvedeva and Shemiankin [1961] have examined the propagation of spherical waves in a medium satisfying a Coulomb yield condition, for a compaction law written in the form

$$-\sigma = K\left[1 - \left(\frac{\rho_0}{\rho}\right)^n\right],$$

where K and n are positive constants. The model is applied to various rocks.

The <u>locking medium</u> was considered by many authors. The main idea is: once the density of the medium reaches a certain limit value, characteristic of the medium, no further density increase is possible. During unloading, the density remains constant and equal to the limiting density. There has also been proposed a model known as an ideal locking medium, which is defined by most authors as follows. The material, of a certain initial density, does not at first offer any resistance to compression, but once the limiting density ρ^* is reached, this remains constant during any subsequent motion and Coulomb's yield condition is satisfied. In a certain sense, the ideal locking medium is a model corresponding to the perfectly plastic/rigid model of plasticity theory. It seems that the idea of locking medium was first used in the paper by Ishlinskii *et al.* [1954] and Salvadori *et al.* [1955]. These authors start from the experimental observation that if a cylindrical hole some ten meters deep and of diameter 10 to 15 cm is made in a soil (of clay type) and if an explosion is produced in this hole, the density of the soil to a distance of 1 m around the hole increases considerably (approximately 30 per cent). This experiment is described mathematically.

Ishlinskii [1954] established the equations of plane motion of certain materials like sand. He made four hypothesis: the medium is non-elastic and incompressible, the stress do not depend on the rate of strain, the directions of principal axes of stresses coincide at each point with the directions of the principal axes of strain, the state of stress satisfies the same yield condition as in the static case. The propagation of spherical shock waves in sandy type medium was considered by Kompaneets [1956]. He assumed that initially the medium is of density ρ_0 and has a small resistance to shear. When reaching the density ρ_1, the medium becomes incompressible and plastic, and satisfies a plasticity condition of the Coulomb form

$$\sigma_r - \sigma_\theta = k + m(\sigma_r + \sigma_\theta + \sigma_\varphi).$$

He also assumes that shock waves are produced by an explosion detonated in a small sphere. Andriankin and Koryavov [1959] consider that the soil compaction along the shock wave front is not constant. They assumed that the compaction produced by the shock wave front depends on the amplitude of the wave; a power law of compaction of the form

$$p = p_0 \beta^{1-n} \varepsilon^n \,,$$

is proposed. Here n and β are constants, p_0 is the pressure ahead of the wave front and $\varepsilon = 1 - \rho_0/\rho$.

A viscous model for plug formation in plates is due to Pytel and Davids [1963]. The paper presents the analysis of a viscous model for the study of impact of plates by projectiles under conditions which would lead to failure of the plate by the formation of a plug. The impact is represented by a velocity uniformly distributed over a circular area on the plate surface. Only the vertical shearing stress is considered and it is assumed to depend only on the radial coordinate. The stress, velocity and displacement profiles are calculated for the viscous model. The calculated displacement profiles are compared with experimental profiles.

Spherical elastic-plastic boundaries due to loading were considered by Srinivasan and Ting [1974]. They have considered spherical waves in an elastic-plastic, isotropically work-hardening medium generated by radial stress uniformly applied at a spherical cavity. The radial stress and its time derivative at the cavity may be discontinuous at time t_0. If the applied radial stress is continuous while its time derivative is not, the discontinuity propagates into $r > r_0$ along the characteristics and/or the elastic-plastic boundaries. If the applied radial stress itself is discontinuous, the discontinuity may propagate into $r > r_0$ in the form of a shock wave, or a centered simple-wave, or a combination of both. In any case, the solutions in the neighborhood of (r_0, t_0) are obtained for all possible combinations of discontinuous loadings applied at $r = r_0$. This is a systematic study on the nature of the solution near (r_0, t_0) where the applied load is discontinuous. Solutions for specials materials, such as linearly work-hardening or ideally-plastic ones, and for special applied loadings at the cavity obtained by other workers, in which the nature of the solutions near (r_0, t_0) are assumed a priori rather than determined, are compared with the results obtained here.

Wave propagation in elastic circular membrane is considered by Hutton and Counts [1974]. One is starting from the problem given in Cristescu [1967] Ch. 5. The equations of motion for meridian line of the membrane are

$$\eta_0 \frac{\partial^2 \eta}{\partial \tau^2} = \frac{p}{p_0} \eta \frac{\partial \omega}{\partial \eta_0} + \frac{\partial}{\partial \eta_0} \left(\frac{\xi \eta}{1 + \varepsilon_1} \frac{\sigma_1}{p_0} \frac{\partial \eta}{\partial \eta_0} \right) - (1 + \varepsilon_1) \frac{\sigma_2}{p_0} \xi$$

$$-\eta_0 \frac{\partial^2 \omega}{\partial \tau^2} = \frac{p}{p_0} \eta \frac{\partial \eta}{\partial \eta_0} - \frac{\partial}{\partial \eta_0} \left(\frac{\xi}{1 + \varepsilon_1} \frac{\sigma_1}{p_0} \frac{\partial \omega}{\partial \eta_0} \right)$$

where η and η_0 are the actual and initial radial coordinates, ω is the axial coordinate, $\varepsilon_1, \varepsilon_2, \varepsilon_3$ are the meridional, circumferential, normal strains, and $\sigma_1, \sigma_2, \sigma_3$ are the corresponding stress components, p/p_0 is the ratio of instantaneous pressure to maximum pressure, τ is the nondimensional time, and $\xi = 1 + \varepsilon_2$. The totally hyperbolic equations are integrated along characteristics. Strain and inertia waves propagate with constant velocities during elastic deformation.

The buckling problem. Dynamic axisymmetric buckling of circular cylindrical shells struck axially by a mass is studied by Karagiozova and Jones [2000] in order to clarify the initiation of buckling and to prove some insight into the buckling mechanism as a transient process. It is assumed that the material is elastic-plastic with linear strain hardening and displaying the Bauschinger effect. The deformation process is analyzed by a numerical simulation using a discrete model. It is found that the development of the buckling shape depends strongly on the inertia properties of the striker and on the geometry of the shell. A new concept is presented by Karagiozova and Jones [2001], [2002] for the dynamic elastic-plastic axisymmetric buckling of circular cylindrical shells under axial impact. The phenomena of dynamic plastic buckling and dynamic progressive buckling are analyzed from the viewpoint of stress wave propagation resulting from an axial impact. The conditions for the development of dynamic plastic buckling are obtained. It is concluded that shells made of strain rate insensitive materials can respond either by dynamic plastic buckling or dynamic progressive buckling, depending on the inertia properties of the shell, while those shells made of strain rate sensitive materials respond always by dynamic progressive buckling. It is shown that the prediction for the peak load depends on the particular yield criterion used in the analysis. The speeds of the stress waves that can propagate in an elastic-plastic medium with isotropic linear strain hardening in a plane stress state are obtained by Karagiozova [2004] to analyze the influence of the transient deformation process on the initiation of buckling in square tubes under axial impact. The kinematic conditions across a surface of discontinuity were employed to obtain the stress wave propagation speeds for an initial condition. It is shown that the plastic wave speeds depend on the stress state and on the direction of wave propagation. The material hardening properties have a stronger effect on the speed of the slow plastic wave, while the shear stress affects both the speeds of the fast and slow plastic waves. The magnitude of the instantaneously applied in-plane load at $t = 0$, which results from an impact in the direction "one", is obtained as a function of the initial velocity when using the differential relationship along the characteristics.

A general non-dimensional formula based on the dynamic cavity-expansion model is proposed buy Chen and Li [2002] to predict penetration depth into several mediums subjected to a normal impact of a non-deformable projectile. The proposed formula depends on two dimensionless numbers and shows good agreement with penetration tests on metal, concrete and soil for a range of nose shapes and impact velocities. The validity of the formula requires that the penetration depth is larger than the projectile diameter and the projectile nose length while projectile remains rigid without noticeable deformation and damage.

A theoretical investigation of plane waves in granular soils is presented by Osinov [1998]. Dynamic equations are derived with the use of the hypoplasticity theory for granular materials. For numerical calculations the material parameters of Karlsruhe sand are used. Wave speeds slopes of characteristics of dynamic equations are calculated for various stresses and densities.

8.3 Perforation with Symmetries

For perforation first to be seen is the paper by Backman and Goldsmith [1978]. This big paper presents the problem until 1978, characteristics of the projectile and the targets, semi-infinite targets, thin plate penetration and perforation, and thick plates.

The problem was considered also by Demiray and Eringen [1978]. By means of a nonlocal viscous fluid model, an investigation is carried out of the problem of penetration of a cylindrical projectile into a plate leading to a failure of the plate by a plug formation. The effect of impact is represented by a uniform initial velocity distribution over a circular region on the surface of the plate. The behavior of this plate material is assumed to be viscous and spatially nonlocal, and only the effects of vertical shearing stress are considered. The expression of stress, velocity and displacement are obtained and the calculated displacement profiles are compared with some existing experimental profiles.

The problem of impact of a liquid cylinder on an elastic-plastic solid was considered by Lush [1991]. He has modified the model to incorporate the unloading of the plastic wave which is produced by multiple reflections of the elastic release wave. In general the agreement between penetration rate, contact pressure and depth of penetration is good.

Afterwards, the ballistic strengths of composites are considered mainly in Cheeseman and Bogetti [2003]. First is described a single fiber subjected to transverse impact (see Ch. 5). Then is described the mechanisms influencing ballistic performance. The fibers are Kevlar or Spectra laminates which are impacted with impact velocities up to 1000 m/s. If the cover factors of the fabrics are from 0.6 to 0.95 they are effective in ballistic applications. The crimp that is the undulation on the yarns due to their interlacing in the woven structure is also discussed. Generally is discussed all the factors which have collaboration in the perforation.

An oblique penetration was considered by Macek and Duffey [2000]. A spherical cavity expansion is assumed with the local velocity normal equal to the penetrator surface V_N (see Fig. 8.3.1). Incompressible kinematics is also assumed. The incompressible condition is

$$r^{*3} - r_i^{*3} = r^3 - r_i^3 , \qquad (8.3.1)$$

where $*$ refer to the deformed configuration, the other radii to the undeformed configuration. The subscript i refer to the inner surface of the ith layer. The radial displacement is

$$u_r = r^* - r = (r_i^{*3} + r^3 - r_i^3)^{1/3} - r . \qquad (8.3.2)$$

The radial velocity distribution is obtained from (8.3.2)

$$V_r = \frac{r_i^{*2}}{r^{*2}} V_i , \qquad (8.3.3)$$

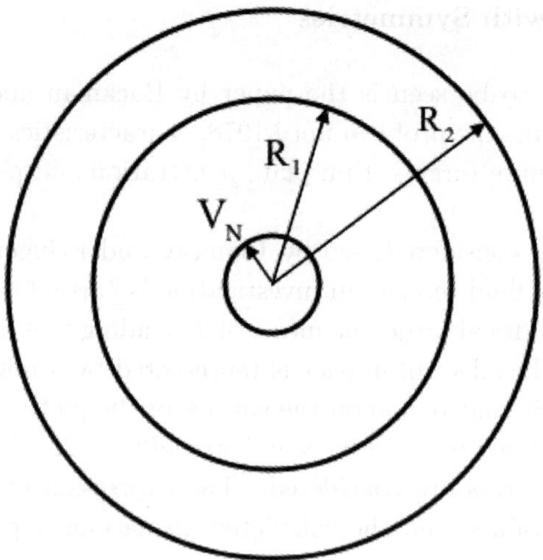

Fig. 8.3.1   Spherical cavity expansion idealization.

with V the velocity. The radial acceleration is determined by differentiation:

$$a_r = \left(\frac{r_i^*}{r^*}\right)^2 \dot{V}_i + 2r_i^* V_i^2 \left(\frac{1}{r^{*2}} - \frac{r_i^{*3}}{r^{*5}}\right). \tag{8.3.4}$$

The corresponding spherically symmetric strain field as a result of incompressibility is obtained from (8.3.2):

$$\varepsilon_\theta = \frac{u_r}{r} = \frac{r^*}{r} - 1 = \frac{r^*}{(r^{*3} - r_i^{*3} + r_i^3)^{1/3}} - 1, \tag{8.3.5}$$

$$\varepsilon_r = \frac{du_r}{dr} = \frac{(r^{*3} - r_i^{*3} + r_i^3)^{2/3}}{r^{*2}} - 1. \tag{8.3.6}$$

The approximated effective strain field is

$$\varepsilon_\theta - \varepsilon_r = \frac{R^3}{r^3} = \frac{r_i^{*3} - r_i^3}{r^{*2}(r^{*3} - r_i^{*3} + r_i^3)^{1/3}} \approx \frac{r_i^{*3} - r_i^3}{r^{*3}}. \tag{8.3.7}$$

For a Hooke's material with spherically symmetry and incompressible, the effective stress is

$$\sigma_e = \sigma_\theta - \sigma_r = \frac{2}{3} E(\varepsilon_\theta - \varepsilon_r). \tag{8.3.8}$$

If the material has yielded and a Mohr–Coulomb yield criterion with damage is used then

$$Y = \sigma_\theta - \sigma_r = Y_0 + \lambda P, \quad P \geq -\frac{Y_0}{\lambda}, \quad \varepsilon_\theta \leq \varepsilon_{ut}, \tag{8.3.9}$$

$$Y = \sigma_\theta - \sigma_r = \lambda_D P, \quad \varepsilon_\theta > \varepsilon_{ut}, \tag{8.3.10}$$

where ε_{ut} is the tensile failure strain, P is the pressure, Y_0 is the yield stress at zero pressure, λ is the Mohr–Coulomb slope, and λ_D denotes the slope after failure.

The equilibrium equation in the deformed configuration is

$$\frac{d\sigma_r}{dr^*} + \frac{2}{r^*}(\sigma_r - \sigma_\theta) = \rho a_r = \rho \dot{V}_r .\tag{8.3.11}$$

Combining (8.3.4) and (8.3.11)

$$\frac{d\sigma_r}{dr^*} + \frac{2}{r^*}(\sigma_r - \sigma_\theta) = \rho \left(\frac{r_i^*}{r^*}\right)^2 \dot{V}_i + 2\rho r_1^* \left(\frac{1}{r^{*2}} - \frac{r_i^{*3}}{r^{*5}}\right).\tag{8.3.12}$$

The governing equation for an elastic sub-layer is found from (8.3.7), (8.3.8) and (8.3.12):

$$\frac{d\sigma_r}{dr^*} = \rho \left(\frac{r_i^*}{r^*}\right)^2 \dot{V}_i + 2\rho r_i^* V_i^2 \left(\frac{1}{r^{*2}} - \frac{r_i^{*3}}{r^{*5}}\right) + \frac{4E}{3r^{*4}}(r_i^{*3} - r_i^3).\tag{8.3.13}$$

The effective stress is

$$\sigma_e = \frac{2}{3} E \frac{r_i^{*3} - r_i^3}{r^{*3}} .\tag{8.3.14}$$

Integrating (8.3.13) with $\sigma_r = \sigma_{r0}$ at $r^* = r_0^*$ on the outer surface layer,

$$\sigma_r(r^*) = \sigma_{r0} + 2\rho r_i^* V_i^2 \left(\frac{1}{r_0^*} - \frac{1}{r^*} + \frac{r_i^{*3}}{4r^{*4}} - \frac{r_i^{*3}}{4r_0^{*4}}\right) + \rho r_i^* \dot{V}_i \left(\frac{r_i^*}{r_0^*} - \frac{r_i^*}{r^*}\right)$$

$$+ \frac{4E}{9} \left(\frac{r_i^{*3}}{r_0^{*3}} - \frac{r_i^{*3}}{r^{*3}}\right)\left(1 - \frac{r_i^3}{r_i^{*3}}\right).\tag{8.3.15}$$

Putting $\bar{r} = r_i^*/r^*$ and $\bar{r}_0 = r_i^*/r_0^*$ this equation becomes

$$\sigma_r(\bar{r}) = \sigma_{r0} + 2\rho V_i^2 \left(\bar{r}_0 - \bar{r} + \frac{\bar{r}^4 - \bar{r}_0^4}{4}\right) + \rho r_i^* \dot{V}_i(\bar{r}_0 - \bar{r})$$

$$+ \frac{4E}{9}(\bar{r}_0^3 - \bar{r}^3)\left(1 - \frac{r_i^3}{r_i^{*3}}\right).\tag{8.3.16}$$

The equivalent pressure is

$$P(\bar{r}) = -\frac{\sigma_r + 2\sigma_\theta}{3} .\tag{8.3.17}$$

Using (8.3.7), (8.3.8), (8.3.16) in (8.3.17) the pressure for an elastic layer is

$$P(\bar{r}) = -\sigma_r - \frac{4}{9} E\bar{r}^3 \left(1 - \frac{r_i^3}{r_i^{*3}}\right)$$

$$= -\left\{\sigma_{r0} + 2\rho V_i^2 \left(\bar{r}_0 - \bar{r} + \frac{\bar{r}^4 - \bar{r}_0^4}{4}\right) + \rho r_i^* \dot{V}_i(\bar{r}_0 - \bar{r})\right\}$$

$$- \frac{4E\bar{r}_0^3}{9} \left(1 - \frac{r_i^3}{r_i^{*3}}\right).\tag{8.3.18}$$

The radial stress distribution and pressure depend also on the acceleration term.

For a plastic sub-layer the effective stress σ_e can be written using (8.3.8), (8.3.9), and (8.3.17) as

$$\sigma_e = \sigma_\theta - \sigma_r = Y_0 + \lambda P = \frac{\alpha}{\lambda} Y_0 - \alpha \sigma_r \,, \tag{8.3.19}$$

where $\alpha = 3\lambda/(3 + 2\lambda)$. From (8.3.19) we get also

$$P = -\frac{\alpha}{\lambda} \left(\frac{2}{3} Y_0 + \sigma_r \right). \tag{8.3.20}$$

Using (8.3.19) in (8.3.12) we get for the plastic sub-layer:

$$\frac{d\sigma_r}{dr^*} + \frac{2\alpha}{r^*} \sigma_r = \rho \left(\frac{r_i^*}{r^*} \right)^2 \dot{V}_i + 2\rho r_i^* V_i^2 \left(\frac{1}{r^{*2}} - \frac{r_i^{*3}}{r^{*5}} \right) + \frac{2\alpha Y_0}{\lambda r^*} \,. \tag{8.3.21}$$

Multiplying with the integrating factor $r^{2\alpha}$ we get:

$$\sigma_r r^{*2\alpha} = 2\rho R V_i^2 \left(\frac{r^{*2\alpha-1}}{2\alpha - 1} - \frac{r_i^{*3} r^{*2\alpha-4}}{2\alpha - 4} \right) + \frac{\rho r_i^{*2} \dot{V}_i}{2\alpha - 1} r^{*2\alpha-1} + \frac{Y_0}{\lambda} r^{*2\alpha} + K \,. \tag{8.3.22}$$

Applying the boundary condition $\sigma_r = \sigma_{rx}$ at the elastic-plastic boundary $r^* = r_x^*$ we get:

$$\sigma_r(r) = \frac{Y_0}{\lambda} \left[1 - \left(\frac{r_x^*}{r^*} \right)^{2\alpha} \right] + \left(\frac{r_x^*}{r^*} \right)^{2\alpha} \sigma_{rx} + 2\rho r_i^* V_i^2 \left[\frac{1}{r^*(2\alpha - 1)} - \frac{r_i^{*3}}{r^{*4}(2\alpha - 4)} \right]$$

$$- 2\rho r_i^* V_i^2 \left(\frac{r_x^*}{r^*} \right)^{2\alpha} \left[\frac{1}{r_x^*(2\alpha - 1)} - \frac{r_i^{*3}}{r_x^{*4}(2\alpha - 4)} \right] + \frac{\rho r_i^{*2} \dot{V}_i}{2\alpha - 1} \left(\frac{1}{r^*} - \frac{r_x^{*2\alpha-1}}{r^{*2\alpha}} \right). \tag{8.3.23}$$

By non-dimensionalizing the radius $\bar{r}_x = r_i^*/r_x^*$ we obtain

$$\sigma_r(\bar{r}) = \frac{Y_0}{\lambda} \left[1 - \left(\frac{\bar{r}}{\bar{r}_x} \right)^{2\alpha} \right] + \left(\frac{\bar{r}}{\bar{r}_x} \right)^{2\alpha} \sigma_{rx}$$

$$+ 2\rho V_i^2 \left[\frac{\bar{r} - \bar{r}_x (\bar{r}/\bar{r}_x)^{2\alpha}}{2\alpha - 1} - \frac{\bar{r}^4 - \bar{r}_x^4 (\bar{r}/\bar{r}_x)^{2\alpha}}{2\alpha - 4} \right]$$

$$+ \frac{\rho r_i^* \dot{V}_i}{2\alpha - 1} \left[\bar{r} - \bar{r}_x \left(\frac{\bar{r}}{\bar{r}_x} \right)^{2\alpha} \right]. \tag{8.3.24}$$

Three possible stress states are possible: elastic $\sigma_e < Y_0 + \lambda P$, plastic $\sigma_e = Y_0 + \lambda P$ or pressure-failed $P < -Y_0/\lambda$. To determine if the material yields first or pressure-fails first, the expanded radius at which yielding occurs R_y can be compared to the expanded radius at which failure occurs R_f. Setting (8.2.14) equal to the yield stress produces

$$\sigma_e = \frac{2}{3} E \left(\frac{R_y}{r^*} \right)^3 = Y_0 + \lambda P \,. \tag{8.3.25}$$

Using (8.3.18) in (8.3.25) yields

$$R_y^3 = \frac{9(Y_0 - \lambda\sigma_r)r^{*3}}{(6+4\lambda)E} \, . \tag{8.3.26}$$

The failure radius R_f is found by setting the pressure given by (8.3.18) equal to the failure pressure:

$$P = -\sigma_r - \frac{4ER_f^3}{9r^{*3}} = -\frac{Y_0}{\lambda} \, . \tag{8.3.27}$$

Solving for R_f^3:

$$R_f^3 = \frac{9r^{*3}}{4\lambda E}(Y_0 - \lambda\sigma_r) \, . \tag{8.3.28}$$

From (8.3.26) and (8.3.28) the material will yield before pressure-failing $R_y^3 < R_f^3$ if

$$\frac{1}{6+4\lambda} < \frac{1}{4\lambda} \tag{8.3.29}$$

because $\sigma_r < 0$. Any real Mohr–Coulomb material $0 < \lambda < \infty$ will yield before pressure-failing. The materials will not pressure-fail after yielding if

$$P = -\frac{\alpha}{\lambda}\left(\frac{2}{3}Y_0 + \sigma_r\right) > -\frac{Y_0}{\lambda} \tag{8.3.30}$$

or if

$$\sigma_r < \frac{Y}{\lambda} \, . \tag{8.3.31}$$

Since $\sigma_r < 0$ pressure-failure cannot occur.

The elastic-plastic interface $\bar{r}_x$ is found from the condition that the effective stress σ_e in the elastic region equals the current yield stress, $Y = Y_0 + \lambda P$. From (8.3.14) and (8.3.18)

$$\frac{2}{3}E\bar{r}_x^3\left(1 - \frac{r_i^3}{r_i^{*3}}\right) = Y_0 + \lambda P = Y_0 + \lambda\left[-\left\{\sigma_{r0} + 2\rho V_i^2\left(\bar{r}_0 - \bar{r}_x + \frac{\bar{r}_x^4 - \bar{r}_0^4}{4}\right)\right.\right.$$

$$\left.\left. + \rho r_i^* \dot{V}_i(\bar{r}_0 - \bar{r}_x)\right\} - \frac{4E\bar{r}_0^3}{9}\left(1 - \frac{r_i^3}{r_1^{*3}}\right)\right] \, . \tag{8.3.32}$$

Simplifying (8.3.32) we find the following equation for $\bar{r}_x$:

$$C_1\frac{\bar{r}_x^4}{4} - \bar{C}_1\bar{r}_x + C_2\bar{r}_x^3 - C_3 = 0 \, , \tag{8.3.33}$$

where

$$C_1 = 2\lambda\rho V_i^2 \, , \quad \bar{C}_1 = 2\lambda\rho V_i^2 + \lambda\rho r_i^*\dot{V}_i \, , \quad C_2 = \frac{2}{3}E\left(1 - \frac{r_i^3}{r_i^{*3}}\right) \, ,$$

$$C_3 = Y_0 - \lambda\sigma_{r0} + 2\lambda\rho V_i^2\left(\frac{\bar{r}_0^4}{4} - \bar{r}_0\right) - \lambda\rho r_i^*\dot{V}_i\bar{r}_0 - \frac{4\lambda E\bar{r}_0^3}{9}\left(1 - \frac{r_i^3}{r_i^{*3}}\right) \, .$$

Equation (8.3.33) is solved by Newton iteration for $0 < \bar{r}_x < 1$.

Further is described a damaged sub-layer, damage-undamaged interface location, stress-based failure, multilayer systems, etc.

A comparison with the experiments (data by Longcope and Forrestal [1983]): 162 kg penetrator, 1.56 m long, 0.156 m diameter ogival nose shape (6.0 caliber radius head), impact velocity 520 m/s, impact angle 90°, angle of attack of 0°, $\rho = 1.62 \times 10^{-9}$ Mg/mm^3, $E = 3.192 \times 10^3$ MPa, $Y = 10.0$ MPa, $\lambda = 1.00$, $\lambda_d = 1.00$. The comparison is very good. Some other comparisons with experimental data are also given.

Another spherical cavity-expansion and comparison with experiment is due to Forrestal and Tzou [1997]. A spherically symmetric cavity is expanded from zero initial radius at constant velocity V. For slow enough V, there are three regions of response: an elastic region, a region with radial cracks (the material reaches its tensile strength), and a plastic region (the material reaches its shear strength). The plastic region is bounded by $r = Vt$ and $r = ct$, the cracked region is bounded by $r = ct$ and $r = c_1 t$, and the elastic region is bounded by $r = c_1 t$ and $r = c_d t$ where r is the radial Eulerian coordinate, t is time and c, c_1, and c_d are velocities. In the plastic region a linear pressure-volumetric strain relation and a Mohr–Coulomb yield criterion are satisfied:

$$pK(1 - \rho_0/\rho) = K\eta$$

$$p = (\sigma_r + \sigma_\theta + \sigma_\phi)/3; \quad \sigma_\theta = \sigma_\phi \qquad (8.3.34)$$

$$\sigma_r - \sigma_\theta = \lambda p + \tau; \quad \tau = [(3 - \lambda)/3]Y$$

with Y the uniaxial compressive strength, and σ_r and σ_θ are the Cauchy stress, positive in compression. In the cracked region $\sigma_\theta = 0$.

For the elastic incompressible material we have:

$$\frac{\partial v}{\partial r} + \frac{2v}{r} = 0, \quad \frac{\partial \sigma_r}{\partial r} + \frac{2}{r}(\sigma_r - \sigma_\theta) = -\rho_0 \left(\frac{\partial v}{\partial t} + v \frac{\partial v}{\partial r} \right) \qquad (8.3.35)$$

with v the particle velocity. From (8.3.34) and (8.3.35) we have

$$\frac{\partial \sigma_r}{\partial r} + \frac{\alpha \lambda \sigma_r}{r} + \frac{\alpha \tau}{r} = -\rho_0 \left(\frac{\partial v}{\partial t} + v \frac{\partial v}{\partial r} \right) \qquad (8.3.36)$$

where $\alpha = 6/(3 + 2\lambda)$. With dimensionless variables $S = \sigma_r/\tau$, $U = v/c$, $\varepsilon = V/c$ and the similarity transformation $\xi = r/ct$ where c is he elastic-plastic interface velocity we have:

$$\frac{dU}{d\xi} + \frac{2U}{\xi} = 0, \quad \frac{dS}{d\xi} + \frac{\alpha \lambda S}{\xi} = -\frac{\alpha}{\xi} + \frac{\rho_0 c^2}{\tau} \left(\xi \frac{dU}{d\xi} - U \frac{dU}{d\xi} \right). \qquad (8.3.37)$$

The boundary condition at the cavity surface is $U(\xi = \varepsilon) = \varepsilon$ and (8.3.37a) is $U = \varepsilon^3/\xi^2$. Introducing in (8.3.37b) and multiplying with $\xi^{\alpha\lambda}$, we get:

$$\frac{d}{d\xi}(\xi^{\alpha\lambda} S) = -\alpha \xi^{\alpha\lambda-1} - \frac{2\rho_0 c^2 \varepsilon^3}{\tau} \xi^{\alpha\lambda-2} + \frac{2\rho_0 c^2 \varepsilon^6}{\tau} \xi^{\alpha\lambda-5} \qquad (8.3.38)$$

which can be integrated

$$S = -\frac{1}{\lambda} - \frac{2\rho_0 c^2 \varepsilon^3}{\tau(\alpha\lambda - 1)}\frac{1}{\xi} + \frac{2\rho_0 c^2 \varepsilon^6}{\tau(\alpha\lambda4)}\frac{1}{\xi^4} + C\xi^{-\alpha\lambda} \tag{8.3.39}$$

where C is the integration constant. Particle velocity is obtained by differentiating displacement:

$$S = \frac{2}{3\xi^3} + \frac{3\rho_0 c^2}{E\xi}, \quad U = \frac{3\tau}{2E\xi^2}. \tag{8.3.40}$$

At the elastic and plastic interface the Hugoniot jump conditions hold:

$$\rho_2(v_2 - c) = \rho_1(v_1 - c), \quad \sigma_2 + \rho_2 v_2(v_2 - c) = \sigma_1 + \rho_1 v_1(v_1 - c) \tag{8.3.41}$$

where 1 and 2 represent the quantities in elastic and plastic regions. For incompressible material $\rho_1 = \rho_2 = \rho_0$ and radial stress and particle velocity are continuous.

From (8.3.41b) we get for

$$\xi = 1 : \varepsilon = \frac{V}{c} = \left(\frac{3\tau}{2E}\right)^{1/3}$$

and

$$S = \frac{2}{\alpha\lambda}\xi^{-\alpha\lambda} - \frac{1}{\lambda} + \frac{2\rho_0 V^2}{\tau}\left[\frac{\varepsilon}{(1 - \alpha\lambda)\xi} - \frac{\varepsilon^4}{(4 - \alpha\lambda)\xi^4}\right]$$
$$+ \frac{\rho_0 c^2}{\tau}\left[\frac{3\tau}{E} + \frac{2\varepsilon^3}{\alpha\lambda - 1} - \frac{2\varepsilon^6}{\alpha\lambda - 4}\right]\xi^{-\alpha\lambda}.$$

Or the radial stress at the cavity surface:

$$S(\varepsilon) = \frac{2}{\alpha\lambda}\varepsilon^{-\alpha\lambda} - \frac{1}{\lambda} + \frac{\rho_0 V^2}{\tau}\left[\frac{6}{(1 - \alpha\lambda)(4 - \alpha\lambda)} - \frac{2\alpha\lambda}{1 - \alpha\lambda}\varepsilon^{1-\alpha\lambda} + \frac{2\varepsilon^{4\alpha\lambda}}{4 - \alpha\lambda}\right] \tag{8.3.42}$$

In the elastic-cracked-plastic response $\sigma_\theta = \sigma_\phi = 0$ and the radial stress is

$$S = \frac{3 + 2\lambda}{\lambda(3 - \lambda)}\xi^{-\alpha\lambda} - \frac{1}{\lambda} - \frac{2\rho_0 c^2}{\tau}\left[\frac{\varepsilon^3}{(\alpha\lambda - 1)\xi} - \frac{\varepsilon^6}{(\alpha\lambda - 4)\xi^4}\right]$$
$$+ \frac{2\rho_0 c^2}{\tau}\left(\frac{\varepsilon^3}{\alpha\lambda - 1} - \frac{\varepsilon^6}{\alpha\lambda - 4}\right)\xi^{-\alpha\lambda}. \tag{8.3.43}$$

For a compressible target the equations are:

$$\rho\left(\frac{\partial v}{\partial r} + \frac{2v}{r}\right) = -\left(\frac{\partial \rho}{\partial t} + v\frac{\partial \rho}{\partial r}\right), \quad \frac{\partial \sigma_r}{\partial r} + \frac{2(\sigma_r - \sigma_\theta)}{r} = -\rho\left(\frac{\partial v}{\partial t} + v\frac{\partial v}{\partial r}\right) \tag{8.3.44}$$

Eliminating σ_θ and ρ we get

$$\frac{\partial v}{\partial r} + \frac{2v}{r} = -\frac{\alpha}{2K(1 - \eta)}\left(\frac{\partial \sigma_r}{\partial t} + v\frac{\partial \sigma_r}{\partial r}\right),$$
$$\frac{\partial \sigma_r}{\partial r} + \frac{\alpha\lambda\sigma_r}{r} + \frac{\alpha\tau}{r} = -\frac{\rho_0}{1 - \eta}\left(\frac{\partial v}{\partial t} + v\frac{\partial v}{\partial r}\right) \tag{8.3.45}$$

where

$$\alpha = \frac{6}{3 + 2\lambda}, \quad \eta = \frac{\alpha\tau}{2K}\left(\frac{\sigma_r}{\tau} - \frac{2}{3}\right).$$

Introducing the dimensionless variables

$$S = \sigma_r/\tau, \ U = v/c, \ \varepsilon = V/c, \ \beta = c/c_p, \ \beta_1 = c_1/c_p, \ \beta\varepsilon = V/c_p,$$

$$c_p^2 = K/\rho_0, \ \xi = r/ct$$

the system of equations becomes

$$\frac{dU}{d\xi} = \frac{2U/\xi + (\tau\alpha^2/2K\xi)[(\xi - U)/(1 - \eta)](\lambda S + 1)}{(\alpha\beta^2/2)[(\xi - U)/(1 - \eta)]^2 - 1},$$

$$\frac{dS}{d\xi} = \frac{\alpha/\xi + \alpha\lambda S/\xi + (2\beta^2 KU/\tau\xi)[(\xi - U)/(1 - \eta)]}{(\alpha\beta^2/2)[(\xi - U)/(1 - \eta)]^2 - 1}$$

(8.3.46)

which are to be solved with the Runge–Kutta method. The radial stress and particle velocity are continuous at the elastic-plastic interface. Thus at $\xi = 1$ we have

$$S_1 = S_2 = \frac{2[(1 - 2v)(1 + \gamma\beta) + (1 + v)(\gamma\beta)^2]}{3(1 - 2v)(1 + \gamma\beta) - 2\lambda(1 + v)(\gamma\beta)^2},$$

$$U_1 = U_2 = \frac{3\tau(1 + v)(1 - 2v)(1 + \gamma\beta)}{E[3(1 - 2v)(1 + \gamma\beta) - 2\lambda(1 + v)(\gamma\beta)^2]},$$

(8.3.47)

$$\gamma^2 = \left(\frac{c_p}{c_d}\right)^2 = \frac{1 + v}{3(1 - v)}, \quad c_d^2 = \frac{E(1 - v)}{(1 + v)(1 - 2v)\rho_0}$$

with S_2 and U_2 known, one can calculate the dimensionless radial stress and particle velocity in the plastic region.

Further the elastic-cracked-plastic response is calculated and a cavity-expansion numerical results is given for $K = 6.7$ GPa, $Y = 130$ MPa, $\lambda = 0.67$, $E = 11.3$ GPa, $v = 0.22$, $f = 13$ NPa, $\rho_0 = 2260$ kg/m^3. The penetration equations and comparison with data are reasonable.

Dynamic cavity expansion of ceramic materials when subjected to cavity expansion at constant velocity was considered by Satapathy and Bless [2000], Satapathy [2001]. All five zones are considered: cavity, comminuted zone, radially cracked zone, elastic zone and undisturbed zone (Fig. 8.3.2). For spherical symmetry the conservation equations for mass and momentum in Eulerian coordinates are:

$$\frac{\partial\rho}{\partial t} + \frac{1}{r^2}\frac{\partial}{\partial r}(\rho r^2 v) = 0,$$

$$\frac{\partial\sigma_r}{\partial r} + 2\frac{\sigma_r - \sigma_\theta}{r} = -\rho\left(\frac{\partial v}{\partial t} + v\frac{\partial v}{\partial r}\right).$$

(8.3.48)

The stresses are positive in compression.

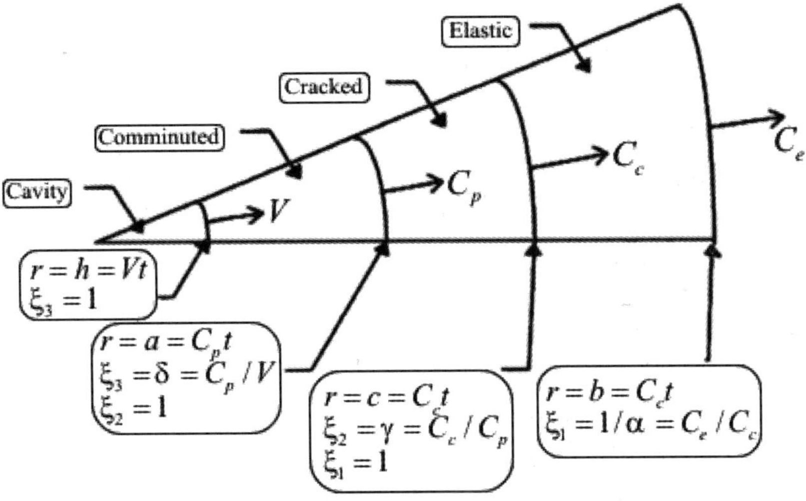

Fig. 8.3.2   Dynamic response regions in a spherical cavity.

In the elastic region due to spherical symmetry the stresses are

$$\sigma_r = -\frac{E}{(1+v)(1-2v)}\left[(1-v)\frac{\partial u}{\partial r} + 2v\frac{u}{r}\right],$$

$$\sigma_\theta = -\frac{E}{(1+v)(1-2v)}\left[v\frac{\partial u}{\partial r} + \frac{u}{r}\right]. \tag{8.3.49}$$

Since the particle velocity is negligible in the elastic region, it is ignored and the equation to be integrated is

$$\frac{\partial^2 u}{\partial r^2} + \frac{2}{r}\frac{\partial u}{\partial r} - \frac{2u}{r^2} = \frac{1}{C_e^2}\frac{\partial^2 u}{\partial t^2},$$

where C_e is the dilatational elastic wave seed, and with $\xi_1 = r/C_c t$ and $\bar{u}_1 = u/C_c t$ this equation becomes

$$\frac{d^2\bar{u}}{d\xi_1^2} + \frac{2}{\xi_1}\frac{d\bar{u}}{d\xi_1} - \frac{2\bar{u}}{\xi_1^2} = \alpha^2\xi^2\frac{d^2\bar{u}}{d\xi_1^2}, \tag{8.3.50}$$

with $\alpha = C_c/C_e$. Forrestal and Luk [1988] have shown that this equation has the solution

$$\bar{u} = A\alpha\xi_1 - B\frac{1 - 3\alpha^2\xi_1^2}{3\alpha^2\xi_1^2},$$

where A and B are constants of integration, with

$$B = -\frac{3}{2}A, \qquad A = \frac{2\sigma_f}{\rho_0 C_c^2}\frac{\alpha^2(1-v)}{2v(\alpha^3 - 1) + 2\alpha^3 - 3\alpha^2 + 1}.$$

In the cracked region only radial stresses are transmitted

$$\frac{\partial\sigma_r}{\partial r} + \frac{2\sigma_r}{r} = -\rho\frac{\partial^2 u}{\partial t^2}.$$

Since

$$\sigma_r = -E\frac{\partial u}{\partial r}, \quad \text{and} \quad \xi_2 = \frac{r}{C_p t}, \quad \bar{u}_2 = \frac{u}{C_p t},$$

this equation becomes

$$\frac{d^2\bar{u}_2}{d\xi_2^2} + \frac{2}{\xi_2}\frac{d\bar{u}_2}{d\xi_2} = \beta^2\xi_2^2\frac{d^2\bar{u}_2}{d\xi_2^2}, \quad \beta = \frac{C_p}{C_{cr}}, \quad C_{cr} = \sqrt{\frac{E}{\rho}}$$

with the following solution

$$\bar{u}_2 = -c_1\frac{1+\beta^2\xi_2^2}{\xi_2} + c_2$$

and

$$c_1 = \frac{Y}{E(\beta^2-1)}, \quad c_2 = \frac{Y}{\gamma E}\frac{\beta^2\gamma^2+1}{\beta^2-1} + \frac{\sigma_f\gamma}{\rho_0 C_e^2}\frac{(1-v)(2\alpha^3-3\alpha^2+1)}{2v(\alpha^3-1)+2\alpha^3-3\alpha^2+1}.$$

For the passage to the next domain the conditions

$$\rho_1(v_1-C) = \rho_2(v_2-C), \quad \sigma_2-\sigma_1 = \rho_1(C-v_1)(v_2-v_1)$$

are used. Denoting the volumetric strain in the elastic side

$$\eta = \frac{P}{K} = 1 - \frac{\rho_0}{\rho} = -\frac{6\sigma_f}{\rho C_e^2}\frac{\alpha^2(\alpha-1)(1-v)}{2\alpha^3(1+v)-3\alpha^2+1-2v},$$

the following relation between α and β is obtained

$$Y(1-\beta^2\gamma^2)\left[1-\frac{\rho_0}{3K}\left(\frac{C_c-V_1}{1-\eta_1}\right)^2\right] + \sigma_1\beta^2\gamma^2 - \gamma^2$$

$$\gamma^2 = \left(\frac{\alpha}{\beta}\frac{C_e}{C_{cr}}\right)^2$$

$$= \frac{Y(1-\beta^2\gamma^2)[1-\frac{\rho_0}{3K}(\frac{C_c-V_1}{1-\eta_1})^2] + \sigma_1\beta^2\gamma^2 - \eta_1\beta^2\gamma^2\rho_0[(C_e-V_1)/(1-\eta)]^2}{\sigma_1 - \eta_1\rho_0[(C_e-V_1)/(1-\eta_1)]^2}.$$

The product $\beta\gamma$ is function of α only.

One assumes that the shear strength increases linearly with the confining pressure

$$\frac{\sigma_r-\sigma_\theta}{2} = \lambda_1\frac{\sigma_r+2\sigma_\theta}{3}.$$

Using this equation in the equation of motion $(8.3.48)_2$ we obtain

$$\frac{\partial\sigma_r}{\partial r} + 2\bar{\alpha}\frac{\sigma_r}{r} = -\rho\left(\frac{\partial v}{\partial t} + v\frac{\partial v}{\partial r}\right), \tag{8.3.51}$$

where $\bar{\alpha} = 6\lambda_1/(3+4\lambda_1)$. Using the similarity transformation

$$\xi_3 = \frac{r}{Vt}, \quad \bar{u}_3 = \frac{u}{Vt} \quad \text{and} \quad U = \frac{v}{V},$$

in the comminuted region, (8.3.51) becomes

$$\frac{d\sigma_r}{d\xi_3} + 2\bar{\alpha}\frac{\sigma_r}{\xi_3} = -\rho V^2 \frac{dU}{d\xi_3}(U - \xi_3). \tag{8.3.52}$$

One is disregarding dilatancy and compaction and from the mass conservation

$$v = \frac{D_1}{r^2},$$

where D_1 is the constant of integration. This is in non-dimensional quantities

$$U = \frac{1}{\xi_3^2}. \tag{8.3.53}$$

Inserting it in the Eq. (8.3.52) and integrating, we get

$$\sigma_r = -\frac{\rho V^2}{\xi_3^{2\bar{\alpha}}}\left(-\frac{2}{2\bar{\alpha}-1}\xi_3^{2\bar{\alpha}-4} + \frac{2}{2\bar{\alpha}-1}\xi_3^{2\bar{\alpha}-1}\right) + \frac{D_2}{\xi_3^{2\bar{\alpha}}},$$

where D_2 is the integration constant. The density for this region is the same as that at the cracked-comminuted boundary $\rho = \rho_0/(1 - Y/3K)$.

Denoting the pressure at the cavity surface by P_c, the stress required to maintain the constant cavity expansion velocity is:

$$P_c \equiv \sigma_r\bigg|_{\xi_3=1} = Y\delta^{2\bar{\alpha}} + \rho_0 V\left(\frac{\delta^{2\bar{\alpha}-4}}{2-\bar{\alpha}} - \frac{2\delta^{2\bar{\alpha}-1}}{1-2\bar{\alpha}} + \frac{3}{(1-2\bar{\alpha})(2-\bar{\alpha})}\right). \tag{8.3.54}$$

The quantity δ is related to α and β by

$$\delta^3 = \frac{1}{3[c_2 - c_1(1 + \beta^2)]}.$$

The equation in the saturated region becomes

$$\frac{d\sigma_r}{d\xi_3} + \frac{4\tau}{\xi_3} = -\rho V^2 \frac{dU}{d\xi_3}(U - \xi_3).$$

Integrating this equation with (8.3.53) yields

$$P_c = \frac{2\tau}{\bar{\alpha}} + 4\tau\ln\delta_1 + \frac{\rho V^2}{2}\left(3 + \frac{1}{\delta_1^4} - \frac{4}{\delta_1}\right).$$

The application is for the alumina ceramic Coors AD995, for which we have $E = 373.14$ GPa, $K = 231.8$ GPa, $\rho = 3890$ kg/m, and the quasi-static compressive strength $Y = 2.62$ GPa, and the tensile strength $\sigma_f = 0.262$ GPa. The tungsten projectile for impact velocities up to 3.5 km/s are shown in Fig. 8.3.3.

A dynamic spherical cavity expansion in a pressure sensitive elastoplastic medium was considered also by Durban and Masri [2004]. The material behavior is described by the hypoelastic model of the Drucker–Prager material with a non-associated flow rule, with arbitrary strain-hardening. Simple analytical solutions are given for the fully incompressible elastic/perfectly plastic material with a non-associated flow rule. A numerical analysis is given for the fully incompressible strain-hardening solid with a non-associated flow rule.

In another paper by Gao *et al.* [2004] one considers the same problem and comparison with experiments, perforation of concrete. The following nine assumptions are given and argued. The medium of concrete can be considered as ideal fluid for high-velocity impact, where the shear modulus is zero, which can satisfy

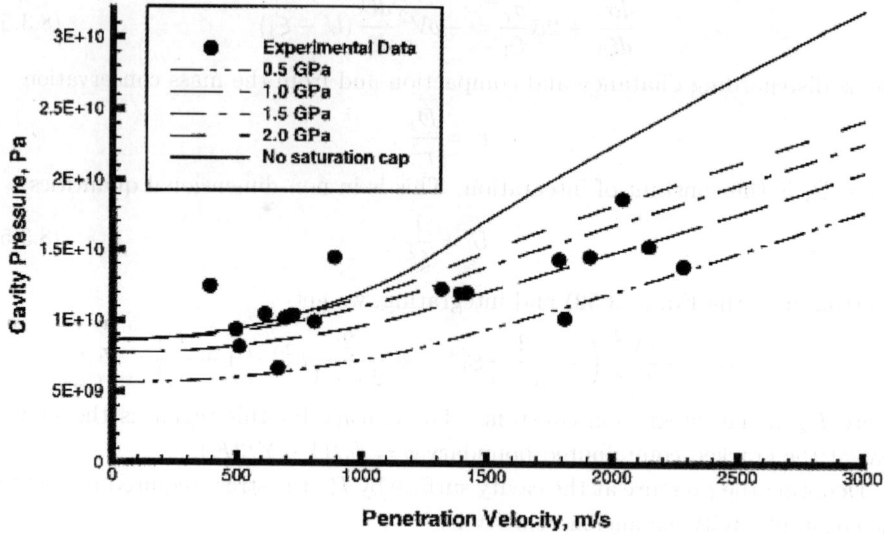

Fig. 8.3.3   Cavity expansion pressure versus penetration velocity for different shear-saturation levels.

the needs of engineering analysis for the case of high-velocity and high-pressure. To be compared with the compressive strength, the tensile strength is very little. For high-velocity and high-pressure impacting, the tensile strength can be neglected. During the impact or penetration, the elastic ultimate stress of concrete can be neglected while the wave velocity of elastic stress is very high. There is an ultimate density for concrete during the impact pressing, to which the density of concrete medium can not be increased again for further compressing. A series of experiments had proved that the ultimate density had almost no changes for the higher pressure. During the compressing until ultimate density, the concrete medium subjects to ideal plastic deformation. There is no change of stress. In this process, the air voids are gradually compressed out of the concrete. The wave velocity of stress is zero. The region of medium response will expand in the external normal direction of the projectile surface. In the compressed region, the density and volume of the concrete material have not any more changes. Therefore, the stress wave is constant-volume wave. The thermal conduction can be neglected during impacting. The projectile is non-deformable.

The dynamic penetration of the projectile is studied by writing the mass conservation, the momentum conservation and the energy conservation. One is studying the penetration equation and the scabbing produced. The criteria for scabbing is $\sigma^* - \sigma \geq \sigma_s$ where σ^* is the front peak value of the attenuation stress wave, σ is the compressing stress where crack is caused and σ_s is the tensile stress of concrete.

For a parabola nose projectile and concrete with density of $\rho = 2400$ kg/m^3, with ultimate (the highest) density of $\rho^* = 2640$ kg/m^3, the compressive strength

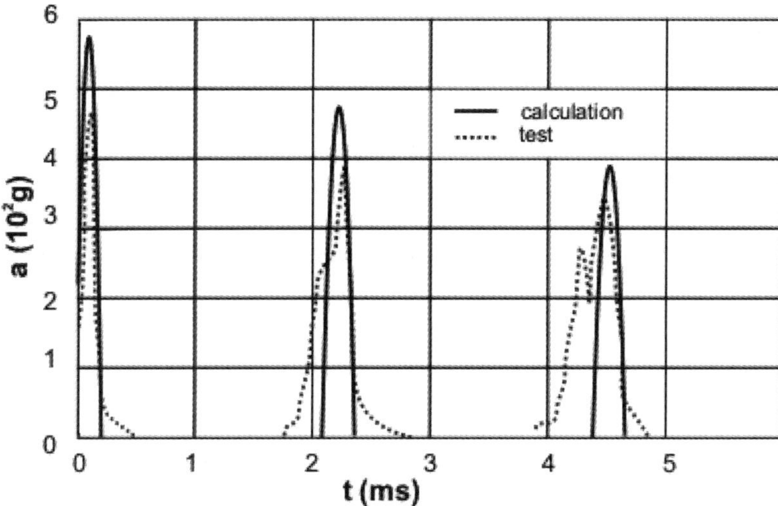

Fig. 8.3.4   The measurement and calculation deceleration curves of a projectile in the second test.

$\sigma_c = 3.0 \times 10^7$ N/m^2, mass of the projectile $m_p = 7.63$ kg, the moment of inertia $J_p = 0.162$ kg/m^2, the friction coefficient $\mu_f = 0.1$ and the thickness of one target plate is 0.3 m. The plates were 1 m apart and the striking velocity of the projectile is 532 m/s in the first series of test and 580 m/s in the second. One has measured the deceleration curve and also the scabbing. The case of three cement plates is shown in Fig. 8.3.4.

A new constitutive model for cold compaction of metal powders is developed by Gu *et al.* [2001]. The plastic flow of metal powders at the macroscopic level is assumed to be representable as a combination of a distortion mechanism, and a consolidation mechanism. For the distortion mechanism the model employs a pressure-sensitive, Mohr–Coulomb type yield condition, and a new physically based non-associated flow rule. For the consolidation mechanism the model employs a smooth yield function which has a quarter-elliptical shape in the mean-normal pressure and the equivalent shear stress plane, together with an associated flow rule. The constitutive model has been implemented in a finite element program. The material parameters in the constitutive model have been calibrated for MH-100 iron powder by fitting the model to reproduce data from true triaxial compression experiments, torsion ring-shear experiments, and simple compression experiments. The predictive capability of the constitutive model and computational procedure is checked by simulating two simple powder forming processes: (i) a uniaxial strain compression of a cylindrical sample, and (ii) forming of a conical shaped-charge liner. In both cases the predicted load-displacement curves and density variations in the compacted specimens are shown to compare well with corresponding experimental measurements.

The transverse cracks in glass/epoxy cross-ply laminates impacted by projectiles was considered by Takeda *et al.* [1981]. They started from the glass/epoxy cross-ply laminates impacted by cylindrical projectiles with different impactor nose shapes and lengths, a sequential delamination mechanism is dominant, initiated by a generator strip, of with approximately equal to the impactor diameter, cut from the first lamina by two through-the-thickness cracks parallel to the fibers of the first lamina Cristescu *et al.* [1975]. They have continued the tests and the impactor velocity for the development of the transverse cracks appears to be independent of the impactor type and to be about 23 m sec^{-1}. Above this threshold velocity, the mean transverse crack distance decreases sharply as the impactor velocity increases. The curves are eventually flattened out at higher velocities. Another paper about damage of composites is due to Nemes and Spéciel [1995]. The problem is mathematically well-posed wave propagation problem, with an unique and stable solution.

A big new paper, devoted to a new membrane model for ballistic impact response and V_{50} performance of multi-play fibrous systems was written by Phoenix and Porwal [2003]. They started with Rakhmatulin and Dem'yanov [1961] equations

$$\frac{1}{a_0}\frac{\partial^2 u}{\partial t^2} = \frac{1}{r}\frac{\partial}{\partial r}(\varepsilon_t r \cos\gamma) - \left[\frac{u}{r^2}\right]^{\oplus} \tag{8.3.55}$$

and

$$\frac{1}{a_0}\frac{\partial^2 v}{\partial t^2} = -\frac{1}{r}\frac{\partial}{\partial r}(\varepsilon_t r \sin\gamma) \tag{8.3.56}$$

where a_0 is reduced to $a_0 = \sqrt{E/\rho}$, the symbol $[\ ]^{\oplus}$ means the quantity is kept only if positive, and

$$\varepsilon_t = \sqrt{\left(1 + \frac{\partial u}{\partial r}\right)^2 + \left(\frac{\partial v}{\partial r}\right)^2} - 1$$

is the strain. These equations are discussed and integrated. Ultimately one expresses the threshold projectile velocity V_{50} above which the membrane fails by perforation 50% of the time, in terms of Cunniff's dimensionless scaling parameters

$$V_{50}/\sqrt[3]{\Omega} = 2^{1/3}\varepsilon_{y\,\text{max}}^{1/12}(1 + \theta^2\Gamma_0)/K_{\text{max}}^{3/4}$$

where K_{max} is accurately represented by

$$K_{\text{max}} \approx \exp\left\{-\frac{4\theta^2\Gamma_0}{3(1+\theta^2\Gamma_0)}(\psi_{\text{max}}^2 - 1)\right\}$$

$$\times \psi_{\text{max}}^{1/3}\left\{\frac{\sqrt{\psi_{\text{max}}/\varepsilon_{y\,\text{max}}}(\psi_{\text{max}} - 1)}{\ln[1 + \sqrt{\psi_{\text{max}}/\varepsilon_{y\,\text{max}}}(\psi_{\text{max}} - 1)]}\right\}^{2/3}$$

and

$$\psi_{\text{max}} \approx \sqrt{(1 + \theta^2\Gamma_0)/(2\theta^2\Gamma_0)}$$

with $\sqrt[3]{\Omega}$ a normalizing velocity and Ω is expressed in yarn (fiber) properties as

$$\Omega = (1/2)(\sigma_{y\,\mathrm{max}}\varepsilon_{y\,\mathrm{max}}/\rho_y)\sqrt{E_y/\rho_y}\,.$$

The comparison with Kevlar 29, Kevlar KM2, PBO and Spectra 1000 are good. For Kevlar 29, we have $\sigma_{y\,\mathrm{max}} \approx 2.9$ GPa, $\varepsilon_{y\,\mathrm{max}} \approx 0.034$, $E_y \approx 74$ GPa, $\rho_y = 1440$ kg/m^3 and producing the normalizing velocity $\sqrt[3]{\Omega} \approx 624$ m/s..

Warren *et al.* [2004], document the results of a combined experimental, analytical, and computational research program that investigates the penetration of steel projectiles into limestone targets at oblique angles. The striking velocities were 0.4 to 1.3 km/s.

In several papers, one have studied the effect of preliminary strain hardening on the flow stress of titanium and a titanium alloy during shock compression by Razoreniv *et al.* [2005]. The effect of preliminary strain hardening of titanium and a titanium alloy on their mechanical properties under quasi-static and high-rate ($> 10^5$ s^{-1}) loading is studied. Preliminary hardening is accomplished using equal-channel angular pressing and shock waves. High-rate deformation is attained via shock-wave loading of samples. The experimental results show that structural defects weaken the dependence of the yield strength on the strain rate. The difference in the rate dependence can be so high that the effect of these defects on the flow stress can change sign when going from quasi-static to high-rate loading. The flyer plates were launched at velocities of 0.1–1.23 km/s.

8.4 Modeling of the Taylor Cylinder Impact Test. Anisotropy

The anisotropy of the impact bar was also considered, starting from a simplified description of the Taylor impacting bar. One has considered the simplified theory by Hawkyard *et al.* [1968], Hawkyard [1969], Hutchings [1979] and Jones *et al.* [1987]. The projectile is deformed by the impact and, from geometrical measurements made before and after impact, the yield stress of the specimen is deduced. Copper and mild steel were first considered at various temperatures. The profiles predicted of the specimen, after the tests, were afterwards approximated in various ways. A comparison with the measured shape of the specimen after the test is afterwards done. Polymers were also considered with projectiles measured before and after impact. New equation was developed. But again with a schematic illustration of the projectile. In another paper Jones *et al.* [1991] a simple theoretical analysis of the old problem is presented. The analysis is more complete, but retains the mathematical simplicity of the earlier versions. The major thrust is to separate the material response into two phases. The first phase is dominated by strain rate effects and has a variable plastic wave speed. The second phase is dominated by strain hardening effects and has a constant plastic wave speed. Estimates for dynamic yield stress, strain, strain-rate, and plastic wave speed during both phases are given. In the paper an example of a cooper specimen impacting a steel anvil with an initial speed of 187 m/s, is given.

Experimental techniques are described and illustrated by Nemat-Nasser *et al.* [1994] for direct measurement of temperature, strain-rate, and strain effects on the flow stress of metals over a broad range of strains and strain rates. Taylor anvil tests are performed, accompanied by high-speed photographic recording of the deformation, and the results are compared with those obtained by finite-element simulations, leading to fine tuning of parameters in the material's flow stress.

Constitutive models and solution algorithms for anisotropic elastoplastic material strength are developed by Maudlin and Schiferl [1996] for use in the high-rate explicit multi-dimensional continuum mechanics codes. The constitutive modeling is posed in an unrotated material frame using the polar decomposition theorem. Continuous quadratic and discontinuous piecewise yield functions obtained from polycrystal simulations for metallic hexagonal-close-paked and cubic crystal structures are utilized. Associative flow formulations incorporating these yield functions are solved using various solution algorithms; explicit, semi-explicit and geometric normal return schemes are assessed for stability, accuracy and efficiency. Isotropic scaling (hardening) of the anisotropic yield surface shapes is included in the modeling. An explosive forming application involving large strain was selected to investigate the effect of using anisotropic materials. Axisymmetric two-dimensional forming simulations were performed for both crystal structures producing resultant formed shapes that are unique to the material's initial yield anisotropy, and are distinct from isotropic results. Initial axisymetric geometry of a simple explosive forming problem for titanium simulated with the EPIC code, is given.

In another paper a new one-dimensional simplified analysis of the Taylor impact test is presented by Jones *et al.* [1996]. This analysis differs from any previously presented in that the wave mechanics are separated from the calculation of dynamic stress. The new results utilize post test measurements to estimate key parameters in the plastic wave propagation. However, these measurements are incorporated into the analysis in a very unconventional way. A comparison with continuum code calculations shows very good agreement has been achieved.

Simple conservation relationships (jump conditions) in conjunction with postulated material constitutive behavior are applied by Maudlin *et al.* [1997] to steady plastic strain waves propagating in problems of uniaxial stress and Taylor Cylinder Impact. These problems are simulated with a two-dimensional Lagrangian continuum mechanics code for the purpose of numerically validating the jump relationships as an accurate analytical representation of plastic wave propagation. The constitutive behavior used in this effort assumes isotropy and models the thermodynamic response with a Mie-Grunisen Equation-of-State and the mechanical response with the rate-dependent Johnson-Cook and MTS flow stress models. The jump relationships successfully replicate the results produced by continuum code simulations of plastic wave propagation and provide a methodology for constructing mechanical constitutive models from experimental plastic wave speed information.

Comparison are also presented between experimental speeds from Taylor Cylinder Impact tests with jump relationships and continuum code predictions, indicating that the above mentioned flow stress models may not accurately capture plastic wave propagation speeds in annealed and hardened copper.

Taylor cylinder impact testing is used by Maudlin *et al.* [1999] to validate anisotropic elastoplastic constitutive modeling by comparing polycrystal-computed yield-surface shapes (topography) with measured shapes from post-test Taylor specimens and quasi-static compression specimens. Measured yield-surface shapes are extracted from the experimental post-test geometries using classical *r*-value definitions modified for arbitrary stress state and specimen orientation. Rolled tantalum (body-centered-cubic metal) plate and clock-rolled zirconium (hexagonal-close-packed metal) plate are both investigated. The results indicate that an assumption of topography invariance with respect to strain rate is well justified for tantalum. However, a strong sensitivity of topography with respect to strain rate for zirconium was observed, implying that some accounting for a deformation mechanism rate dependence associated with lower-symmetry materials should be included in the constitutive modeling. Discussion of the importance of this rate dependence and texture evolution in formulating constitutive models appropriate for finite-element model applications is provided.

In another paper by Maudlin *et al.* [1999] Taylor impact tests using specimens cut from a rolled plate of tantalum were conducted. The tantalum was experimentally characterized in terms of flow stress and crystallographic texture. A piece-wise yield surface was interrogated from an orientation distribution function corresponding to this texture assuming two slip system modes, in conjunction with an elastic stiffness tensor computed from the same orientation distribution function and single crystal elastic properties. This constitutive information was used in a code of a Taylor impact test.

A forged and round-rolled pure tantalum bar stock was observed by Maudlin *et al.* [2003] to exhibit large asymmetry in bulk plastic flow response when subjected to large strain Taylor cylinder impact testing. This low-symmetry behavior was analyzed experimentally investigating both the initial stock and the impact-deformed material via x-ray crystallographic texture measurements and auto-mated electron back-scatter diffraction scans to establish spatial micro structural uniformity. Polycrystal simulations based upon the $\langle 110 \rangle$ measured duplex texture and experimentally inferred deformation mechanisms were performed to project discrete yield surface shapes. Subsequent least squares fitting and eigensystem analysis of the resulting quadratic forth-order tensors revealed strong normal/shear coupling in the yield surface shape. This mixed-mode coupling produces a shearing deformation in the 1–2 impact plane of a Taylor specimen whose axis is coincident with the compressive 3-axis. The resultant deformation generates an unusual rectangular-shaped impact footprint that is confirmed by finite-element calculations compared to experimental post-test geometries.

8.5 Analysis of the Steady-State Flow of a Compressible Viscoplastic Medium Over a Wedge (Cazacu *et al.* [2006])

Penetration mechanics has a long and rich history. Information concerning the stress-time response in the target during penetration is lacking. A comprehensive review of empirical equations for maximum penetration depth in rock, concrete, soil, ice, and marine sediments can be found in Batra [1987], Heuze [1990], Zukas *et al.* [1990], etc. The maximum penetration depth is expressed as a function of initial impact velocity, penetrator cross-sectional area, penetrator weight, and/or nose geometry. However, in these equations the only target material properties considered are the initial density and unconfined strength.

A large number of semi-analytical models where penetration is idealized as uniform expansion of a spherical or cylindrical cavity into a semi-infinite target have been proposed. These cavity expansion analyses provide the radial stress at the cavity surface as a function of the cavity velocity and acceleration. This radial stress is considered to be the load on the penetrator and subsequently used in conjunction with equations of motion to calculate the maximum depth of penetration. As emphasized in recent contributions, good agreement between model and data could be expected only if the compressibility and rate sensitivity of the target material are accounted for.

In the present section (Cazacu *et al.* [2006]) I present a new model for calculating the resistance to penetration into geological or geologically derived materials. In the analyses it is assumed that the rate of penetration and all flow fields are steady as seen from the tip of the penetrator. This hypothesis is supported by penetration tests into cementitious materials (grout, concrete) for impact velocities below 1000 m/s. The shear response of the target material is modeled by a modified Bingham type viscoplastic equation proposed by Cazacu and Cristescu [2000], Cristescu *et al.* [2002]. The rationale for adopting such a model is that it accounts for both strain rate and compaction effects on yielding, which are key properties of any porous/brittle material. In the law-pressure regime, a non-linear pressure-volume relationship is considered. Based on experimental observations which show that in the high-pressure regime a very large increase in pressure is necessary to produce even a very small change in density, the hypothesis of a "locking medium" is adopted: the density cannot exceed a critical value. The penetrator is wedge-shaped with high length-over-diameter ratio. Contact between the projectile nose and target is considered to be of Coulomb type with constant friction coefficient, whereas the frictional contact between the remainder of the projectile and target is considered to be slip-rate dependent. Resistance to penetration is calculated for different interface conditions between the target and wedge. It is shown that for low to intermediate impact velocities, accounting for friction alters the optimal wedge geometry for penetration performance by blunting the nose. Furthermore, the higher the velocity, the greater the influence of the nose geometry (wedge semi-angle) on penetration.

The deviatoric response of the target material is modeled by a non-homogeneous Bingham type rigid/viscoplastic equation:

$$\boldsymbol{D'} = \begin{cases} 0 & \text{if } II_S \leq k^2(\rho)\,, \\[2mm] \dfrac{1}{2\eta} \left\langle 1 - \dfrac{k(\rho)}{\sqrt{II_s}} \right\rangle \mathbf{s} & \text{if } II_S > k^2(\rho)\,, \end{cases} \tag{8.5.1}$$

with $\langle\ \rangle$ the Macauley bracket. The influence of the degree of compaction on the behavior is modeled through the dependence of the yield limit in shear, k, on the current density ρ. Thus, for stress states satisfying $II_S \leq k^2(\rho)$ the model response is rigid, otherwise the model response is viscoplastic. A power law variation of the yield limit with the current density is assumed

$$k(\rho) = k_0 + \beta \left(1 - \frac{\rho_0}{\rho}\right)^n, \tag{8.5.2}$$

where ρ_0 and $k_0 = k(\rho_0)$ are the density and yield stress of the undeformed medium, respectively, while n and β are material constants. The yield stress k is given in stress units and η in Poise. The stress-volume relationship adopted reflects the following experimental observations (see Schmidt [2003]):

- in the low to moderate pressure regime, most cementitious materials show a highly non-linear mean stress-volumetric strain response, the reversible decrease in volume being very small,
- in the high-pressure regime, a very large increase in pressure is necessary in order to produce even a very small change in volume.

Thus, the hypothesis of a "locking medium" applies, i.e., the density cannot exceed a critical value. This critical density, called locking density, is the density at which no volume change occurs under hydrostatic conditions. It will be denoted by ρ_* and the pressure level at which this density is first reached, called locking pressure, is denoted by p_*. Hence, $\rho = \rho_0$ for $p > p_*$ while for $\rho \leq \rho_*$, the pressure vs. density relationship $p = p(\rho)$ is assumed to be of the form

$$p(\rho) = p_* \left(\frac{\rho}{\rho_*}\right)^m \tag{8.5.3}$$

where m is a material constant. The relationships (8.5.1)–(8.5.3) could be inverted, to give:

$$\boldsymbol{\sigma} = \begin{cases} p(\rho)\mathbf{I} + \left(2\eta + \dfrac{k(\rho)}{\sqrt{II_{D'}}}\right) \boldsymbol{D'} & \text{for } \boldsymbol{D'} \neq 0\,, \\[3mm] p(\rho)\mathbf{I} & \text{for } \boldsymbol{D'} = 0\,. \end{cases} \tag{8.5.4}$$

Note that the model accounts for both rate dependency and compaction effects on yielding, which are key properties of any cementations or geologic materials.

Generally post-test observations of penetration experiments in geologic or cementations materials (grout, concrete) indicate that there are four stages of the

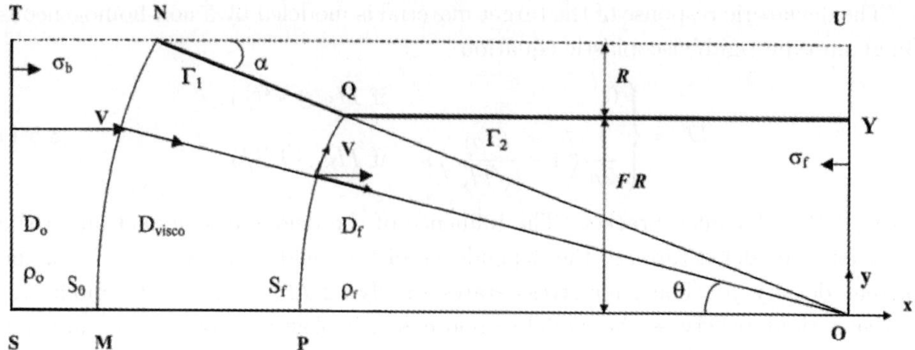

Fig. 8.5.1   Schematic of the flow over the wedge.

penetration event. The first stage corresponds to nose penetration when the target material is restrained from large movements; the second stage corresponds to chipping and cratering at the point of impact; the third stage occurs when the projectile is fully embedded and the pusher plate hits the target and strips off while the forth stage occurs when the full length of the projectile is in the penetration channel or tunnel created by the penetrator.

The analyses presented here concerns the steady-state flow of the target material over a projectile fully embedded in the target (i.e., forth stage of penetration). The penetrator is considered to be rigid and wedge-shaped. Post-test observations indicate that for impact velocities up to 1000 m/s, penetration paths are relatively straight and stable with regard to the original shotline. Thus, we can assume that the problem is axi-symmetric with respect to the wedge centerline.

Let us denote by $D = D_0 \cup D_{\text{visco}} \cup D_f$ the domain occupied by the penetrator and the target material around it (see Fig. 8.5.1). The lateral extent of this domain is $F \times R$, where R is the wedge semi-height and F is a number estimated from post-test observations. In the domains D_0 and D_f (i.e., in front of and behind the penetrator), it is assumed that the target material is in rigid body motion while in the domain D_{visco} the material undergoes viscoplastic deformation. We further suppose that in D_{visco} the flow lines are centered at a certain pole O. Hence, the viscoplastic domain D_{visco} is bounded by the nose surface, Γ_1, MP, and the two surfaces S_0 and S_f. The surface S_f is of radius $OP \equiv r_f = (R/\sin\alpha)F$, while the surface S_0 is of radius $OM \equiv r_0(R/\sin\alpha)(1 + F)$ (see Fig. 8.5.1). We assume that there exists velocity discontinuities tangent to the surfaces S_0 and S_f, while the components of velocity normal to these surfaces are continuous. Since across S_0

$$[\vec{V}] \cdot \vec{n} = 0, \tag{8.5.5}$$

with $\vec{n}$ the outward normal unit vector, it follows that in the viscoplastic domain, D_{visco}, the only non-zero velocity component is the radial component,

$$V_r = V\cos\theta, \quad 0 \le \theta \le \alpha \tag{8.5.6}$$

and consequently the only non-zero components of the rate of deformation are

$$D_{\theta\theta} = \frac{V\cos\theta}{r}$$

and

$$D_{r\theta} = -\frac{V\sin\theta}{2r} \qquad (8.5.7)$$

The continuity equation reduces to:

$$\frac{\partial}{\partial r}(r\rho) = 0. \qquad (8.5.8)$$

Since $\rho = \rho_0$ on S_0, in the viscoplastic domain,

$$\rho = \frac{r_0}{r}\rho_0. \qquad (8.5.9)$$

Obviously, the compaction law (8.5.9) could also be obtained by imposing that the velocity be the same in the domains D_0 and D_f (debit compatibility). In the following, we estimate the resistance to penetration based on energy considerations. The wedge semi-angle that minimizes the resistance to penetration is then determined for different interface conditions between the wedge and the target.

Along the surface of the wedge significant *friction* arises. The mechanics of friction at high sliding speeds is very complex. Much of the work reported is at lower speeds or pressures than those occurring during a penetration event. In view of this, in our analysis we assume that Coulomb friction law applies along Γ_1, the surface area of contact between the nose of the rigid projectile and the target, i.e.

$$\tau = \mu\sigma_n, \qquad (8.5.10)$$

where τ is the shear stress in the deformed material, μ is a friction coefficient taken as constant, and σ_n is the normal pressure between the nose and the target material.

In plastic forming of metals, when one body is fully plastic and the other is rigid, it is often assumed a friction law of the form (see Avitzur [1968]):

$$\tau = \mu_r\sigma_Y/\sqrt{3}, \qquad (8.5.11)$$

where τ is the shear stress in the deformed material, μ_r is a friction coefficient taken constant and σ_Y is the yield limit of the material. In general, $0 \leq \mu_r \leq 1$, the case $\mu_r = 0$ corresponds to "no friction", while $\mu_r = 1$ corresponds to adherence of the plastic body to the rigid die wall. Hence, according to the friction law (8.5.11) irrespective of the normal pressure between the two bodies, the shear stress is constant. The density of the target material behind the penetrator is: $\rho = \rho_f$ where $\rho_f = \rho|_{r=r_f}$ while the state of stress is such that $\sqrt{II_S} = k(\rho_f)$ (i.e., on the yield surface) (see Cristescu [1975]). Thus, it can be assumed that the contact between the rigid penetrator and the compacted material behind the penetrator is described by a law similar to (8.5.11), i.e.

$$\tau = \mu_r k(\rho_f). \qquad (8.5.12)$$

The theorem of power expanded is used to *compute the resistance to penetration.* In the domains D_0 and D_f no deformation occurs and therefore no internal power dissipation is involved. In the viscoplastic domain, the stress power per unit volume, $\boldsymbol{\sigma} : \boldsymbol{D}$ can be expressed as

$$\boldsymbol{\sigma} : \boldsymbol{D} = \sigma_{ij} D_{ij} = 4\eta II_{D'} + 2k(\rho)\sqrt{II_{D'}} + p \cdot \operatorname{tr} \boldsymbol{D}. \tag{8.5.13}$$

Using the work-hardening law (8.5.2) in conjunction with the radial compaction law (8.5.8), we obtain

$$\boldsymbol{\sigma} : \boldsymbol{D} = \eta \frac{V^2}{r^2}\left(\frac{3 + \cos^2\theta}{3}\right) + \left[k_0 + \beta\left(1 - \frac{r}{r_0}\right)^n\right]\frac{V}{r}\sqrt{\frac{3 + \cos^2\theta}{3}}$$

$$+ p_*\left(\frac{\rho_0}{\rho}\right)^m \left(\frac{r_0}{r}\right)^m \frac{V\cos\theta}{r}. \tag{8.5.14}$$

Integrating over D_{visco}, we arrive at:

$$\dot{W} = \int_{D_{\text{visco}}} \boldsymbol{\sigma} : \boldsymbol{D}\, dV = \int_{r_f}^{r_0}\int_0^\alpha (\boldsymbol{\sigma} : \boldsymbol{D})\, r\, d\theta dr\,,$$

or

$$\dot{W} = \eta\frac{V^2}{6}(7\alpha + \sin\alpha\cos\alpha)\ln\left(\frac{r_0}{r_f}\right)$$

$$+ \frac{V}{\sqrt{3}}F(\alpha)\left[k_0(r_0 - r_f) + \beta\frac{r_0}{n+1}\left(1 - \frac{r_f}{r_0}\right)^{n+1}\right]$$

$$+ p_*\frac{V\sin\alpha}{m-1}\left(\frac{\rho_0}{\rho_*}\right)^m r_0\left[\left(\frac{r_0}{r_f}\right)^{m-1} - 1\right] \tag{8.5.16}$$

where $F(\alpha) = \int_0^\alpha \sqrt{3 + \cos^2\theta}\, d\theta$. The surfaces S_0 and S_f are surfaces of velocity discontinuity. The power dissipated at the crossing of these surfaces is: $\dot{W}_{S_0, S_f} = \int_{S_0}\vec{t}\cdot[\vec{v}]dA + \int_{S_f}\vec{t}\cdot[\vec{v}]\, dA$, where $\vec{t}$ is the stress vector while $[\vec{v}]$ is the jump in velocity across the respective surface. Since the magnitude of the tangential jump is $V\sin\theta$, we obtain:

$$\dot{W}_{S_0} = \int_{S_0}\vec{t}\cdot[\vec{v}]\, dA = \int_0^\alpha \sigma_{r\theta}V\sin\theta r_0 d\theta$$

$$= -\frac{\eta V^2}{2}(\alpha - \sin\alpha\cos\alpha) - r_0 k_0 V\sqrt{3}F_1(\alpha)\,,$$

$$\dot{W}_{S_f} = \int_{S_0}\vec{t}\cdot[\vec{v}]\, dA = \int_0^\alpha -\sigma_{r\theta}V\sin\theta r_f d\theta$$

$$= \frac{\eta V^2}{2}(\alpha - \sin\alpha\cdot\cos\alpha) + \left[k_0 + \beta\left(1 - \frac{r_f}{r_0}\right)^n\right]V\sqrt{3}F_1(\alpha) \tag{8.5.16}$$

where

$$F_1(\alpha) = \int_0^\alpha \frac{\sin^2\theta}{\sqrt{3 + \cos^2\theta}}d\theta\,.$$

Assuming Coulomb friction [see (8.5.10)] along Γ_1, the surface area of contact between the nose of the rigid projectile and the target, the dissipation due to friction is:

$$\dot{W}_{\Gamma_1} = \mu \int_{r_0}^{r_f} \left(V \cos \alpha \cdot \sigma_{\theta\theta}|_{\theta=\alpha} \right) dr$$

$$= \mu V \cos \alpha \left\{ \frac{4}{3} V \cos \alpha + \eta \ln \left(\frac{r_0}{r_f} \right) + \frac{4}{3} \frac{\cos \alpha}{\sqrt{3 + \cos^2 \alpha}} \right.$$

$$\left. \times \left[k_0 (r_0 - r_f) + \beta \frac{r_0}{n+1} \left(1 - \frac{r_f}{r_0} \right)^{n+1} \right] \right\}$$

$$+ \mu V \cos \alpha \frac{p_*}{m-1} \left(\frac{\rho_0}{\rho_*} \right) r_0 \left[\left(\frac{r_0}{r_f} \right)^{m-1} - 1 \right]. \qquad (8.5.17)$$

Assuming that the friction law (8.5.11) applies along Γ_2, the surface area of contact between the penetrator and the compacted target material behind it (see Fig. 8.5.1), then

$$\dot{W}_{\Gamma_2} = \mu_r \int_{\Gamma_2} k(\rho_f) V ds = \mu_r k(\rho_f) V L, \qquad (8.5.18)$$

where L is the length of the shank. Next, the resistance to penetration can be computed from energy balance. Consider the control volume $V = D_0 \cup D_{\text{visco}} \cup D_f$ i.e., in the domain comprised between the planes $x = 0$ and $x = OS$ and $y = 0$ and $y = F \cdot R$, (see Fig. 8.5.1). At steady state, the stress power theorem writes:

$$\int_{D_{\text{visco}}} \boldsymbol{\sigma} : \boldsymbol{D} \, dV + \dot{W}_{S_0} + \dot{W}_{S_f} + \dot{W}_{\Gamma_1} + \dot{W}_{\Gamma_2}$$

$$= \int_{\partial V} (\sigma_b V - \sigma_f V) \, dA. \qquad (8.5.19)$$

In (8.5.19), σ_b denotes the magnitude of the stress vector acting along the direction of the velocity on the boundary of the domain D_0 (i.e., in front of the penetrator) while σ_f is the magnitude of the stress vector acting on the boundary of D_f. We may assume that $\sigma_f = k(\rho_f)$, where ρ_f is the density of the compacted target material behind the penetrator. Substituting (8.5.14) to (8.5.18) into (8.5.19), we obtain the expression of σ_b, the resistance to penetration, as:

$$\sigma_b = \frac{1}{V(F+1)R} (\dot{W} + \dot{W}_{S_0} + \dot{W}_{S_f} + \dot{W}_{\Gamma_1} + \dot{W}_{\Gamma_2} + k(\rho_f) V F R). \qquad (8.5.20)$$

To illustrate the predictive capabilities of the model, we will apply it to *concrete* for impact velocities of 300 m/s to 1000 m/s. In an attempt to study the influence of the frictional contact at the penetrator/target interface on the resistence to penetration for various impact velocities, the friction coefficient μ between the tip of the projectile and target was varied between 0.001 and 0.2. The coefficient of friction between the body of the wedge and the target was set to $\mu_r = 0.6$. The target material parameters involved in Eqs. (8.5.1)–(8.5.3) were assigned numerical values for conctrete material of initial density $\rho_0 = 2$ g/cm^3 and unconfined compressive strength is of 42.2 MPa. This implies that the pressure range of interest varies from tents of Mega-Pascals (MPa) to the order of a Giga-Pascal (GPa) for a case involving a high-strength steel penetrator. Thus, to characterize the behaviour of the concrete material for pressures representative for the in situ material, quasi-static compression tests were conducted for confining pressures in the range 0–450 MPa using a standard fully automated MTS testing machine at a strain rate of approximately $0.77 \cdot 10^{-6}$/s.

To better characterize the time and history effects on the behaviour of the material, in each quasi-static test the loading path considered of several loading-creep-unloading-reloading cycles. As already mentioned, before passing from loading to unloading, the load was held constant for 10–20 minutes in order to separate viscous effects from unloading. As an example, in Figs. 8.5.2(a) and (b) are shown the results of a tests under $\sigma_3 = 200$ MPa confinement. The variation of the axial strain rate, $\dot{\varepsilon}_1$ with time during the third creep cycle is presented in Fig. 8.5.3 showing that within the load was held constant until the time rate of change of the axial strain approached zero. Note that hysteresis effects were very much reduced thus allowing a rather accurate determination of the elastic parameters from the slopes of the unloading curves. No Hugoniot shock test data were available for

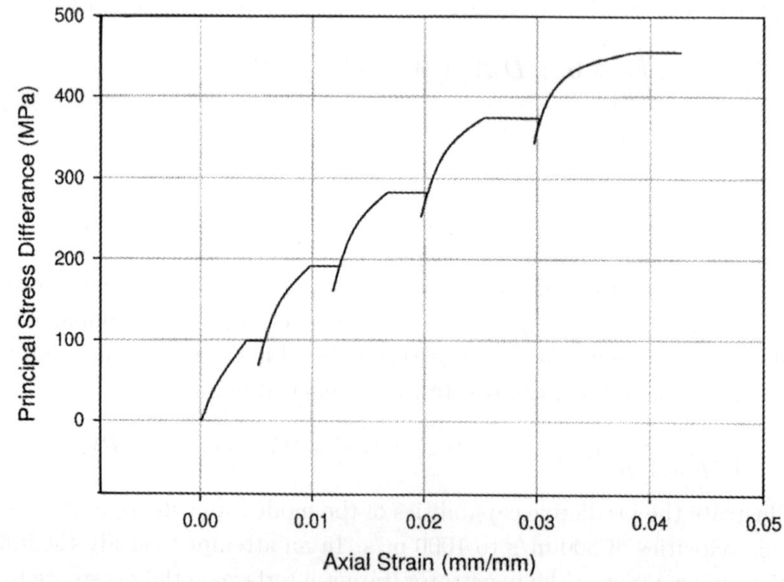

Fig. 8.5.2 Cyclic quasi-static test results under $\sigma_3 = 200$ MPa: (a) principal stress difference versus axial strain; (b) principal stress difference versus radial strain.

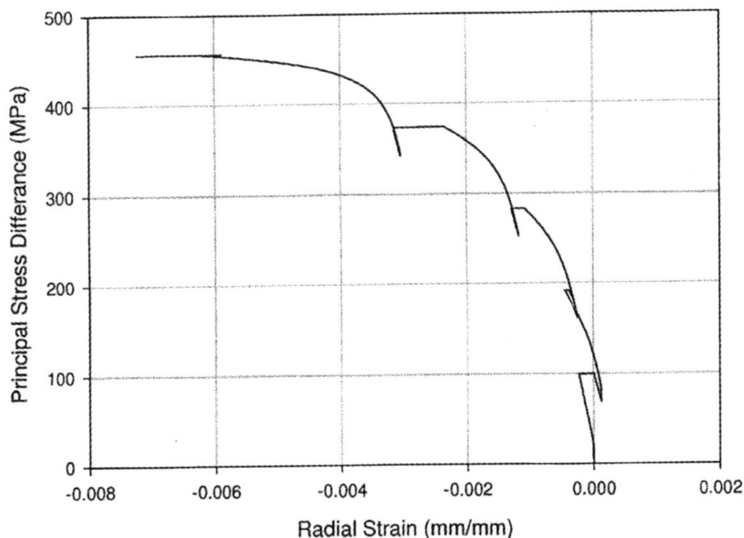

Fig. 8.5.3 Axial strain rate versus time during this 3rd creep cycle under $\sigma_3 = 300$ MPa confining pressure.

this material. Thus, the licking pressure was considered to be $p_* = 0.5$ GPa, a value which is based on post-test data on cementious materials. The corresponding locking density ρ_* is estimated using the relationship

$$\rho_* = \rho_0 \frac{1}{1 - \varepsilon_V^f}$$

where $\varepsilon_V^f = 0.068$ corresponds to the volumetric deformation at failure under confinement 450 MPa. For sake of simplicity, we assume a quadratic dependence of pressure on density [i.e., we set $m = 2$ in (8.5.3)]. We take the yield limit of the undisturbed medium $k(\rho_0)$ to be equal to the unconfined yield limit, i.e., $k(\rho_0) = 10$ MPa.

One of the challenges associated with modeling penetration is that the only information that could be gathered during penetration concerns the projectile trajectory and deceleration. To date, there is no in-target instrumentation for measuring deformation during penetration, there is no data from which the yield limit $k(\rho_f)$ may be inferred. Thus, we will assume that $k(\rho_f) = FF \cdot k(\rho_0)$, FF being a parameter, which depends on the impact velocity. Setting $n = 2$ in the law of variation of $k(\rho)$, we estimate the parameter β as

$$\beta = \frac{k(\rho_0)(FF - 1)}{(1 - \rho_0/\rho_f)^2} \, .$$

The wedge dimensions are set at: $R = 6.35$ mm, $L = 89$ mm.

As an example, in Fig. 8.5.4 is shown (σ_b/p_*), the resistance to penetration normalized by the locking pressure, versus the projectile semi-angle α for an impact

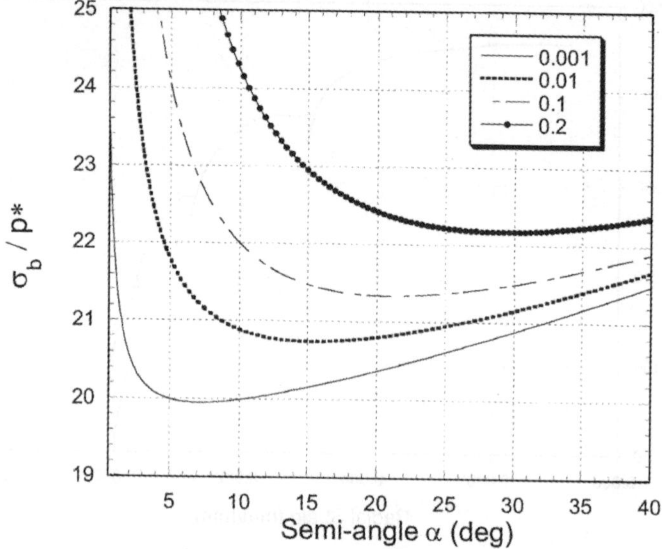

Fig. 8.5.4   Resistance to penetration σ_b normalized by the locking pressure p_* as a function of projectile semi-angle α for different values of the friction coefficient between the target and the wedge tip; impact velocity $V = 300$ m/s.

velocity $V = 300$ m/s. Calculations were done for 4 different values of the coefficient of friction between the wedge tip and target material. The yield stress of the compacted target $k(\rho_f)$ was considered to be 2.5 $k(\rho_0)$.

Note that irrespective of the contact conditions between the wedge and target, there exists a critical angle for which the resistance to penetration is minimal.

All the others parameters being kept the same, the more friction that is present, the blunter the nose required to achieve minimum resistance to penetration (of course, assuming that no erosion takes place and that the nose does not fail). This trend was observed for all impact velocities.

The plots of the optimal wedge-angle vs. impact velocity for different values of the friction coefficient between the tip and target are presented in Fig. 8.5.5. Since compaction ratios were not available, in the calculations it was assumed that for an impact velocity of 1000 m/s $k(\rho_f)/k(\rho_0)$ is double than at 300 m/s impact velocity. The results indicate that at higher impact velocities a sharper nose is required for optimum performance. However, for modest friction ($\mu = 0.01$), the sharpening is minimal. Figure 8.5.6 shows the resistance to penetration corresponding to optimum wedge angle versus the coefficient μ between the wedge tip and target for different impact velocities. Irrespective of the contact conditions, an increase in resistance to penetration occurs with increased impact velocity.

In an attempt to study the influence of the frictional contact at the penetrator/target interface on the resistance to penetration for various impact velocities,

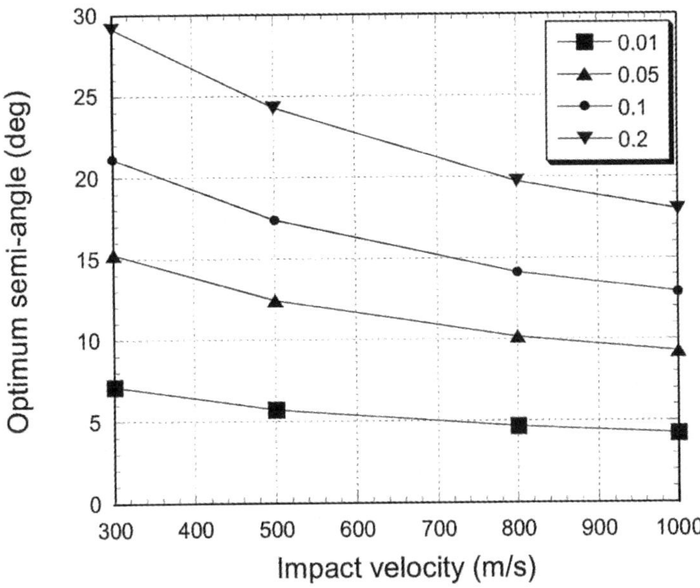

Fig. 8.5.5   Optimum projectile semi-angle versus impact velocity for various contact conditions between the wedge tip and viscoplastic target.

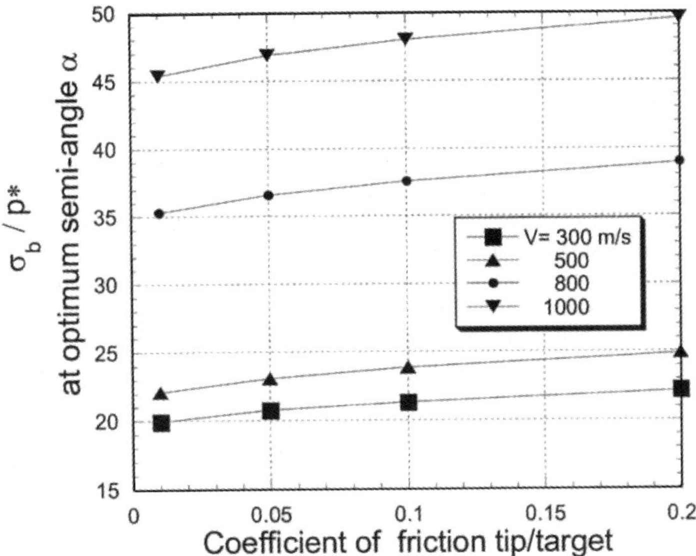

Fig. 8.5.6   Normalized resistance to penetration corresponding to the optimum projectile semi-angle versus the coefficient of friction between the wedge tip and viscoplastic target for various impact velocities.

the friction coefficient μ between the tip of the projectile and target was varied between 0.001 and 0.2. The coefficient of friction between the body of the wedge and the target was set to $\mu_r = 0.6$. It was found that irrespective of the contact conditions between the penetrator and target, there exists a critical angle for which the resistance to penetration is minimal. All the other parameters being kept the same, the more friction that is present, the blunter the nose required to achieve minimum resistance to penetration (of course, assuming that no erosion takes place and that the nose does not fail). This trend was observed for all impact velocities in the range considered 300–1000 m/s. The resistance to penetration at optimal nose angle was found to be greater for higher impact velocity.

An theory of optimal nose geometry for a rigid penetrator is due to Jones *et al.* [1998]. A net force on the nose of a rigid projectile normally penetrating a compliant target is given. The frictional effects are neglected.

A comparison of experimental results with simulations over a wide range of stiking velocities (less than 1000 m/s) is giving accurate prediction (Waren and Tabbara [2000]).

Bibliography

Aliev H. M., 1963, *Dokl. Akad. Nauk. SSSR* **151**, 80–83 (in Russian).

Andriankin E. I. and Koryavov V. P., 1959, Shock wave in a plastic medium of variable density, *Dokl. Akad. Nauk. SSSR* **128**, 257–260 (in Russian).

Avitzur B., 1968, *Metal Forming: Processes and Analysis*, Mc Graw Hill, New York.

Backman M. E. and Goldsmith W., 1978, The mechanics of penetration of projectile into targets, *Int. J. Engng. Sci.* **16**, 1–99.

Batra R. C., 1987, Steady-state penetratration of viscoplastic targets, *Int. J. Engng. Sci.* **24**, 9, 1131–1141.

Béda G., 1997, Constitutive equations and nonlinear waves, *Nonlinear Analysis, Theory, Methods & Applications* **30**, 1, 397–407.

Burke J. J. and Weiss V. (eds.), 1971, *Shock Waves and the Mechanical Properties of Solids*, Syracuse University Press, 417 pp.

Campbell J. D. 1972, *Dynamic Plasticity of Metals*, Springer-Verlag, Wien, 92 pp.

Cazacu O. and Cristescu N. D., 2000, Constitutive model and analysis of creep flow of natural slopes, *Italian Geo. Journal* **34**, 3, 44–54.

Cazacu O., Ionescu I. R. and Perrot T., 2004, Penetration of a rigid body into a viscoplastic compressible fluid, *ECCOMAS 2004*, Jyväskylä, 24–28 July, 2004.

Cazacu O., Cristescu N. D. and Schmidt M. J., 2006, Analysis of steady-state flow of a compressible viscoplastic medium over a wedge, *Int. J. Num. Anal. Meth. Geomechanics* **30**, 489–499.

Chadwick P., Cox A. D. and Hopkins H. G., 1964, *Phil. Trans. Roy. Soc. London A* **256**, 235–300.

Cheeseman B. A. and Bogetti T. A., 2003, Ballistic impact into fabric and compliant composite laminates, *Comp. Structures* **61**, 161–173.

Chen X. W. and Li Q. M., 2002, Deep penetration of a non-deformable projectile with different geometrical characteristics, *Int. J. Impact Engng.* **27**, 619–637.

Chernyshov A. D., 1966, On the character of strong jumps in certain complex media, *PMM J. Appl. Math. Mech.* **30**, 6, 1226–1232 (in Russian).

Cristescu N., 1967, *Dynamic Plasticity*, North-Holland, 614 pp.

Cristescu N., 1975, Plastic flow through conical converging dies, using a viscoplastic constitutive equation, *Int. J. Mech. Sci.* **17**, 425–433.

Cristescu N., 1980, On the optimum die angle in fast wire drawing, *J. Mech. Working Tech.* **3**, 275–287.

Cristescu N. D., 2000, A model of stability of slopes, in *Slope Stability 2000. Proceedinds of Sessions of Geo-Denver 2000*, Griffiths D. V., Fenton G. A. and Martin T. R. (eds.), *Geotechnical Special Publication* **101**, 86–98.

Cristescu N., Malvern L. E. and Sierakowski R. L., 1975, Failure mechanisms in composite plates impacted by blunt-ended penetrators, foreign object impact damage to composites, *American Society for Testing and Materials, ASTM, STP* **568**, 159–172.

Cristescu N. D., Craciun E. M. and Soós E., 2003, *Mechanics of Elastic Composites*, Chapman & Hall/CRC, 682 pp.

Cristescu N. D., Cazacu O. and Cristescu C., 2002, A model for slow motion of natural slopes, *Canadian Geotechnical Journal* **39**, 4, 924–937.

Davids N. (ed)., 1960, *International Symposium on Stress Wave Propagation in Materials*, Interscience Pub., 337 pp.

Davids N. (ed)., 1969, *Stress Wave Propagation in Materials, Intern. Sympo., Pennsylvania*, Inter. Sci., New York–London.

Demiray H. and Eringen A. C., 1978, A nonlocal model for plug formation in plates, *Int. J. Engng. Sci.* **16**, 287–297.

Durban D. and Masri R., 2004, Dynamic spherical cavity expansion in a pressure sensitive elastoplastic medium, *Int. J. Solids Struct.* **41**, 5697–5716.

Ezra A. A., 1973, *Principles and Practice of Explosive Metalworking*, Industrial Newspapers Limited, 270 pp.

Fomin V. M. and Kiselev S. P., 1997, *Elastic-Plastic Waves in Porous Materials, in High-Pressure Shock Compression of Solids IV*, Davison L., Horie Y. and Shahinpoor M. (eds.), Springer, 205–232.

Forrestal M. J. and Luk V. K., 1988, Dynamic spherical cavity-expansion in a compressible elastic-plastic solid, *J. Appl. Mech.* **55**, 2, 275–279.

Forrestal M. J. and Tzou D. Y., 1997, A spherical cavity-expansion penetration model for concrete targets, *Int. J. Solids Struct.* **34**, 31–32, 4127–4146.

Gao S., Jin L. and Liu H., 2004, Dynamic response of a projectile perforating multi-plate concrete targets, *Int. J. Solids Struct.* **41**, 4927–4938.

Grigorian S. S., 1959, On the general equations of the dynamics of soils, *Dokl. Acad. Nauk. SSSR* **124**, 285–288 (in Russian).

Grigorian S. S., 1960a, On the final definition on the motion of soils, *Prikl. Mat. Mech.* **24**, 651–662 (in Russian).

Grigorian S. S., 1960b, On the fundamental representation of the dynamics of soils, *Prikl. Mat. Mech.* **24**, 6, 1057–1072 (in Russian).

Grigorian S. S., 1967, Some problems of mathematical theory of deformation and failure of solid soils, *Prikl. Mat. Mech.* **31**, 643–669 (in Russian).

Gu C., Kim M. and Anand L., 2001, Constitutive equations for metal powders: application to powder forming processes, *Int. J. Plasticity* **17**, 147–209.

Hawkyard J. B., 1969, A theory for the mushrooming of flat-ended projectiles impinging on a flat rigid anvil, using energy considerations, *Int. J. Mech. Sci.* **11**, 313–333.

Hawkyard J. B., Eaton D. and Johnson W., 1968, The mean dynamic yield strength of cooper and low carbon steel at elevated temperatures from measurements of the "mushrooming" of flat-ended projectiles, *Int. J. Mech. Sci.* **10**, 929–948.

Heuze F. E., 1990, An overview of projectile penetration into geological-materials, with emphasis on rocks, *Int. J. Rock Mech. Min. Sci.* **27**, 1–14.

Hutchings I. M., 1979, Estimation of yield stress in polymers at high strain-rates using G. I. Taylor's impact technique, *J. Mech. Phys. Solids* **26**, 289–301.

Hutton D. V. and Counts J., 1974, Wave propagation in an elastic circular membrane subjected to impulsive pressure loading, *J. Appl. Mech. Trans. ASME* **41**, 1, 290–291.

Ishlinskii A. Yu., 1954, *Ukr. Mat. Zh.* **6**, 430–441 (in Russian).

Ishlinskii A. Yu., Zvolinskii N. V. and Stepanenko I. A., 1954, *Dokl. Akad. Nauk. SSSR* **95**, 729–731 (in Russian).

Johnson G. R. and Holmquist T. J., 1988, Evaluation of cylinder-impact test data for constitutive model constants, *J. Appl. Phys.* **64**, 8, 3901–3910.

Jones S. E., Gillis P. P. and Foster J. C. Jr., 1987, On the equation of motion of the undeformed section of a Taylor impact specimen, *J. Appl. Phys.* **61**, 2, 499–502.

Jones S. E., Gillis P. P., Foster J. C. Jr and Wilson L. L., 1990, A one-dimensional, two-phase flow model for Taylor impact specimens, *Trans. ASME, J. Engn. Mat. Tech.* **113**, 228–235.

Jones S. E., Maudlin P. J. and Foster J. C. Jr, 1996, An engineering analysis of plastic wave propagation in the Taylor test, *Int. J. Impact Engng.* **19**, 2, 95–106.

Jones S. E., Rule W. K., Jerome D. M. and Klug R. T., 1998, On the optimal nose geometry for a rigid penetrator, *Comp. Mech.* **22**, 413–417.

Karagiozova D., 2004, Dynamic buckling of elastic-plastic square tubes under axial impact — I: stress wave propagation phenomenon, *Int. J. Impact Engng.* **30**, 143–166.

Karagiozova D. and Jones N., 2000, Dynamic elastic-plastic bucling of circular cylindrical shell under axial impact, *Int. J. Solids Struct.* **37**, 2005–2034.

Karagiozova D. and Jones N., 2001, Influence of stress waves on the dynamic progressive and dynamic plastic buckling of cylindrical shell, *Int. J. Solids Struct.* **38**, 6723–6749.

Karagiozova D. and Jones N., 2002, On dynamic buckling phenomena in axially loaded elastc-plastic cylindrical shells, *Int. J. Non-Lin. Mech.* **37**, 1223–1238.

Koiter W. T., 1953, Stress–strain relations, uniqueness and variational theorems for elastic-plastic materials with a singular yield surface, *Quart. Appl. Math.* **11**, 350–354.

Kolsky H. and Prager W., 1964, *Stress Waves in Anelastic Solids*, Springer-Verlag, 342 pp.

Kompaneets A. S., 1956, *Dokl. Akad. Nauk. SSSR* **109**, 49–52 (in Russian).

Lee L. H. H., 1975, Bifurcation and uniqueness in dynamics of elastic-plastic continua, *Int. J. Engng. Sci.* **13**, 1, 69–76.

Longcope D. B. and Forrestal M. J., 1983, Penetration of targets describet by a Mohr Coulomb failure criterion with a tension cutoff, *J. Appl. Mech.* **50**, 327–333.

Lush P. A., 1991, Comparison between analytical and numerical calculations of liquid impact on elastic-plastic solid, *J. Mech. Phys. Solids* **39**, 1, 145–155.

Macek R. W. and Duffey T. A., 2000, Finite cavity expansion method for near-surface effects and layering during earth penetration, *Int. J. Impact Engng.* **24**, 239–258.

Maudlin P. J. and Schiferl S. K., 1996, Computational anisotropic plasticity for high-rate forming applications, *Comput. Methods Appli. Mech. Engrg.* **131**, 1–30.

Maudlin P. J., Foster J. C. Jr. and Jones S. E., 1997, A continuum mechanics code analysis of steady plastic wave propagation in the Taylor test, *Int. J. Impact Engng.* **19**, 3, 231–256.

Maudlin P. J., Gray III G. T., Cady C. M. and Kaschner G. C., 1999, High-rate material modelling and validation using the Taylor cylinder impact test, *Phil. Trans. R. Soc. Lond. A* **357**, 1707–1729.

Maudlin P. J., Bingert J. F., House J. W. and Chen S. R., 1999, On the modeling of the Taylor cylinder impact test for orthotropic textured materials: experiments and simulations, *Int. J. Plasticity* **15**, 139–166.

Maudlin P. J., Bingert J. F. and Gray III G. T., 2003, Low-symmetry plastic deformation in BCC tantalum: experimental observations, modeling and simulations, *Int. J. Plasticity* **19**, 483–515.

Medvedeva N. S. and and Shemiankin E. I., 1961, *J. Appl. Mech. Tech. Phys.* **6**, 78–87.

Nemat-Nasser S., Li Y.-F. and Isaacs J. B., 1994, Experimental/computational evaluation of flow stress at high strain rates with application to adiabatic shear banding, *Mech. of Mat.* **17**, 111–134.

Nemes J. A. and Spéciel E., 1995, Use of a rate-dependent continuum damage model to describe strain-softening in laminated composites, *Comp. Structures* **58**, 6, 1083–1092.

Osinov V. A., 1998, Theoretical investigation of large-amplitude waves in granular soils, *Soil Dynamics Earthquake Engn.* **17**, 13–28.

Phoenix S. L. and Porwal P. K., 2003, A new membrane model for the ballistic impact response and V50 performance of multi-play fibrous systems, *Int. J. Solids Struct.* **40**, 6723–6765.

Prager W., 1953, On the use of singular yield conditions and associated flow rules, *J. Appl. Mech.* **20**, 3, 317–320.

Pytel A. and Davids N., 1963, A viscous model for plug formation in plates, *J. Franklin Inst.-Engineering Appl. Math.* **276**, 5, 394–406.

Razorenov S. V., Savinykh A. S., Zaretsky E. B., Kanel G. I. and Kolobov Yu. R., 2005, Effect of preliminary strain hardening on the flow stress of titanium and titanium alloy during shock compression, *Physics of the Solid State* **47**, 4, 663–669.

Rakhmatulin Kh. A. and Dem'yanov Yu. A., 1961, *Strength under High Transient Loads* (English translation, *Israel Program for Scientific Translations*, 1966).

Rakhmatulin H. A. and Stepanova L. I., 1958, On the propagation of explosion shock waves in sols, in: *The Problems of Destruction of Rocks by Explosion* (Moscow) 149–159 (in Russian).

Rakhmatulin H. A., Sagomonian A. Yu. and Alekseev N. A., 1964, *The Problems of Dynamics of Soils* (Izv. MGU, Moscow) (in Russian).

Rinehart J. S., 1975, *Stress Transients in Solids*, HyperDynamics, 230 pp.

Sagomonian A. Yu., 1961, One-dimensional motion of soils with spherical, cylindrical and plane waves, in: *Dynamics of Soils*, No. 44 (Moscow), 43–87 (in Russian).

Salvadori M. G., Skalak R. and Weidlinger R., 1955, *Trans. N.Y. Acad. Sci. Ser. II* **21**, 427–434.

Satapathy S., 2001, Dynamic spherical cavity expansion in brittle ceramics, *Int. J. Solids Struct.* **38**, 5833–5845.

Satapathy S. S. and Bless J. S., 2000, Cavity expansion resistance of brittle materials obeying a two-curve pressure-shear behavior, *J. Appl. Phys.* **88**, 7, 4004–4012.

Schmidt M. J., 2003, *High Pressure and High Strain Rate Behaviour of Cementitious Materials: Experiments and Elastic/Viscoplastic Modeling*, PhD dissertation, University of Florida.

Schmidt M. J., Cazacu O., Ross C. A. and Cristescu N. D., 2001, Dynamic behaviour of mortar: experimental data and modeling, *Proceedings of the 10th International Conference on Computer Methods and Advances in Geomechanics*, Tucson, Arizona.

Shemiankin E. I., 1961, *J. Prikl. Mech. Tech. Fiz.* **5**, 91–99 (in Russian).

Shewmon P. G. and Zackay V. F., 1961, *Response of Metals to High Velocity Deformation*, Intersience Publishers, 491 pp.

Srinivasan M. G. and Ting T. C. T., 1974, Initiation of spherical elastic-plastic boundaries due to loading at a spherical cavity, *J. Mech. Phys. Solids* **22**, 415–435.

Takeda N., Sierakowski R. L. and Malvern L. E., 1981, Transverse cracks in glass/epoxy laminates impacted by projectiles, *J. Material Sci.* **16**, 7, 2008–20011.

Warren T. L. and Tabbara M. R., 2000, Simulations of the penetration of 6061-T6511 aluminum targets by spherical-nosed VAR 4340 steel projectiles, *Int. J. Solids Struct.* **37**, 4419–4435.

Warren T. L., Hanchak S. J. and Poormon K. L., 2004. Penetration of limestone targets by ogive-nosed VAR 4340 steel projectiles at oblique angles: experiments and simulations, *Int. J. Imp. Engineering* **30**, 1307–1331.

Zhu G., Goldsmith W. and Dharan C. K. H., 1992, Penetration of laminated Kevlar by projectiles — I. Experimental investigation. *Int. J. Solids Struct.* **29**, 4, 399–420.

Zukas J. A., Nichols T., Greszczuk L. B. and Curran D. R., 1990, *Impact Dynamics*, Wiley, New York, 155–214.

Zukas J. A., 2004, *Introduction to Hydrocodes*, Elsevier.

Chapter 9

Hypervelocity Impact (Information)

9.1 Introduction

Several papers were published in the last years about hypervelocity impact. It seems that the beginning is a big paper by Al'tshuler [1965]; it is a review paper containing everything which was done up to this year (shock adiabats and their experimental registration, methods of obtaining semiempirical equation of state, shock adiabats, speed of sound and isentropic elasticity of shock-compressed bodies, on the composition of the earth's core and the mantle, dynamic strength of materials, etc.). A paper by Alekseevskii [1966] is devoted to penetration of a rod with high velocity; the main ideas and schemes are given an example is computed for the velocity of 1470 m/s. Then is the paper by Tate [1967]. A modified hydrodynamic theory which takes some account of strength effects is used to predict the deceleration of a long rod after striking a target. The results are then compared with experimental data from X-ray observation. The Bernoulli equation is used. The theory of long rod penetration as given in a previous paper is extended by Tate [1969] to take into account of the deformation of a soft rod against a rigid target and the penetration of a rigid projectile into a soft target. It is shown that it is theoretically possible to have a decrease in depth of penetration with increasing impact velocity, and a method for deducing the average radius of the hole is given. The Bernoulli equation is again given for materials which behave as a fluid. A semi-infinite solenoid in a uniform velocity flow field is used by Tate [1986], as a model for the quasi-steady primary phase of long rod penetration. Assumptions: the materials to be incompressible, to obey a Mises yield criterion, when yielding the materials are perfectly plastic, the Levy–Mises flow law applies, the materials are eitherelastic or plastic, a modified Bernoulli equation is established. In another paper Tate [1990] is proposing a segmented rod to explain penetration; the projectile is fired up to 1800 m/s.

Allen and Rogers [1961] study the penetration of metal rods into semi-infinite metal targets. They have investigated experimentally at velocities up to 0.3 cm/μs. Results are compared with predictions from the hydrodynamic theory of jet penetration. Experimental results indicate that the effective yield strength is relatively

425

independent of the strength of the rod. The hydrodynamic theory of penetration was determined to be generally acceptable except where the density of the jet is much greater than that of the target.

The impact flash jet initiation phenomenology was considered by several authors. Shaped charge theory predicts that a jet will form when the impact angle between two colliding surfaces is greater than an initiation angle criterion which is a function of material properties and impact velocity. Because the effective impact angle continuously sweeps over a range of values as a spherical particle impacts a flat target, this jet initiation angle criterion is satisfied at some point during the penetration process. A simple model has been developed by Ang [1990] which predicts the jet initiation from either the target or the particle material.

The hypervelocity impact of a projectile upon a thin metal plate and subsequent formation of back-surface debris is reviewed by Anderson *et al.* [1990]. At sufficiently high impact velocities, roughly greater than 3.0 km/s, shock formation and interaction dominate and control the overall response of both the projectile and the target plate. The authors focus upon the importance of shock heating, melting, and vaporization in the application. They also assess the current status of computational modeling of this kind of impact event, specifically addressing recent work bearing on the sensitivity of such modeling to the equations of state and certain numerical issues. More accurate data in multi-dimensional experiments may be able to resolve the question of what are the major causes of differences between calculations and experiments.

The theory is summarized and the origins of the target resistance term are examined by Anderson and Walker [1991]. Numerical simulations were performed of a tungsten-alloy, long-rod projectile into a semi-infinite hardened steel target at three impact velocities sufficiently high to result in projectile erosion (impact velocity apt to 1.5 km/s). The constitutive responses of the target and projectile were varied parametrically to assess the effects of strain hardening, strain-rate hardening, and thermal softening on penetration response. The results of one of the numerical simulations were selected to compare and contrast in detail with the predictions of the Tate model.

A computational study was performed by Anderson *et al.* [1993a] to quantify the effects of strain rate on replica-model experiments of penetration and perforation. The impact of a tungsten-alloy, long-rod projectile into an armor steel target at 1.5 km/s was investigated. It was found that over a scale factor of 10, strain-rate effects change the depth of penetration, for semi-infinite targets, and the residual velocity and length of projectile, for finite-thickness targets, by an order of 5%. Although not modeled explicitly in the present study, the time-dependence of damage was examined. Damage accumulation is a strong function of absolute time, not scaled time. At homologous times, a smaller scale will have less accumulated damage than a larger scale; therefore, the smaller scale will appear stronger, particularly in situation where the detailes of damage evolution are important.

The penetration resistance of hard layers, such as ceramics and hardened steels, struck by high velocity long rod projectiles can be characterized by the depth of penetration tests (Bless and Anderson [1993]). The depth of penetration tests can be used to calculate average penetration resistance. The tests can also be used to compute differential efficiency. Implications for the effectiveness of hypervelocity penetration (for 2–5 km/s) are that the optimum velocity for energy efficient penetration will be much higher for hard materials than for conventional armor steel. Furthermore, ceramics will continue to substantially outperform armor steels, while high hardness steels will lose their relative advantages against long rod projectiles above 3 km/s.

Numerical simulations are used by Anderson *et al.* [1993b] to examine long-rod penetration as a function of impact velocity. Similarities and differences between the penetration histories are analyzed, including penetration and tail velocities, penetration depths, crater radii, centerline interface pressures, and the extends of plastic flow in the projectile and target. The Tate's Bernoulli equation is used

$$\frac{1}{2}\rho_p(v-u)^2 + Y_p = \frac{1}{2}\rho_t u^2 + R_t$$

where Y_p is the flow stress of the projectile, and R_t is defined as the target resistance in the one-dimensional formulation, v is the projectile (tail) velocity, u is the penetration (interface) velocity. The one-dimensional modified Bernoulli theory is often used to examine long-rod penetration into semi-infinite targets, and integral to the theory is a term that describes the resistance of the target to penetration. It is shown that the hydrodynamic theory gives smaller penetration velocities. The impact velocities are 1.5–4.5 km/s. One is also observing that the target resistance decreases with impact velocity, and it is shown that this is a consequence of both the residual phase of penetration and variations in the size of the plastic zone field.

Calculations of steel target penetration by $L/D \leq 1$ tungsten and tungsten alloy projectiles have been extended to $L/D = 1/32$ over the velocity range 1.5 to 5 km/s by Orphal *et al.* [1993]. L is the projectile length and D the projectile diameter. The ratio of crater to projectile diameter tends to 1 as L/D decreases over this entire velocity range. For impact velocities of 1.5 and 3 km/s, penetration depth normalized by projectile length increases with decreasing projectile L/D up to a maximum value and then decreases for still lower L/D. Experiments at impact velocities of 2 and 3 km/s confirm these results. For 5 km/s impact velocity, the calculations show P/L increasing (P depth of penetration) with decreasing projectile L/D over the entire range $1/32 \leq L/D \leq 1$. Comparison with the experiments by Bjerke *et al.* [1992] is also done.

Walker and Anderson [1995] found that in general the time histories of penetration predicted by the Tate model can be in good agreement with those computed from numerical simulations. However, discrepancies exist between the model and numerical simulations at the beginning and at the end of penetration. From insights provided by numerical simulations, assumptions are made concerning the velocity and stress profiles in the projectile and the target. Using these assumptions, the time-dependent, cylindrically-symmetric, axial momentum equation is explicitly integrated along the centerline of the projectile and target to provide the equation of motion. The model requires the initial interface velocity — which can be found, for example, from the shock jump conditions — and material properties of the projectile and target to compute the time history of penetration. Agreement between the predictions of this one-dimensional, time-dependent penetration model is in good agreement with experimental results and numerical simulations. Analytic penetration modeling usually relies on either a momentum balance or an energy-rate balance to predict depth of penetration by a penetrator based on initial geometry and impact velocity. Based on the flow field and constitutive assumptions it is then possible to derive a momentum or an energy-rate balance. In another paper Walker *et al.* [1995] is presented the results of impact tests, as well as discusses the shaped charge design modifications for the nickel and molybdenum launchers. Radiographs are presented of impacting projectiles, as are post test photographs of various targets. The data are unique in that they represent low L/D projectile impacts into both monolithic blocks and spaced plates at velocities above 10 km/s. The aluminum projectiles are being launched at 11.25 ± 0.20 km/s, the molybdenum projectiles at 11.72 ± 0.10 km/s, and the nickel projectiles at 10.81 ± 0.10 km/s. The impact data for aluminum agree well with trends seen at lower velocity light gas gun testing and scaled approaches.

The paper by Walker [2001] examines the use of assumed flow fields within a target created by impact and then examines the resulting predicted behavior based on either momentum conservation or energy conservation. It is shown that for the energy rate balance to work, the details of energy transfer mechanisms must be included in the model. In particular, how the projectile energy is initially transferred into target kinetic energy and elastic compression energy must be included. As impact velocity increases, more and more energy during the penetration event is temporarily deposited within the target as elastic compression and target kinetic energy. This energy will be dissipated by the target at a later time, but at the time of penetration it is this transfer of energy that defines the forces acting on the projectile. Thus, for an energy rate balance approach to successfully model penetration, it must include the transfer of energy into kinetic energy within the target and the storage of energy by elastic compression. Understanding the role of energy dissipation in the target clarifies the various terms in analytic models and identifies their origin in terms of the fundamental physics. Understanding the modes of energy transfer also assists in understanding the hypervelocity result that penetration depth only slowly increases with increasing velocity even through the kinetic energy increases as the square of the velocity.

At ordnance velocities (1.0–1.9), there is a pronounced decrease in penetration efficiency, as measured by P/L, when projectiles of larger L/D are used, as mentioned by Anderson *et al.* [1995b]. The influence of L/D on penetration is refered to the L/D *effect*. They numerically examine the L/D effect at higher velocities, from 2.0 km/s to 4.5 km/s. It is found that as the velocities increases, there is a change in mechanism for the L/D effect. At ordnance velocities the L/D effect is mostly due to the decay in penetration velocity during the steady-state region of penetration. At higher velocities, the steady-state region of penetration shows no L/D dependence, and the L/D effect is due primarily to the penetration of the residual rod at the end of the penetration event. This change in mechanism is related to the change in slope of the penetration-versus-impact velocity for eroding projectiles.

Anderson *et al.* [1995a] study in a paper the penetration behavior of tungsten alloy, long-rod penetrators into high-hard steel, at two impact velocities 1.25 km/s and 1.70 km/s. The positions of the nose and tail of the projectile were measured by means of a 600 kV flash X-ray system at different times during penetration. The experiments were numerically simulated. The computational results are in very good agreement with the experimental position-time data.

The computational results by Lawrence *et al.* [1995] are discussed and compared with new experimental observations obtained at an impact velocity of ~ 10 km/s. In the experiment, the debris cloud was generated by the impact of a plate-shaped titanium projectile with a thin titanium shild.

When 0.76 mm diameter tungsten rods impact AD995 Alumina at velocities up to 3.5 km/s, Subramanian and Bless [1995] show that penetration and consumption velocities are approximately linear functions of impact velocity. At 3.5 km/s, the penetration falls below the hydrodynamic limit by 15%. Neither the primary penetration nor the total penetration (i.e., final depth of penetration) reached the limiting value of 2.23 suggested by hydrodynamic theory. The residual penetration, which is the difference between the total and primary penetration, was almost zero at 3 to 3.5 km/s. This suggests that the ceramic retains considerable strength ahead of the penetrator/target interface. Primary penetration approaches 75% of the hydrodynamic limit.

A systematic study is decribed by Chhabildas *et al.* [1995], has led to the successful launch of thin flier plates to velocities of 16 km/s. The energy required to launch a flier plate to 16 km/s is approximately 10 to 15 times the energy required to melt and vaporize the plate.

In the experiments by Brannon and Chhabildas [1995], a thin plate of zinc is impacted by a tantalum flier plate at speeds ranging from 8 to 10.1 km/s, producing pressures from 3 Mbar to over 5,5 Mbar and temperatures as high as 39000 K. Such high pressures produce essentially full vaporization of the zinc.

Trucano and Chhabildas [1995] used the Sandia Hypervelocity Launcher to get velocities in excess of 10 km/s. In the first experiment the launch of an intact 0.33 gram titanium alloy chunk flier to a velocity of 10.2 km/s.

In a paper by Anderson *et al.* [1996a] one is considering the ballistic impact experiments which were performed on ceramic laminate targets at three scale sizes, normally 1/3, 1/6 and 1/12, to quantify the effects of scale on various responses, in particular, the ballistic limit velocity. Some of the measured quantities showed little or no dependence on scale size, whereas other quantities, particularly the ballistic limit velocity, were found to vary with scale size. The impact velocities are up to 2.69 km/s. The ballistic limit velocity V_{BL} is

$$V_r = \begin{cases} 0, & 0 \le V_s \le V_{BL} \\ a(V_s^p - V_{BL}^p)^{1/p}, & V_s > V_{BL} \end{cases}$$

where V_r, V_s and V_{BL} are the residual, striking (impact), and limit velocities.

In another paper by Orphal *et al.* [1996] a series of 26 terminal ballistics experiments was performed to measure the penetration of simple confined aluminum nitride targets by a long tungsten rod. Impact velocities ranged from 1.5 to about 4.5 km/s. Penetrator diameter D was 0.762 mm. From the data $u = dp/pt =$ speed of penetration into the target and $v_c = d(L - L_r)/dt =$ speed of "consumption" of the long rod were obtained. In a paper by Anderson *et al.* [1996], a common measure of penetration efficiency is given by the depth of penetration P into a semi-infinite target normalized by the original length of the projectile L. It has been known for many years that P/L depends upon the aspect ratio L/D for projectiles with relatively small aspect ratios, e.g. $1 \le L/D \le 10$. This influence of L/D on penetration is referred to as the L/D *effect*. Although observed, the L/D effect for large aspect ratio rods is not as well documented. Further, published penetration equations have not included the L/D effect for high aspect ratio rods. The authors have compiled a large quantity of experimental data that permits the quantification of the L/D effect for projectiles with aspect ratios of $10 \le L/D \le 30$. Numerical simulations reproduce the observed experimental behavior; thus, no new physics is required to explain the phenomenon. The numerical simulations allow investigation of the fundamental mechanics leading to a decrease in penetration efficiency with increasing aspect ratio.

In a paper by Piekutowski *et al.* [1996] one is studying the perforation experiments of 26.3 mm thick aluminum plates and 12.9 mm diameter, 88.9 mm long, ogive-nose steel rods. For normal and oblique impacts with striking velocities between 280 and 860 m/s, one is measuring the residual velocities and displayed the perforation process with X-ray photographs. They have developed perforation equations that accurately predict the ballistic limit and residual velocities.

Forty terminal ballistics experiments were performed by Orphal *et al.* [1997] to measure the penetration of simple confined boron carbide targets by long tungsten roads. Impact velocities ranged from 1.5 to about 5.0 km/s. For tests with velocities $1.493 \le v \le 2.767$ km/s, the penetrator diameter was 1.02 mm. For tests with impact velocities $v \ge 2.778$ km/s the penetrator diameter was 0.762 mm. For tests in the velocity range $2.335 < v < 2.761$ km/s both penetrator sizes were used. The penetration velocity u is found to depend linearly on the impact velocity $u = 0.757v - 0.406$.

The shock-wave compression of brittle solids was studied by Grady [1998]. Extensive experimental investigation in the form of large-amplitude, nonlinear wave-profile measurements which manifest the shock strength and equation-of-state properties of brittle solids has been performed.

The results of experiments performed to determine a complete set of independent elasic constants and the effect of shock-induced shear strain on the delamination strength of a glass-fiber-reinforced plastic composite is due to Dandecar *et al.* [1998].

Two analytical models for the crater size generated by long-rod and thick-walled tube projectiles are presented by Lee and Bless [1998]. The first is based on energy; in a steady-state penetration, the kinetic energy loss of a projectile is related to the total energy deposited in the target. This simple approach provides an upper bound for the crater size. The second approach is based on the observation that two mechanisms are involved in cavity growth due to long projectiles: flow of projectile erosion products, which exerts radial stress on the target and opens a cavity, and radial momentum of the target as it flows around the projectile nose. The speed of impact is less than 5 km/s. This analysis includes the centrifugal force exerted by the projectile, radial momentum of the target, and the strength of the target. Thus, it can estimate the extent of cavity growth due to projectile mushrooming, which cannot be predicted by other analyses. This model is shown to be in good agreement with experimental data.

Yaw has been known to generally influence the penetration performance of long-rod projectiles. Experiments have shown that even small angles of yaw can significantly degrade performance. The experiments by Bless *et al.* [1999] have shown that even small angles of yaw can significantly degrade performance. The authors show that a critical feature of a yawed impact is the transverse load on the penetrator. Transverse loads tend to decrease the misalignment of rod axis and velocity vector. They use classical cavity expansion theory to quantify the impact transients and determine the magnitude of the transverse load. Then, a steady-state slot-cutting model is used to calculate the shape and orientation of a projectile that exits a finite plate. They found that this is contrary to the findings of some previous studies considered. The strength pf the projectile may be ignored compared to the inertial loads even at the relatively high impact velocities. The theory agrees well with reverse impact experimental data on finite plate.

Two and three-dimensional numerical simulations have been conducted by Anderson *et al.* [1999b] to help better understand the penetration and perforation of chemical submunition targets by tungsten-alloy long-rod projectiles. In particular, the computational results are analyzed to assess modeling assumptions in the application of the modified Bernoulli model of Tate to this class of problems. For the study, the chemical submunitions were treated as long steel cylinders filled with water. Impacts near the submunitions ends were neglected. Many of the simulations conducted looked at rod penetration through two successive submunitions. For purposes of the computational study, the cylinders were idealized as flat plates. The study considered three impact obliquities (60°, 70° and 80°) and two impact velocities (2.0 and 4.0 km/s). Penetration velocities, erosion rates, and the effects of a finite projectile diameter were investigated as a function of obliquity and impact velocity.

Experimental data by Anderson *et al.* [1999a] show that penetration velocities and instantaneous penetration efficiencies fall below that expected from hydrodynamic theory, even at impact velocities as high as 4.0 km/s. For steel rod impacting an aluminum target, the hydrodynamic penetration velocity is achieved only at impact velocity of about 4 km/s. Numerical simulations, using appropriate strength values, are in excellent agreement with the experimental data. Parametric studies demonstrate that both projectile and target strength have a measurable effect even at such high impact velocities.

The ballistic performance of 17 penetrator materials, representing 5 distinct steel alloys treated to various hardnesses along with one tungsten alloy, has been investigated by Anderson *et al.* [1999]. Residual length and velocities, as well as the ballistic limit velocities, were determined experimentally for each of the alloy types for length to diameter (L/D) ratio 10 projectiles against finite-thick armor steel targets. An equation for the residual velocity V_R in terms of the impact velocity V_P and ballistic limit is

$$\frac{V_R}{V_{BL}} = a \left[\left(\frac{V_P}{V_{BL}} \right)^m - 1 \right]^{1/m}$$

were V_{BL} is the ballistic limit velocity, was used. The target thickness normalized by the projectile diameter (T/D) was 3.55. For some of the projectile types, a harder target, with the same thickness, was also used. It was found that the ballistic limit velocity decreases significantly when the projectile hardness exceeds that of the target. Numerical simulations are used to investigate some of the observed trends. It is shown that the residual projectile length is sensitive to projectile hardness; the numerical simulations reproduce this experimental observation. The results of experiments showed that the residual velocity either increased or remained approximately constant as the projectile strength was increased. It is assumed that the plastic work per unit volume is approximately a constant, that is, there is a trade off between strength and ductility. An Eulerian wave code was used by Anderson *et al.* [1999d] to simulate the impact, penetration, and detonation of a 23-mm high-explosive projectile into a water-filled tank. The pressure-time response is compared with results from an experiment. Computed peak pressures and impulses compare well with the experimental values.

Orphal and Anderson [1999] report a direct observation of the streamline reversal of eroded rod material proposed by Allen and Rogers [1991]. Allen and Rogers suggested that the turning of high-velocity long-rod penetrator material at the target interface could be viewed as a reversal of the direction of a streamline with only a change of sign in the velocity. Allen and Rogers' streamline reversal model has two important consequences. First the eroded debris has a speed of $v_d = 2u - v$ relative to the target, where v is the impact velocity and u is the speed of penetration of the rod relative to the target. Secondly, a consequence of this formula is that the length of the rod debris, is given by the difference in the initial length of the rod and the remaining length of the rod. Results of a time-resolved experiment for a tungsten penetrator into a polycarbonate target at 3.61 km/s and a corresponding numerical simulation are consistent with streamline reversal. Numerical simulations are then used in a parametric study to investigate the effects of various density relations between penetrator and target materials.

The dimensions of holes produced by the impact of aluminum spheres with various thicknesses and aluminum sheets are presented and analyzed by Piekutowski [1999]. The sphere diameters ranged from 3.18 mm to 19.05 mm and the bumper sheet thickness to projectile diameter ratio, ranged from 0.026 to 0.504 with the majority of the tests having this ratio of less than 0.234. Impact velocities ranged from 1.98 km/s to 7.23 km/s. As this ratio is increased the holes tended to be less circular.

Piekutowski *et al.* [1999] performed a series of depth of penetration experiments using 7.11 mm diameter, 71.12 mm long ogive-nose steel projectiles and 254 mm diameter aluminum targets. The projectiles had a nominal mass of 0.021 kg and were launched with striking velocities between 0.5 and 3.0 km/s. Post test radiographs of the targets showed three different regions of penetrator response as the striking velocity increased: (1) the projectiles remained rigid and visibly undeformed; (2) the projectiles deformed during penetration without nose erosion, deviated from the target centerline, and exited the side of the target of turned severely within the target; and (3) the projectiles eroded during penetration and lost mass. Similar experiments have been done with spherical nose steel projectiles by Forestal and Piekutowski [2000].

Goldsmith [1999] has published a very big review paper on the subject. Everything published until 1999 is mentioned: background and scope, oblique impact, yaw impact, impact with yaw and obliquity, moving targets, rotating penetrators, tumbling impact, impingement of jets or long rods with relative target motion, ricochet, and conclusions-status on non-standard projectile impact. He is also mentioning the rate of the impact velocities: subordnance $v_0 < 500$ m/s, ordnance $500 < v_0 < 1500$ m/s, ultraordnace $1.5 < v_0 < 3$ km/s, and hypervelocity $v_0 > 3$ km/s. The references are also quite big, starting with Euler and continuing to 367 references. See also Batra and Lin [1988].

Powder guns or a two-stage, light-gas launched the projectiles at normal impacts to striking velocities between 0.4 and 1.9 km/s, by Frew *et al.* [2000]. In addition, an analytical penetration equation that described the target resistance by its density and a strength parameter determined from depth of penetration versus striking velocity data, is also done.

9.2   Further Studies

The process of long-rod penetration into thick metallic targets is examined by Rosenberg and Dekel [2001] through a series of two-dimensional simulations. The aim of the research presented, is to uncover the inherent material similarities in this process. In particular, the search is for non-dimensional parameters which account for the depth of penetration, such as the density ratio, and the relative strengths of penetrator and target. The velocity of impact is less than 7 km/s. The simulation results are in accord with existing empirical data, shedding more light on the penetration process and emphasizing the difficulties in achieving an overall normalization procedure for this process.

Target hole sizes and geometries were measured for a series of highly oblique hypervelocity impacts of steel spheres against thin laminated targets by Orphal and Anderson [2001]. The impact velocity was nominally 4.6 km/s for most of the experiments with a few tests conducted at 7.3 km/s. Impact obliquity ranged from 60° to 80° from the normal to the target plane. Projectiles were stainless steel spheres with masses of 222 g, 25 g, and 1 g. Targets were laminated MX-2600 silica phenolic bonded to a 2024-T3 substrate. Target thickness was varied to give thickness to projectile diameter 0.6 and 0.3 for each projectile.

The normal penetration of a deformable projectile into an elastic-plastic target is considered by Roisman *et al.* [2001]. The force imposed on the projectile by the target is generally a complex function of the strength of the target material, the projectile velocity, its diameter and shape, as well as the instantaneous penetration depth. When this force exceeds a certain critical value the projectile begins to deform. At moderate to high values of the impact velocity, the projectile's tip material flows plastically with large deformations causing the formation of a mushroom-like configuration. This process is accompanied by erosion of the projectile material. In the rear, elastic, part of the projectile the deformation remain small and the region can be approximated as a rigid body being decelerated by the projectile's yield stress. The general model allows one to predict the penetration depth, the projectile's eroded length and the crater diameter. It has been shown that in the limit of very high impact velocities the present model reduces to the well-known form of the hydrodynamic theory of shaped-charge jet. Also, a simplified asymptotic formula for the crater radius has been derived which includes the effect of the target's yield stress and compares well with experimental data for very high impact velocities.

The paper by Lynch *et al.* [2001] presents scale size firings of two novel shape KE penetration into a steel/ceramic/steel target at four velocities between 1.8 and 2.9 km/s. The two novel shapes were a three piece segment rod and a telescopic rod/tube. Two unitary rod designs were also included in the assessment. All the penetrators had a similar mass of 60 grams. Test data against semi-infinite rolled homogeneous armor was used to obtain the mass effectiveness of the ceramic target for each rod shape and velocity. The performance rankings of the penetrators against the ceramic target were found to be similar to those for semi-infinite rolled homogeneous armor. Hydrocode analysis of the experiments gave some valuable insights into the penetration processes of the two novel penetrator designs. Predicted depth of penetration compared very well with experimental values, but enhancements to the physics of the ceramic model are needed in order to simulate cover plate effects.

Tests surpassing 11 km/s have been reported by Reinhart *et al.* [2001]. The temperature, and hence the density of the flyer-plate is also well known prior to impact.

In a paper by Hari Manoj Simha *et al.* [2002] is described the computational modeling of the penetration response of a high-putity ceramic, namely the AD-99.5 alumina. This material is the most widely investigated ceramic, and extensive materials testing and ballistic data are available. The model development is based on constitutive relationships inferred from bar impact and plate impact data. A novel element removal scheme for ceramics is presented, and the code is then used to investigate the penetration response of AD-99.5 alumina in the depth of penetration and semi-infinite configurations. The computations are found to be in excellent agreement with the experimental results. The impact velocities are smaller than 3.5 km/s.

A pulse shaping techniques to obtain compressive stress–strain data for brittle materials with the split Hopkinson pressure bar apparatus is presented by Frew *et al.* [2002]. The conventional split Hopkinson pressure bar apparatus is modified by shaping the incident pulse such that the samples are in dynamic stress equilibrium and have nearly constant strain rate over most of the test duration. A thin disk of annealed or hard copper is placed on the impact surface of the incident bar in order to shape the incident pulse. Analytical models and data that show a wide variety of incident strain pulses are produced by varying the geometry of the copper disks and the length and striking velocity of the striker bar. A model that corrects for the added sabot mass and shows good agreement with experiments conducted with high-strength steel striker and incident bars is presented by Forrestal *et al.* [2002]. They point out that this added sabot mass effect will be even more pronounced with titanium, aluminum or magnesium bars. In another paper Piekutowski [2003] study the impact tests using aluminum bumper sheets ranging from 0.076 mm to 4.80 mm thick and impact velocities ranging from 1.98 to 7.19 km/s used to determine the fragmentation initiation threshold for aluminum spheres impacting at hypervelocity. Tests using 0.53 mm and 12.70 mm diameter aluminum spheres indicated that the fragmentation initiation threshold velocity scaled with the bumper-thickness projectile diameter ratio. When a conventional split Hopkinson pressure bar is used to investigate the dynamic flow behavior of ductile metals, Chen *et al.* [2003] observe that the results at small strain ($\varepsilon <\sim 2\%$) are not considered valid owing to fluctuations associated with the early portion of the reflected signal and the nonequilibrated stress state in the specimen. Using a pulse-shaping technique, the dynamic elastic properties can be determined with a split Hopkinson pressure bar, as well as the dynamic plastic flow. They present a description of the experimental technique and the experimental results for a mild steel. The dynamic compressive stress–strain curve is composed of a lower strain-rate elastic portion and a high strain-rate plastic flow portion.

An experimental method for determining failure and fragmentation properties of metals in catastrophic breakup events has been pursued by Grady and Kipp [1995]. Spherical samples of the test metals were launched at high velocities with a two-sage light-gas gun. Fragmentation and motion of the debris were diagnosed with multiple flash radiographies. Grady and Kipp [1997] in a study describe an experimental fracture material property test method specific to dynamic fragmentation. A spherical metal sample is subjected to controlled impulsive stress loads by acceleration to a high velocity in a light-gas launcher with subsequent normal impact onto a thin plate. Motion, deformation and fragmentation of the test sample are diagnosed with multiple flash radiographies. Strain to failure and fragmentation property data for several steels, copper, aluminum, tantalum and titanium are reported. In a study by Chhabildas *et al.* [2001] is reported well-controlled experiments conducted to determine the fracture resistant properties of AerMet 100 steels. One of the objectives of this study is to determine the influence of fracture toughness properties on the fracture and fragmentation process. Both sphere impact tests and cylinder expansion test geometry were used to determine the dynamic fracture resistant coefficients. These experiments were conducted at strain rates of $\sim 14 \times 10^3$/s for the cylinder expansion tests; the strain rate for the sphere impact tests varied over 50 to 100×10^3/s. A glass-fiber-reinforced polyester composite subjected under shock loading to 20 GPa was considered by Dandekar *et al.* [2003]. These experiments show that: (1) the material deforms elastically in compression to at least 1.3 GPa; (2) the deformation coordinates of shocked and re-shocked material lie on the deformation locus of initially shocked to 4.3 GPa; (3) and the release path of material shocked to varying magnitudes of stresses indicate that the material expands such that its density when stresses are released in the range of 3–5 GPa from a peak compressive stress of 9 GPa and above is lower than the initial density of the material.

Forrestal and Hanchak [2002] conducted depth-of-penetration experiments with ogive-nose steel projectiles and limestone targets to determine the penetration limit velocity. The penetration limit velocity is defined as the minimum striking velocity required to embed the projectile in the target. For striking velocities smaller than the penetration limit velocoity, the projectile rebounds from the target. The penetration limit velocity for these experiments was found to be about 300 m/s.

Hugoniot measurements were performed by Knudson *et al.* [2003], on aluminum in the stress range of 100–500 GPa using a magnetically accelerated flyer plate technique. The impact velocities exess 20 km/s. The improved accuracy enhances the understanding of the response of aluminum to 500 GPa.

An application studied by Lemke *et al.* [2003a] [2003b] in greit detail, involves the intense magnetic field to accelerate flyer plates to very high velocities, over 20 km/s, for use in shock loading experiments.

In a paper by Chocron *et al.* [2003] a blended model is presented: momentum balance is used to calculate the semi-infinite portion penetration, and the Ravid-Bodner failure modes are used to determine projectile perforation. In addition, the model has been extended to handle multiple plate impact. Numerical simulations show that after target failure the projectile still continues to erode for some microseconds. This time has been estimated and incorporated into the model. Agreement with results from numerical simulations is quite good.

In another paper by Chhabildas *et al.* [2003] are presented experiments in which a target plate of aluminum is impacted by a titanium-alloy flier plate at speeds ranging from 6.5 to 11 km/s, producing pressures from 1 Mbar to over 2.3 Mbar and temperatures as high as 15000 K. The aluminum plate is totally melted at stresses above 1.6 Mbar. An investigation of strength of AD995 alumina in the shocked state was assessed over the stress range of 26–120 GPa by Reinhart and Chhabildas [2003]. Velocity interferometry was used to measure loading, unloading, reloading and cycling loading profiles from the initial shocked state. These results show that alumina retains considerable strengths at pressures exceeding 120 GPa. An important observation, as with some metals, is that there is a substantial increase in strength during reloading and well-defined elastic behavior is observed. The unloading and reloading technique described also yields data to estimate a dynamic shock hydrostat, i.e., the mean pressure compression curve. In a report Vogler *et al.* [2003] is described recent developments on a technique for the expansion and fragmentation of tubes using a two-stage light gas gun. Results of experiments on two materials, which provide insight into the physics of the process, are presented. A series of plate impact experiments was performed by Vogler *et al.* [2004] that also included restock and release configuration. Hugoniot data were obtained from the elastic limit (15–18 GPa) to 70 GPa and were found to agree reasonably well with the somewhat limited data in the literature. This time no phase transition can be conclusively demonstrated. The experimental data are consistent with a phase transition at a shock stress of about 40 GPa.

The exact solution to the long-rod penetration equations is revisited by Segletes and Walters [2003], in search of improvements to the solution efficiency, while simultaneously enhancing the understanding of the physical parameters that drive the solution. Improvements are offered in these areas. The presentation of the solution is simplified in a way that more tightly unifies the special- and general-case solutions to the problem. Added computational efficiencies are obtained by expressing the general-case solution for penetration and implicit time in terms of a series of Bessel functions.

The terminal phase, or Phase 3, of penetration is investigated using numerical simulations by Anderson and Orphal [2003]. Results of the first set of simulations, for zero-strength tungsten-alloy projectiles into armor steel at velocity of 1.5, 3.0, and 6.0 km/s are reported. For projectiles of $L/D \geq 3$, steady-state penetration is achieved. For $L/D \geq 3$, the deceleration of both the nose and tail of the projectile are essentially independent of L/D.

The paper by Mullin *et al.* [2003] describes a formal similitude analysis for the problem of hypervelocity impact into shielded targets. The analysis indicates that the dissimilar material model approach can be used to investigate the response of prototype impact situations using models that impact at lower velocities. The significance of this approach lies in the ability to infer debris cloud damage for impact velocities that are higher than obtainable with gas-gun technology. Six model/prototype pairs of experimental tests were performed, with the response of one-half of each pair intended to simulate the response of the other. The experiments consisted of symmetric material impacts of a sphere into a thin shield plate separated from an instrumented witness plate by a vacuum gap. Data collected include debris structure, damage area, debris momentum, shield hole size, and ejecta momentum for different debris cloud phases using aluminum, zinc, and cadmium at various impact velocities. The results obtained showed excellent correlation for all debris crowd structure, materials, and velocities. The scaled data from each pair of tests showed less than 10 percent deviation, with most data exhibiting less than 5 percent deviation.

Impact of the order of 2 km/s to 30 km/s or more, are considered as hypervelocity impact (Eftis *et al.* [2003]). Shock pressures caused by impact at lower speeds vary approximately between 10 and 200 GPa. The volume reduction caused by such pressures can vary between 10 to 35%. The temperature is also increasing with 400-5000 K. The deforming material strain-rate can be of the order 10^6–10^8 s^{-1}. All thermal changes are adiabatic and the material behavior is ductile. Impact pressures are as high as 10^3–10^4 GPa, with associated volume reductions as high as 50%. Launch velocities of approximately 4–8 km/s are currently used with metallic projectiles of 0.6–0.9 cm diameter. The strain rate are additive, and the elastic one satisfies again the Jaumann stress rate. The target are 1100 rectangular target plates. Three ratios of the projectile diameter to the target thickness are chosen for the simulations, providing a wide range of damage features. The simulated impact damage is compared with experimental damage of corresponding test specimens, illustrating the capability of the model.

In a paper by Kozachuk *et al.* [2003] one is studying the deviations of the penetration from hydrodynamic calculations due to an interaction of some jet with the crater walls. The jet velocities ranged from 6.5 to 1.4 km/s. Quantitative estimations were made of a reduction of the jet length resulted from the interaction with the crater walls in targets of various nature. The strength resistance of brittle materials to penetration was studied experimentally in a wide range of penetration velocities from 10^{-5} to 10^4 m/s, by Kozhushko and Sinani [2003]. The penetration into brittle materials is characterized by a progressive decrease in the strength resistance due to growing fracture ahead of a penetrator. The averaged strength resistance is about 0.2–0.5 hardness of the target material. With the increase of impact velocity Chen and Li [2003] show that penetration mechanism may change at a transition point from nondeformable projectile penetration regime to semi-hydrodynamic penetration regime. A series of penetration experiment using steel projectiles with ogive and hemispherical noses were conducted. Aluminum targets with various hardness were hit at striking velocities between 0.5 and 3.0 km/s.

2D numerical modeling of impact cratering has been utilized to quantify an important depth-diameter relationship for different crater morphologies, simple and complex. It is generally accepted (Wünnemann and Ivanov [2003]) that the final crater shape is the result of a gravity-driven collapse of the transient crater, which is formed immediately after the impact. Numerical models allow a quantification of the formation of simple craters, which are bowl-shaped depressions with a lens of rock debris inside, and complex craters, which are characterized by a structural uplift. The computation of the cratering process starts with the first contact of the impactor and the planetary surface and ends with the morphology of the final crater. Using different rheological models for the sub-crater rocks, the authors quantify the influence on crater mechanics. To explain the formation of complex craters in accordance to the threshold diameter between simple and complex craters, they utilize the Acoustic Fluidization model. They carried out a series of simulations over a broad parameter range with the goal to fit the observed depth/diameter relationships as well as the observed diameters on the Moon, Earth and Venus.

The paper by Anderson [2003] is a review paper, presenting all the things until 2003. As dictated by precedence, the article highlights some of the significant events and activities in his career. Besides other things he mentioned also his contribution in penetration/armor mechanics. He is mentioning all the Symposia on hypervelocity impact, starting from 1955. Also he is mentioning his first papers on hypervelocity impact. Then he is mentioning the main points of the hypervelocity impact: armor/anti-armor initiative, penetration mechanics (with Walker–Anderson penetration model) etc. He is also mentioning hypervelocity impact symposia and the hypervelocity impact society. A commonality of the areas is that they all deal with characterizing and understanding the response of materials or structures to intense loads.

The paper by Leigh Phoenix and Porwal [2003] develops an analytical model for the ballistic impact response of fibrous of interest in body armor applications. It focuses on an un-tensional 2D membrane impacted transversely by a blunt-nosed projectile, a problem that has remained unsolved for a half a century. Membrane properties are assumed characteristic of the best current body armor materials, which have very high stiffness and strength per unit weight, and low strain-to-failure. Under constant projectile velocity they first develop self-similar solution forms for the tensile implosion wave and the curved cone wave that develops in its wake. Through matching boundary conditions at the cone wave front, they obtain an accurate approximate solution for the membrane response including cone wave speed and strain distribution. Then they consider projectile deceleration due to membrane reactive forces, and obtain results on cone velocity, displacement and strain concentration versus time. Other results obtained are the membrane ballistic limit, and the residual velocity when penetrated above this limit. They then derive an exact functional representation of a V_{50} master curve to reduce data for a wide variety of fabric systems impacted by blunt cylindrical projectiles. This curve is given in terms two dimensionless parameters based only on fiber mechanical properties and the ratio of the fabric areal density to the projectile mass divided by its area of fabric contact. Their functional representation has no fitting parameters beyond one reflecting uncertainty in the effective diameter of the impact zone relative to the projectile diameter, and even then the values are consistent across several experimental systems. Their extremely successful comparison of their analytical model to experimental results in the literature raises fundamental questions about many long-held views on fabric system impact behavior and parameters thought to be important.

Reisman *et al.* [2003] performed quasi-isentropic compression experiments on radiation-damaged stainless steel. They were loaded with a ramp compression wave. The velocity histories suggest a sudden volume reduction of the material above 40 kbar caused by the collapse of nanosized voids.

A review paper is published by Field *et al.* [2004]. A variety of techniques used to obtain the mechanical properties of materials at high rates of strain (≥ 10 s^{-1}) are summarized. These include dropweight machines, split Hopkinson pressure bar, Taylor impact and shock loading by plate impact. A significant literature is mentioned.

Using intense magnetic pressure macroscopic aluminum or titanium flyer plates have been launched by Knudson *et al.* [2004] to velocities in excess of 22 km/s.

An analysis of stress uniformity in split Hopkinson bar test is done by Yang and Shim [2005]. They are based on the method of characteristics for one-dimensional waves. They show that the shortest time to achive uniformity of stress in a specimen requires the incident pulse to have a specific profile.

Klepaczko and Hughes [2005] consider the terminal velocity up to 1300 m/s. Ballistic experiments indicate that below hydrodynamic transition a significant erosion of steel penetrators is observed with concrete targets. It was found that a proper normalization of penetration experiments performed with different scales leads to the universal parameters defined as the rate of wear and the rate sensitivity of wear.

In another paper Brara and Klepaczko [2005] publish the result of tensile tests on cylindrical specimens made of wet and dry concrete, performed at different strain rates from 10 s^{-1} to 120 s^{-1} by means of a special experimental technique. The experiments clearly demonstrated a significant increase of tensile strength occurring in the range of strain rates above 10 s^{-1}.

A coupled constitutive model is proposed for anisotropic damage and permeability variation in britle rocks under deviatoric compressive stresses by Shao *et al.* [2005]. Darcy's law is used for fluid flow at the macroscopic scale whereas laminar flow is assumed at the microcrack scale.

Bibliography

Alekseevskii V. P., 1966, Penetration of a rod into a target at high velocity, *Combustion, Explosion, and Shok Waves* **2**, 2, 99–106.

Allen W. A. and Rogers J. W., 1961, Penetration of a rod into a semi-infinite target, *J. Franklin Ins.* **272**, 4, 275–284.

Al'tshuler L. V., 1965, Use of shock waves in high-pressure physics, *Sovit Phys. Uspekhi.* **8**, 1, 52–91.

Anderson C. E. Jr., Trucano T. G. and Mullin S. A., 1990, Debris cloud dynamics, *Int. J. Impact Engng.* **9**, 1, 89–113.

Anderson C. E. Jr. and Walker J. D., 1991, An examination of long-rod penetration, *Int. J. Impact Engng.* **11**, 4, 481–501.

Anderson C. E. Jr., Mullin S. A. and Kuhlman C. J., 1993a, Computer simulation of strain-rate effects in replica scale model penetration experiments, *Int. J. Impact Engng.* **13**, 1, 35–52.

Anderson C. E. Jr., Littlefield D. I. and Walker J. D., 1993b, Long-rod penetration, target resistance and hypervelocity impact, *Int. J. Impact Engng.* **14**, 1–12.

Anderson C. E., Hohler V., Walker J. D. and Stilp A., 1995a, Time-resolved penetration of long rods into steel targets, *Int. J. Impact Engng.* **16**, 1, 1–18.

Anderson C. E. Jr., Walker J. D., Bless S. J. and Sharron T. R., 1995b, On the velocity dependence of the L/D effect for long-rod penetrators, *Int. J. Impact Engng.* **17**, 13–24.

Anderson C. E. Jr, Mullin C. A., Piekutowski A. J., Blaylock N. W. and Poormon K. L., 1996a, Scale model experiments with ceramic laminate targets, *Int. J. Impact Engng.* **18**, 1, 1–22.

Anderson C. E. Jr., Walker J. D., Bless S. J. and Partom Y., 1996b, On the L/D effect for long-rod penetrators, *Int. J. Impact Engng.* **18**, 3, 247–264.

Anderson C. E. Jr., Orphal D. L., Franzen R. R. and Walker J. D., 1999a, On the hydrodynamic approximation for long-rod penetration, *Int. J. Impact Engng.* **22**, 23–43.

Anderson C. E. Jr., Wilbeck J. S. and Elder J. S., 1999b, Long-rod penetration into highly oblique, water-filled targets, *Int. J. Impact Engng.* **23**, 1–12.

Anderson C. E. Jr., Hohler V., Walker J. D. and Stilp A. J., 1999c, The influence of projectile hardness on ballistic performance, *Int. J. Impact Engng.* **22**, 619–632.

Anderson C. E. Jr., Sharron T. R., Walker J. D. and Freitas C. J., 1999d, Simulation and analysis of a 23-mm HEI projectile hydrodynamic ram experiment, *Int. J. Impact Engng.* **22**, 981–997.

Anderson C. E. Jr., 2003, From fire to ballistics: a historical retrospective, *Int. J. Impact Engng.* **29**, 13–67.

Anderson C. E. Jr. and Orphal D. L., 2003, Analysis of the teminal phase of penetration, *Int. J. Impact Engng.* **29**, 69–80.

Ang J. A., 1990, Impact flash jet initiation phenomenology, *Int. J. Impact. Engng.* **10**, 23–33.

Batra R. C. and Lin P.-R., 1988, Steady state deformations of a rigid perfectly plastic rod striking a rigid cavity, *Int. J. Engng. Sci.* **26**, 2, 183–192.

Bjerke T. W., Zucas J. A. and Kinsey K. D., 1992, Penetration performance of disk shaped penetrators, *Int. J. Impact Engng.* **12**, 2, 263–280.

Bless S. J. and Anderson C. E. Jr., 1993, Penetration of hard layers by hypervelocity rod projectiles, *Int. J. Impact. Engng.* **14**, 85–93.

Bless S. J., Satapathy S. and Normandia M. J., 1999, Transverse loads on a yawed projectile, *Int. J. Impact Engng.* **23**, 77–86.

Brannon R. M. and Chhabildas L. C., 1995, Experimental and numerical investigation of shock-induced full vaporization of zinc, *Int. J. Impact Engng.* **17**, 109–120.

Brara A. and Klepaczko J. R., 2005, Experimental characterization of concrete in dynamic tension, *Mech. Mater.* (in print).

Chen W., Song B., Frew D. J. and Forrestal M. J., 2003, Dynamic small strain measurements of a metal specimen with a split Hopkinson pressure bar, *Experimental Mechanics* **43**, 1, 20–23.

Chen X. W. and Li Q. M., 2004, Transition from nondeformable projectile penetration to semihydrodynamic penetration, *J. Engn. Mechanics, ASCE* **130**, 1, 123–127.

Chhabildas L. C., Kmetyk L. N., Reinhart W. D. and Hall C. A., 1995, Enhanced hypervelocity launcher — capabilities to 16 km/s, *Int. J. Impact. Engng.* **17**, 183–194.

Chhabildas L. C., Thornhill T. F., Reinhart W. D., Kipp M. E., Reedal D. R., Wilson L. T. and Grady D. E., 2001, Fracture resistant properties of aermet steels, *Int. J. Impact Engng.* **26**, 77–91.

Chhabildas L. C., Reinhart W. D., Thornhill T. F., Bessette G. C., Saul W. V., Lawrence R. J. and Kipp M. E., 2003, Debris generation and propagation phenomenology from hypervelocity impacts on aluminum from 6 to 11 km/s, *Int. J. Impact Engng.* **29**, 185–202.

Chocron S., Anderson C. E. Jr., Walker J. D. and Ravid M., 2003, A unified model for long-rod penetration in multiple metallic plates, *Int. J. Impact Engng.* **28**, 391–411.

Dandekar D. P., Boteler J. M. and Beaulieu P. A., 1998, Elastic constants and delamination strength of a glass-fiber-reiforced polymer composite, *Comp. Sci. Technology* **58**, 1397–1403.

Dandekar D. P., Hall C. A., Chhabildas L. C. and Reinhart W. D., 2003, Shock response of a glass-fiber-reinforced polymer composite, *Composiote Structures* **61**, 51–59.

Eftis J., Carrasco C. and Osegueda R. A., 2003, A constitutive-microdamage model to simulate hypervelocity projectile-target impact, material damage and fracture, *Int. J Plasticity* **19**, 1321–1354.

Field J. E., Walley S. M., Proud W. G., Goldrein H. T. and Siviour C. R., 2004, Review of experimental techniques for high rate deformation and shock studies, *Int. J. Impact Engng.* **30**, 725–775.

Forestal M. J. and Piekutowski A. J., 2000, Penetration experiments with 6061-T6511 aluminum targets and spherical-nose steel projectiles at striking velocities between 0.5 and 3.0 km/s, *Int. J. Impact Engng.* **24**, 57–67.

Forrestal M. J., Frew D. J. and Chen W., 2002, The effect of Sabot mass on the striker bar for split Hopkinson pressure bar experiments, *Experimental Mechanics* **42**, 2, 129–131.

Forrestal M. J. and Hanchak S. J., 2002, Penetration limit velocity for ogive-nose projectiles and limestone targets, *J. Appl. Mech.* **69**, 853–854.

Frew D. J., Forrestal M. J. and Chen W., 2002, Pulse shaping techniques for testing brittle materials with a split Hopkinson pressure bar, *Experimental Mechanics* **42**, 1, 93–106.

Frew D. J., Forrestal M. J. and Hanchak S. J., 2000, Penetration experiments With limestone targets and ogive-nose steel projectiles, *J. Appl. Mech.* **67**, 841–845.

Goldsmith W., 1999, Non-ideal projectile impact on targets, *Int. J. Impact Engng.* **22**, 95–395.

Grady D. E. and Kipp M. E., 1995, Experimental measurement of dynamic failure and fragmentation properties of metals, *Int. J. Impact Engng.* **32**, 17/18, 2779–2791.

Grady D. E. and Kipp M. E., 1997, Fragmentation properties of metals, *Int. J. Impact Engng.* **20**, 293–308.

Grady D. E., 1998, Shock-wave compression of brittle solids, *Mech. Mater.* **29**, 181–203.

Hari Manoj Simha C., Bless S. J. and Bedford A., 2002, Computational modeling of the penetration response of a high-purity ceramic, *Int. J. Impact Engng.* **27**, 65–86.

Klepaczko J. R. and Hughes M. L., 2005, Scaling of wear in kinetic energy penetrators, *Int. J. Impact Engng.* **31**, 435–459.

Knudson M. D., Lemke R. W., Hayes D. B., Hall C. A., Deeney C. and Asay J. R., 2003, Near-absolute Hugoniot measurements in aluminum to 500 GPa using a magnetically accelerated flyer plate technique, *J. Appl. Phys.* **94**, 7, 4420–4431.

Knudson M. D., Hanson D. L., Bailey J. E., Hall C. A., Asay J. R. and Denney C., 2004, Principal Hugoniot, reverberating wave and mechanical reshock measurements of liquid deuterium to 400 GPa using plate impact techniques, *Phys. Rev.* **69**, 144209–144219.

Kozachuk A. I., Kozhushko A. A., Rumyantsev B. V., Sinani A. B., Vlasov A. S. and Zilberbrand E. L., 2003, On interaction of shaped-charge jets with the crater walls in penetrating metals and brittle materials, *Int. J. Impact Engng.* **29**, 385–390.

Kozhushko A. A. and Sinani A. B., 2003, Hypervelocity impact for brittle targets, *Int. J. Impact Engng.* **29**, 391–396.

Lawrence R. J., Kmetyk L. N. and Chhabildas L. C., 1995, The influence of phase changes on debris-cloud interactions with protected structures, *Int. J. Impact Engng.* **17**, 487–496.

Lee M. and Bless S. J., 1998, Cavity models for solid and hollow projectiles, *Int. J. Impact Engng.* **21**, 10, 881–894.

Leigh Phoenix S. and Porwal P. K., 2003, A new membrane model for the ballistic impact response and V_{50} performance of multi-ply fibrous systems, *Int. J. Solids Struct.* **40**, 6723–6765.

Lemke R. W., Knudson M. D., Hall C. A., Haill T. A., Desjarlais P. M., Asay J. R. and Mehlhorn T. A., 2003a, Characterization of magnetically accelerated flyer plates, *Phys. Plasmas* **10**, 4, 1092–1099.

Lemke R. W., Knudson M. D., Robinson A. C., Haill T. A., Struve K. W., Asay J. R. and Mehlhorn T. A., 2003b, Self-consistent, two-dimensional, magnetohydrodynamic simulations of magnetically driven flyer plates, *Phys. Plasmas* **10**, 5, 1867–1874.

Lynch N. J., Bless S. J., Brissenden C., Berry B. and Pedersen B., 2001, Novel KE penetrator performance against a steel/ceramic/steel target at 0° over the velocity range 1800 to 2900 m/s, *Int. J. Impact Engng.* **26**, 475–486.

Mullin S. A., Anderson C. E. Jr. and Wilbeck J. S., 2003, Dissimilar material velocity scalimg for hypervelocity impact, *Int. J. Impact Engng.* **29**, 468–485.

Orphal D. L., Anderson C. E. Jr., Franzen R. R., Walker J. D., Schneidewind P. N. and Majerus M. E., 1993, Impact and penetration by $L/D \leq 1$ projectiles, *Int. J. Impact Engng.* **14**, 551–560.

Orphal D. L., Franzen R. R., Piekutowski A. J. and Forrestal M. J., 1996, Penetration of confined aluminum nitride targets by tungsten long rod at 1.5–4.5 km/s, *Int. J. Impact Engng.* **18**, 4, 355–268.

Orphal D. L., Franzen R. R., Charters A. C., Menna T. L. and Piekutowski A. J., 1997, Penetration of confined boron carbide targets by tungsten long rods at impact velocities from 1.5 to 5.0 km/s, *Int. J. Impact Engng.* **19**, 1, 15–29.

Orphal D. L. and Anderson C. E. Jr., 1999, Streamline reversal in hypervelocity penetration, *Int. J. Impact Engng.* **23**, 699–710.

Orphal D. L. and Anderson C. E. Jr., 2001, Target damage from highly oblique hypervelocity impacts of steel spheres against thin laminated targets, *Int. J. Impact Engng.* **26**, 567–578.

Piekutowski A. J., Forrestal M. J., Poormon K. L. and Warren T. L., 1996, Perforation of aluminum plates with ogive-nose steel roads at normal and oblique impacts, *Int. J. Impact Engng.* **18**, 7–8, 877–887.

Piekutowski A. J., 1999, Holes produced in thin aluminum sheets by the hypervelocity impact of aluminum spheres, *Int. J. Impact Engng.* **23**, 711–722.

Piekutowski A. J., Forrestal M. J., Poormon K. L. and Warren T. L., 1999, Penetration of 6061-T6511 aluminum targets by ogive-nose steel projectiles with striking velocities between 0.5 and 3.0 km/s, *Int. J. Impact Engng.* **23**, 723–734.

Piekutowski A. J., 2003, Fragmentation-initiation threshold for spheres impacting at hypervelocity, *Int. J. Impact Engng.* **29**, 563–574.

Reinhart W. D. and Chhabildas L. C., 2003, Strength properties of Coors AD995 alumina in the shocked state, *Int. J. Impact Engng.* **29**, 601–619.

Reinhart W. D., Chhabildas L. C., Carroll D. E., Bergstresser Y. K., Thornhill T. F. and Winfree N. A., 2001, Equation of state measurements of materials using a three-stage gun to impact velocities of 11 km/s, *Int. J. Impact Engng.* **26**, 625–637.

Roisman I. V., Yarin A. L. and Rubin M. B., 2001, Normal penetration of an eroding projectile into an elastic-plastic target, *Int. J. Impact Engng.* **25**, 573–597.

Reisman D. B., Wolfer W. G., Elsholz A. and Furnish M. D., 2003, Isentropic compression of irradiated stainless steel on the Z accelerator, *J. Appl. Phys.* **93**, 11, 8952–8957.

Rosenberg Z. and Dekel E., 2001, Material similarities in long-rod penetration mechanics, *Int. J. Impact Engng.* **25**, 361–372.

Satapathy S. S. and Bless J., 2000, Cavity expansion resistance of brittle materials obeying a two-curve pressure-shear behavior, *J. Appl. Phys.* **88**, 7, 4004–4012.

Segletes S. B. and Walters W. P., 2003, Extensions to the exact solution of the long-rod penetration/erosion equations, *Int. J. Impact Engng.* **28**, 363–376.

Shao J. F., Zhou H. and Chau K. T., 2005, Coupling between anisotropic damage and permeability variation in brittle rocks, *Int. J. Num. Anal. Meth. Geomec.* **29**, 1231–1247.

Subramanian R. and Bless S. J., 1995, Penetration of semi-infinite AD995 Alumina targets by tungsten long rod penetrators from 1.5 to 3.5 km/s, *Int. J. Impact Engng.* **17**, 807–816.

Tate A., 1967, A theory for the deceleration of long rods after impact, *J. Mech. Phys. Solids* **15**, 387–399.

Tate A., 1969, Further results in the theory of long rod penetration, *J. Mech. Phys. Solids* **17**, 141–150.

Tate A., 1986, Long rod penetration models-part I. A flow field model for high speed long rod penetration, *Int. J. Mech. Solids* **28**, 8, 535–548.

Tate A., 1990, Engineering modelling of some aspects of segmented rod penetration, *Int. J. Impact Engng.* **9**, 3, 327–341.

Trucano T. G. and Chhabildas L. C., 1995, Computational design of hypervelocity launchers, *Int. J. Impact Engng.* **17**, 849–860.

Vogler T. J., Thornhill T. F., Reinhart W. D., Chhabildas L. C., Grady D. E., Wilson L. T., Hurricane O. A. and Sunwoo A., 2003, Fragmentation of materials In expanding tube experiments, *Int. J. Impact Engng.* **29**, 735–746.

Vogler T. J., Reinhart W. D. and Chhabildas L. C., 2004, Dynamic behavior of boron carbide, *J. Appl. Phys.* **95**, 8, 4173–4183.

Walker J. D., 2001, Hypervelocity penetration modeling: momentum vs. energy and energy transfer mechanisms, *Int. J. Impact Engng.* **26**, 809–822.

Walker J. D. and Anderson C. E. Jr., 1995, A time-dependent model for long-rod penetration, *Int. J. Impact Engng.* **16**, 1, 19–48.

Walker J. D., Grosch D. J. and Mullin S. A., 1995, Experimental impacts above 10 km/s, *Int. J. Impact Engng.* **17**, 903–914.

Wünnemann K. and Ivanov B. A., 2003, Numerical modelling of the impact crater depth-diameter dependence in an acoustically fluidized target, *Planetary and Space Science* **51**, 13, 831–845.

Yang L. M. and Shim V. P. W., 2005, An analysis of stress uniformity in split Hopkinson bar test specimens, *Int. J. Impact Engng.* **31**, 129–150.

Author Index

Subject Index